HYDROGEN AND ITS FUTURE AS A TRANSPORTATION FUEL

Other related SAE books:

Hydrogen Fuel For Surface Transportation
(Order No. R-160)

For information on these or other related books, contact SAE by phone at (724) 776-4970, fax (724) 776-0790, e-mail: publications@sae.org, or the SAE website at www.sae.org

Hydrogen and Its Future as a Transportation Fuel

PT-95

Published by:
Society of Automotive Engineers, Inc.
400 Commonwealth Drive
Warrendale, PA 15096-0001
USA
Phone: (724) 776-4841
Fax: (724) 776-5760
February 2003

For permission and licensing requests contact:

SAE Permissions
400 Commonwealth Drive
Warrendale, PA 15096-0001-USA
Email: permissions@sae.org
Fax: 724-772-4891
Tel: 724-772-4028

Global Mobility Database®

All SAE papers, standards, and selected books are abstracted and indexed in the Global Mobility Database.

For multiple print copies contact:

SAE Customer Service
Tel: 877-606-7323 (inside USA and Canada)
Tel: 724-776-4970 (outside USA)
Fax: 724-776-1615
Email: CustomerService@sae.org

ISBN 0-7680-1128-0
Library of Congress Catalog Card Number: 2002115467
SAE/PT-95
Copyright © 2003 SAE International

Positions and opinions advanced in this publication are those of the author(s) and not necessarily those of SAE. The author is solely responsible for the content of the book.

SAE Order No. PT-95

Printed in USA

Preface

Hydrogen as a Transportation Fuel

The six chapters in this Progress in Technology Series look at the progress the transportation industry is achieving toward making hydrogen the fuel of the future. The 43 SAE technical papers presented by various authors at SAE technical conferences from 1997 through 2002 provide a slice in time of the progress. Hydrogen has many advantages but some challenges that must be overcome for it to be a successful transportation fuel.

Introduction

Fossil fuels, mainly gasoline and diesel fuel, have been the fuel of choice for the past 100 years. Today with an emphasis on emissions and the desire to find an alternative fuel, hydrogen is being looked at and touted by many as the fuel of the future.

Hydrogen is one of the simplest and lightest elements in the Universe, it is made up of a single proton with one electron. It is also the most abundant element in the universe, being found in 90% of all matter. Since it is a major component of water that covers 70% of the Earth's surface and is found in most organic matter, hydrogen is the third most abundant element on Earth.

Thus the abundance of hydrogen makes it a candidate as a transportation fuel as far as making sure that it can be obtained in sufficient quantities. It can be obtained from electrolysis of water by using a number of energy sources such as nuclear, solar, and fossil fuels. With the interest in hydrogen as a future transportation fuel a lot of research has dealt with producing it from biomass sources. Everything from food scraps to sugar has been investigated as a source of hydrogen. The reforming of hydrogen from hydrocarbon sources such as gasoline, methane, ethanol, natural gas, and other sources appears to be a promising approach. In fact, 95% of the hydrogen that is used today comes from reforming natural gas.

The second biggest draw for hydrogen to be used as a transportation fuel is the lack of emissions when burned. When hydrogen is burned with oxygen only, heat and water are released. However, when burned with air, which is mostly nitrogen, some oxides of nitrogen (NOx) are produced. However, this level of NOx is less than that obtained when using hydrocarbon fuels. Also, without the hydrocarbon fuel, no carbon monoxide or carbon dioxide is created. This lack of exhaust emissions is the big draw for hydrogen to be used in internal combustion engines.

Hydrogen and the Automotive Industry

The automotive industry is taking a couple of different directions to use hydrogen to reduce exhaust emissions and reduce the transportation industry's dependence on oil. The first is to use hydrogen as a fuel in the internal combustion engine or to basically replace gasoline as the vehicle's fuel. Researchers have also looked at using hydrogen in conjunction with diesel-fuel powered engines. A number of manufacturers have been working on engines that will maintain the performance and range expected by the consumer.

A slightly different track is to use hydrogen to power a small internal combustion engine in a hybrid vehicle or one that uses an electric motor to drive the wheels and batteries to provide the electric power. The internal combustion engine is used to recharge the batteries or to supply electrical energy to the electric motors.

The use of hydrogen as a fuel to power a fuel cell is currently the most promising to produce a true zero emissions vehicle. A fuel cell uses an electrochemical reaction that produces water vapor, heat, and electrical energy. The electrical energy from the fuel cell can be used to power an electric motor for propulsion. The abundance of electrical energy from a fuel cell and the different package requirements, cooling requirements, and fuel system makes a fuel cell vehicle a design challenge. The challenge is to place the fuel cell components so that they work efficiently in a vehicle originally designed for an internal combustion engine.

The use of hydrogen in the transportation industry has not been restricted to automobiles. A number of buses and sport utility vehicles have been conceived to verify the technology. In some cases it is easier to package the fuel cell components into a bus since there is more available room.

The Challenges

Whether hydrogen fuel is used to power an internal combustion engine or a fuel cell, there are a number of challenges that must be overcome before hydrogen can be as prevalent a transportation fuel as gasoline. Some of the reasons hydrogen is being considered as a fuel also causes some of the challenges.

The immediate challenge is to find a way to store enough hydrogen on-board a vehicle in a quantity to have a sufficient driving range, 200-300 miles, before having to refuel. The most common storage method has been to compress the hydrogen under 3000-5000 psi in canisters. Researchers have also experimented with storing the hydrogen under cryogenic temperatures on-board the vehicle. A number of other storage methods such as hydrides and slurries are under consideration.

After the on-board storage problem is solved the largest challenge is the hydrogen refueling infrastructure. Consumers expect to find a refueling area anywhere and everywhere, as is the case with gasoline refueling sites. The cost of creating the infrastructure and who finances it has been a matter under much discussion. Also in the mix is the choice of using reformers on-board the vehicles, at homes, or at refueling stations to convert hydrocarbon fuels to hydrogen. Many researchers have suggested that there will be a transition period where fossil fuels may be reformed into hydrogen to lower emissions while the mix of gasoline- and hydrogen-powered vehicles adjust to consumer demand. Not only is the future fuel in question but also the powerplant -- internal combustion engine, hybrid, or fuel cell. All of which can and may use hydrogen as the fuel.

TABLE OF CONTENTS

Hydrogen Issues

Internal Combustion Engines

Diesel Engines

Hybrid Vehicles

Fuel Cells

Hydrogen Storage and Generation

HYDROGEN ISSUES

2002-01-1927

Hydrogen: On the Horizon or Just a Mirage?

Marianne Mintz, Stephen Folga, Jerry Gillette and John Molburg
Argonne National Lab

ABSTRACT

This paper contrasts "well-to-tank" costs of supplying gaseous hydrogen (GH_2) transportation fuel to light-duty vehicles via steam-methane-reforming (SMR) of natural gas. One decentralized pathway (i.e., station reforming) and two centralized pathways (i.e., production in market demand centers, and production at or near resource supply centers) are considered. All pathways are likely to cost in excess of $600 billion ($18-22/GJ) at high production volumes. With limited production, the decentralized path holds promise as an initial strategy. However, unless substantial cost reductions can be achieved either in the reforming process itself or via economies of scale, it is unlikely to be cost-competitive at high volumes and may preclude other policy objectives like carbon sequestration. With increasing production, centralized production becomes the less costly. Our analyses suggest three possibilities for reducing the cost of hydrogen infrastructure: (1) development of small, low-cost station reforming technologies via partial oxidation, SMR or other processes; (2) centralized GH_2 production at very large plants, perhaps using site-specific feedstock and production processes; and (3) pipeline improvements, including lower-cost dedicated pipelines, designing new natural gas pipelines to levels allowing future transport of hydrogen, and the development of gaseous product pipelines.

INTRODUCTION

With interest in hydrogen-fueled transportation systems mounting, can we expect to see the development of a hydrogen fuel infrastructure in the foreseeable future? There certainly are compelling reasons to move in that direction. Imports accounted for 51% of U.S. petroleum use in 2000, a share that is expected to rise to 64% by 2020 (EIA 2000). About a quarter of those imports originated in the Persian Gulf and, given the worldwide distribution of petroleum reserves, that share may be expected to rise over the next 20 years. Transportation is 97% dependent on petroleum (Davis 2000), much of it imported. Today, the sector accounts for approximately two-thirds of U.S. petroleum consumption; in 20 years, it is expected to account for 72% (EIA 2000). With continued growth in transportation energy demand, dwindling domestic oil reserves, and an increasing concentration of remaining reserves in potentially unstable parts of the world, it is becoming critical to identify a transition path for the transportation sector. While it will take decades for a successor to gain the kind of market dominance now held by imported oil, transition planning must begin well before large numbers of vehicles requiring non-petroleum fuel are on the road. A key element of that planning must include developing the necessary infrastructure to produce, transport and deliver large quantities of non-petroleum fuel.

Unlike other demand sectors that have shifted from oil to natural gas, electricity and renewable fuels, the transportation sector's long-standing commitment to the internal combustion engine (ICE) and the need for mobile energy storage has limited the potential for fuel shifts. With its high energy density, ease of handling, and chemical properties fine-tuned to the demands of the ICE, petroleum products have become "ideal" fuels for most transportation applications. Unseating that market dominance is not an easy task.

Yet it is a task that cannot be avoided, at least in the long term. Experts argue over whether the transition will occur in 20, 30, or more years. Though the timeframe will undoubtedly affect the timing of investment decisions, it will not affect the broad outline of the ultimate path or the need to identify and plan for the required infrastructure.

The paths that have been identified to date tend to be expensive, experimental or both, and at this time there is no clear-cut "winner". Meanwhile, vehicle manufacturers have announced plans to introduce hydrogen-fueled vehicles by 2003, largely in response to California's zero-emission-vehicle (ZEV) mandate. Although a few refueling facilities have been built (most notably in Las Vegas and Sacramento) and a skeleton infrastructure exists to serve the refining industry (primarily around Houston and Chicago), the supply of hydrogen transportation fuel for these ZEVs is likely to be quite limited.

Today, most hydrogen is used as an industrial chemical, to produce ammonia and methanol, and to increase refinery yields. Although the U.S. consumed 3.2 trillion cubic feet (tcf) of hydrogen in 1999 (equivalent to 1.1 quadrillion Btu on an energy basis), very little was used as a fuel or energy carrier (Suresh 2001).

METHODOLOGY

The well-to-tank [1]costs reported here were estimated by modeling pathways from extraction of the primary energy resource (or feedstock), through its conversion into a vehicular fuel, to its dispensing into a vehicle tank. Figure 1 illustrates the three pathways considered.

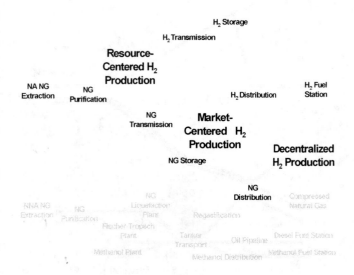

Figure 1. Well-to-Tank Pathways Considered in This Analysis

Our cost modeling approach may be described in terms of the following four steps.

- Detail potential pathways from resource extraction to fuel dispensing, both to describe the path and to bound the analysis.

- Determine the "tank-in" total quantity of fuel that must be dispensed to all hydrogen-fueled vehicles expected to be on the road in a given year.

- Size pathway stages. Working backward from the "tank-in" fuel requirement and efficiency assumptions for the vehicle and each stage in the pathway, the quantity of fuel needed at each stage is estimated and facilities are sized accordingly.

- Calculate capital and annual operating cost of pathway stages. Based on facility sizes and unit costs, costs are estimated by pathway stage and summed. The model permits unit costs to vary in response to "learning", scenario-related changes in infrastructure, and financial assumptions.

The analysis reported here used the IMPACTT model (Mintz and Saricks 1998) as a pre-processor to compute

---

[1] Exploration and production costs upstream of the wellhead are not explicitly considered. Pathways originate at the wellhead, with resource price set at wellhead price.

the "tank-in" requirement and the NICC model to compute capital and operating costs of each pathway.

IMPACTT ANALYTICAL APPROACH

The Integrated Market Penetration and Anticipated Cost of Advanced Transportation Technologies (IMPACTT) model estimates annual energy use and emissions produced by conventional and advanced vehicles as they move through the light-duty-vehicle fleet. IMPACTT incorporates a vehicle-stock module that adds new vehicles (advanced or conventional) and retires old vehicles from an initial vehicle population profile to produce annual profiles of the auto and light-truck population by age and technology; a usage module to compute vehicle travel, oil displacement, and fuel use by technology; and an emissions module to compute upstream and operational emissions of criteria pollutants and greenhouse gases for autos and light trucks, again by technology. The usage module computes the quantity of petroleum that would have been consumed by conventional vehicles in the absence of advanced vehicles, the quantity of "petroleum equivalent" consumed by advanced vehicles, and the net savings due to the presence of advanced vehicles in the fleet. In this application, only the annual quantity of hydrogen consumed by advanced vehicles is input to the cost model. That quantity corresponds to the fuel use by hydrogen-fueled vehicles under two different market penetration scenarios (see Fig. 2). In both scenarios, hydrogen-fueled vehicles are assumed to achieve 2.0 times the fuel economy of conventional vehicles.

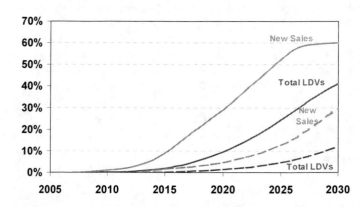

Figure 2. Hydrogen-Fueled Vehicles as a Percent of New Sales and Total Light Duty Vehicles (LDVs) under High and Low Market Penetration Cases

NICC ANALYTICAL APPROACH

The NICC (Natural gas Infrastructure Component Cost) model computes the type and cost of the infrastructure required to bring natural-gas-based transportation fuels from the wellhead to the vehicle tank. It includes the three hydrogen pathways considered here as well as pathways detailing the conversion of natural gas to CNG or LNG for transportation use, to Fischer-Tropsch liquids, or to methanol.

HYDROGEN PRODUCTION

Although hydrogen can be produced from a number of resources and processes, this study focused on steam reforming of methane (SMR) because it is currently the lowest cost, most widely used process for producing hydrogen. From an engineering viewpoint, the main processing steps are feed compression, feed purification, steam reforming, shift conversion, steam generation, purification by pressure swing adsorption (PSA) and product compression. A simplified process flow diagram of a typical steam-methane reformer, based on natural gas, is shown in Fig. 3.

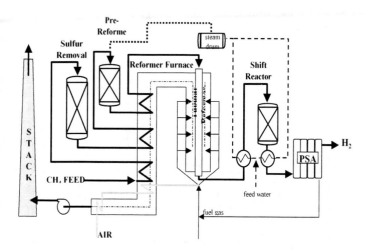

Figure 3. Process Flow Diagram of Stand-Alone H_2 Production from Steam Reforming of Natural Gas (Adapted from Aasberg-Petersen et al. 1998)

The SMR process generally has an efficiency of 70-75% in large-scale applications. In this study, we assume efficiencies of 72% for large-scale facilities and 67% for small-scale, decentralized facilities (Wang 2001). Other options for hydrogen production, including partial oxidation and water electrolysis, were not considered.

In this study, natural gas was selected as the primary energy resource not only because of the relatively attractive economics of SMR conversion, but also because natural gas is widely available and has an extensive, in-place production, transportation, and distribution infrastructure. Although there is considerable uncertainty over the adequacy of North American supplies to meet rising demands for natural gas in power generation and residential and commercial use (let alone conversion to hydrogen transportation fuel), we assume that all natural gas originates in continental or offshore North American fields.

Once produced at either a market- or resource-centered location, hydrogen must be transported to markets. Today's hydrogen distribution system is extremely limited: there are only 450 miles of hydrogen pipeline in the United States, compared with 200,000 miles of oil pipeline and 1.3 million miles of natural gas pipeline (Suresh 2001, Association of Oil Pipelines 2001).

Nonetheless, hydrogen pipeline distribution is a firmly established technology. The key obstacle to making the fuel widely available is the scale of expansion needed to serve transportation markets.

In theory, natural gas can be transported via pipeline in a blend containing up to 20% hydrogen, offering a cleaner fuel without the need to modify natural gas pipelines (Oney et al. 1994). Modifying the same pipelines to carry pure hydrogen, however, requires addressing a number of issues, including the potential for embrittlement of some steels and sealing difficulties at fittings that are tight enough to prevent natural gas, but not hydrogen, from escaping. Regardless of these materials issues, construction of new pipelines to carry hydrogen could benefit from joint use of existing rights-of-way for natural gas distribution. In this analysis we assume that new steel pipelines will be needed for hydrogen transmission and distribution, but that existing rights-of-way will be available.

DECENTRALIZED HYDROGEN PRODUCTION

As shown in Fig. 4, the initial portions of this pathway are the existing natural gas production and delivery infrastructure. After recovery and processing, natural gas is transported from producing regions to local natural gas utilities along the transmission pipeline network. Compressor stations, located at approximately 150-mile intervals, boost the pressure lost as a result of the friction of gas moving through steel pipe. Underground storage facilities, located near market demand centers, store natural gas for use during peak demand periods and ensure reliable service. At the "city gate" natural gas enters smaller-diameter distribution pipelines (or "mains") and service lines for delivery to most homes and nearly 5 million businesses, including refueling stations where it is converted to hydrogen.

Figure 4. Decentralized Hydrogen Pathway

In this pathway, H_2 refueling stations consist of a series of skid-mounted, modular, portable units containing:

- compressors to pressurize natural gas from an assumed suction pressure of 30 psig to the delivery pressure of 200 psig required for the SMR process, and hydrogen output from 200 psig to 6,000 psig;
- hydrogen production including SMR and PSA units;
- three-bank cascade storage capable of storing approximately 100,000 scf of compressed H_2;

- compressed hydrogen dispensers, to fast-fill vehicles at 6,000 psig; and
- associated control equipment.

The modular design of the H_2 refueling station is designed to permit one or more dispensing stations that afford vehicle-fueling times comparable to gasoline refueling stations.

Hydrogen Compressors. It is assumed that multi-stage, positive-displacement reciprocal compressors are used to boost the pressure of the hydrogen leaving the PSA unit from 200 psig to approximately 6,000 psig.[2] Compressors are assumed to be non-lubricated (oil free) to reduce the potential for oil contamination of the hydrogen product. To minimize potential explosion hazards, equipment meeting Class I, Div. 2 standards is assumed (RIX Industries 2001).

The Theoretical Horsepower Equation is used to estimate the power requirements of the compressors. The number of compressor stages is estimated as a function of inlet and outlet pressures (McAllister 1990).

Standard production compressors and custom-designed compressors for pressurizing hydrogen are available at sizes ranging from less than 1 HP to 500 HP, and for both low- and high-pressure applications (up to 10,000 psia). A parametric study was performed to estimate economies-of-scale associated with the capital cost of new H_2 compressors as a function of power rating (brake horsepower), based upon Turnquest (2000). As shown in Fig. 5, capital cost is highly dependent on brake horsepower.

Figure 5. Capital Cost of H_2 Compressors (2000 $)

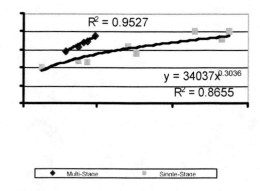

The costs shown in Fig. 5 are for currently available compressor packages, which are skid-mounted and include control systems. They do not include pouring a concrete pad, electrical and cooling water installation, and connection of gas pipelines (Turnquest 2000).

Based on recent developments, innovative compressor designs could be available in the near term at 1/2 to 1/3 the cost of conventional compressors. Thus, a 50% reduction in the capital cost estimated in Fig. 5 is assumed in this analysis.

Hydrogen Storage. Because the demand for hydrogen is unlikely to be constant throughout the day, on-site storage will be needed. A number of advanced hydrogen storage methods -- including metal hydrides, high-pressure tanks, carbon nanotubes, and glass microspheres -- are currently under development. While encouraging results are being obtained in the laboratory, these methods are unproven commercially and costing is highly speculative. Thus, compressed GH_2 storage in high-strength, pressure tanks certified to 6,000 psig is assumed in this analysis.

The number of DOT-certified hydrogen storage tanks is estimated as a function of pressure, required hours of product hydrogen storage, and hydrogen flow rate (227,000 SCFD, equivalent to 50,000 gasoline gal equivalent (GGE) per month).

Hydrogen Dispensers. Hydrogen dispensers specially designed for public refueling applications are currently available and allow vehicles to be refueled in three to five minutes, much like typical gasoline pumps (Kraus Group Inc. 2001). Standard features include flow rates of up to 10,000 SCFM (17,000 Nm^3/hr) and fill pressures of 3,000 or 3,600 psig at 70°F. They are designed to meet ASME requirements for pressure systems.

Hydrogen Production (Refueling Stations with Onsite SMR). Today, steam methane reformers are commercially available, primarily as custom-built units for large industrial customers. A parametric study was performed to estimate scale economies associated with the capital cost of "conventional" hydrogen generation (including the reformer, shift reactor, PSA system, and NO_x control) as a function of hydrogen output, using data reported by Ogden (1997, 1998) and in the open literature. In these data, installed capital costs for "conventional" hydrogen generation varied from $1.5 million for a plant producing 100,000 SCFD to $3 million for a plant producing 1 million SCFD. Results of this analysis are discussed below, under the market-centered pathway.

Small-scale SMRs are not yet commercially available. Thus, our analysis is more indicative than conclusive. However, manufacturers claim that new, small, on-site hydrogen plants using steam reforming of commercially available natural gas, can offer cost savings of 20% or more over conventional SMRs (Salm 1997). For this reason, results of our parametric analysis were decreased by 20%.

---

[2] Neither centrifugal nor piston compressors can achieve pressures of 6,000 psig (ORNL 1996). On-board storage is assumed to be at 5,000 psig.

We use activity-based cost (ABC) estimating[3] to estimate the "balance of plant" (i.e., capital cost of a hydrogen refueling station supplying 50,000 GGE/month with onsite reforming). Direct costs include not only equipment, but also initial spare parts to ensure process operation in the event of a failure of a major piece of installed equipment. Indirect costs include distributables (general conditions) and overhead, as well as construction management, architect/ engineering (A/E) fees and program management.

This analysis assumes an outdoor H_2 refueling station that must be sheltered and built of non-combustible materials, with storage and dispensing equipment that must be above ground (not beneath power lines) and beyond a minimum distance from streets, structures and adjacent properties. Capital cost is estimated at approximately \$2.4 million.[4] The station is assumed to be available 24 hours per day, 365 days per year.

Given the high capital cost of the hydrogen refueling station, it is assumed that the station will be continuously manned, to reduce the likelihood of vandalism. A fully burdened cost of \$80,000 per person-year is assumed. Since labor is the largest component of annual costs (58% of the total), a fully automated and unmanned refueling station would reduce annual costs significantly. Annual costs also include maintenance and repair of the hydrogen refueling station (based on the installed costs of the individual major station components), property taxes, and insurance. The annual cost of a 227,000 SCFD hydrogen refueling station is estimated to be approximately \$686,000, assuming a natural gas cost of \$3/MMBtu and a power cost of \$0.06/kWh (EIA 2000).

MARKET-CENTERED HYDROGEN PRODUCTION

In this path (Fig. 6), individual steam reforming plants are located in demand regions with pipeline connections to local refueling stations. The pipeline network might also include industrial outlets to help equalize demand loads, although this was not considered in this analysis.

Hydrogen Production. The number of market-area hydrogen plants is estimated as a function of total U.S population, which is projected to be 350 million in 2030 (Bureau of the Census 2001), of which 65% are assumed to live in large metropolitan areas.

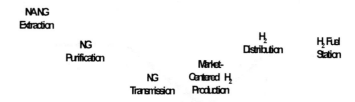

Figure 6. Market-Centered Hydrogen Pathway

Assuming one hydrogen plant per million people, a total of 228 regional hydrogen plants (i.e., 350 million / 1 million x 65% = 228) are needed in the year 2030.

The daily hydrogen production rate from the market-area plants is estimated as a function of the "tank-in" hydrogen requirement for vehicles in 2030 and the number of plants. Under our low market penetration case (Fig. 2), the production rate from a single hydrogen plant is 67.7 million SCFD. Under our high market penetration case, the production rate from a single hydrogen plant is 218.1 million SCFD.

As mentioned above, a parametric study was performed to determine economies-of-scale associated with the capital cost and annual operating and maintenance (O&M) costs of "conventional" hydrogen generation (including the reformer, shift reactor, PSA system, and NO_x control) as a function of hydrogen output, based upon various sources (Thomas et al. 1998, APC 2002, PR Newswire 2000, BOC Gases 2000). Figures 7 and 8 show the dependence of capital cost and annual O&M cost on hydrogen production rate.

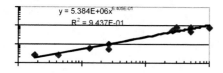

Figure 7. Capital Cost of a Market-Centered Hydrogen Production Plant

The relationship between the units in Figs. 7 and 8 is:

$$\text{Ton/day} = \text{MSCFD} \times (1{,}000 \text{ scf/Mscf}) \times (2 \text{ lb } H_2/\text{lb-mole}) / (359 \text{ scf/lb-mole}) / (2{,}000 \text{ lb/ton})$$

Hydrogen Pipelines. This analysis assumes that the design of the hydrogen distribution pipeline network will be similar to that of natural gas main pipelines, where service pipelines connect the end-user (the H_2 refueling stations in this case) with a larger-diameter pipeline ring which encompasses the community being served. Figure 9 provides a conceptual illustration of the proposed hydrogen distribution pipeline network for a market-area plant supporting 54 refueling stations. The assumed characteristics of the various pipeline components shown

---

[3] In ABC estimating, work is broken down into discrete, quantifiable activities, the costs of which are then estimated in terms of requirements for labor and materials on a per unit basis. The cost for each activity is equal to labor hours required times labor costs, plus material and subcontracting costs.

[4] Note that the parametric scaling approach taken in this analysis estimates a capital cost of \$4.6 million for a 750,000 SCFD refueling station, which is in agreement with the estimate of \$4.5 million in ORNL (1996).

in Fig. 9 are provided in Table 1. Distribution pipelines in the Resource-Centered pathway (see below) have the same characteristics.

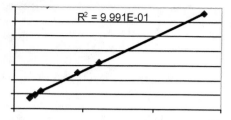

R² = 9.991E-01

Figure 8. O&M Costs of a Market-Centered Hydrogen Production Plant ($/yr)

The pressure drop in the network shown in Fig. 9 was estimated using the "Panhandle A" equation for steady-state flow of compressible gases in pipelines (McAllister 1990). The inlet pressure at the H_2 refueling stations is on the order of 200 psig.

Because a large fraction of pipeline cost is for installation, natural gas pipeline construction prices were used to estimate hydrogen pipeline costs. This analysis assumes that the installed cost of hydrogen pipelines is 80% higher than that of natural gas pipelines, based on Veziroglu and Barbir (1998). Table 2 provides unit costs of hydrogen pipelines assumed in this analysis.

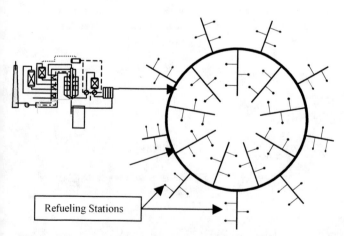

Figure 9. Conceptual Representation of a Hydrogen Pipeline Network Connecting a Market-Centered Production Plant with Refueling Stations

Refueling Stations. The number of H_2 refueling stations in a given year is estimated as a function of the "tank-in" hydrogen requirement and the refueling capacity of a single station, assumed to be 50,000 gasoline gallon

equivalents (GGEs) per month. In this analysis, each market-area hydrogen plant supports an average of approximately 180 refueling stations. The H_2 refueling station design is simpler than in the Decentralized pathway since SMR and PSA units are not needed. The power requirements for the compressors supplying hydrogen at 5,000 psig and the number of DOT-certified hydrogen storage tanks are estimated in a manner similar to that for the Decentralized pathway.

Table 1. Characteristics of H_2 Distribution Network in Market-Centered and Resource-Centered Pathways

Component	Radius (miles)	Diameter (inch)	Length (miles)
H_2 Pipeline Connecting Pipeline Ring with H_2 Production Plant	N/A	12	15
H_2 Pipeline Ring Encompassing Community	25	12	157
H_2 Pipeline Connecting H_2 Refueling Stations with H_2 Pipeline Ring	N/A	3	900 [a]

[a] Estimated assuming a total of 180 refueling stations, a service pipeline unit length of 15 miles, and 3 refueling stations per service pipeline.

Table 2. Unit Capital Cost of Hydrogen Pipelines as a Function of Inside Diameter

Diameter (inch)	Capital Cost of Natural Gas Pipeline ($/mile)	Assumed Cost Increase for H_2 Pipelines	Estimated Capital Cost of H_2 Pipeline ($/mile)
3	$200,000	80%	$400,000
9	$500,000	80%	$900,000
12	$600,000	80%	$1,000,000
14	$800,000	80%	$1,400,000

As in the Decentralized pathway, the unit cost of a hydrogen refueling station is computed using activity-based cost (ABC) estimating. For the Market-Centered pathway, the costs of natural gas compression and steam methane reforming are excluded, resulting in a capital cost of $560,000 for a single H_2 refueling station dispensing 50,000 GGE/month (vs. $2.4 million for a station that also produces H_2 by the SMR process). For comparison purposes, the cost of upgrading an existing gasoline refueling station to dispense 30,000 GGE/month of methanol has been quoted at $62-$70,000 (American Methanol Foundation 2001).

The annual cost of a H_2 refueling station supplying 50,000 GGE/month (227,000 SCFD) is estimated to be approximately $110,000 (versus $686,000 for a refueling station with on-site production of H_2). The station is assumed to be operated 24 hours per day, 365 days per year, and to be intermittently manned (a fully burdened cost of $80,000 per person-year is assumed). The estimate includes maintenance and repair (based on installed costs of individual major station components), property taxes and insurance.

RESOURCE-CENTERED HYDROGEN PRODUCTION

This pathway utilizes many of the same assumptions as the Market-Centered path, but with different assumptions for the size and location of hydrogen production facilities and the consequent need for gaseous hydrogen transmission pipelines. This path (Fig. 10) assumes that large, dedicated H_2 plants using SMR and PSA technology produce and purify hydrogen into a pipeline-quality product. H_2 plants are assumed to be located near major sources of domestic natural gas, which are geographically distributed throughout the U.S. as shown in Fig. 11.

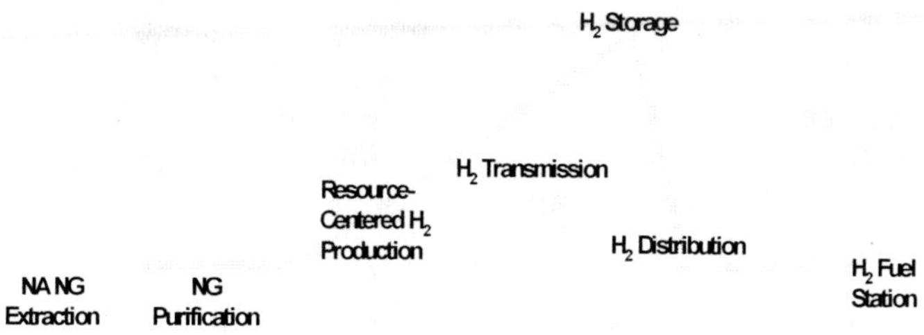

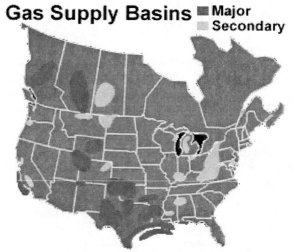

Figure 10. Resource-Centered H_2 Pathway

Gas Supply Basins ■ Major ■ Secondary

Figure 11. Natural Gas Supply Basins in North America (NGSA 2001)

Hydrogen Production. The scale and geographic distribution of Resource-Centered hydrogen plants are assumed to be analogous to those of the U.S. petroleum refining industry, which uses a complex series of processes to manufacture finished petroleum products from crude oil and other hydrocarbons. While refining began as simple distillation, refiners currently must use more sophisticated additional processes and equipment in order to produce the mix of products that the U.S. market demands. U.S. refining capacity, as measured by the daily processing capacity of crude oil distillation units alone, has been nearly constant in recent years. As of 2000, U.S. refining capacity totaled 16,540,990 bbl/day

(EIA 2001), with a maximum refining capacity of about 437,000 bbl/day at a single site (BP Amoco's Texas City, TX refinery). The minimum number of refineries needed to achieve current capacity is therefore about 38 (i.e., 16,540,900 bbl/day divided by 437,000 bbl/day). This analysis assumes an equal number of Resource-Centered H_2 plants produce transportation fuel in 2030.

Again, the daily production rate from Resource-Centered plants is a function of the "tank in" requirement under each of the two market penetration cases and the number of plants. Under the low market penetration case (4,060,000 MMscf/yr) hydrogen production at a single plant is 292.7 MM SCFD. Under the high market penetration case (13,100,000 MMscf/yr), hydrogen production at a single plant is 942.2 MM SCFD.

Hydrogen Pipelines. This analysis assumes that the hydrogen transmission pipeline network functions much like the network for long-distance delivery of refined petroleum products, where a limited number of petroleum refineries deliver product to metropolitan areas. Pipelines are the primary option for transcontinental transportation of refined petroleum products because they are at least an order of magnitude cheaper than rail, barge, or road alternatives. As of 1998, the U.S. had approximately 86,500 miles of product trunk lines and 114,000 miles of crude trunk and gathering lines (AOPL 2001). The average length of product trunk lines for transmission purposes is on the order of 600 miles per refinery (i.e., 86,500 miles divided by 152 petroleum refineries). This analysis assumes the same average length of H_2 transmission trunk lines connecting resource-centered production plants with metropolitan areas (in six separate lines for an average length of 100 miles per line).

Because of the low pressure-drop per unit-mile associated with high-pressure H_2 transmission pipelines, compressor stations are not required to boost the pressure along the H_2 transmission network. For moderate pipeline diameters like those assumed in this study (See Tables 2 and 3) pressure declines from 1,000 psig at the inlet of the transmission pipeline to 500 psig at the inlet of the hydrogen distribution network, to approximately 200 psig at the inlet of the refueling station.

Table 3.   Unit Cost of Hydrogen Pipelines: Market-Centered and Resource-Centered Pathways

Transport Segment	Pipeline Segment	Unit Cost ($/mile)
Transmission	H_2 transmission pipeline, 9" d, Low Market Penetration Case	$900,000
Transmission	H_2 transmission pipeline, 14" d, High Market Penetration Case	$1,400,000
Distribution	H_2 pipeline connecting pipeline ring with H_2 production plant	$1,000,000
Distribution	H_2 pipeline ring encompassing community	$1,000,000
Distribution	H_2 pipeline connecting H_2 refueling stations with H_2 pipeline ring	$400,000

Hydrogen Storage. H₂ leaving the transmission pipeline may be augmented by stored supplies to enhance deliverability. Storage acts as a buffer between the transmission pipeline and the distribution system, enhancing reliability by augmenting supply during peak use periods. Local storage of H_2 also reduces the time needed for the delivery system to respond to increased demand and allows continuous service if production or transportation services are interrupted.

Depending on the geology of an area, underground storage of hydrogen gas may be possible (Zittel 1996). Underground storage of hydrogen-natural gas mixtures occurs in France and Germany. Underground storage of helium, which diffuses faster than hydrogen, has been practiced successfully in Texas (Hart 1997).

Underground storage of hydrogen requires a large cavern or area of porous rock with an impermeable cap rock above. A porous layer of rock saturated with water is an example of a good cap rock layer. Other options include abandoned natural gas wells, solution-mined salt caverns, and manmade caverns. Figure 12 shows the different types of underground storage (Tobin and Thomson 2001).

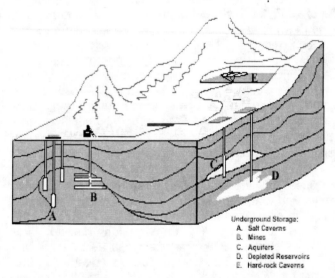

Figure 12. Types of Underground Storage (Tobin and Thomson 2001)

Underground storage is the most inexpensive means of storing large quantities of hydrogen (Amos 1998). Capital costs vary depending on whether there is a suitable natural cavern or rock formation, or whether a cavern must be mined. Using abandoned natural gas wells is the cheapest alternative, followed by solution salt mining and hard rock mining.

We assume that the number of underground H_2 storage fields is equal to the number of resource-centered H_2 production plants (i.e., 38). The mass of H_2 contained in each underground storage field is a function of the assumed days of gas storage and the daily hydrogen production rate. In this analysis, a seven-day supply of

underground hydrogen is assumed adequate to ensure continuous supply.

The capital cost of underground storage of gaseous H_2, has two components, one associated with development of the underground cavern to contain the H_2 and the second associated with the compressor station required to increase inlet H_2 pressure to levels necessary to maintain the integrity of the storage field (Amos 1998).

Operating costs for underground storage are limited to energy and maintenance costs related to compressing the gas for storage (and possibly boosting the pressure coming back out). In this analysis, annual costs also include water-cooling of the compressor station.

Refueling Stations. Capital and annual costs ($560,000 and $110,000/year, respectively) for a station dispensing 50,000 GGE/month are the same as in the Market-Centered pathway.

RESULTS

Total Investment. Figure 13 contrasts well-to-tank capital costs of the three natural gas-to-hydrogen pathways for two hydrogen production levels. The lower bars correspond to relatively low penetration of the LDV market coupled with high hydrogen-vehicle fuel efficiency, which translates into modest demand for H_2 transportation fuel. Such demand could also be representative of the early stages of a transition to hydrogen transportation fuel. The higher bars correspond to high market penetration and lower vehicular fuel efficiency, which translates into greater demand for H_2 transportation fuel. All costs are in undiscounted year 2000 dollars.

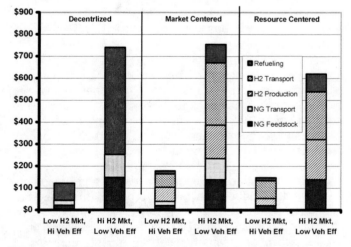

Figure 13. Fuel Infrastructure Total Investment by Pathway Component, Vehicle Efficiency and Market Penetration ($10^6$$)

As indicated by the lower series of bars for each pathway, well-to-tank costs to produce and deliver 5.2

billion GJ of H_2, enough to fuel a fleet of 20 million vehicles, range from $120-180 billion. Decentralized production is at the low end of this range, with no hydrogen transport component, but $78 billion for refueling.[5] Centralized production is 20-40% higher (of which $60-80 billion is for H_2 transport).

At higher cumulative production volumes (35.1 GJ, fueling a fleet of approximately 130 million vehicles), the well-to-tank costs of all pathways are similar but the distributions differ. Refueling accounts for more than two-thirds the cost of the Decentralized pathway, while hydrogen transport accounts for 35-40% of the cost of centralized pathways.

Unit Costs. In terms of unit cost, the comparison is quite different. At lower production volumes, the Decentralized pathway has the lowest unit cost ($23/GJ), declining to $21/GJ at higher production volumes. As shown in Fig. 14, both centralized pathways have higher costs at low production volumes, but comparable costs at higher volumes. Again, refueling represents approximately two-thirds the cost of the Decentralized pathway, while hydrogen transport represents 40% or more of the cost of the other two alternatives.

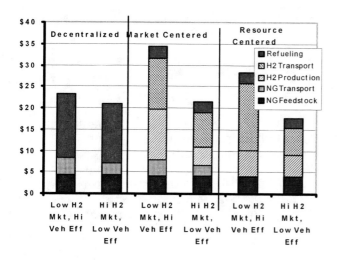

Figure 14. Unit Cost of Hydrogen by Pathway, Vehicle Efficiency and Market Penetration (2000 $/GJ)

CONCLUSIONS

For each of the three natural gas-based pathways considered in this study, total investment costs on the order of $600-800 billion have been estimated. Unit costs were found to vary not only by pathway, but also by volume of hydrogen delivered. For relatively small volumes of hydrogen transportation fuel, the Decentralized pathway has the lowest unit costs ($23/GJ); for higher production volumes, centralized pathways can have lower unit costs ($18-22/GJ). These

compare with a delivered gasoline cost of less than $7/GJ.[6] Even assuming higher crude oil prices and increased hydrotreating to further reduce sulfur content, gasoline is unlikely to exceed $10/GJ. Bringing hydrogen unit costs into this range will not be an easy task. Although the Decentralized pathway entails lower hydrogen transport costs, it more than makes up for those savings with higher refueling station costs. Reducing those costs is a major focus of the US Department of Energy's Hydrogen Program. Centralized pathways have much lower refueling station costs but higher hydrogen transport costs. Reducing those costs could bring these alternatives close to the range of gasoline.

Station reforming of hydrocarbons via partial oxidation or other processes is one possibility for reducing the cost of the Decentralized pathway. Another is reducing the cost of small-scale SMR through process improvements and volume production of components. For centralized pathways, cost reductions could come from production at very large plants, perhaps using site-specific feedstocks and production processes, and from distribution via lower cost pipelines. Since hydrogen transport is the largest single component of unit cost, potential cost-reduction strategies include materials research to reduce the cost of dedicated hydrogen pipelines, designing new natural gas pipelines to levels allowing future transport of hydrogen, and investigation of the limits of new and existing natural gas pipelines to transport combined mixtures. The latter two possibilities are also potential transition strategies. Transition issues, the perennial Achilles heel of hydrogen-fueled transportation, will be a key factor in evaluating potential pathways for delivering hydrogen transportation fuel. While this paper does not address transition issues per se, our findings can provide important inputs to that work.

ACKNOWLEDGMENTS

The authors gratefully acknowledge the assistance of Bob Kirk, Ed Wall, Pete Devlin, and Steve Chalk of the U.S. Department of Energy's Office of Advanced Automotive Technologies, as well as that of Larry Johnson of Argonne National Laboratory's Transportation Technology Research and Development Center; without their support, this work could not have been completed.

---

[5] In station reforming, the "refueling" category includes not only dispensing but also onsite H_2 production and 12-hr storage.

---

[6] Assuming $0.90 per gallon, excluding taxes and markups.

REFERENCES

Air Products and Chemicals, Inc. (APC). *Air Products To Build/Own/Operate World-Class Hydrogen Plant In Carson, California To Serve Equilon/Arco And Other Refining Customers In The Area*, available at http://www.airproducts.com/ corp/rel/98157.htm (accessed on Feb. 14, 2002).

American Methanol Foundation, *Methanol Refueling Station Costs*, prepared by EA Engineering, Science, and Technology, Inc. for the American Methanol Foundation, Feb. 1, 1999, available at http://www.methanol.org/fuelcell/special/statiocosts.pdf (accessed on March 21, 2001).

Amos, W.A. *Costs of Storing and Transporting Hydrogen*, National Renewable Energy Laboratory, NREL/TP-570-25106, Nov. 1998, available at http://www.nrel.gov/docs/fy99osti/25106.pdf

Association of Oil Pipe Lines (AOPL). *Fact Sheet, U.S. Oil Pipe Line Industry*, available at http://www.aopl.org/pubs/pdf/fs2000.pdf (accessed on Sept. 12, 2001).

Aasberg-Petersen, K. et al, *Membrane Reforming for Hydrogen*, Catalysis Today 46: 193-201, 1998.

BOC Gases. *BOC Gases/Foster Wheeler Bring on Stream Largest Hydrogen Facility in South America*, available at http://www.boc.com/news/gasnews/1-6-97.htm (accessed on Sept. 15, 2000).

Davis, Stacy. *Transportation Energy Data Book, Ed. 20*, Oak Ridge National Laboratory, ORNL-6959, Oct. 2000.

Hart, D. *Hydrogen Power: The Commercial Future of 'the Ultimate Fuel.'* London: Financial Times Energy, 1997.

Hydrogen Technical Advisory Panel (HTAP). *Realizing a Hydrogen Future*, National Renewable Energy Laboratory, Aug. 1999, available at http://www.eren.doe.gov/hydrogen/pdfs/brochure.pdf (accessed on Feb. 15, 2002).

Kraus Group Inc. *Fast Fill Hydrogen Dispensers*, available at http://www.krausind.mb.ca/kraus/hydrogen/overview.html (accessed on Aug. 16, 2001).

McAllister, E.W., editor. Pipeline Rules of Thumb Handbook, 2nd Edition, Gulf Publishing Company, Houston, TX, 1990.

Mintz, M. and C. Saricks. *IMPACTT5A Model: Enhancements and Modification Since December 1994*, Argonne National Laboratory, ANL/ESD/TM-154, Sept. 1998.

Natural Gas Supply Association (NGSA). *How Much Natural Gas Is There?* available at http://www.naturalgas.org/SUPPLY.HTM (accessed on Sept. 12, 2001).

Oak Ridge National Laboratory (ORNL). *Hydrogen Fueling Station's Time Has Come, Researchers Conclude*, Feb. 12, 1996, available at http://www.ornl.gov/Press_Releases/archive/mr19960212-00.html (accessed on Aug. 16, 2001).

Ogden, J., M. Steinbugler, and T. Kreutz, *Hydrogen Energy System Studies*, Proceedings of the U.S. DOE Hydrogen Program Review Meeting, Alexandria, VA, April 28-30, 1998.

Ogden, J., M. Steinbugler, and T. Kreutz. *Hydrogen as a Fuel for Fuel Cell Vehicles: A Technical and Economic Comparison*, presented at the National Hydrogen Assn., 8th Annual Meeting, Arlington, VA, March 11-13, 1997.

Oney, F., T. Veziroglu and Z. Dulger. *Evaluation of Pipeline Transportation of Hydrogen and Natural Gas Mixtures*, Intl. J of Hydrogen Energy, 19(10):813-822, 1994.

PR Newswire. Air *Products Brings Carson, California Hydrogen Plant On-Stream*, available at http://www.prnewswire.com/cgi-bin/stories.pl?ACCT=104&STORY=/www/stpry/02-02-2000/0001131007&EDATE= (accessed on Sept. 15, 2000).

RIX Industries. *Hydrogen Compressors*, available at http://www.rixindustries.com/products_hydr.html (accessed on Aug. 16, 2001).

Salm, W. *Hydrogen "Filling Station" May Pave the Way for Fuel Cells*, available at http://www.us-tech.com/Jan97/news/news001.htm (accessed on Aug. 27, 2001).

Suresh, B. *CEH Product Review*, Chemical Economics Handbook, SRI International, 2001.

Terrible, G. Shahani, C. Gagliardi, W. Baade, R. Bredehoft and M. Ralston. *Consider Using Hydrogen Plants to Cogenerate Power Needs*, Hydrocarbon Processing, 78(12), Dec. 1999, available at http://www.hydrocarbonprocessing.com/archive/archive_99-12/99-12_consider-terrible.htm.

Thomas, C.E., James, B., Lomax, Jr., F., Kuhn, Jr. *Integrated Analysis of Hydrogen Passenger Vehicle Transportation Pathways*, Draft Final Report, National Renewable Energy Laboratory, Subcontract AXE-6-16685-01, March 1998.

Tobin, J. and J. Thompson. *Natural Gas Storage in the United States in 2001: A Current Assessment and Near-Term Outlook*, U.S. Department of Energy, Energy Information Administration, 2001, available at

http://www.eia.doe.gov/pub/oil_gas/natural_gas/feature_
articles/2001/storage_outlook_2001/storage.pdf
(accessed on Sept. 12, 2001).

Turnquest, R. Sales and Marketing Manager, RIX
Industries, personal communication, Oct. 3, 2000.

US Department of Commerce, Bureau of the Census,
*Annual Projections of the Total Resident Population as of
July 1: Middle, Lowest, Highest, and Zero International
Migration Series, 1999 to 2100*, available at
http://www.census.gov/population/projections/nation/sum
mary/np-t1.pdf

U.S. Department of Energy, Energy Information
Administration (EIA). *Annual Energy Outlook 2001*,
DOE/EIA-0383(2001), Dec. 2000.

U.S. Department of Energy, Energy Information
Administration, *Oil Market Basics, Refining*, 2001
available at
http://www.eia.doe.gov/pub/oil_gas/petroleum/analysis_p
ublications/oil_market_basics/default.htm (accessed on
Sept. 12, 2001).

Veziroglu, T. and F. Barbir. *Hydrogen Energy
Technologies*, Emerging Technology Series, United
Nations Industrial Development Organization, Vienna,
1998.

Wang, M.Q. *Development and Use of GREET 1.6 Fuel-
Cycle Model for Transportation Fuels and Vehicle
Technologies*, Argonne National Laboratory,
ANL/ESD/TM-163, June 2001 available at
http://greet.anl.gov/pdfs/greet1-6summary.pdf.

Zittel, W. and R. Wurster. *Hydrogen in the Energy
Sector*, Ludwig-Bölkow-Systemtechnik GmbH, Aug. 7,
1996

2001-01-2528

Hydrogen Vehicle Fueling Alternatives: An Analysis Developed for the International Energy Agency

Susan M. Schoenung
Longitude 122 West, Inc.

ABSTRACT

Vehicles fueled by hydrogen produce virtually no pollutant emissions and are projected to become a serious alternative to hydrocarbon-fueled vehicles in the future. Current vehicle designs produce power by consuming hydrogen in either a fuel cell or a spark-ignition engine. Hydrogen can be stored on-board as either a compressed gas or a cryogenic liquid. Hydrogen-fueled vehicles need a ready source of fuel for routine use. The infrastructure to provide convenient fueling for passenger vehicles must be put in place in the near future, both in the US and internationally. Options for providing hydrogen fuel include electrolysis of water, reforming of hydrocarbon fuels at the fueling station; and transport of bulk hydrogen. This paper presents results of comparative analysis for passenger vehicle fueling options using either liquid or gaseous hydrogen.

INTRODUCTION

The comparison of options for hydrogen production, storage and utilization in the transportation sector is of great interest both in the US and in many other countries around the world, as a move to cleaner fuels is contemplated. The International Energy Agency (IEA) under its Hydrogen Implementing Agreement (Annex 13, "Design and Optimization of Integrated Systems") is sponsoring comparative analysis of hydrogen fueling alternatives. Two other projects are addressing remote power generation (on a Norwegian island) and residential power and heating (in a Netherlands suburban community).

The goal of these projects is to address specific hydrogen demonstration opportunities, with respect to energy independence, improved domestic economies and reduced emissions. These development activities are selected to provide both specific findings to the immediate region and also generic conclusions to the hydrogen energy community. Rigorous analysis aids both the specific project and can be extended to additional opportunities in participating countries.

The transportation analysis is based primarily on current U.S. and European experience with hydrogen fueling infrastructure. It is meant to contribute to the ongoing discussion, both in the U.S. and internationally, on the preferred choice for fueling options and hydrogen distribution alternatives.

PROJECT SCOPE

The overall scope of the transportation analysis includes a comparison of hydrogen passenger vehicle fueling options, including:

- Fueling alternatives, primarily various sources of gaseous or liquid hydrogen.
- Vehicle configuration alternatives, primarily various hydrogen storage and vehicle power plant selections, e.g. fuel cell or internal combustion engine (ICE).
- Driving cycle implications.
- Cost variations for electricity, natural gas and hydrogen with conditions and over international boundaries.

Figures of merit for the overall project include:

- Costs, both capital and operating, which contribute to the cost of hydrogen fuel as delivered to the vehicle
- Efficiency, both in terms of vehicle fuel economy, and also in terms of overall energy conversion efficiency
- Footprints for the fueling station alternatives
- Emissions, for each alternative system

This paper specifically addresses fueling station alternatives, where the primary considerations are:

- Liquid or gas hydrogen storage on-site
- On-site or off-site hydrogen production
- The utilization factor of the fueling station, i.e., how many vehicles "tank up" per day relative to capacity

CASE STUDIES - Six specific fueling station cases analyzed are described below and illustrated in Figure 1.

- Bulk liquid hydrogen transported to the fueling station by truck, stored as a cryogenic liquid and dispensed to the vehicle as a liquid. This case is suitable for a BMW or Ford hydrogen ICE vehicle.

The remaining cases represent fuel cell vehicles.

- Bulk liquid hydrogen transported to the fueling station by truck, stored as a cryogenic liquid and dispensed to the vehicle as a gas at 5000 psi.
- Bulk gaseous hydrogen transported to the fueling station by existing pipeline, stored as a compressed gas and dispensed to the vehicle as a gas.
- Gaseous hydrogen generated at the fueling station from natural gas by steam methane reforming, stored as a compressed gas and dispensed to the vehicle as a gas.
- Gaseous hydrogen generated at the fueling station from natural gas by a partial oxidation process, stored as a compressed gas and dispensed to the vehicle as a gas.
- Gaseous hydrogen generated at the fueling station by electrolysis, stored as a compressed gas and dispensed to the vehicle as a gas. (Grid electricity is assumed to power the electrolyzer. Renewable electricity will be considered later.)

FUELING STATION OPERATION - All fueling station alternatives assume the station is open for business 24 hours per day. Hydrogen is supplied to vehicles from the storage system. The storage is filled intermittently (once per week) in the case of truck delivery. Storage can be refilled continuously in the cases of pipeline gas and on-site production, or scheduled for optimum times, based on the cost of electricity, or other consumables.

The liquid-to-liquid system is assumed to operate similarly to that in the Munich airport [1] by interconnect to the vehicle and simple pumping of cryogenic liquid into the on-board storage tank. The liquid-to-gas system requires a pump and vaporizer, which are available technologies. The gas-to-gas system requires that pressure to the vehicles be maintained at 5000 psi, even though pressure in the storage tank will drop as it is emptied. Therefore a boost compressor is used to provide gas at suitable pressure to the vehicle.

TECHNOLOGY DESCRIPTIONS – Because this IEA Annex is addressing near-term opportunities, only commercial hydrogen production technologies were considered. The large reformer is assumed to be existing technology, centrally located with existing distribution networks of trucking routes or pipelines.

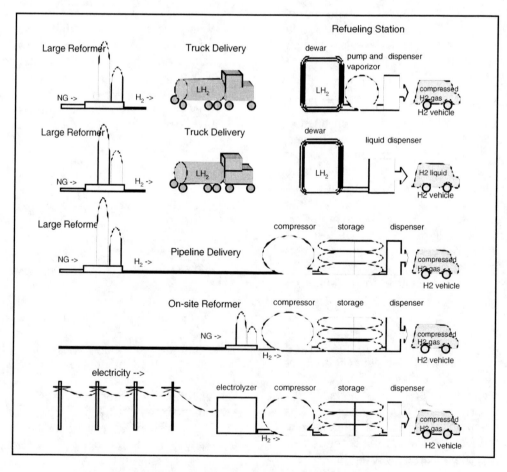

Figure 1. Fueling station alternatives

The smaller, on-site generator could be either a scaled-down steam methane reformer (SMR), an autothermal reformer (ATR), or a partial oxidation reformer (also called an under-oxidized burner, or UOB). For this analysis, both a small SMR and a UOB were considered. Small electrolyzers are commercial products, although improvements in performance and cost are projected with a growing market.

Storage technologies are also current, commercial types: a cryogenic liquid dewar for liquid hydrogen, and pressurized tanks for gaseous hydrogen. Other storage types, such as metal hydrides were not considered for bulk on-site storage. Storage compressors are commercial types with suitable flowrate to 5000 psi. A boost compressor is required to dispense at 5000 psi to the vehicle, once gas pressure in the storage tank drops as hydrogen is dispensed. The dispensers are still developmental, with limited commercial practice both for gaseous and liquid hydrogen. The liquid - to - gas system requires a pump and vaporizer, which is an available technology.

ANALYSIS

ASSUMPTIONS - A number of initial assumptions were made regarding operation of the transportation system. These include:

- Only passenger vehicles are currently under consideration.
- The vehicles are assumed to be fueled with hydrogen (either liquid or gas), i.e., fuel processing is NOT done on board. (Later analysis will look at on-board processing.)
- The base case calls for capacity to refuel 100 vehicles per day.
- Each vehicle fueling event requires 4 kg of hydrogen. This is a "consensus" value of multiple other studies and matches the current value for PNGV targets and for the Ford hydrogen ICE vehicle. (Later analysis will look at variations in this requirement based on variations in vehicle weight and configuration, due to storage type and power plant, and to driving cycle assumptions.) The resulting daily delivery at 100% utilization is 400 kg hydrogen per day.
- Available dispensing hours are 24 hours per day, 365 days per year.
- On-site hydrogen production capacity is sized to fill the required storage once per day. (A trade-off between production capacity or rating and operating hours was considered for the electrolyzer and UOB cases to minimize the combined capital cost and electricity cost.)
- Liquid delivery is scheduled once per week. Thus the dewar is sized for one week's dispensing plus 30%, to maintain proper conditions in the tank. [2]
- Compressed gas storage capacity is oversized by 40% to maintain adequate pressure for dispensing via boost compressor.

A fueling station consists of the hydrogen production unit or receiving area, storage and its associated facilities, and the dispensing area with two dispensing units. A representative layout is shown in Figure 2.

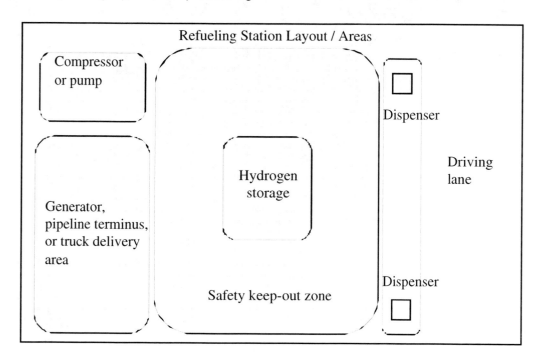

Figure 2. Fueling station layout for footprint analysis

17

COST ANALYSIS - The cost analysis consists of computing both capital cost and the delivered cost of hydrogen for each alternative station case.

Capital Costs - The capital cost components include:

- Hydrogen generator (for on-site cases)
- Storage system and auxiliaries
- Storage compressor (for gaseous cases)
- Boost compressor (for gaseous cases)
- Dispensers and auxiliaries

For costing purposes, we chose a commercial compressor for which performance and cost are known. Likewise, we chose a known boost compressor and a known pump and vaporizer system that fits the size parameters of this study. The fill time per vehicle is just over 4 minutes. Table 1 lists the capital cost assumptions for the base cases.

Table 1. Capital Cost Assumptions (base case)

Item	Units	Cost	Ref.
Pipeline terminus	$ per each	10,000	estimate
Small SMR	$/scfd	10	[3], [4]
UOB	$/scfd	4.6	[5]
Electrolyzer	$/kW H2	600	[6]
Liquid dewar	$/gal	9.5	[2]
Compressed gas cylinders	$/scf	2.2	[7], [8]
Storage compressors (254 scfm at 5000 psi)	$ per each	170,000	[3], [9]
Boost compressors	$ per each	80,000	[3]
Gas dispenser	$ per each	25,000	[3]
Liquid dispenser	$ per each	100,000	[1]
Storage regulator valve	$ per each	10,000	[3]
Pump and vaporizer	$ per each	36,000	[3]

Capital costs for the 400 kg / day station are additive:

Total Capital Cost =

Cost of generator + Cost of storage system + Cost of compressors + Cost of dispensers

Operating Costs - The operating cost components include:

- Capital charge
- Natural gas, if purchased
- Hydrogen, if purchased
- Electricity
- Catalysts or other consumables
- Operation and maintenance (O&M) expenses
- Labor

Table 2 lists the operating cost assumptions for the base cases. The cost of bulk liquid hydrogen is dependent on the delivery distance, although the delivery cost is almost independent of the amount, above a certain value. We assumed an average 800 mile delivery distance (round trip) and a minimum delivery of 10,000 gal LH2. On-site cases assume labor for three full-time persons (3-shifts), while delivery cases assume two.

Table 2. Operating Cost Assumptions (base case)

Item	Units	Value	Ref.
Capital charge rate	% of capital cost/yr	15	[3], [9]
Natural gas	$/MMBTU	6	Base case
Liquid Hydrogen (bulk)	$/GJ	26	[2]
Gaseous Hydrogen (bulk)	$/GJ	10	[8], [10]
Electricity – on peak	¢/kWh	7	[11]
Electricity – off peak	¢/kWh	2.5	[11]
O&M	% of capital cost	4	[3]
Labor	$/yr/person	50,000	[3]
Catalysts	$/1000 scf H2	.65	[3], [9]

Operating costs are additive for each applicable item in each case, and for the number of vehicles served.

Annual Operating Cost =

Capital charge + Cost of consumables + Cost of electricity + O&M + Labor

The capital charge, O&M and labor are independent of the number of vehicles served, whereas the cost of consumables and cost of electricity are proportional to the number of cars served. These assumptions were used in determining the impact of utilization factor on the cost of delivered hydrogen.

The cost of hydrogen dispensed, served or "delivered" to the vehicle is calculated based on the annual operating costs divided by the amount of hydrogen delivered, in GJ. The cost per driving cycle can be converted into an equivalent $/gal gasoline.

Delivered cost of hydrogen ($/GJ) =

Annual operating cost / GJ hydrogen dispensed per year

Each of the station components has an associated efficiency and/or power requirement for operation. Efficiency is applied to the conversion to hydrogen, on a HHV basis. Electricity costs are calculated on the basis of power required during the time of operation. Whether the operation occurs on-peak (6 hours per day) or off-peak (18 hours per day) is also taken into consideration. The values assumed in this analysis are tabulated in Table 3.

Table 3. Operating parameters of fueling station components

Item	Effi-ciency %	Power required	Ref.
Small SMR	67	0.6 kWh / 1000 scf H2	[3],[12]
UOB	69	17.38 kWh/1000 scf H2	[5]
Electrolyzer	80		[6]
Storage compressor (254 scfm at 5000 psi)		82 kW	[3]
Boost compressor (100 hp)		75 kW	[3]
Liquid pump (20 l/min)		1 kW	estimate
Pump and vaporizer (20 l/min)		22.4 kW	[3]

FOOTPRINT ANALYSIS - The footprint of the fueling station is of interest for comparing land area requirements for the various alternatives. The footprint was calculated for the following components, shown previously in Figure 2:

- Hydrogen generator (if applicable)
- Delivery area (if applicable)
- Hydrogen storage (including safety keep-out zone) and associated facilities
- Dispensing area, including driving lane.

The safety, keep-out zone or perimeter around the stored hydrogen is set by fire safety codes and standards, and depends on the state and volume of hydrogen as well as the fire-proofing provided for near-by equipment and buildings. Footprint assumptions for the various components are listed in Table 4.

Table 4. Footprint areas of fueling station components

Item	Units	Value	Ref.
Truck delivery area	m² per each	39	Estimate
Pipeline terminus	m² per each	20	[2]
Liquid dewar (a standing cylinder)	m² / kg stored	.206	[2]
Pressurized cylinders - stacked (5000 psi)	m² / kg stored	.826	[13]
Liquid dispenser area	m² per each	9	[9]
Safety perimeter	m	5	[14], [15]
Liquid pump	m² per each	negligible	Estimate
Evaporator and pump	m² per each	4	Estimate
Storage compressor	m² / scfm	.0236	[5]
Boost compressor	m² / scfm	4	[3]
Small SMR	m² / scfd	6.4×10^{-4}	[3], [12]
UOB Hydrogen generator	m² / scfd	3.13×10^{-4}	[5]
Electrolyzer	m² / scfd	4.9×10^{-4}	[6]
Gas dispenser area	m² per each	9	Estimate
Driving lane	m² per each	12	Estimate

Footprints of the individual components for each case are additive:

Fueling station footprint area =

Hydrogen source + Storage system (including safety zone) + Delivery area + Dispensing area

RESULTS AND DISCUSSION

This section contains results of the study to date, including capital costs, delivered hydrogen costs, sensitivity analyses, and fueling station footprints.

FUELING STATION CAPITAL COSTS - Figure 3 presents the components of capital costs for the six alternative cases. The results show that, on a capital cost basis, the on-site production systems are much more expensive.

COST OF DELIVERED HYDROGEN - The delivered cost of hydrogen is determined from the annual operating costs divided by the total GJ of hydrogen delivered per year. Figure 4 presents a comparison of delivered cost of hydrogen ($/GJ) for the 6 base cases assuming 100% utilization factor for the fueling station, i.e., 100 cars are filled every day. The components of the fuel cost for each case are shown. Unlike the capital cost results, these results show that, on an operating cost basis, all the options produce hydrogen at a somewhat comparable price.

SENSITIVITY STUDIES - One very important consideration is the impact of under-utilization of the fueling station, as may be the case in the early years of operation. Figure 5 shows a comparison of the cost of delivered hydrogen for 100% (100 cars per day) and 50% utilization (50 cars per day) for the 6 base cases. Capital charges, O&M and labor are independent of utilization, whereas consumables (i.e. natural gas or bulk hydrogen) and electricity use depend on the number of vehicles served. These results show that under-utilization more severely penalizes the on-site generation systems because of their greater capital cost.

FOOTPRINTS - The footprint area results for each of the six base cases are presented in Figure 6. The components of the footprints for each case are shown. Note that the liquid storage results are based on storing one week's mass of liquid hydrogen on-site, whereas only a day's worth of gaseous hydrogen is stored on-site. The footprints are dominated by the safety keep-out zone to a perimeter of 15 ft (5 m).

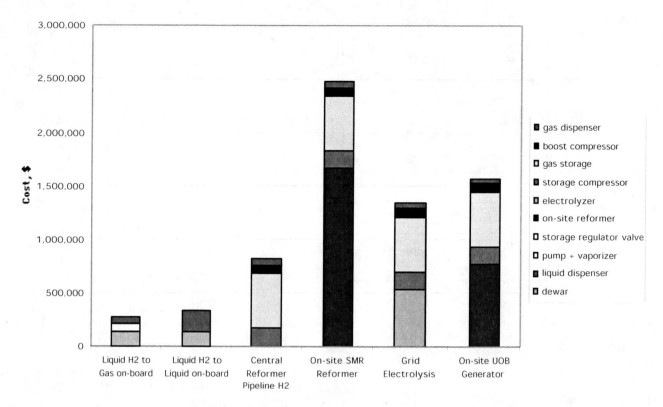

Figure 3. Components of Capital Costs for Base Cases

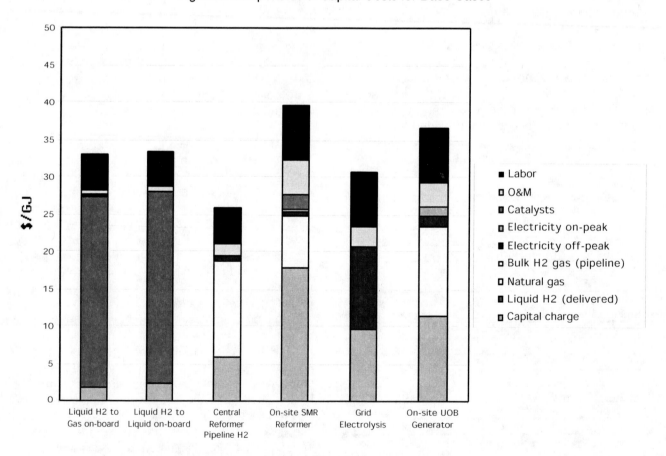

Figure 4. Components of Delivered Hydrogen Cost for Base Cases, 100% Utilization Factor

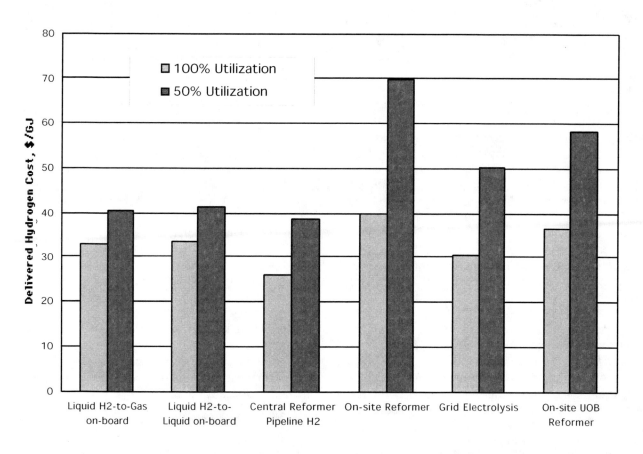

Figure 5. Cost of Delivered Hydrogen for 100% and 50% Utilization Factors

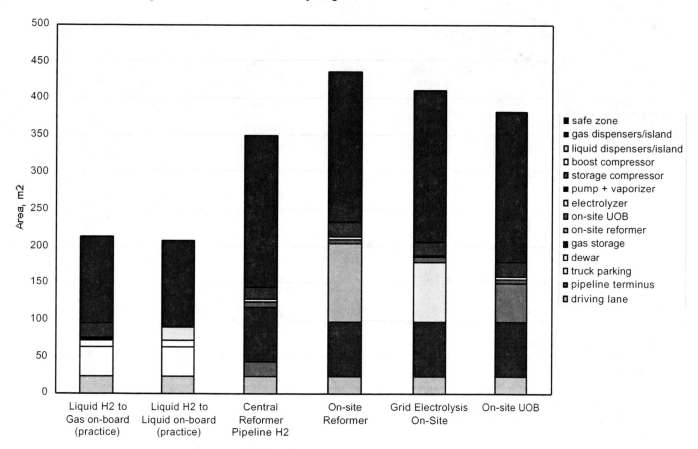

Figure 6. Footprint components for the 6 base cases, 100 vehicles served per day.

CONCLUSIONS AND FUTURE ANALYSIS

- From the analysis done to date, the major conclusions are:
- The capital cost of the on-site generation options is greater than bulk delivery of hydrogen to the fueling station. This assumes existing delivery infrastructure. The cost trend for the on-site generation technologies is downward.
- The small steam methane reformer is the most costly option; however there is growing commercial development of this technology and lower costs are projected.
- The served or delivered cost of hydrogen is generally lowest where existing hydrogen pipelines can provide gaseous hydrogen to the refueling station.
- When the station is underutilized, the delivered cost of hydrogen from all sources is always greater because the capital, O&M, and labor charges are independent of the utilization factor.
- For on-site generation alternatives, where the capital investment is higher, the delivered cost of hydrogen is particularly higher when the utilization factor is less than 100%.
- Footprint areas for on-site hydrogen generation alternative are somewhat greater than for bulk hydrogen delivered by truck or pipeline.
- Footprint areas for liquid storage options depend on the safety keep-out zone requirements.

This analysis suggests that while early stations may be based on the use of delivered hydrogen, on-site generation will be competitive as costs of these technologies decrease and full utilization of the stations is realized.

This project is ongoing; the following analyses are in progress:
- Calculations of fuel economy and system efficiency
- Implications of alternative on-board storage alternatives, especially with regard to weight
- Implications of alternative power plant selections, i.e., fuel cell compared with ICE
- Implications of on-board processing, with regard to both weight and cost per driving cycle
- Investigation of driving cycle variations
- A comparison of overall emissions
- Sensitivity analysis to natural gas and electricity prices
- Consideration of renewable sources of grid electricity
- Analysis and impact of upstream technology developments, especially hydrogen pipelines and large-scale SMR production

All of these analyses are being coordinated with other efforts of the IEA Hydrogen Programme, so that cost and performance assumptions are consistent. The emissions analysis, is being undertaken in conjunction with the efforts of the Paul Scherrer Institute in Switzerland.

ACKNOWLEDGEMENT

This work was performed under the sponsorship of the International Energy Agency and the Department of Energy, contract no. DE-FC36-00GO10607.

A COMMENT ON DATA SOURCES

An attempt has been made in this analysis to use current or near-term values for costs and equipment sizes. In many cases, vendors were approached for these data because no published literature exists. Some vendors and manufacturers were fairly cooperative in providing data; others less so, for competitive reasons. Some information was available from commercial web sites. When necessary, estimates from other studies have been used. The most significant are those of Ogden, et al, Thomas, et al, and various other Department of Energy laboratory sources.

REFERENCES

1. BMW / ARAL Munich Airport literature, 2000

2. Jeff Richards, Praxair, personal communication, based on Praxair product literature, February 2001

3. J. M. Ogden, et al. "Hydrogen Energy System Studies," DOE Report, Contract No. XR-11265-2, 1995

4. J. Keller, Sandia National Laboratories, personal communication, based on DOE Hydrogen Program test results, February 2001

5. Hydrogen Burner Technology, technical specifications, based on HBT product literature. HBT web site: www.hydrogenburner.com; personal communication, Greg Hummel, HBT, 2001

6. M. J. Fairlie and P. B. Scott, "Filling Up with Hydrogen 2000," Proceedings of the 2000 DOE Hydrogen Program Review, 2000; Stuart web site: www.electrolyser.com

7. B. D. James, F. D. Lomax, and C. E. Thomas, " Integrated Analysis of Hydrogen Passenger Vehicle Transportation Pathways," Activity Report, NREL Subcontract No. AXE-6-16685-01, 1997.

8. W. A. Amos, "Costs of Storing and Transporting Hydrogen," NREL Report No. NREL/TP-570-25106, 1998

9. C. E. Thomas and B. D. James, "Analysis of Utility Hydrogen Systems," DOE Report, Contract No. ACG-8-18012-01, 1998

10. J. M. Ogden, Hydrogen Energy System Studies," Proceedings of the 1999 U. S. DOE Hydrogen Program Review, 1999

11. J. Iannucci, Distributed Utility Associates, personal communication, based on Energy Information Agency data, 2000

12. D. Edlund, "A Versatile, Low-Cost and Compact Fuel Processor for Low Temperature Fuel Cells," IdaTech web site: www.idatech.com

13. A. R. Abele and A. P. Niedzwiecki, "Quantum's Experience – Leveraging DOE Funding to Accelerate Development and Commercialization of Advanced Hydrogen Storage Technologies." *Proceedings of the 12th Annual U.S. Hydrogen Meeting.* Washington D.C. National Hydrogen Association, 2001.

14. NFPA. Standard 50A: Standard for Gaseous Hydrogen Systems at Consumer Sites and Stand and 50B: Standard for Liquified Hydrogen Systems at Consumer Sites. National Fire Protection Association, Inc. 1999.

15. O. Weinmann, Safety analysis of the liquid hydrogen storage system in Hamburg, 1994, (in German) and personal communication, HEW-AG (Germany), 2001.

Safety Considerations in Retailing Hydrogen

R. F. Cracknell, J. L. Alcock, J. J. Rowson, L. C. Shirvill and A. Üngüt
Shell Global Solutions

ABSTRACT

To be used in public, untrained people must be able to handle hydrogen with the same degree of confidence and with no more risk than conventional liquid and gaseous fuels. Physical properties relevant to the safety of hydrogen as a fuel are reviewed and compared to gasoline, LPG and methane. The key parameters are flammability, detonability, ignition energy, materials compatibility, buoyancy and toxicity. For many years, Shell has conducted an experimental programme on gas safety, which has recently been extended to include hydrogen. A selection of results from this programme is presented.

INTRODUCTION

If hydrogen is to be a fuel used by the general public, untrained people must be able to handle hydrogen with the same degree of confidence and with no more risk than conventional liquid and gaseous fuels. In this context, risk should be regarded as the product of the probability of an incident or accident occurring and the magnitude of its hazardous consequences.

Prevention and control of accidental formation and ignition of large volumes of fuel-air mixtures are crucial to the safe operation of hydrogen systems. Adequate understanding of the overpressures generated in an accident situation is essential for the protection of the public and also of operating plant and safety equipment.

The safe handling and use of hydrogen requires an appreciation of its physical properties in each of the forms in which its use as a vehicle fuel is considered. These include as a gas, liquid and adsorbed to another material, e.g. metal powders, carbon nanofibres, glass beads.

Fire and explosion hazards must be carefully assessed to determine the relative safety of a fuel for each potential application. Hydrogen can be safer than conventional fuels in some situations, and more hazardous in others. The relative safety of hydrogen compared to other fuels must therefore take into consideration the particular circumstances of its accidental release. Several reviews (DTI [1], Barbir[2], Cadwallader & Herring[3], Ringland [4]) have been published that consider the safety of hydrogen as a vehicle fuel. These

have concentrated primarily on hydrogen safety related to the vehicle itself rather than the wider context of a fuelling infrastructure. In this wider context it is vital to understand the risks associated with fuel delivery to forecourt or on-site manufacture as well as the risks associated with releases from on-site storage and dispensing operations.

2. CHARACTERISTICS OF HYDROGEN

2.1 Propensity to Leak

Hydrogen gas has the smallest molecule and has a greater propensity to escape through small openings than liquid fuels or other gaseous fuels.

For releases from pressurised systems the flow is likely to be choked. For the same pressure and hole size, hydrogen would leak approximately 2.8 times faster than natural gas and 5.1 times faster than propane on a volumetric basis. However the energy density of hydrogen is lower than that of methane or propane such that its energy leakage rate would be 0.88 times that of methane and 0.61 times that of propane.

2.2 Hydrogen Embrittlement

Prolonged exposure to hydrogen of some high strength steels can cause them to lose their strength, eventually leading to failure. Proper choice of materials to avoid these risks is required.

2.3 Dispersion

Hydrogen gas is more diffusive and under most conditions more buoyant than gasoline, propane or methane and hence tends to disperse more rapidly if released. The one exception is for cryogenic releases of hydrogen where the very cold vapour cloud initially formed can be denser than the surrounding air.

At low concentrations the effects of buoyancy become less significant because the density of the fuel-air mixture is similar to that of air. Buoyancy effects are also less significant for high momentum releases. For these releases the orientation of the release will determine the direction in which the hydrogen cloud forms. These releases are the most likely to occur for the high-pressure systems probable for hydrogen storage.

2.4 Flammability and Ignition

Hydrogen has much wider limits of flammability in air than methane, propane or gasoline and the minimum ignition energy is about an order of magnitude lower than for other combustibles, Table 1.

Table 1: Flammability and Ignition Characteristics

	Hydrogen	Methane	Propane	Gasoline
Flammability limits (vol. % in air)				
Lower limit (LFL)	4	5.3	2.1	1
Upper limit (UFL	75	15	9.5	7.8
Minimum ignition energy (mJ)	0.02	0.29	0.26	0.24

The wide range of flammability of hydrogen-air mixtures compared to other combustibles is in principle a disadvantage with respect to potential risks. A hydrogen vapour cloud could potentially have a greater volume within the flammable range than a methane cloud formed under similar release conditions.

On the other hand there are only minor differences between the lower flammable limits (LFLs) of hydrogen and methane, and those of propane and gasoline are even lower. In many accidental situations the LFL is of particular importance as ignition sources of sufficient energy are often present to ignite a fuel-air mixture once a flammable concentration has been reached. In some circumstances (e.g. low momentum releases) the dispersion characteristics of hydrogen may make it less likely that a flammable mixture will form than for the other fuels. In addition the 4 vol.% LFL for hydrogen only applies to upward propagating flames. For downward propagating flames experiments have shown that between 9 and 10 vol.% hydrogen is required [5,6]. For methane the difference between LFLs for upward and downward propagating flames is less, 5.3 versus 5.6 vol.%.

In practical release situations the lower ignition energy of hydrogen may not be as significant a differentiation between the fuels as it first seems. The minimum ignition energy tends to be for mixtures at around stoichiometric composition (29 vol.% for hydrogen). At the LFL the ignition energy for hydrogen is similar to that of methane, Figure 1. In addition many so called weak ignition sources such as electrical equipment sparks, electrostatic sparks or sparks from striking objects involve more energy than is required to ignite methane, propane and other fuels. A weak electrostatic spark from the human body releases about 10 mJ.

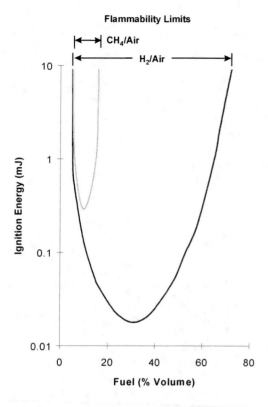

Figure 1: Minimum Ignition Energies

Static electricity generation has been implicated as causing ignition in hydrogen venting situations[7]. However there are many spurious ignition sources and phenomena that could cause ignition and this is an area that is poorly understood. Among these is "diffusion ignition" whereby a shock wave from expansion of high-pressure gas into air is postulated to cause local auto-ignition [8,9].

The minimum autoignition temperature of hydrogen at ambient pressure is higher than that of methane, propane or gasoline, Table 2. However the autoignition temperature depends on the nature of the source. The minimum is usually measured in a heated glass vessel, however if a heated air jet or nichrome wire is used the autoignition temperature of hydrogen is lower than the other fuels.

Table 2: Autoignition Temperatures

Autoignition Temperatures (°C)	Hydrogen	Methane	Propane	Gasoline
Minimum	585	540	487	228-471
Heated air jet (0.4 cm diameter)	670	1220	885	1040
Nichrome wire	750	1220	1050	

To summarise, in the event of a fuel spill you could expect hydrogen to form a flammable mixture more readily than methane due to its higher buoyancy promoting its rapid mixing in air and due to its slightly lower flammable limit and larger flammable range. Gasoline would be orders of magnitude slower than hydrogen or methane at forming a flammable mixture of the same size and propane would be somewhere in between.

Although the rapid mixing property of hydrogen leads to the rapid formation of flammable mixtures, they also lead to its ready dispersal and thus generally shorter duration of the flammable hazard than for the other fuels (on an equal volume basis). Despite the UFL of hydrogen being much higher than that of methane, its higher buoyancy leads to it dispersing to concentrations below the LFL more quickly. However this does not apply to spills of cryogenic liquid. At its boiling point the density of hydrogen vapour approaches that of air, while for methane it is greater. This can lead to the formation of transiently non-buoyant flammable mixtures extending considerable distances from the spill.

2.5 Combustion characteristics

Hydrogen gas can burn as a jet flame with combustion taking place along the edges of the jet where it mixes with sufficient air. In the open, flammable mixtures undergo slow deflagration (also known as a cloud or flash fire). Where the flame speed is accelerated e.g. by extreme initial turbulence, turbulence from obstacles, or confinement, the result is an explosion. An extreme example is a detonation where the flame speed is supersonic. Once initiated a detonation is self-sustaining (i.e. turbulence or confinement are not required) as long as the combusting mixture is within the detonable range.

Hydrogen flames are different to hydrocarbon flames in that there is little or no soot formation and the lower radiation from the flame makes the flame itself hotter than hydrocarbon flames. Objects engulfed by a hydrogen jet flame tend to heat up faster than when in the same size methane flame because the convective component is considerably greater. However the lower radiative component means that there is less radiation transferred to objects (or people) outside the flame.

The quenching gap for hydrogen is smaller than for methane, propane and gasoline (Table 3). The quenching gap is the largest passage that can prevent propagation of a flame through that passage when it is filled with a flammable fuel-air mixture. When the dimension of a passage is less than a critical width a flame front is extinguished because heat transfer and/or free radical loss become great enough to prevent flame propagation. The quenching gap depends on gas composition, temperature, pressure and passage geometry. The design of flame arrestors and flame traps are dependent on quenching gap measurements. The small quenching gap for hydrogen requires tighter tolerances, which makes equipment capable of containing hydrogen flames more difficult to build than equipment for hydrocarbon flames.

Table 3: Quenching Gap

	Hydrogen	Methane	Propane	Gasoline
Quenching gap at NTP (mm)	0.6	2	2	2

The *laminar burning velocity* of a gas-air mixture is the velocity of the cold reactants (without flame stretch) relative to the plane comprising the flame. The *flame speed* is the speed of the flame front as would be observed by a stationary observer outside the flame. This will be higher than the laminar burning velocity because the position of the flame front is driven by the expansion of hot combustion products, (particularly where the hot combustion products can not be vented). The flame speed may increase further with the effect of turbulence, from the either the gas release mechanism or from the interaction of the flame front with obstacles.

In extreme cases of flame acceleration, the mode of combustion may switch from deflagration to detonation. A detonation explosion is more severe than a deflagration explosion, the overpressures generated are higher (in the region of 20 to 1 versus up to 8 to 1 [10]) and hence much greater physical damage is possible.

Generally, the propensity for a combustible mixture to support the transition from deflagration to detonation (DDT) is related to the laminar burning velocity (Table 4). The higher the laminar burning velocity is, the greater is, the tendency for DDT to occur.

Table 4: Explosion Characteristics

	Hydrogen	Methane	Propane	Gasoline
Detonability limits (vol. % in air) Lower limit (LDL) Upper limit (UDL)	11-18 59	6.3 13.5	3.1 7	1.1 3.3
Maximum laminar burning velocity (m/s)	3.46	0.43	0.47	
Concentration at maximum (vol. %)	42.5	10.2	4.3	
Laminar burning velocity at stoichiometric (m/s)	2.37	0.42	0.46	0.42
Concentration at stoichiometric (vol. %)	29.5	9.5	4.1	1.8

A deflagration can make the transition to a detonation if the concentrations in the flammable cloud are within the detonable range and the flame front can accelerate to a speed above the sonic velocity in air. The flow driven by the expansion of hot combustion products means that DDT is a particular concern for one-dimensional cases, such as pipelines. However, even for spherical gas clouds, DDT can occur in principle if the dimensions of the cloud are large enough to provide sufficient run-up distance for the flame to accelerate, and if there are turbulence promoting structures to accelerate the flame or there are pressure wave reflecting bodies such as walls.

A further concern is whether the turbulence in an emerging high pressure hydrogen gas jet release, coupled with its exceptionally high burning velocity, may also provide the conditions for detonation rather than deflagration to occur on ignition. Direct detonation of a hydrogen gas cloud is less likely than a deflagration as the ignition energy required is in the 10 kJ range (c.f. figure 1), the minimum concentration is higher and the detonation range is narrower than the flammable range.

3. Shell Hydrogen Experiments

As was discussed in the last section, hydrogen has a much higher laminar burning velocity than conventional hydrocarbon fuels such as gasoline, methane and propane. It is much more buoyant than these other fuels and so the build-up of a significant flammable gas cloud is less likely. Nevertheless, following an accidental release occurring on a retail forecourt if a flammable gas cloud can build up and find an ignition source, then one might expect to see much higher overpressures for hydrogen-air explosions in comparison to the other fuels discussed. Moreover, the risk of a DDT occurring cannot be completely discounted *a priori*.

The question as to whether there are any credible scenarios associated with a forecourt release of hydrogen which could give rise to detonation is vital therefore to properly assess the risks associated with retailing hydrogen as a fuel. Shell Hydrogen has sponsored an experimental programme, conducted by Shell Global Solutions (UK) to address these concerns.

3.1   Explosion experiments with quiescent gas mixtures

Figure 2 Test rig with example configuration

Mixtures of hydrogen and air were constrained within an open box containing obstacle grids (Figure 2). This was achieved by wrapping polythene sheets to the side of the box and affixing them with magnetic strips. The polythene sheets come away immediately following ignition. The largest dimension of the rig is 1.2m and the rig may be considered to be representative of lengthscales of dispensing equipment on a forecourt. The flame front is accelerated by obstacles in the array. Experiments were conducted with varying levels of obstacle "congestion".

For these experiments, hydrogen was compared to ethylene and acetylene, which are known to be more reactive gases then methane or propane. The laminar burning velocity reported in the literature [11,12] for each fuel is presented in Figure 3. As can be seen from the figure and from table 4, the maximum in the hydrogen laminar burning velocities occurs

for somewhat rich fuel air mixtures. Nevertheless for the whole range of fuel-air ratio the laminar burning velocity of hydrogen is higher than acetylene or ethylene. Based on the discussion in section 2.5, one might expect therefore that explosion severity would be in the rank order Hydrogen>Acetylene>Ethylene.

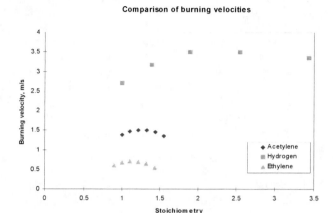

Figure 3 Laminar Burning velocities of hydrogen, acetylene and ethylene as a function of stoichiometry

Figure 4 shows a representative hydrogen explosion test in progress. Flame speeds and explosion overpressures were recorded as a function of time. Experiments were conducted for each of the fuels at various values of stoichiometry and levels of congestion. For a given set of conditions, the repeatably observed rank order for explosion severity was found to be Acetylene>Hydrogen>Ethylene.

It appears therefore that for the length scales we have investigated, the severity of hydrogen explosions is not as great as indicated by the magnitude of the laminar burning velocity of hydrogen relative to other hydrocarbons. Also for length scales representative of dispensing equipment, there appear to be no credible events in which ignition of a quiescent hydrogen-air cloud could lead to a deflagration to detonation transition.

We are currently looking for joint industry partners to investigate unconfined hydrogen explosions in length scales that would be representative of onsite hydrogen manufacturing equipment.

Figure 4 Example of hydrogen explosion experiment 28 ms after ignition

3.2 Jet Release Experiments.

The previous section described the combustion of quiescent hydrogen air mixtures. Although the situation allows an experiment to be properly characterised and understood, it does not represent a completely realistic scenario, since a gas cloud will most likely have been formed by the mixing with air of a jet of pressurised hydrogen or the evaporation of a liquid spill.

We have therefore undertaken a series of experiments releasing a hydrogen jet into a congested region. This serves as a "demonstration" rather than an experiment since it is difficult to characterise. Nevertheless for congested regions of size 1-1.5m, although local high overpressures were observed on jet ignition, there was no evidence of any event likely to lead to a deflagration to detonation transition from a 25 bar release. We are in the process of confirming this conclusion at pressures up to 150 bars. We are also seeking joint industry funding to investigate the hazards associated with jet releases of hydrogen at 350 bar and 700 bar.

CONCLUSION

The comparative safety of hydrogen can only be judged based on the particular circumstances in which it will be used. In some instances hydrogen's propensity to dissipate quickly, relatively high LFL and low energy density may make it a safer fuel than the alternatives considered. In other cases hydrogen's wide flammable range, small quenching gap and propensity to detonate might make it less safe.

Detailed safety analyses will be required to establish the relative safety of different fuels for each specific application and potential accident scenario.

The high laminar burning velocity of hydrogen compared to other hydrocarbon fuels is a reason to be concerned about the possible severity of an explosion following ignition of an accidental hydrogen release in a congested region. In particular there has been a concern as to whether there are any credible forecourt accident scenarios in which a detonation could occur. Our experimental programme suggests a detonation is unlikely for gas clouds of the length scale of dispensing equipment. Moreover for the length scales that we have investigated, the severity of hydrogen explosions is not as great as indicated by the magnitude of the laminar burning velocity of hydrogen relative to hydrocarbons.

ACKNOWLEDGMENTS

The author thanks Shell Hydrogen for permission to publish.

CONTACT

Roger.F. Cracknell
Shell Global Solutions (UK)
Cheshire Innovation Park
P.O. Box 1
Chester
CH1 3SH
United Kingdom
Email: roger.f.cracknell@opc.shell.com

REFERENCES

1. Directed Technologies Inc. Direct-hydrogen-fuelled proton-exchange-membrane fuel cell system for transportation applications, Hydrogen safety report, DOE/CE/50389-502, May 1997.
2. Barbir, F. Safety issues of hydrogen in vehicles, International Association for Hydrogen Energy, Technical Papers, http://www.iahe.org/hydrogen%20safety%20issues.htm
3. Cadwallader, L.C. & Herring, J.S., Safety issues with hydrogen as a vehicle fuel, INEEL/EXT-99-00522, 1999.
4. Ringland, J., Safety issues for hydrogen-powered vehicles, Sandia National Laboratories, SAND94-8226, March 1994.
5. Berman, M., A critical review of recent large-scale experiments on hydrogen/air detonations, Nuclear Science and Engineering, Vol. 93, pp 321-347, 1986.
6. Swain M.N. et al, Gaseous fuel transport line leakage - natural gas compared to hydrogen, pp 161-170 in Alternative Fuels: Alcohols, hydrogen, natural gas and propane (SP-982), Proceedings of 1993 Society of Automotive Engineers' Future transportation technology conference, San Antonio, TX, Aug 9-12, 1993.
7. Edeskuty, F.J. & Stewart, W.F., Safety in the handling of cryogenic fluids, Chapter 7, Plenum Publishing Corporation, New York, 1996.
8. Bond J., Sources of Ignition Flammability Characteristics of Chemicals and Products, Butterworth Heinemann, Oxford, 1991.
9. Wolanski, P. and Wojcicki, S. Investigation into the mechanism of the diffusion ignition of a combustible gas flowing into an oxidizing atmosphere. In 14th Symposium (International) on Combustion, The Combustion Institute pp 1217-1223, 1974.
10. The SPFE Handbook of Fire Protection Engineering, Ed. DiNenno, P et al, 2nd Edition, National Fire Protection Association, June 1995.
11. Scholte, T.G. and Vaags, P.B., Combustion and Flame, 3, 495 (1959).
12 Dixon-Lewis, G., Garside,J.E. Kilham, J.K., Roberts, A.L. and Williams, A., Inst. Gas Eng. Comm. 861 (1971).

2000-01-1538

Overview of Hybrid Electric Vehicle Safety and the Potential for Hydrogen Ignition by Static Electricity

Vahid Motevalli and Kartik Venkat Bulusu
Transportation Research Institute, The George Washington University

ABSTRACT

Hybrid Electric vehicles (HEVs) and Fuel Cell vehicles (FCVs) are showing promise of success as a commercial product as they are being developed by the industry. It is only prudent to closely consider safety issues for both post-crash and failure (non-crash) scenarios.

A review of most relevant technologies being considered for HEVs was performed to identify potential hazard conditions and interactions between systems and sub-systems within these vehicles. Energy storage, propulsion systems and fuel storage were examined for different configurations of such vehicles.

It is anticipated that plastics, composites and other nonconductive materials will be used more widely in future cars. This can result in an increased propensity to generate substantial static charge levels. Furthermore, the presence of high-voltage and high-current lines, batteries, electric motors and other components not present in conventional vehicles with alternative fuels or hydrogen justifies this examination. Areas of potential hazards in post-crash as well as nominal operating conditions are discussed. Special emphasis has been placed on investigating ignition of hydrogen-air mixtures by electro-static discharges.

INTRODUCTION

A number of automakers have announced and have developed FCVs using compressed hydrogen. The controversy about hydrogen safety and whether hydrogen is anymore unsafe than gasoline (or CNG, etc.) may never be resolved to the satisfaction of everyone and numerous work can be cited on this topic [1-8]. A more extensive bibliography on this topic can be found in reports by Motevalli[9] and Thomas[10]. Regardless, it is clear that all technologically feasible must be done to prevent a hydrogen-induced fire or explosion in these future vehicles under conditions that would have not been life-threatening in a conventional vehicle. A loss of life in a HEV or FCV due to a fire or explosion that would not have readily occurred in a conventional vehicle under similar circumstances can potentially result in a great setback to the technologies being developed. The potential sources of ignition in a HEV/FCV are dominated by the electrical drivetrain and energy storage (e.g. batteries). Aside from the use of alternative fuels, existence of high voltage and current sources and lines is another major difference with conventional vehicles. Although systems may not be fully designed yet, electric vehicles and concept designs are good guides for potential sources of ignition. Following are some of the possible sources of ignition in HEVs:

1. Hot surfaces
2. Exhaust gas leak (for IC or Gas Turbine engines, fuel reformer burner or heat exchanger)
3. Low or high voltage arc
4. Electric spark across wires (loose wire, worn insulation, damaged wire)
5. Any heating element (current limiter failure)
6. Broken halogen light bulb
7. Electric contacts, switches or relays

Given these and many other potential ignition sources in a post-crash condition, issues such as hydrogen detectability, invisibility of a Hydrogen flame and a large flammable range would be of particular concern in terms of safety.

OVERVIEW

Safety is a subjective measure in that its boundaries are frequently defined by regulation, which are mostly prescriptive in nature and not performance-based [9]. Often, the probability of an event that could have an adverse affect on safety is used to make decisions about the potential hazards. This is especially challenging if the systems being evaluated are not fully developed and certainly with no historical data. Perhaps, it is important to first define "safety" for our purposes. Any condition that poses a life-threatening hazard to a vehicle occupant is a safety concern. Such a situation may arise in a crash, where the vehicle structure and safety systems (i.e. seat belt, airbags) initially protects the occupant, but other hazards, such as a fire, may develop. The evaluation of safety can be organized in the following fashion:

1. Hazard assessment for operation in extreme conditions.
2. Unpredictable component interactions leading to a hazardous condition.
3. Sub-system component level failures leading to catastrophic failures.
4. Hazards developed due to crash situations.

The focus of this paper is to identify the areas of potential safety concerns with emphasis on Fuel Cell vehicles based on the current state of technologies being considered for the PNGV concept vehicles. For the purpose of clarity, when referred to HEV here, it is implied that FCV is a sub-set and is included in the consideration, while the reverse is not true. The hybrid vehicle, by design, will have high voltage electric motors and wiring in compact areas and perhaps in proximity to IC engines, fuel lines and fuel storage areas. This provides potential fire problems, especially when the vehicle is to be used over 5 to 10 years resulting in deviation from design performance of components, degradation and improper maintenance. Concern about electric and HEV safety arising from conditions itemized in 1-4 is not new. One of the first reviews of this issue was published by Brown and Hall [11] who mainly examined the safety standards, and lack of them, for future "electric and hybrid vehicles". Their paper noted many potential fire, explosion and electrical hazards mostly emanating from concern with hydrogen generation in batteries and electrolyte spillage. The important role that static electricity plays is developed later in this paper.

HAZARD ASSESSMENT

In order to arrive at a reasonable evaluation of such complex systems as the proposed HEV/FCV, critical systems and sub-system components need to be identified and evaluated. Furthermore, critical possible hazardous events need to be identified and evaluated to devise appropriate counter-measures. Currently, a number of potential and prototype approaches exist for these vehicles. Examination of all potential vehicle designs with possible event scenarios and resulting hazards forms a very large matrix. To organize this process, clear definition and appropriate tools were devised.

Figure 1 shows the process used to perform the hazard assessment for the HEV. The process is initiated by selecting the vehicle configuration, which includes selecting the propulsion unit (e.g. fuel cell or CIDI), the type of energy storage system, fuel type and system and the electric motor/controller. The characteristics and specifications of each component and sub-component is provided in this step [1].

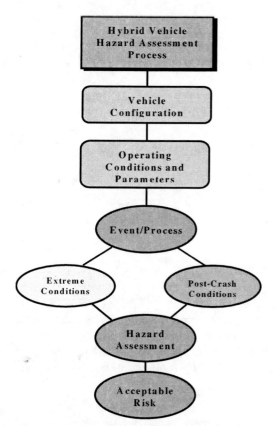

Figure 1.   Hazard Assessment Process [9]

In general, the configurations possible for a HEV (i.e., type of energy storage, propulsion unit and drivetrain design) are numerous. For each configuration, there would be numerous conditions that can lead to a hazard. In order to identify the areas of potential safety concerns, an interactive software program, Hybrid Vehicle Hazard Assessment (HVHA), was developed. Figure 2 shows the first screen of the program, which shows the three major portions for the vehicle as the a) propulsion system, b) Energy storage system and c) fuel system. Depending on the choices at each level made, the software would guide the user to system configuration selection [9]. The vehicle configuration is divided into three segments: propulsion system, energy storage and fuel system. The top level of the HVHA chart as implemented in the HVHA program is shown in Figure 2. The chart is comprised of different branches that can be considered in selection of the propulsion system, energy storage and fuel systems.

It is the potential for fire and explosion from fuels used in the propulsion system and oxidant (normally air) which is obviously present is a major concern. In the HEVs the added challenge is that the electric motor(s), energy storage systems and power electronics and power controllers all provide potential sources of ignition. These sources of ignition are not normally present in conventional vehicles in such abundance and with such high level of energy discharge potentials. The presence of high temperature sources and surfaces as an ignition source (such as exhaust gases and systems) vary and often would not pose a higher potential risk than conventional vehicles (and they may even pose a reduced risk). Given the highly competitive nature of this industry and challenges ahead and given the substantial investment by the industry and the government, introduction of the new technologies, particularly fuel cell systems must be done in a manner that meets the public safety expectations.

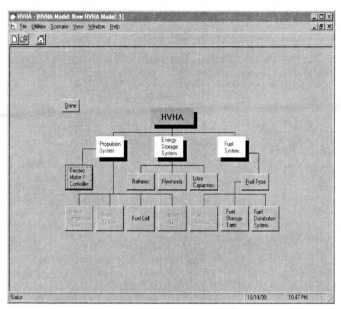

Figure 2.   HVHA First Screen [9]

DISCUSSION OF POTENTIAL SAFETY ISSUES

A report published in February 1997 [12] listed over 11 major safety issues among which the following categories are perhaps the most relevant to hybrid vehicles.

1. Fire (and explosion)
2. Heat Burns
3. Steam Burns
4. Chemical Burns
5. Refueling hazards
6. Electric Shock and static electricity discharge
7. Toxic fumes
8. Collision
9. Rollover
10. Noise
11. Manufacturing

The focus in this section will be limited to Fuel Cells, which have emerged, in the last decade, as a potential replacement for internal combustion vehicles. The low to no emission for a FCV makes the cost-benefit assessment for the potential additional hazards an important study. Some of the key issues are listed below.

1. Hydrogen storage, distribution within the vehicle and even small quantity storage of generated hydrogen from fuel reforming may pose safety hazards. The basic problem will be the highly diffusive property of hydrogen, the intensity of combustion (high flame temperature and high energy per unit mass), and low flammable limits. There may be a certain need for hydrogen sensors placed in such a vehicle. Refueling of hydrogen is an ongoing issue addressed in report published recently [10].

2. Hydrogen gas release as a result of battery damage or malfunction.

3. Hydrogen gas release as a function of the chemical processes used by the battery to produce electricity (discharge) and/or the amount of current applied across the battery cells during the charging process.

4. Fuel Reformers – different types of reformers and variations in fuels can produce hazardous conditions. Component failures can lead to a leak of primary fuel or generated hydrogen. Hot surfaces, especially the burner unit and its exhaust can serve as ignition sources.

5. The fuel cell stacks need to be durable and reliable for long-term operation in a vibrating environment. Some high pressure and temperature operations create potential thermal management and fire hazard concerns.

6. Electrolyte flammability is of concern for the phosphoric acid fuel cells. Chemical burns can be an additional concern for post-crash and maintenance conditions.

7. Increased pressure (i.e. water vapor pressure) due to increased temperature can produce a marginal hazard.

The quantity of hydrogen flowing through the fuel lines can be easily determined in the fuel cell system. The fuel cell thermodynamics can be used to develop a simple relationship for the quantity of hydrogen needed to power a fuel cell as shown by Eqn. (1). The three types of fuel cells currently used and under development, the Proton Exchange Membrane (PEM), Solid Oxide(SO) and the Phosphoric Acid (PA), are considered and the hydrogen flow rate for each is calculated by Eqns 3-5. The three relations are simply different based on the operating temperature of these fuel cells. Figure 3 shows the mass flow rate of hydrogen based on the power rating and fuel cell efficiency. In hazard assessment, one must consider the worst case. That is when the power is maximized and the efficiency is at a minimum. In that case, for the 50% efficiency, all three fuel cells approximately require the same amount of hydrogen flow. At a peak power of 60 kW, about 1.2 g/s of hydrogen would be flowing. If the main fuel line entering the fuel cell stack were to be ruptured in a post-crash scenario, it is reasonable to assume that an ignition source (hot surface, spark or electrostatic discharge) would be available. This would result in a 140 kW flame based on the heat of combustion of hydrogen. If the line has sufficient pressure, the flame may act as a torch. Thus, quick shut-off would be extremely important. However, should a pinhole develop in the fuel line, the resulting flow (assume 10% of the total) would result in a very weak flame (about 15 kW) and may not pose a high risk given the low probability of the event.

$$\dot{n}_{H2} = \frac{\dot{W}_{act}}{W_{rev}\eta}$$

(1)

where;

$$W_{rev} = \Delta H_T - T\Delta S_T$$

(2)

and

$\dot{W}_{act}$ = actual power (kW)

η = fuel cell efficiency

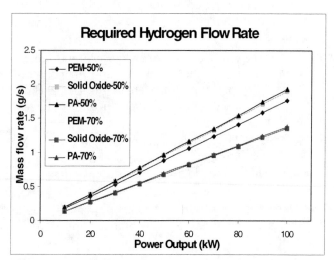

Required Hydrogen Flow Rate

Figure 3. Fuel flow rate for the candidate Fuel Cell types as a function of power and efficiency [9]

For PEM fuel cell, T = 80 °C

$$\dot{n}_{H_2} = \frac{\dot{W}_{act}}{22618\eta}$$

(3)

For Solid Oxide fuel cell, T = 500 °C

$$\dot{n}_{H_2} = \frac{\dot{W}_{act}}{210468\eta}$$

(4)

For Phosphoric Acid (PA) fuel cell, T = 500 °C

$$\dot{n}_{H_2} = \frac{\dot{W}_{act}}{207000\eta}$$

(5)

Another critical case in the context of this discussion is a leak from a compressed hydrogen tank in an enclosed space, such as a residential garage. Although, this is a very low-probability event, the high-risk associated with it and the tremendous challenge it poses, make it a good candidate to demonstrate the extreme difficulties that may be posed.

The rate of mass flow through a hole in a tank under pressure and assuming an isentropic expansion is:

$$\dot{m}(t) = m_o(t)\frac{(RT_o)^{1/2}}{V_o}A_eC_r$$

(6)

Where the terms are defined in details in [9] and $m_o(t)$ is the instantaneous mass of the compressed gas in the tank [10].

$$m_o(t) = \left[m_o^{\frac{(1-\gamma)}{2}} + K_oA_oC_rt\frac{(\gamma-1)}{2}\right]^{\frac{2}{1-\gamma}}$$

(7)

For comparison, in addition to hydrogen, CNG, which has been considered as an alternative fuel is used here to calculate the flammability of the gas mixture. The total mass of hydrogen or CNG (based on methane) needed to achieve the PNGV range of 608 km (380 miles) can be calculated based on the 3X fuel efficiency set as a goal. Using the above for a full tank, the following values were used:

Hydrogen: m=2.309 kg, P=34.5 MPa

CNG: m=5.537 kg, P=10.43 MPa

In an actual system, a higher quantity of CNG may be needed since the PEM FCV efficiency would be much higher.

Calculation of this scenario using equations 6 and 7 and assuming a total enclosure volume of 140 m³ is shown in Figure 4. Since both hydrogen and natural gas are much lighter than air, it is assumed that a flammable mixture would first form near the ceiling of the enclosure and therefore the scenario should consider a smaller volume (a more severe case than here). In Figure 4, the full volume is used and it clearly shows that a hole diameter of 5 mm in case of hydrogen, leaves no time for sensing and response. The 1 mm hole is a pin head size that can develop in many ways. Approximately 45 seconds is needed for the entire volume to reach a flammable limit (i.e. a well-ventilated area). It is clear that a full tank with a sizable hole would pose a grave safety hazard for either gas and sensing may not be a solution. This result also shows that a hydrogen leak in an open space (perhaps post-crash), can also produce hazard conditions since local concentration around the vehicle and within crevasses of the vehicle can easily reach the lower flammability of hydrogen and CNG, 4% and 5%, respectively.

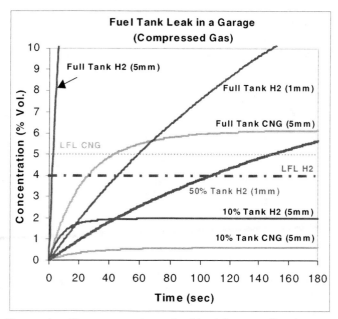

Figure 4. Compressed gas fuel tank in a confined space [9]

STATIC ELECTRICITY AS A SOURCE OF IGNITION

The development of electrical charges might not be in itself a potential fire or explosion hazard [13]. The rapid discharge of energy in the form of a spark or a sudden recombination of separated positive and negative charges are typically the cause for concern. In order for static discharge to be a source of ignition, four conditions must be fulfilled:

1. There must be an effective means of static generation.

2. There must be a means of accumulating the separate charges and maintaining a suitable difference of electrical potential.

3. There must be a spark discharge of sufficient energy, and

4. Spark must occur in a flammable mixture.

Electrostatic discharges are produced when electric fields, produced by high-accumulated charges, break down. Some or all of the stored electrical energy is converted to heat during the breakdown and this heat may provide the thermal energy needed to initiate the ignition of a flammable mixture. The energy available for ignition depends on the type of discharge and its origin. Ignition will occur when the energy of discharge exceeds the Minimum Ignition Energy (MIE). Furthermore, the ions produced in the process affect the kinetics of the ignition chemistry.

Gas discharge phenomena can generally be divided into those that occur at low pressures (partial vacuum) («$1.33*10^{-2}$ MPa) such as glow discharges and those that occur at atmospheric pressures and higher pressures (»0.1 MPa) such as corona or spark discharges [14].

Corona discharges are relatively low power electrical discharges that take place at or near atmospheric pressure. The corona is invariably generated by strong electric field using small diameter wires, needles, or sharp edges on an electrode. Corona discharge exists in several forms, depending on the polarity of the field and the geometrical configuration of the electrodes. For positive corona in the needle plate electrode configuration, discharge starts with the burst pulse corona and proceed to the streamer corona, glow corona, and spark discharge as the applied voltage increases (see Figure 4). For negative corona in the same geometry, the initial form of discharge will be the Trichel pulse corona, followed by the pulseless corona and spark discharge as the applied voltage increases (Figure 5).

The continuity of arc or glow discharge is guaranteed by the presence of an adequate source of current or high voltage. If the total energy available for a discharge is limited, for example by the presence of a large capacitor, the electrical discharge tends to manifest itself in the form of rapid impulse type filament discharges known as sparks at pressures above atmospheric pressure. The time and space dependent development of such a discharge represents a complex physical phenomena that depends on numerous parameters such as pressure (if above atmospheric), electrode geometry, and electrode gap.

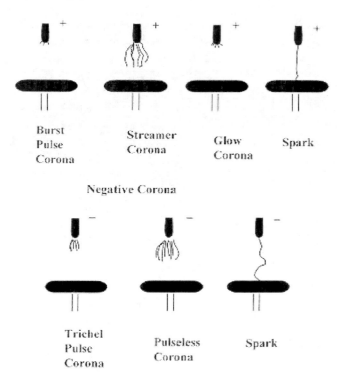

Figure 5. Corona Discharges [14]

STATIC ELECTRICITY DISCHARGE IN HYBRID ELECTRIC VEHICLES – Any automobile has the potential for electrostatic generation. As plastics and other nonconductive materials are used more widely in cars, it becomes increasingly easy to generate substantial static charge levels. As electronic controls grow in use, the sensitivity of vehicle components to electrostatic discharge damage increases. Symptoms range from annoying spark discharges between a driver and car body to failures of electronic control modules or even discharges across plastic sections of fuel lines, storage system, electric motor, etc.

At least four attributes of prototype hybrid electric vehicles make their Electro Static Discharge (ESD) properties different, and more critical, than in conventional automobiles.

1. The hybrid vehicles will have to make extensive use of plastic and composites to minimize weight and energy consumption.

2. Low rolling resistance (for dynamic friction) tires are important for these vehicles. They use less carbon black filler, and tend to be significantly less conductive than conventional tires.

3. The electrical architectures being considered for hybrid vehicles tend to provide multiple opportunities for charge storage. Isolated battery systems can experience ESD arcs through instrumentation and through electronic components. The hybrid and electrical vehicles contain high voltage lines, electric motor and electrical controllers, which provide opportunity for ESD in extreme conditions as well as failure modes and post-crash scenarios.

4. Highly ignitable fuels, such as hydrogen can be present, in the vehicle for propulsion system or evolved from batteries, especially during fast recharging or over-discharging. Aside from factors specific to HEVs some of the issues present in conventional vehicles may be exacerbated in HEVs. A section to fuel flow induced static electricity is included here due to high velocity flow and use of alternative fuels in HEVs.

Hydrogen Safety – Extensive work has been done on hydrogen safety such that by Thomas[10]. The key issues that need to be considered for hydrogen safety are:

1. Wide flammability range of H_2.
2. High diffusion rate.
3. Buoyant/odorless and hard to detect.
4. Explosive potential (high energy per unit mass).
5. Very low ignition energy is required.
6. Invisible flame.

Minimum Ignition Energies (MIE), have been measured for flammable mixtures using standard test methods [15,16], which employ a spark as a source for ignition energy. The spark is produced using two electrodes. Tests have shown that approximately 0.25 mJ of stored energy can ignite the optimum mixture of most gaseous hydrocarbons (like propane, toluene, acrolein etc.) in air (see Table 1). Unsaturated hydrocarbons may have lower

ignition energies. Hydrogen and acetylene have the lowest MIE values, making hydrogen use in fuel cell vehicles a particular concern.

Among the 6 issues noted above, difficulties in detection, low required ignition energy and the high flammability range (in combination) pose the highest challenges.

Table 1. Minimum Ignition energy for air/fuel mixtures[9]

Substance	Minimum Ignition Energy (mJ)
Ethyl acetate	1.42
Acetone	1.15
Methane	0.30
Propane	0.25
Toluene	0.24
Gasoline Vapor	0.24
Cyclopropane	0.18
Acrolein	0.13
Ethylene	0.08
Acetylene	0.017
Hydrogen	0.017

CONDITIONS FOR EXPLOSIONS – Static charge is usually found on non-conductive material where high resistivity prevents the movement of the charge. There are at least two situations where the static charge can move quickly and be dangerous in a combustible atmosphere. The first is where a grounded object intensifies the static field until it overcomes the dielectric strength of the air and allows current to flow in the form of a spark. The second case is where the charge is on a floating conductor such as an isolated metal plate. Here the charge is very mobile and will flash to a proximity ground at the first opportunity.

Previous research has shown that even ESD discharge from the body can ignite a flammable mixture [17]. In this work and work by Rose and Priede [18] and Felstead, et al.[19], the effect of electrode characteristics have been investigated. These researchers have determined that great variations in the MIE and propensity of ignition can be achieved by varying the shape and material of the electrodes. A plot from the work of Rose and Priede is shown in Figure 6. In these works, the electrode pairs were the same shape and mixing of the shapes (e.g. sphere with a sharp pointed electrode) were not examined. Furthermore, the MIE measurements essentially rely on spark type discharges and other ESD mechanisms have not been examined. These other conditions could be present in a HEV environment and a fuller understanding is warranted.

Experimental Set-up – Apparatus for the Standard Test Method for Minimum Ignition Energy and Quenching Distance in Gaseous Mixtures (ASTM E582) [16] was modified and a 5-liter spherical reaction vessel has been constructed for the study of the behavior of ESD for variations in electrode geometry's and diameters. This method covers the determination of minimum energy for ignition (initiation of deflagration) and associated flat – plate ignition quenching distances [21]. The results of this investigation will be reported in a later publication. Rose and Priede have shown [18] that MIE decreased with a corresponding decrease in the diameter (Figure 6). Similar study was conducted with electrodes of different material. Results indicated that MIE decreased as electrical conductivity increased. It was also found that variations in electrode geometry from hemispherical to flanged did not produce a detectable effect on MIE. However, they could clearly define the quenching distance of the particular mixture below which it was impossible to achieve ignition.

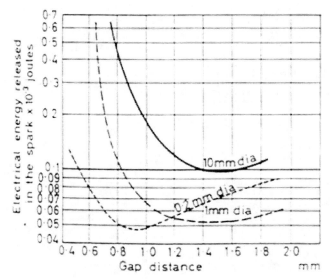

Figure 6. Effect of Electrode Configuration on the Minimum Ignition Energy in the Spark the 50 per cent Hydrogen – Air mixture for various gap distances. [18]

Static electricity discharges do happen very frequently without giving rise to ignitions or explosions. The reason is that most discharges do not have enough energy (or rather energy density) to start the necessary interactions between enough molecules of the vapor and the oxygen of the air. Or, may be more precisely, that the incidental ratio between the concentrations of the vapor and the oxygen requires more energy dissipated per unit volume than can be delivered by a likely static discharge.

The dependence of ignition energy on mixture for hydrogen and methane is shown in Table 1 [9]. At the lower fuel concentrations, which are applicable to open-air conditions, methane and hydrogen are quite comparable. Even so, there are situations where hydrogen's very low ignition energies do have a practical importance. A sudden discharge of hydrogen through a relief valve or burst disk has, in some cases, ignited. Electrostatic ignition of gases in vent stacks has also been observed [20]. These issues require a further study of ignition of hydrogen by ESD and effect of electrode geometry.

CONCLUSION

The large number of potential HEV designs, complexity of the systems and the large number of non-conventional systems and sub-systems, warrant this and other safety reviews. It is clearly shown here that the presence of high voltage and current devices and lines in the HEV along with hydrogen or other alternative fuel (particularly gaseous fuels) will increase the potential for a fire and explosion compared to conventional vehicles. This requires FMEA type analysis of these designs and redundant gas detection and safety systems. Furthermore, a simple enclosure gas leak case, demonstrates that research FCV should consider the use of on-board fuel reformers from the safety prospective.

This paper also examined the electro-static discharge as a source of ignition, particularly with regard to hydrogen. The effect of electrode shape and material and types of ESD warrants further investigation for hydrogen-air mixtures.

ACKNOWLEDGMENTS

This work has been supported by a grant from DOE Office of Advanced Automotive Technologies.

REFERENCES

1. Hord, J., "Is Hydrogen A Safe Fuel", International Journal Of Hydrogen Energy, V. 3. pp. 157-176, 1978.

2. Karim, G. A., "Some Considerations of the Safety of Methane, as an Automobile Fuel Comparison with Gasoline, Propane and Hydrogen Operation", SAE Technical Paper 830267, Feb. 28 - Mar. 4, 1983.

3. Hansel, J. G., Mattern, G. W., Miller, R.N., "Safety Considerations in the Design of Hydrogen-Powered Vehicles", Int. J. Hydrogen Energy, V. 18, No. 9, pp. 783-790, 1993.

4. Reider, R., Edeskuty, F., "Hydrogen Safety Problems", Int. J. of Hydrogen Energy, V. 4, pp. 41-45, 1979.

5. Ringland J. T., "Safety Issues for Hydrogen Powered Vehicles", Sandia Report 94-8226 UC-407, Sandia National Laboratory, 1994.

6. Shooter, D., Kalelicar, A., " Benefits and Risks Associated with Gaseous Fueled Vehicles", Arthur D Little Report To the Massachusetts Turnpike Authority, Boston, MA, May 1972.

7. Swain, M. R., "A Comparison of H2, CH4 and C3H8 Fuel Leakage in Residential Settings" Int. J. Hydrogen Energy, V. 17, No. 10, pp. 807-815, 1992.

8. Das, L. M., "Safety Aspects Of A Hydrogen-Fuelled Engine System Development", Int. J. Hydrogen Energy, V. 16, No. 9, pp. 619-624, 1991.

9. Motevalli, V., An Approach to Hazard Assessment for Hybrid Electric /PNGV Vehicle, Phase-1 Report, Submitted to Office of Advanced Automotive Technologies, Office of Transportation Technologies, Office of Energy Efficiency and Renewable energy, DOE, 1999.

10. Thomas, C.E., "Direct-Hydrogen-Fueled Proton-Exchange-Membrane Fuel Cell System for Transportation Application: Hydrogen Vehicle Safety Report", Prepared for DOE under contract DE-AC02-94CE50389, Directed Technologies, Inc., May 1997.

11. Brown, P. J. and Hall, R. T., "Safety Considerations for Electric and Hybrid Vehicles", SAE Technical Paper Series, pp. 291-315, Paper # 821164, 1982.

12. The Environmental Impacts and Safety of Electric Vehicles, Int'l Center for Technology Assessment, Rpt. #2, Feb. 26, 1997.

13. NFPA 77, "Static Electricity", 1993 Ed., ANSI/NFPA 77.

14. Chang, J. S. & Kelly, A.J. "Gas Discharge Phenomena" Handbook of Electrostatic Processes, Eds. N.Y. Marcel-Dekker, 1995, pp. 152-156.

15. Jonassen, N., "Explosions and Static Electricity", Electrical overstress/electrostatic discharge Symposium Proceedings, pp. 7.1.1-7.1.7, 1995.

16. Additional Sources ASTM Std. # E 582 - 88, " Standard Test Method for Minimum Ignition Energy and Quenching Distance in Gaseous Mixtures", American Society for Testing and Materials, 1988.

17. Wilson, N., "The Ignition of Natural Gas by Spark Discharges from the Body", Inst. Phys. Conf. Ser. No. 66, Session 1, Electrostatics 1983, pp. 21-27, Oxford.

18. Rose, H.E., Priede, T., "Ignition Phenomena in Hydrogen – Air Mixtures", Seventh Symposium (International) on Combustion, The Combustion Institute, 28 August – 3 September, 1958, pp. 436 – 445.

19. Felstead, D.K., Roger, R.L., and Youn, D.G., "The Effect of Electrode Characteristics on the Measurement of the Minimum Ignition Energy of Dust Clouds", Inst. Phys. Conf. Ser. No. 66, Session IV, Electrostatics 1983, pp. 105-110, Oxford.

20. Pidoll, U.V., Kramer, H. and Bothe, H., Research Report 508, "Avoidance of Ignition of Gasoline/Air mixture during refueling of Motor vehicles at filling stations". Feb 1996, DGMR Division, German Society for Petrol, Coal Science and Technology.

21. Litchfield, E.L., Hay, M.H., Kubala, T.S., and Monroe, J.S., "Minimum Ignition Energy and Quenching Distance in Gaseous Mixtures," BuMines. R.L. 70009, August 1967, 11 pp.

CONTACT

Vahid Motevalli, Ph.D., P.E.
GW Transportation Research Institute,
The George Washington University,
20101 Academic Way, Ashburn, VA 20147

INTERNAL COMBUSTION ENGINES

2002-01-0243

Ford Hydrogen Engine Powered P2000 Vehicle

Steven J. Szwabowski, Siamak Hashemi, William F. Stockhausen, Robert J. Natkin, Lowell Reams, Daniel M. Kabat and Curtis Potts
Ford Motor Co.

ABSTRACT

The first known, North American OEM vehicle powered exclusively by a hydrogen fueled internal combustion engine (H$_2$ICE) has been developed and tested. This production viable, low cost, low emission vehicle is viewed as a short term driver for the hydrogen fueling infrastructure ultimately required for fuel cell vehicles. This vehicle features a highly optimized hydrogen IC engine, a triple redundant hydrogen safety system, and a dedicated gaseous hydrogen fuel system. The vehicle and its test results are presented in this paper.

INTRODUCTION

A Zetec based 2.0 liter H$_2$ICE was mapped in the Ford Scientific Research Laboratory engine dynamometer facility [1,2], at both a constant fuel air equivalence ratio of 0.55, when throttled, and over a range from 0.12-0.70, when unthrottled; thus, providing a starting point for vehicle calibration development. The engine was tested with a 14.5:1 compression ratio, fixed cam timing, but without EGR or after-treatment system. Following the completion of the engine dynamometer development, the hydrogen engine team developed a plan to build and test the engine in a vehicle. This port fuel injected (PFI) four valve per cylinder engine was integrated, together with a five speed manual transaxle into a P2000 [3], an aluminum intensive five passenger family sedan developed by Ford to support PNGV work, Figures 1 and 2.

FUEL PROPERTIES - Before proceeding further to describe the vehicle control system, perhaps it might be beneficial to highlight some hydrogen properties that directly impact the engine control system and its calibration [4]. First, the stoichiometric air fuel ratio for pure hydrogen is 34.2:1, thus an equivalence ratio of 0.55 represents an operating air fuel ratio of 62.2:1. The combustion properties of hydrogen, its low ignition energy and wide flammability limits make operation at

Figure 1: P2000 powered by hydrogen fueled internal combustion engine.

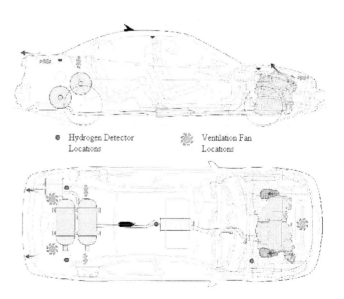

Figure 2: Vehicle wire frame showing locations of fuel and safety system within the vehicle.

this equivalence ratio and those as low as 0.12 possible. Hydrogen flammability limits range from 4 to 75 volume % in air (1 to 7.8 volume % for gasoline), while its minimum ignition energy is 0.02 mJ, roughly one tenth that of gasoline. Finally, hydrogen flame velocities of 2.37 m/sec compared to 0.415 m/sec for gasoline, result in ignition timing that is significantly more retarded than the gasoline Zetec.

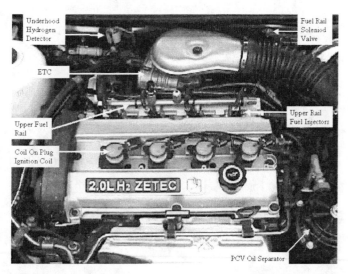

Figure 3: Hydrogen internal combustion engine, based on Ford 2.0L Zetec.

CONTROL SYSTEM – The vehicle control system consisted of an electronic throttle operated in a pedal follower configuration, an air meter based open loop fuel control, and electronically controlled coil on plug ignition. In addition, a PCV coalescing oil separator was added to prevent recirculation of oil into the combustion chamber reducing the potential for preignition or backflash [1,5,6,7].

FUEL SYSTEM – The fuel system, see Figure 4, was designed for safety as well as functionality. All of the major components of the hydrogen fueling system were located in the vehicle trunk except for the fuel rail solenoid and the fuel injectors. Figure 5 shows a photograph of the hydrogen fuel system and plumbing in the trunk. The fuel line feeding hydrogen gas from the trunk to the engine was located under the floor pan. The major components of the hydrogen fuel system are:

- Two carbon fiber wrapped aluminum gaseous fuel tanks for a total volume of 87 liters. The tanks are rated for a maximum fill pressure of 278.5 bar (4040 psig) and a nominal operating pressure of 248.2 bar (3600 psig). This configuration of tanks was used because of the hydrogen embrittlement problem with steel. Steel tanks are not rated for hydrogen.
- Both tanks have in-tank isolation solenoid valves with internal check valve, thermal pressure relief device vent port (PRD), and a pressure transducer, all located at one end of the tank. At the other end of the tanks are end caps with a temperature probe for assessing accurate in-tank fuel quantity.

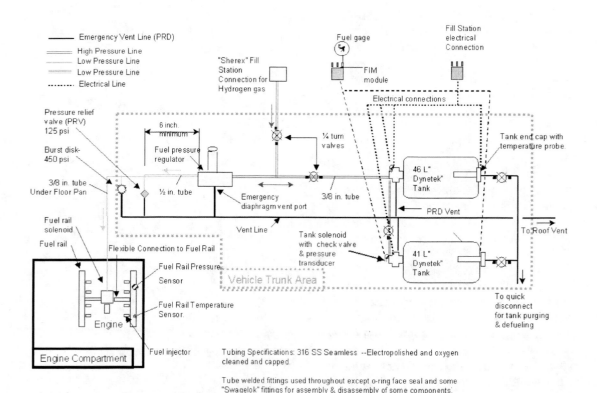

Figure 4: Vehicle fuel charging system schematic diagram.

- A single stage pressure reducing regulator set at 5.2 bar (75 psig) with an emergency diaphragm vent port.
- A pressure relief valve (PRV) downstream of the pressure regulator set at 8.6 bar (125psig).
- An over pressure burst-disk downstream of the pressure regulator set at 31 bar (450psig).
- A hydrogen fuel, fill station receptacle with cover and internal check valve. Fill station electrical connector for vehicle grounding and to supply the vehicle's fuel tank pressure and temperature information to the filling station for fill monitoring is separate.
- A fuel rail isolation solenoid valve located in the engine compartment feeding the port injector fuel rail.
- Quarter turn lockdown valves are located in the plumbing system for tank and fill station connector isolation.

All of the high-pressure hydrogen plumbing utilized 9.5 mm (3/8 inch) diameter by 1.24 mm (0.049 inch) wall thickness type 316 seamless stainless steel tubing with weld fittings throughout except for some O-ring face seal and double-tapered compression type fittings installed to simplify assembly and service. Low-pressure plumbing and the fuel system vent line utilized a 12.7 mm (1/2 inch) diameter by 1.65 mm (0.065 inch) wall thickness type 316 seamless stainless steel tubing. The vent line, connected to all vented fuel system components, was routed so the exit was through a look alike phone antenna located at the center of the vehicle roof just ahead of the rear window glass. Figure 6 shows the detail of the antenna vent location on the vehicle.

The system supplies gaseous hydrogen to the port fuel injectors at a nominal pressure of 5.2 bar (75 psig). A fuel rail solenoid valve, located in the engine compartment, allows the flow of hydrogen from a single stage regulator to the fuel rail when enabled. When disabled the valve isolates the fuel rail from the remainder of the fuel system, thus reducing the potential for hydrogen leakage into the engine compartment. A second control valve located in the fuel tank manages the high pressure flow from the fuel tank to the regulator. This valve like the other is opened while the engine runs, but is closed when the engine is stopped, in response to a hydrogen leak or vehicle crash. Two carbon fiber reinforced aluminum tanks were installed in the trunk with a total volume of 87 liters of gaseous hydrogen, which at 248.2 bar (3600 psig) holds 1.5 kg of fuel. On an energy equivalent basis, these tanks carry an equivalent of 5.7 liters (1.5 US gallons) of gasoline. Finally, pressure relief devices and burst disks are manifolded to a roof mounted vent to protect the fuel system components and enhance safety by venting hydrogen away from passengers and bystanders. More details of this configuration are in a related paper [5].

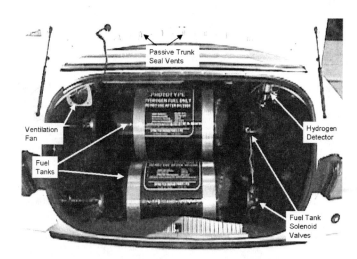

Figure 5: View of vehicle's trunk highlighting fuel tank locations, in addition to ventilation system and hydrogen detection system components.

Figure 6: Hydrogen vehicle high pressure roof vent.

SAFETY SYSTEM - Safety was a fundamental vehicle design consideration for this first prototype. The team built upon the experience of Ford's hydrogen fuel cell and compressed natural gas (CNG) programs, to produce a unique system for this vehicle. This triple redundant system combines active and passive ventilation with hydrogen detectors [8] to enhance safety through redundancy. The heart of this system, the combustible gas detectors, consists of four sensors shown as dots in Figure 2, located in the engine compartment, passenger compartment, and trunk. Alarm conditions are triggered at hydrogen concentrations of 0.6 %, 1 % and 1.6 % (15 %, 25 % and 40 % of lower flammability limit for hydrogen). In addition to the detector, the safety system relies on several ventilation fans, also shown in the figure, to circulate air within the trunk and engine compartments to prevent the formation of high concentration pockets of hydrogen within these areas and create more uniform mixtures at the sensors. Since the hydrogen storage tanks are located in the trunk, two fans were installed there to exhaust the trunk space through the rear spoiler further reducing the potential of trapping hydrogen. Should hydrogen of sufficient concentration be detected, several measures are taken to minimize the potential that it is ignited. These measures include disabling the fuel supply and engine starter, opening the moonroof, and activating all ventilation fans if not already on.

Since hydrogen is significantly less dense than air, it will rise and disperse if it is not trapped. The passive elements of the safety system were designed to take advantage of this behavior. Hood louvers and the trunk seal vents, depressions in the trunk seal (see Figure 5), allow hydrogen to escape to the atmosphere from the engine compartment and trunk respectively. Thus these elements further improve the vehicle's safety providing redundancy without intervention from the other system elements [5].

DEVELOPMENT PLAN – Upon completion of the vehicle build, the vehicle development team began calibration, engine control strategy refinement and testing. Due to the inevitable differences between the dynamometer and vehicle implementations, calibration corrections for environmental, transient, and steady state conditions were developed directly through vehicle testing. This development proceeded over a three month period, through two stages designated as Phase I and Phase II. Each of these and their results will be discussed in what follows.

The primary objective, was vehicle performance, focusing on driveability characteristics such as, acceleration, part throttle responsiveness, and idle quality. Development of an exhaust after-treatment system and the associated calibration was left for subsequent stages of vehicle development. However, since tailpipe emissions are an important metric by which combustion engines are evaluated and compared to one another, emissions tests were performed during each development phase.

All emissions testing of the vehicle was performed at the Ford Motor Company Vehicle Engineering Research Laboratory. The tests consisted of repeated runs of the EPA 75 test procedure, and when time allowed, the highway fuel economy schedule (HWFET). For those unfamiliar with this test procedure, the EPA 75 is the standard U.S. emissions certification test. It consists of a city drive cycle, made up of a 505 second "cold 505", an 876 second bag 2 test, a 10 minute soak and concludes with a "hot 505". When performed, the 765 second HWFET immediately follows the hot 505. The city cycle, the first three parts of the test, covers a drive distance of 17.8 km (11.09 miles) at an average speed of 34.3 km/hour (21.3 miles/hour); the highway fuel economy

portion simulates a drive distance of 16.4 km (10.2 miles) at an average speed of 79.5 km/hour (49.4 miles/hour).

VEHICLE TEST RESULTS

PHASE I CALIBRATION RESULTS - Phase I refers to the initial stage of vehicle calibration development. The results of engine dynamometer mapping [1] were integrated into the vehicle calibration at a constant 0.55 equivalence ratio and MBT spark over all engine speeds and loads. To account for the actual operating conditions observed during vehicle testing, some calibration modifications were made to correct fuel injector flow based on changes in the fuel rail temperature and pressure, and ignition timing based on changes of engine coolant and air charge temperature. Due to the performance limitations of this constant equivalence ratio control, it was decided to document the results as a baseline for future work. Note that Tables 1, 2, and 3 at the end of this paper provide a summary of vehicle test results discussed in detail below.

A series of emissions tests were performed to establish a baseline for all subsequent development. The regulated carbon based vehicle feedgas (FG) emissions, summarized in Figures 7 and 8, are less than the SULEV tailpipe (TP) standards. The CO_2 emissions, Figure 9, although presently unregulated are only 0.4 % of those emitted by the gasoline powered 2.0L Zetec. Since pure hydrogen powers the vehicle, all of these carbon based emissions are produced from the burned and unburned engine oil present in the combustion chamber. As a result, the carbon based emissions remained consistent throughout all of the testing, and will not be discussed any further in this paper.

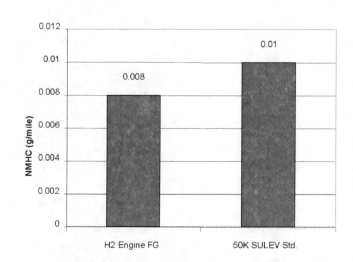

Figure 7: Comparison of vehicle's feedgas HC emissions to SULEV tailpipe emissions standards.

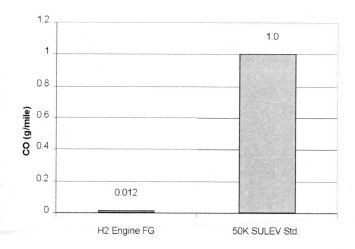

Figure 8: Comparison of vehicle's feedgas CO emissions to SULEV tailpipe emissions standards.

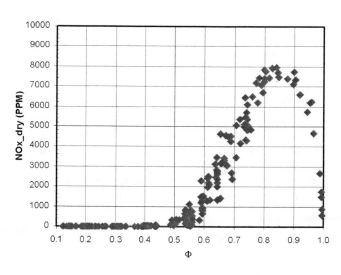

Figure 10: Relationship of fuel air equivalence ratio to NO_x emissions from a H_2ICE [1].

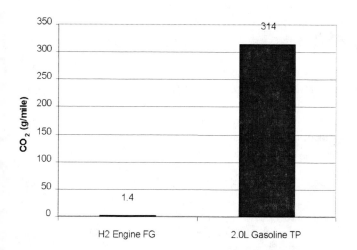

Figure 9: Comparison of vehicle's feedgas CO_2 emissions to gasoline powered Zetec tailpipe CO_2 emissions.

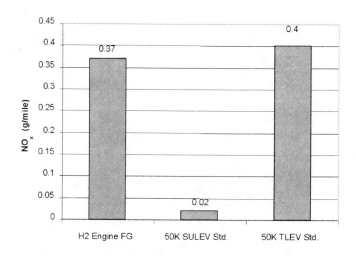

Figure 11: Comparison of vehicle's feedgas NO_x emissions to SULEV and TLEV tailpipe emissions standards for Phase I calibration.

In contrast, tailpipe NO_x emissions are significant for this vehicle as presently configured without an after-treatment system. The NO_x emissions are highly dependent on the equivalence ratio [1] as demonstrated in Figure 10. Based on this figure, the selected operating point is associated with relatively low emissions concentrations. However, since the location is in the knee of this curve, even small fueling errors richer than 0.55 can produce large increases in exhaust NO_x concentrations.

Figure 11, compares the average NO_x emissions produced during the EPA test cycle. These untreated feedgas NO_x emissions are much greater than the EPA's 50 K SULEV tailpipe standard, but just within the TLEV tailpipe standard.

This average, although a valuable figure of merit, does not provide a great deal of insight into the conditions, which create these emissions. Figure 12 represents the contributions to the bag 1 emissions over time. This test includes the cold start and warm-up, from an ambient of 21° C. Note that the NO_x emissions are very small during the cold start and climb as the engine warms up. Compare these results to those from bag 3, Figure 13, which represents the NO_x emissions when the engine is hot. The hot restart produces a pronounced spike in NO_x concentration, after which the concentration maintains levels consistent with those of the final 200 seconds of bag 1. It is conjectured that the cold to hot engine differences in NO_x emissions are a result of the differences in cylinder charge heat transfer rates. In both cases, once the engine is warm, the acceleration and deceleration transients contribute significant amounts to the total emissions. While the increases in NO_x

concentration represent transient fuelling errors rich of the operating point, the magnitude of these peaks implies that the fuelling errors produced by the open loop controller are reasonably small.

In addition to emissions data, this test procedure provided a vehicle fuel economy baseline for the hydrogen powered vehicle. Figure 14, provides a fuel economy comparison for this vehicle when powered by hydrogen relative to the gasoline Zetec engine. Here the hydrogen powered vehicle delivered a 17.9 % fuel economy improvement relative to the gasoline powered vehicle over the city cycle.

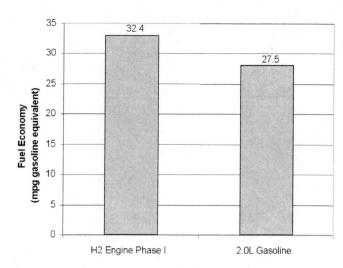

Figure 14: Phase I city cycle fuel economy comparison [9].

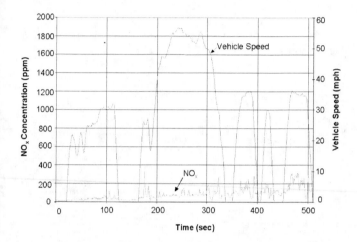

Figure 12: Phase I vehicle NO_x emissions throughout the EPA Cold 505 emissions schedule.

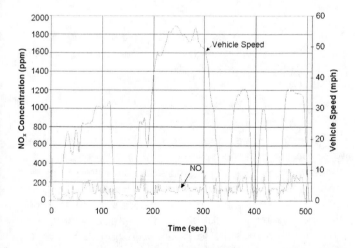

Figure 13: Phase I vehicle NO_x emissions throughout the EPA Hot 505 emissions schedule.

Although the emissions and fuel economy results for this configuration offer a reasonably good starting point, the constant equivalence ratio control resulted in unacceptably poor acceleration performance, summarized in Table 3 and shown in Figure 15. Despite this shortcoming, the Phase I work established a useful vehicle baseline and valuable experience with the hydrogen engine operating characteristics, which were directly applicable to the following work on Phase II.

PHASE II CALIBRATION RESULTS - The principle objective of this work was to optimize the vehicle acceleration performance of the Phase I calibration. As a result, the initial focus was to utilize high load fuel "enrichment" to maximize the vehicle's acceleration performance. A maximum equivalence ratio of 0.7, an increase of 27 % over that of Phase I, was limited by the onset of preignition. In conjunction with the fuel enrichment, it was also necessary to modify the ignition and fuel injection timing in order to reduce the occurrence of preignition. Together, these actions produced a 23 % reduction in the vehicle's 0 to 60 mph times, see Figure 15. While not achieving the performance of the baseline gasoline fueled vehicle, it represents the maximum performance available with this engine's reduced power density.

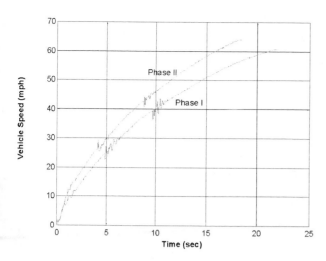

Figure 15: Vehicle acceleration performance comparison, 0 to 60 mph.

In addition to these improvements in acceleration performance, the calibration required refinement in the medium and low load regions to improve driveability and eliminate the intermittent occurrence of intake manifold backflash [6]. In contrast to preignition, which takes place within the cylinder, backflash occurs when the charge ignites within the intake or from the exhaust manifold on valve overlap. The intake manifold backflash observed here was identified by a popping sound and rapid temperature increases within the intake runners. As before these goals were accomplished through the coordination of fuel air equivalence ratio, fuel injection timing and ignition timing. At this stage, the performance objectives for the vehicle's idle speed control quality, launch feel, and throttle responsiveness had been achieved.

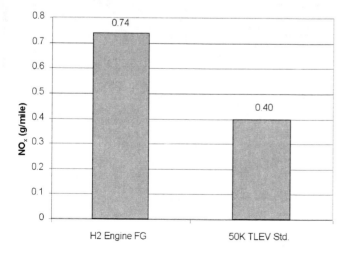

Figure 16: Comparison of vehicle's feedgas NO_x emissions to TLEV tailpipe emissions standards for Phase II calibration.

Having achieved a significantly improved level of vehicle driveability and performance, the vehicle was returned to the emissions laboratory to reevaluate its emissions and fuel economy. High load fuel enrichment, necessary to improve vehicle acceleration performance did not significantly affect the average city fuel economy results. However, it did negatively impact vehicle emissions since operation at an equivalence ratio of 0.70 produces much higher concentrations of NO_x (recall Figure 10). The NO_x emissions, Figure 16, doubled relative to the Phase I results, exceeding the TLEV standard. Comparing Figures 13 and 17, we see that emissions from the new calibration exceed the Phase I levels over the entire test. The NO_x spikes are now very pronounced, relative to Phase I, and occur during every significant acceleration transient of the test. Figure 18 is also included to demonstrate that these levels remain high throughout the highway portion of the drive cycle further contributing to higher overall emissions.

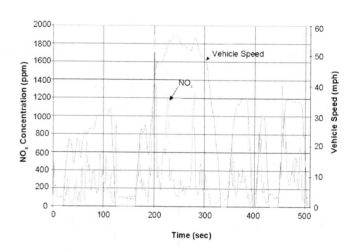

Figure 17: Phase II vehicle NO_x emissions throughout the EPA Hot 505 emissions schedule.

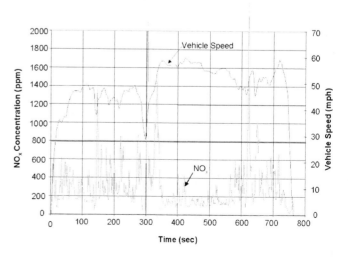

Figure 18: Phase II vehicle NO_x emissions throughout EPA highway fuel economy schedule.

Finally, having just considered the highway cycle NO$_x$ emissions results, it is worthwhile presenting the associated fuel economy results. Figure 19 demonstrates a 6.1 % fuel economy advantage achieved by the hydrogen engine, in addition to the 14.1 % advantage it maintained over the city cycle. The fuel economy advantage of hydrogen engine is smaller over the highway cycle because on average it operates at a higher load, and thus a richer fuel air mixture than required for the city cycle. These richer mixtures are accompanied by greater spark retard, to prevent preignition, thus leading to an overall efficiency reduction.

Driving the vehicle around the engineering center test track resulted in a gasoline equivalent economy of 17 km/liter (40 miles/gallon) and a range of 96 km (60 miles). Future plans call for increasing the fuel tank capacity to 179 liters at 344 bar (5000psig), which would extend the range to approximately 257 to 290 km (160-180 miles).

Time constraints precluded a Phase III, largely unthrottled, calibration development effort. The unthrottled configuration, which was already demonstrated in the dynamometer test cell, was expected to improve both NOx emissions and fuel consumption [1].

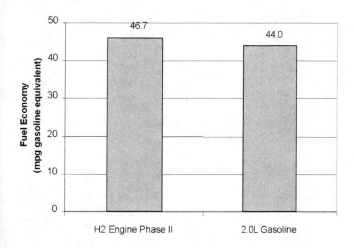

Figure 19: Phase II highway cycle fuel economy comparison [9].

SUMMARY OF RESULTS

EPA 75	NMHC (g/mile)	CO (g/mile)	NO$_x$ (g/mile)	CO$_2$ (g/mile)
Phase I FG	0.0084	0.0117	0.37	1.4
Phase II FG	0.0076	0.0082	0.74	1.4
Gasoline FG	1.97	9.64	1.4	290
Gasoline TP	0.06	0.94	0.03	314
SULEV TP Standard	0.01	1.0	0.02	N/A

Table 1: Summary of H$_2$ICE vehicle emissions results.

EPA Combo	City (mpg)	Highway (mpg)
Phase I	32.4	N/A
Phase II	31.4	46.7
Gasoline	27.5	44.0

Table 2: Summary of H$_2$ICE vehicle fuel economy results [9].

	0-45 mph (seconds)	0-60 mph (seconds)
Phase I	12.0	21.8
Phase II	9.9	16.8

Table 3: Summary of H$_2$ICE acceleration performance.

CONCLUSION

The first production viable, North American OEM H$_2$ICE vehicle has been built and tested. This effort demonstrates a practical first attempt solution for a dedicated hydrogen fueled vehicle incorporating a systems approach to safety and fueling.

Features incorporated into the vehicle include:

- A highly optimized 14.5:1, 2.0 liter H$_2$ICE.
- A gaseous hydrogen fuel system with fuel storage at 248.2 bar (3600 psig) and delivery at 5.2 bar (75 psig).
- A triple redundant hydrogen safety system consisting of gas sensing, active and passive elements.

Results of vehicle testing indicate the following:

- Carbon based engine out emissions testing indicated that HC and CO are less than SULEV standards, and that CO_2 emissions are reduced to about 0.4 % of the tailpipe levels produced by a gasoline fueled engine of the same displacement.
- The engine out NO_x emissions ranged from 0.37 – 0.74 g/mile.
- A metro cycle fuel economy improvement of up to 17.9 % relative to gasoline.
- Smooth, acceptable drive feel in a city setting.
- Acceleration performance would require improvement for full customer acceptability. At equal performance, the metro cycle fuel economy advantage, relative to gasoline, is expected to decrease to about 11 %.

Future testing should investigate technologies such as boosting, EGR, and various configurations of exhaust after-treatment. Combinations of these are expected to substantially improve NO_x emissions, specific power and fuel economy.

ACKNOWLEDGMENTS

The authors wish to express their gratitude to Donald Wilkinson, Allan Kotwicki, Yin Chen, Yong-wha Kim and Woong-chul Yang among many others who contributed to the vehicle development effort.

REFERENCES

1. Tang, X., et. al., "Ford P2000 Hydrogen Engine Dynamometer Development", SAE Technical Paper, 2002-01-0242.
2. Natkin, R. J., et. al., "Ford Hydrogen Engine Laboratory Testing Facility" , SAE Technical Paper, 2002-01-0241.
3. Cornille, H.J., Weishaar, J.C., Young, C.S., "P2000 Body Structure", SAE Technical Paper, 982405.
4. Bain, A., *Sourcebook for Hydrogen Applications*, Hydrogen Research Institute and National Renewable Energy Laboratory, 1998.
5. Stockhausen, W. F., et. al., "Ford P2000 Hydrogen Engine Design and Vehicle Development Program", SAE Technical Paper, 2002-01-0240.
6. Cox, K.E., Williamson Jr., K.D., *Hydrogen: Its Technology and Implications, Utilization of Hydrogen Volume IV,* CRC Press, Boca Raton, 1979.
7. Heywood, J.B., *Internal Combustion Engine Fundamentals*, McGraw-Hill Publishing Company, New York, 1988.
8. "Installation and Service Manual, Part Number 900-096-01 Rev. C", Delphian Corporation, New Jersey, 1999.
9. E.W. Lemmon, M.O. McLinden and D.G. Friend, "Thermophysical Properties of Fluid Systems" in NIST Chemistry WebBook, NIST Standard

Reference Database Number 69, Eds. P.J. Linstrom and W.G. Mallard, July 2001, National Institute of Standards and Technology, Gaithersburg, MD, 20899.

1999-01-0991

Powertrains of the Future: Reducing the Impact of Transportation on the Environment

Jean J. Botti and Carl E. Miller
Delphi Automotive Systems

ABSTRACT

Tomorrow's winning powertrain solutions reside in those technology combinations providing optimized propulsion systems with zero emissions and no cost or performance penalty compared with today's vehicles.

The recent Kyoto Protocol for CO_2 reduction and the California Air Resources Board (CARB) thrust for zero emission vehicles along with the European Regulatory community, motivate car manufacturers to adopt new light body structures with low aerodynamic drag coefficients, low-rolling resistance and the highest efficiency powertrains. The environmental equation expresses car manufacturers aptitude and desire to create zero emission vehicles at acceptable levels of performance unlike limited range electrical powered vehicle products.

The cheapest solution to the environmental equation remains the conventional internal combustion engine ($30 to $50 per kW). However, new system optimization is necessary and will take many forms from alternative fuels enriched with hydrogen to diesel direct injected engines for light duty trucks and sport utility vehicles. Diesel fuel will probably be the favorite mid-term option with direct injection engines coupled with an electric motor. This combination gives higher efficiency, but a new optimized aftertreatment system will be necessary to satisfy emission mandates.

Hybrid vehicles, the Toyota Prius being an excellent example, will increase in popularity with their high efficiency and driving smoothness.

INTRODUCTION

Removing the automobile from the environmental equation is a monumental mission for the automotive industry. Many new technologies are being evaluated to move in this direction and the automotive engineering community is challenged like never before to develop cost effective, customer driven solutions to meet these objectives. There are no areas of current or future technological development that will go unexplored. Everything from vehicle hardware, software and service methods to manufacturing and engineering processes will need to be extensively retooled and redeveloped to focus on the new environmental imperatives. Fundamental changes in powertrain hardware and controls, sensor and actuators, accessories, electrical architecture and power electronics are required for all future vehicle alternatives. Our industry must press very hard for increased fuel economy and lower emissions while maintaining important customer attributes of utility, serviceability and cost effectiveness.

ENVIRONMENT

Global warming has become a very prominent topic worldwide catalyzed by the recent Summit at Kyoto.[1] Although we are using more energy today than ever before, we are also using it more efficiently. The fuel economy of US motor vehicles has doubled over the past twenty-five years, in part from technology advances in engine combustion and powertrain control. Considering well-to-wheel, gasoline from petroleum is a very efficient energy source for vehicle usage compared to other forms and sources of energy. With further anticipated technology advances, there is still significant room for fuel economy improvements in internal combustion engines. This will help extend the remaining world supply of oil further.[2] The latest effort in fuel economy improvement is the result of recent world mandate on the abatement of carbon dioxide. (See Figure 1.) As far as we can ascertain the best and most direct way to reduce carbon dioxide emission from motor vehicles in the next decade is through better fuel economy.

Total CO$_2$ Emissions (in billions of tons)	
Anthropogenic	6
Trees, volcanoes, etc.	200

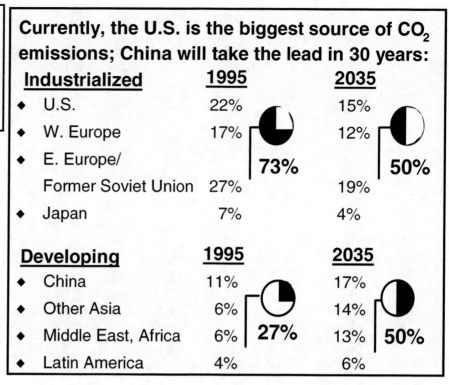

Currently, the U.S. is the biggest source of CO$_2$ emissions; China will take the lead in 30 years:

Industrialized	1995	2035
◆ U.S.	22%	15%
◆ W. Europe	17%	12%
◆ E. Europe/		
Former Soviet Union	27%	19%
◆ Japan	7%	4%
	73%	**50%**

Developing	1995	2035
◆ China	11%	17%
◆ Other Asia	6%	14%
◆ Middle East, Africa	6%	13%
◆ Latin America	4%	6%
	27%	**50%**

Figure 1. Anthropogenic Carbon Dioxide (CO$_2$) emission trends

This translates into cost savings to the consumer and it is believed these economics will entice the consumer to buy "green" products. Short of using the non-carbon alternative fuel, hydrogen, CO$_2$ reduction correlates directly with a gain in fuel economy. Using today's economics, the dollar value in fuel savings accumulated over four years to car buyers in Europe and the U.S. under two proposed Kyoto CO$_2$ reduction levels are summarized in Figure 2. At higher CO$_2$ reduction levels the fuel cost savings can be significant even at moderate fuel prices. As petroleum fuels become scarce, the inevitable increase in fuel prices will come. The rise in fuel price may come sooner than later, perhaps in the form of a carbon surcharge as an incentive towards fuel conservation and "greener" vehicles. The challenge to the technology community is to provide cost effective solutions that do not diminish economic gains for the customer. The key metric for the consumer is not liters or kilojoules but full cost per kilometer![3]

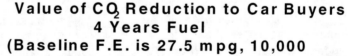
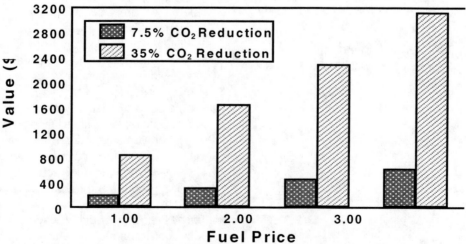

Figure 2. Fuel saving economic value as a function of fuel cost

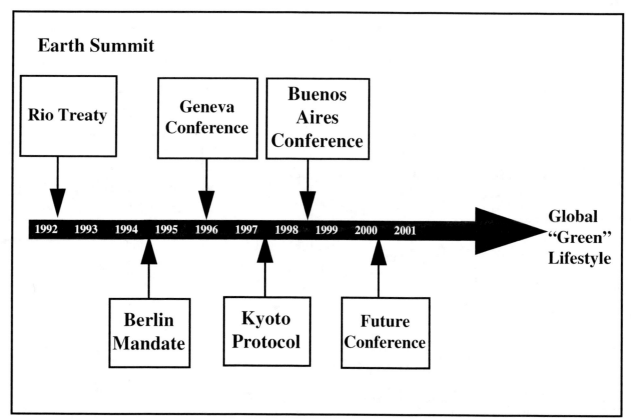

Figure 3. Despite major technical, economic and scientific uncertainties, the global emission policy development process is maturing.

CHARACTERIZING VEHICLE LEGISLATION – Mature, proactive environmental legislating and regulating bodies are found in the U.S., western Europe, and Japan. The U.S. works to preserve individual mobility, the European Union wants balance between mobility and environment, while Japan works to protect their economic system. A developing player in regulating transportation globally is the United Nations via various framework programs and agencies. The Kyoto anthropogenic emission protocol is one of the better known activities in this world forum. (See Figure 3.)

EUROPE – Vehicle environmental standards are set by two primary commissions; the European Commission (EC) via the European Parliament, and 2) the Economic Commission for Europe (ECE). The ECE has a wide membership including the former Soviet bloc countries, members of the European Union (EU) and the USA. Unlike the regulations of the EC and the US EPA incorporated within the relevant countries, legislation standards set by the ECE are adopted to varying extent by its members.

JAPAN – National legislation concerning safety regulations and pollution control for road vehicles is the responsibility of the Ministry of Transport. The declared aim of Japanese emissions legislation is to set standards which go up to the limits of technological feasibility and eventually to apply the same standards to both gasoline and diesel fueled vehicles.

THE UNITED STATES – The EPA is the US federal regulating body. However, the state of California has often modified federal regulations through CARB by adopting more stringent regulations, including requiring progressive introduction of cars with ever more demanding emissions standards.

U.S. Corporate Averaged Fuel Economy (CAFE), as implemented by the EPA, continues to be an important U.S. automotive parameter continuously challenging technology limits.

From the fuel economy CO_2 perspective the CAFE is the premier automotive parameter. (See Figure 4.)

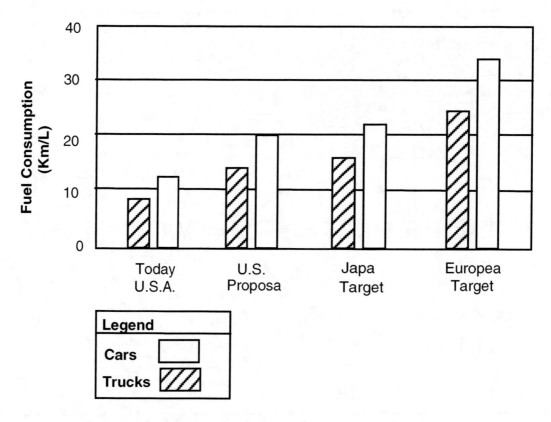

Figure 4. Potential Impact on CAFÉ fuel economy targets

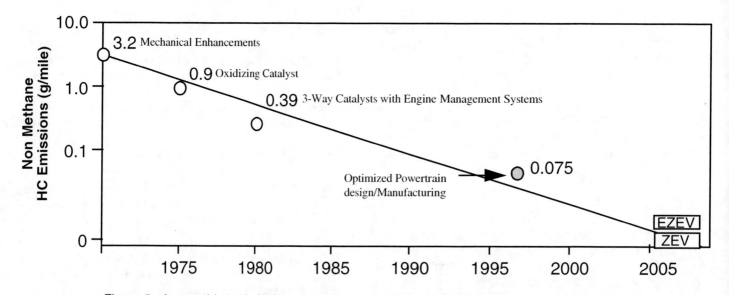

Figure 5. Acceptable technical compromises have halved emission every five to ten years

EMISSIONS REGULATIONS – Air pollution has become a major concern as energy consumption grows along with the population.[4] Transportation vehicles represent a significant source of the total pollutant emissions. The major exhaust constituents identified as pollutants are carbon monoxide, nitrogen oxides, hydrocarbons, and particulates. These pollutants arise from either incomplete or high temperature combustion processes. At higher concentrations carbon monoxide and nitrogen oxides are toxic and nitrogen oxides react with hydrocar-

bons to form photochemical smog proven to be detrimental to human health. Over the past three decades U.S. mandated emission standard trends have decreased logarithmically. Figure 5 shows how car manufacturers and suppliers have responded to these mandates through innovative technologies. This aggressive schedule was satisfied because the basic science was mature and the infrastructure could respond quickly.

ALTERNATIVE POWERTRAINS – In the near term a variety of piston engine powertrains are available to remove the car from the environmental equation. Figure 6 compares various technologies employing alternative fuels. Unfortunately, the most fuel efficient powertrains are not automatically either the most economical or guaranteed to meet all emission requirements.

In a more distant future hybrid vehicles and fuel cells offer substantial fuel economy and emission reductions. (See Figure 7.) Pure electric vehicles will have a hard time to compete economically with alternative fuels because of their cost and their limited range.

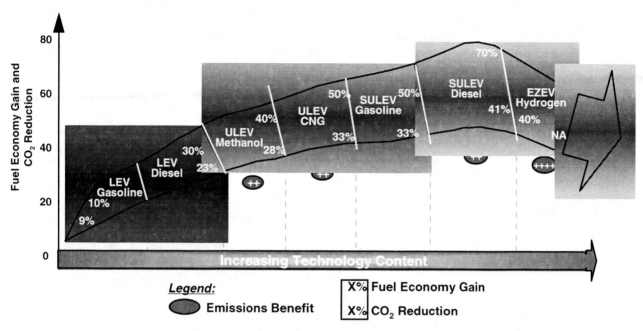

Figure 6. Near term generation powertrain trends.

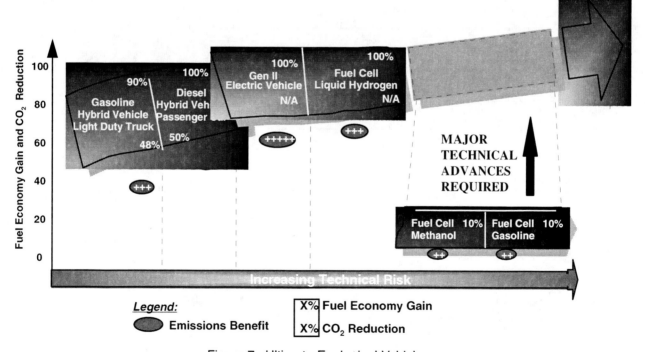

Figure 7. Ultimate Ecological Vehicles.

The gasoline fueled powertrain technology challenge is to develop direct injected engines approaching the powertrain efficiency of a hybrid vehicle especially in the region of low power demand. (See Figure 8.) Such developments could create very efficient, low cost powertrains.

COMPRESSION INJECTED PISTON ENGINES

DIESEL – Diesel engines have higher compression ratios and burn a leaner air-fuel mixture than gasoline engines. The diesel reduces pumping loss and has higher thermal efficiency. Therefore, diesel engines have better fuel economy than any gasoline fueled powertrain including direct injection. (See Figure 9.) However, the diesel exhaust aftertreatment is challenged by high emissions of nitrogen oxides and particulates (black smoke). This situation requires special technology including perhaps a molecular electronic excitation zone coupled to modified passive catalysts for final clean-up. Recent developments

have led us to believe aftertreatment systems employing gas ionization and molecular recombination can reduce particulate levels. Current particulate emission levels will not be acceptable in future diesel powertrains and Europe has established a value of 0.025 gm/km by 2005, an order of magnitude smaller compared with today. Better fuel atomization of the spray aerosol is also directionally correct towards lower emissions and better fuel economy. Fuel delivery pressures will reach 150 to 200 MPa while pilot and variable injection will further enhance combustion and provide noise reduction. Auto manufacturers in the European Union through new diesel V6 and V8 technologies are pathfinders, while Japan is a close follower to European trends. However, in the United States, diesel acceptance remains a major consumer issue to be ameliorated ensuring slow market growth. In the United States, clean diesel powertrains will be required for satisfying anticipated CAFE standards especially for class 1 and 2 trucks and S.U. Vans.

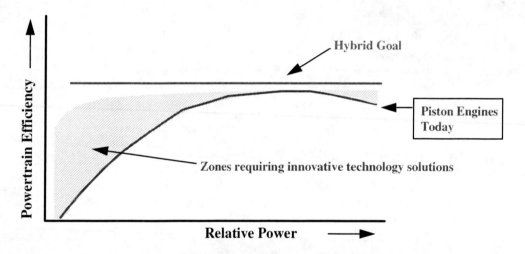

Figure 8. Heuristic map for near term challenges.

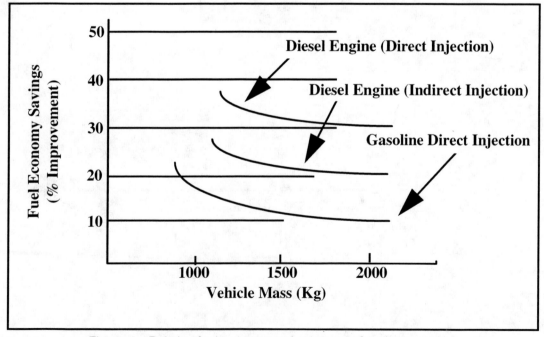

Figure 9. Relative fuel economy enhancement for piston engines.

"CLEAN" DIESEL FUEL – From the consumer's viewpoint, one of the most important diesel fuel properties is density which relates to the energy available to create useful power. However, before the fuel can be used effectively air has to be compressed and fuel injected as a fine spray which ignites and burns with minimum noise and emissions. To satisfy all regulatory, performance and reliability requirements, other fuel characteristics including viscosity, ignition quality, volatility, cleanliness and cold flow must be ensured.

Cetane number is often used to quantify the ignition quality of diesel fuel. (See Figure 10.)

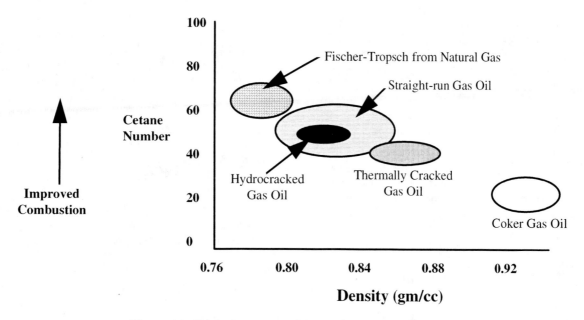

Figure 10. Typical sources of diesel fuels and fuel quality.

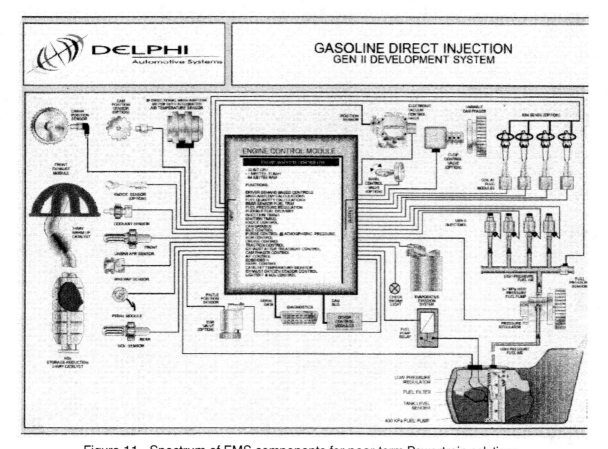

Figure 11. Spectrum of EMS components for near-term Powertrain solutions.

A higher cetane fuel ignites sooner and often reduces combustion noise and emissions although really high cetane numbers may increase smoke and emissions. The refiner's challenge is very regionally dependent where end-use dictates refinery practices. An increase in USA diesel usage would involve relatively long time delays and large refiner capital investment if refineries change their product mix this radically.

Trends suggest Fischer-Tropsch diesel fuel from natural gas will become important in the European Union while refinery evolution will slowly change in Japan and USA to mimic current usage in European Union. Single molecular specie fuels, including dimethyl ether usage, will probably remain very limited.

GASOLINE – Automakers use direct injection gasoline engines to reduce fuel consumption and CO_2 emissions but struggle to simultaneously satisfy NOx emissions regulations.[5] (See Figure 11.) A theoretical advantage of a Gasoline Direct Injection (GDI) engine, compared to current port fuel injection, includes up to a 20% improvement in Brake Specific Fuel Consumption (BSFC), as the result of more precise air/fuel control and less fuel enrichment

on cold starts. New types of aftertreatment technologies are being developed to eliminate NOx from direct injection systems. Investigations globally into alternative aftertreatments have featured Plasma based technologies. Some technologies use gas-phase plasma discharge interactions and appear to have low NOx conversion efficiency and/or high energy consumption. Durability for these aftertreatment technologies also remain an issue.

HYBRID VEHICLES – Hybrid propulsion vehicles combine energy conversion technologies (viz, heat engines, generators, and motors) with energy storage technologies (e.g., lithium ion batteries, ultracapacitors, and flywheels.) Propulsion systems are designed to take advantage of the strengths of each constituent technology resulting in vehicle systems with improved performance and range efficiency as well as positive environmental emission characteristics. Consumer interest in hybrid propulsion does exist, but only if stringent cost, reliability and durability goals are met. Technically, the development of viable energy storage devices and thermally efficient hybrid power units, at a practical systems cost, remain major challenges. (See Figure 12.)

Parameter	Advanced Batteries		Lead-acid battery base-line
	electric vehicles	hybrid vehicles	
Specific energy, Wh/kg	100-200	8-80	25-35
Energy density, Wh/L	130-300	10-100	~ 70
Specific power, W/kg	75-200	625-1600	80-100
Life expectancy, years	5-7	5-7	2-5
Cost, US $/kWh	100-150	150-200	80-100

Figure 12. Realistic technical goals for Advanced Hybrid Batteries.

Hybrid Electric Vehicles (HEV) have several advantages over traditional internal combustion engine vehicles.

- Regenerative braking minimizes the energy lost especially during urban driving.

- Engine is sized to average not peak load, reducing engine mass.

- Fuel efficiency is increased while emissions are reduced.

- Incorporate alternative fuels reducing dependence on liquid petroleum crude derived fuels.

There are two main types of hybrid systems: series and parallel. The series hybrid uses the engine to power a generator producing electricity for the motor driving the

wheels. In a series hybrid, power flow moves in a direct line. A low-output gasoline engine operates almost constantly in its high-efficiency range, charging the battery while it runs. This method requires a motor that is larger and heavier than in the parallel hybrid system.

A parallel hybrid system uses both the engine and the motor to drive the wheels, and allocates the power of each according to driver demands. It is called a parallel hybrid because the power flows in parallel lines viz, the engine can power the wheels while at the same time charging the battery.

Figure 13 summarizes a technical roadmap for hybrid vehicles.

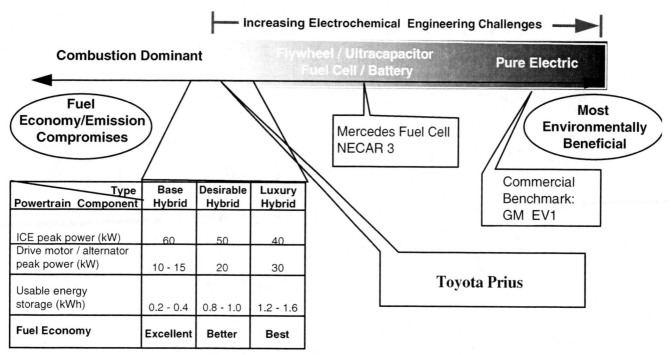

Figure 13. Spectrum of hybrid powertrains.

Powertrain Component \ Type	Base Hybrid	Desirable Hybrid	Luxury Hybrid
ICE peak power (kW)	60	50	40
Drive motor / alternator peak power (kW)	10 - 15	20	30
Usable energy storage (kWh)	0.2 - 0.4	0.8 - 1.0	1.2 - 1.6
Fuel Economy	Excellent	Better	Best

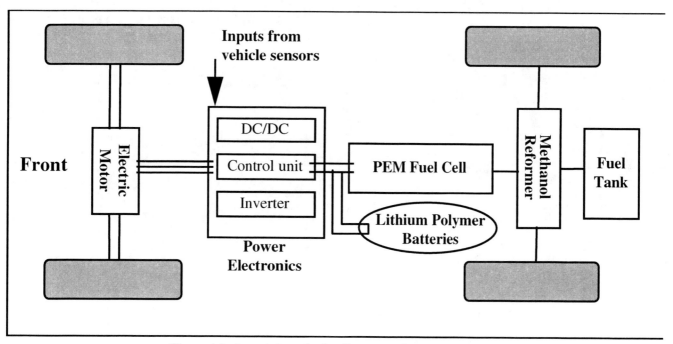

Figure 14. A prototype fuel cell powertrain vehicle system.

The fuel cell has long been considered the ideal energy source to replace piston engines, converting a fuel's chemical energy directly into electricity. A Fuel Cell Powertrain (FCPT) can be fueled with gasoline - or they can be run on methanol, ethanol or natural gas, all of which could replace gasoline at the pumps and keep the existing network of gas stations intact.

A Fuel cell car is powered by the controlled chemical reaction of hydrogen and oxygen in Proton Exchange Membrane (PEM) electrochemical cells. The FCPT converts the hydrogen into electricity which drives the trac-

tion motors. The reaction produces no emissions except pure water and waste heat, and quietly generates onboard electric power. As long as hydrogen and oxygen are available, the cells require no maintenance and would probably last longer than the vehicle. For the near term though the cost and complications of transporting and storing gaseous or liquefied hydrogen make this fuel cell concept impractical for widespread use.

A methanol fueled FCPT reforms methanol to hydrogen as needed providing a reasonable near term commercialization pathway. (See Figure 14.) A mixture of methanol and water is vaporized in the reformer and catalysts reform the vapor into hydrogen and carbon dioxide. Hydrogen and oxygen enter the PEM electrochemical cells releasing electrons which become the electricity that powers the vehicle's motor. On-board conversion of liquid fuels is potentially an attractive hydrogen source, but none exist suitable for on-board automotive applications having a highly integrated reformer with high efficiency, i.e., greater than 85-90%. The challenges are to develop innovative methods for on-board hydrogen-rich fuel feedstream production and more density efficient ways to store gaseous/liquid hydrogen.

Another engineering challenge is the start-up-time for a "cold" reformer PEM system must be reduced from about several minutes to a few seconds. The ideal would be to do it without substantial stored energy required beyond initial traction.[6]

ALTERNATIVE FUELS

HYDROGEN – Among potential alternative energy sources to supplement or replace petroleum fuel is hydrogen. It emerges as the most environmentally friendly fuel. Because of various issues including cost, however, hydrogen is not likely to be widely used in automobiles in the near future. Nevertheless, research and development using hydrogen has been initiated, so when the technical and cost barriers have been removed, it may possibly become as the dominant fuel in the mid 21st century.

CNG/LPG – While the fuel cost and wide availability of Compressed Natural Gas (CNG) give it a significant long term potential to take market share from petroleum fuels, difficulties of on-board storage and range make the most common type of "factory conversions" poorly received by the market-place. A next generation, fully optimized CNG vehicle would look very different from today's bi-fuel or limited range dedicated vehicles. Compression ratio must be raised for best efficiency, turbocharging or direct injection should be used, emissions must be optimized at Equivalent Zero Emission Vehicle (EZEV) levels and, most importantly, vehicles must incorporate single, weight optimized tanks into the vehicle structure. Weight, cost and trunk space make most multiple tank CNG configurations unattractive in the near term. Production and prototype systems that move toward this next generation vision are Honda Civic GX and the GM CNG Impact. Liquefied Petroleum Gas (LPG) is a much more attractive niche fuel than CNG due to larger storage density at lower pressures and development resources should be focused on LPG in the short term.[7]

ULTIMATELY RECOVERABLE PETROLEUM – Although oil prices have been low for a decade, there is a reason to anticipate higher prices when demand begins to outstrip traditional supply. From an economic viewpoint what matters is not when the oil reserves run out but when production starts to decline within the large, low cost reservoirs of oil and increase in new and more expensive smaller reservoirs.

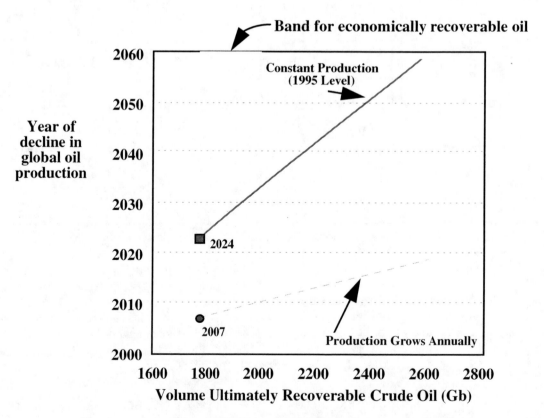

Figure 15. Peak oil projections.

There have been many major oil reserve studies since 1956 when M. King Hubbert, a geologist with Shell Oil Company, predicted oil from the 48 contiguous states would peak sometime around 1968; he was correct. His basic premise was flow from a large oil field starts to fall when about half the recoverable oil is removed. Adding the output of oil fields of various sizes and ages led to a bell-shaped production curve enabling Hubbert to predict oil reserve decline. These metrics have been extended and confirmed in over fifty other global studies. Conservative projections conclude that the decline in the amount of recoverable oil will happen about 2007 if production growth continues. (See Figure 15.) Regardless of the exact timing - decline will take place and the leading automotive suppliers will need a portfolio of alternative powertrain and alternative fuel technologies for the 21st century if they expect to minimize the impact on their business and exploit this opportunity for survival and economic gain.

The illustration shows graphically that at a price - some of the dislocations that otherwise lie ahead could be mitigated for a significant period of time if growth in global oil demand could be reduced or eliminated (viz about one and a half decades at 1800 Gb with production frozen at 1995 levels).

With sufficient preparation the transition to the post-crude oil economy need not be traumatic. For example, advanced methods of producing liquid fuels from natural gas can be scaled up quickly and could become an important source of liquefied transportation fuels. Safer nuclear power to produce hydrogen, cheaper renewable energy, and oil conservation programs could all help postpone the inevitable decline of conventional oil.

The world is not running out of oil -- at least not yet because about 1,000 giga barrels will be extracted using current extraction technologies. What society does face is the end of the abundant and cheap oil on which all industrial nations currently depend.

Figure 16 shows the various energy usage evolution starting from coal and going into the era of non-fossil fuels. These trends will drive alternative fuels and powertrain technologies.

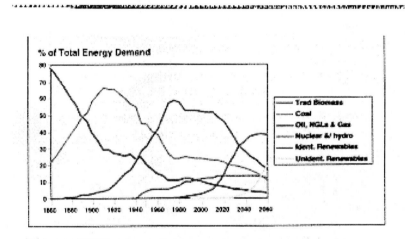

Figure 16. Evolution of Carbon Energy Sources.

SUMMARY

It is very difficult to define the winning powertrain solutions with our inability to predict consumer acceptance. Nevertheless, anticipated powertrain technologies can remove the automobile from the environmental equation. However, performance and cost will remain potential show-stoppers dramatically increasing implementation time. A good example is the electric vehicle where the lack of low cost and high efficient battery technology has blunted market penetration. Continuous monitoring of consumer trends, fuel alternatives and regulations will validate powertrain technology R&D needs. Several technologies have been briefly discussed above and could be applied in sequence and not as substitutes for one another. Again, this will be totally dependent on the adaptation of cost versus performance.

Although we are using more energy today than ever before, we are also using it more efficiently. The fuel economy of USA motor vehicles has doubled over the past 25 years, in part from technology advances in engine combustion and powertrain control. Considering well-to-wheel, gasoline from petroleum is a very efficient energy source for vehicle usage compared to other forms and sources of energy. With further anticipated technology advances, there is still significant room for fuel economy improvements in internal combustion engines. This will help extend the remaining world supply of oil further. The latest effort in fuel economy improvement is the result of recent world mandate on the abatement of carbon dioxide. As far as we know, the best and most direct way to reduce carbon dioxide emission from motor vehicles in the next decade is through better fuel economy.

In the case of hydrogen as a fuel, it will be necessary to obtain consumer acceptance. This paradigm shift will probably be initiated as an on-board partial reformation technology for gasoline, diesel fuel or methanol, within conventional powertrains, before pure hydrogen vehicles begin to dominate the marketplace.

We should not forget the time it takes for technological advancement and basic scientific break through. Governments in cooperation with private enterprise must promulgate policies which create the right incentives for orderly market-based introduction of new technologies. Research and development tax credits and grants would be useful to push the innovation and these will have a synergistic effect with market pull from anticipated carbon taxes on fuel and tax credits for clean and green new vehicles.

REFERENCES

1. Nature / Vol. 390 / 18/25Dec97 page 649

2. Nature / Vol. 388 / 17July97 page 213.

3. Energy in Profile, Shell Briefing Service, Number 2, 1995.

4. World Economic Forum for Automotive and Energy Industries / 2Feb98 R. Watson, Director.

5. Science / Vol 279 / 23Jan98 page 491.

6. Automotive Engineering / Dec. 97 page 81.

7. Ward's Auto World / March, 1998 page 64.

Fuel-Cycle Energy and Emissions Impacts of Propulsion System/Fuel Alternatives for Tripled Fuel-Economy Vehicles

Marianne M. Mintz, Michael Q. Wang and Anant D. Vyas
Argonne National Laboratory

ABSTRACT

This paper presents the results of Argonne National Laboratory's assessment of the fuel-cycle energy and emissions impacts of 13 combinations of fuels and propulsion systems that are potential candidates for light-duty vehicles with tripled fuel economy (3X vehicles). These vehicles are being developed by the Partnership for a New Generation of Vehicles (PNGV). Eleven fuels were considered: reformulated gasoline (RFG), reformulated diesel (RFD), methanol, ethanol, dimethyl ether, liquefied petroleum gas (LPG), compressed natural gas (CNG), liquefied natural gas (LNG), biodiesel, Fischer-Tropsch diesel and hydrogen. RFG, methanol, ethanol, LPG, CNG and LNG were assumed to be burned in spark-ignition, direct-injection (SIDI) engines. RFD, Fischer-Tropsch diesel, biodiesel and dimethyl ether were assumed to be burned in compression-ignition, direct-injection (CIDI) engines. Hydrogen, RFG and methanol were assumed to be used in fuel-cell vehicles.

Impacts were analyzed under alternative scenarios of potential 3X vehicle market penetration. Profiles of 3X and conventional vehicle stocks were then used to estimate fuel supply requirements and emissions produced by all light-duty vehicles (both 3X and conventional) expected to be on the road in each year of the analysis. Energy consumption, and emissions of criteria pollutants and greenhouse gases were estimated for upstream fuel processing/production as well as for vehicle operation. Emissions of criteria pollutants were further disaggregated into urban and nonurban components.

Results show that the fuel efficiency gain by 3X vehicles translates directly into reductions in total energy demand, fossil energy demand, and greenhouse gas (primarily CO_2) emissions. The combination of fuel substitution and fuel efficiency results in substantial petroleum displacement and large reductions in urban emissions of volatile organic compounds and sulfur oxide for all propulsion system/fuel alternatives considered. Although urban emissions of particulate matter smaller than 10 μm increase for CIDI engines operating on RFD, biodiesel, and Fischer-Tropsch diesel, such increases do not occur for CIDI engines operating on dimethyl ether. Fuel-cell vehicles produce large reductions in urban emissions of nitrogen oxide and carbon monoxide; compression-ignition engines operating on RFD, dimethyl ether, Fischer-Tropsch diesel or biodiesel also yield ubstantial reductions in urban emissions of carbon monoxide.

BACKGROUND

This paper summarizes part of the analyses in support of the Partnership for a New Generation of Vehicles (PNGV). Formed as a joint government-industry research and development effort, the PNGV aims to develop vehicles that can achieve up to three times the fuel economy of today's vehicles, about 34.5 km/l or 80 miles per gallon (mpg) for six-passenger automobiles. The PNGV further aims that these three-times-efficient (often called 3X) vehicles will meet the safety and emissions requirements expected to be in place when they are introduced, as well as to maintain the performance, size, utility, and cost of ownership/operation of the vehicles that they replace.

To achieve the 3X goal, the PNGV program is focusing on the development and use of advanced automotive technologies and lightweight materials. To meet the emissions goal or to provide the optimum fuel for new propulsion systems, new fuels (e.g., hydrogen, methanol, ethanol, or dimethyl ether) could also be necessary. New materials and fuels would inevitably require changes in automotive manufacturing, materials production, and fuel production and distribution. Those changes, in turn, will affect energy consumption and emissions.

As part of its oversight of the PNGV, the National Research Council (NRC), a part of the National Academy of Sciences, has created a standing committee to review the PNGV research program. In its report, the NRC Peer Review Committee raised concerns about the potential for "substantial discontinuities" in vehicle manufacturing and the transportation system and identified a need for in-depth assessment of infrastructure consequences including capital requirements, and environmental and safety issues associated with each technology being explored in the PNGV program (NRC 1994). In response to these concerns, the U.S. Department of Energy's Office of Advanced Automotive Technologies commissioned two studies: Argonne National Laboratory's (ANL's) analysis of the additional capital requirements to establish infrastructure for producing and distributing new fuels for 3X vehicles and the fuel-cycle energy and emissions impacts of using new fuels, and Oak Ridge National Laboratory's analysis of light-weight vehicle materials and their infrastructure consequences. This paper summarizes the results of the ANL analysis. A more detailed discussion of ANL's analysis methodology and results is contained in *Assessment of PNGV Fuels Infrastructure Phase 2 Report* (Wang et al. 1998).

SCOPE AND APPROACH

As a point of departure, this analysis assumed that the 3X goal will be achieved for each of the propulsion system/fuel combinations being considered, that each combination will be an equally feasible 3X alternative, and that assessment of the energy and emissions impacts of the 3X alternatives should include an examination of energy use and emissions from upstream fuel production and processing as well from vehicle operation. This latter perspective, known as a full fuel-cycle approach, necessitated the use of two complementary models -- GREET (Greenhouse gases, Regulated Emissions, and Energy use in Transportation) and IMPACTT (Integrated Market Penetration and Anticipated Cost of Transportation Technologies), both of which were developed at ANL. First, full fuel-cycle rates of energy use and emissions produced were estimated for each propulsion system/fuel combination with the GREET model (Wang 1996). These rates were then fed into the IMPACTT model to estimate annual energy use and emissions produced by a mix of vehicles (Mintz et al. 1994). Finally, energy savings and emissions reductions of the given technology were estimated as the difference between a scenario result for each propulsion system/fuel combination and a reference case.

GREET ANALYTICAL APPROACH – The GREET model calculates full fuel-cycle emissions and energy use rates in quantity-per-unit distance for various transportation fuels. For each fuel-cycle activity, GREET first calculates energy use by various process fuels per unit of fuel throughput, and then calculates emissions associated with combustion of process fuels and emissions from chemical processes and other sources. GREET includes both fuel and vehicle cycles and can calculate energy and emissions for either or both (Fig. 1).

For this analysis, GREET was used to calculate energy and emission rates for the fuel cycle only. In this mode, GREET takes into account energy use for primary feedstock recovery and transportation, fuel production, and fuel transportation, storage and distribution. GREET includes emissions caused by process fuel combustion, fuel leakage, and fuel evaporation. Upstream energy use and emissions are calculated in joules (Btu) and g per million units of fuel delivered at the pump.

GREET then calculates operational energy use (which is a member of both the vehicle and the fuel cycle, as shown in Figure 1) from vehicle fuel economy. Vehicular emissions for conventional ICE vehicles fueled with gasoline and diesel are estimated with EPA's Mobile model.[1] Vehicular emissions for other fuels and propulsion systems are calculated from baseline conventional vehicle emissions and anticipated changes between baseline vehicles (gasoline or diesel) and new technologies. Depending on the application, operational energy use may be assigned to either the vehicle or fuel cycle. In this analysis, it was included in the fuel cycle.

Within GREET, fuel-cycle and vehicle-cycle results are converted into rates per unit distance, which may then be reported separately or combined into a total energy-cycle result. Fuel economy is used to convert fuel-cycle energy use and emissions from g/unit energy to g/km (or mi); lifetime vehicle utilization is used to convert vehicle-cycle energy use and emissions from g/vehicle to g/km (or mi).

The key outputs from the GREET model are gram-per-km (g/mi) emissions and joule-per-km (Btu/mi) energy use for various fuel cycles. GREET includes emissions of volatile organic compounds (VOC), carbon monoxide (CO), nitrogen oxides (NO_x), particulate matter smaller than 10 μm (PM_{10}), sulfur oxides (SO_x), methane (CH_4), nitrous oxide (N_2O), and carbon dioxide (CO_2). The three greenhouse gases (GHGs) (CH_4, N_2O, and CO_2) are then weighted by their global warming potentials to estimate CO_2-equivalent GHG emissions. In this study, the global warming potential factors recommended by the Inter-Governmental Panel on Climate Change (IPCC) were used to calculate CO_2-equivalent GHGs. Those factors are 1 for CO_2, 21 for CH_4, and 310 for N_2O.

---

[1] The current version of EPA's Mobile model is Mobile5b. The next version – Mobile6 – is scheduled to be released in late 1999.

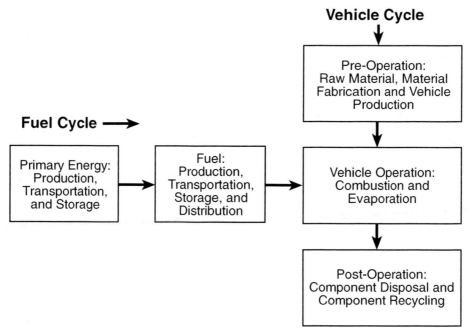

Figure 1. Fuel Cycles and Vehicle Cycles for Transportation Energy and Emissions Analysis

GREET CALCULATIONS OF UPSTREAM EMISSIONS RATES – Although GREET generates both upstream and operational energy use and emissions, only upstream results are fed into the IMPACTT model. A GREET upstream calculations follow these steps. For a given stage in the fuel cycle, energy use (in energy per million units of energy throughput) is calculated and allocated to different process fuels (e.g., NG, residual oil, diesel, coal, and electricity). Fuel-specific energy use, together with fuel-specific emission factors (specific to a particular combination of fuel and combustion technology), is then used to calculate combustion emissions for the stage. GREET has an archive of combustion emission factors for various combustion technologies that use different fuels and are equipped with different emission control technologies. Combustion emission factors for VOC, CO, NO_x, PM_{10}, CH_4, and N_2O were derived primarily from data published by the EPA (EPA AP-42 document). For most fuels, emission factors for sulfur oxide are calculated from sulfur content, assuming that all sulfur contained in process fuels is converted into sulfur dioxide (SO_2). Similarly, carbon dioxide emission factors are calculated by a carbon balance approach (i.e., the carbon contained in the fuel burned, minus the carbon contained in combustion emissions of VOC, CO, and CH_4, is assumed to be converted to CO_2).

In this analysis, emissions of the five criteria pollutants were further separated into total emissions and urban emissions. The major concern with respect to emissions is the effect on human health of exposure to air pollution created by these pollutants. Clearly, emissions that occur in remote, sparsely populated areas pose far less of a health threat than those in densely populated urban areas. Although GREET is not a location-specific model, the issue of human exposure warrants some degree of spatial analysis.[2] Thus, emissions are separated into

total and urban emissions to provide a better indication of health effects from a given combination of 3X fuel and propulsion system technologies.

GREET uses information on facility locations to separate upstream emissions into the two categories. For facilities located inside urban areas, all emissions are considered urban; emissions from all other facilities are considered to be non-urban. In this analysis, "urban" was defined as the 125 metropolitan areas specified in the 1992 Energy Policy Act. On the basis of a general understanding of the geographic location of fuel production facilities (e.g., petroleum refineries, electric power plants, etc.), a set of ratios was approximated. Corresponding to the share of each facility type located in urban areas, the ratios were then used to allocate emissions from each facility type between urban and non-urban areas.

IMPACTT ANALYTICAL APPROACH – The IMPACTT model estimates annual energy consumption and emissions production by conventional and 3X vehicles as they move through the light-duty fleet. IMPACTT incorporates a vehicle stock module that adds new vehicles (3X and/or conventional) and retires old vehicles from an initial vehicle population profile to produce annual profiles of the auto and light-truck population by age and technology; a usage module to compute vehicle travel, oil displacement, and fuel use by technology; and an emissions

---

[2] Ideally, results from GREET could be used in an emissions inventory model to generate an emissions distribution by geographic location. The location-specific inventory could then be fed into an air quality model, the results of which could be combined with a population exposure model to assess human health effects of air pollution.

module to compute upstream and operational emissions of criteria pollutants and GHGs for autos and light trucks, again by technology. The usage module computes the quantity of petroleum that would have been consumed by conventional vehicles in the absence of 3X vehicles, the quantity of petroleum equivalent consumed by 3X vehicles, and the net savings due to the presence of 3X vehicles in the fleet.[3] Upstream energy use is computed post hoc, as a function of operational energy use and a series of GREET-developed rates, which are specific to each potential 3X fuel.

In IMPACTT, operational emissions of NO_x, CO, VOC, and PM_{10} are computed separately for autos and light trucks, using age-based tailpipe emission rates obtained from EPA's MOBILE5b and PART5 models for conventional SI and CI engines operating on gasoline and diesel fuel, respectively, and average operational emission rates for nonconventional engines and fuels estimated with assumptions presented below. Operational emissions of SO_x and CO_2 are computed as a function of fuel consumption and fuel specifications. Operational emissions of N_2O and CH_4 are based on results available from past studies. Upstream emission rates for all fuels (conventional as well as potential 3X fuels) are obtained from GREET.

To evaluate the relative damage (in terms of population exposure to criteria pollutants) associated with alternative 3X power system/fuel technologies, the IMPACTT files created for the high- and low-market-share scenarios (Figure 2) were modified to estimate those portions of operational and upstream emissions likely to occur within urban areas. To do so, urban 3X vehicle sales, survival and utilization had to be estimated in order to generate forecasts of the urban 3X vehicle population and the VMT, energy use, and operational emissions of those vehicles. Urban upstream emissions were then added to these values. Urban upstream emissions were calculated as a function of the GREET-estimated upstream urban emissions rate for each pollutant and the IMPACTT estimate of total fuel use (urban and non-urban) under each scenario.

KEY ANALYTICAL ISSUES

Several key analytical issues had to be addressed before GREET and IMPACTT runs could be completed. These included market penetration of new 3X vehicles, selection of the fuels and propulsion system technologies to be considered in the analysis, specification of fuel pathways, and characterization of inherent operational emissions improvements possible with 3X propulsion system/fuel alternatives.

3X VEHICLE MARKET PENETRATION – Since the impacts of 3X vehicles are dependent not only on engine technology and fuel choice, but also on how quickly and completely they penetrate the light-duty-vehicle market, market penetration was a key issue. In order to explore a range of 3X impacts, three market penetration scenarios were postulated. The scenarios included a reference or base scenario depicting a future without 3X vehicles and two market-share scenarios bracketing a range of 3X vehicle sales (Figure 2).

The reference scenario was taken from the Energy Information Administration's 1997 forecast of transportation energy demand through the year 2015 and extrapolated to 2030 (EIA 1996). This forecast assumes 1.9% per year growth in gross domestic product (GDP), relatively low world oil price (rising from $17.26 in 1995 to only $20.98 per barrel by 2015 [all in 1995 dollars]), continued growth in the number of licensed drivers, and moderate increases in new light-duty-vehicle sales and fuel economy. Under the reference scenario, new car sales increase from 9.31 million with a rated fuel economy of 11.3 km/l (28.3 mpg) in 1995 to 13.13 million rated at 14.7 km/l (34.3 mpg) in 2030; new light-truck sales increase from 5.88 million rated at 8.5 km/l (20.4 mpg) in 1995 to 7.76 million rated at 11.4 km/l (26.3 mpg) in 2030.

The 3X vehicle market-share scenarios retain the basic parameters of the reference scenario but allow such market factors as the level of technology maturity and consumer preferences to vary. Since each of these factors is subject to some uncertainty, two extreme sets of conditions could materialize. Under one set, every factor favorable to 3X vehicles' market success could occur, resulting in rapid consumer acceptance and high sales of new 3X vehicles. Alternatively, some factors may not be as favorable to market success, resulting in slower early acceptance and low-to-moderate sales of new 3X vehicles. Figure 2 illustrates the two market-share scenarios developed for this analysis.

For both market-share scenarios, each candidate PNGV propulsion system/fuel combination is assumed to have the same market penetration, to compete solely with conventional vehicles (not with other PNGV technologies) and, thus, to account for all of the impacts identified. Because competing technologies are set aside for separate fuel/technology comparisons, this assumption provides the basis for analyzing the maximum impact of each technology.

---

[3] Unlike GREET, IMPACTT's fuel-use module computes only downstream or operational energy use. Upstream energy use is then computed as the product of operational energy use and a GREET-supplied rate.

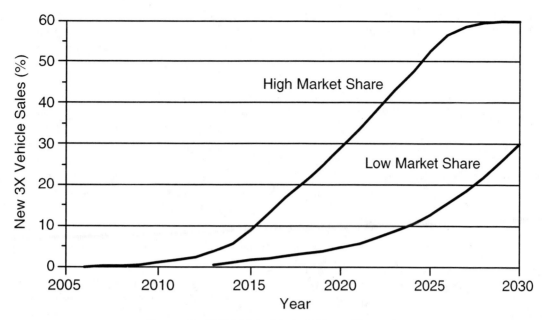

Figure 2. 3X Vehicle Market-Share Scenarios

FUELS AND PROPULSION TECHNOLOGIES – The PNGV is considering three general groups of candidate propulsion technologies: stand-alone and hybrid SI engines, stand-alone and hybrid CI engines, and fuel cells. Direct-injection technologies are likely to be applied to both SI and CI engines.

Throughout this analysis, each candidate propulsion technology was assumed to meet the 3X goal. However, among the three groups, SIDI engines currently achieve the least fuel economy improvement, CIDI engines the next, and fuel cells the most. This implies that if all technologies meet the 3X goal, vehicles equipped with SIDI engines will require the most additional effort (in improving drivetrain efficiency and/or reducing mass, drag, accessory loads, or rolling resistance) to reduce vehicle energy demand, vehicles with CIDI engines will require moderate effort, and vehicles with fuel cells will require the least effort. Investigation of the integrated design of each vehicle type to meet the 3X goal is beyond the scope of this study.

Eleven fuels were analyzed:

- Reformulated gasoline (RFG) which served as the fuel for conventional vehicles in the reference scenario and in both market-share scenarios, and a possible fuel for stand-alone SI, hybrid SI or fuel cell 3X vehicles. Reformulated gasoline was assumed to be federal phase 2 RFG.

- Reformulated diesel (RFD) which was assumed to have sulfur content below 0.01% by weight (as compared to 0.05% in current highway diesel) and to be used in stand-alone CI or hybrid CI vehicles.

- Dimethyl ether (DME) which was assumed to be used in stand-alone CI or hybrid CI vehicles. Although expensive and requiring changes in fuel storage and injection systems, DME may offer significant environmental benefits while exploiting the high thermal efficiency of a CI engine system.

- Fischer-Tropsch (F-T) diesel which is produced from natural gas and has no sulfur, low aromatic content, and high cetane number. To utilize F-T diesel's inherent advantages at a reasonable cost premium, a blend of 50% F-T diesel and 50% conventional diesel (F-T50) was assumed for this analysis.

- Biodiesel which because it is produced from renewable sources, has a large potential for reducing transportation GHG emissions and petroleum use by CI engines. Due to its high production cost and the desire to use the existing diesel distribution system, it is generally suggested that biodiesel be blended with conventional diesel. For this analysis, a blend of 20% biodiesel and 80% conventional diesel (B20) was selected to exploit the benefits of biodiesel at a reasonable cost premium.

- Liquefied petroleum gas (LPG) which was assumed to be used in SI engines primarily for its environmental benefits. LPG also offers some petroleum reduction, since approximately 50-60% of the LPG fraction appropriate for motor fuel use (i.e., propane) currently comes from natural gas (EIA 1997a; EIA 1997b).

- Compressed natural gas (CNG) and liquefied natural gas (LNG) which have the potential for reducing emissions of criteria pollutants and GHGs and petroleum use by SI engines.

- Methanol (MeOH) which was considered in pure form (M100) for SI engines (in a hybrid-electric vehicle or a stand-alone configuration) and as a hydrogen carrier for fuel cells. Methanol fuel cells were assumed to be of the partial oxidation type with on-board reformers. Because of on-board reforming, methanol for fuel cells was assumed to be lower quality, more or less comparable with that burned by SI engines.
- Ethanol (EtOH) which was included because it, alone among the SI fuels considered, is currently made from renewable resources. Pure ethanol (E100) was assumed to be burned in stand-alone SI or hybrid SI engines.
- Hydrogen (H_2) which was considered in gaseous form for use in fuel-cell vehicles.

In the context of the three groups of propulsion technologies, these 11 fuels produced 13 different combinations of propulsion technologies and fuels for which estimates of total fuel-cycle energy consumption and emissions were generated. Note that neither the fuels nor the propulsion technologies considered in this analysis represent a comprehensive picture of all available and potential candidates for application to 3X vehicles. In each case, the selection of the fuel and the propulsion technology with which it was paired was based on a specific advantage of the fuel-vehicle system relative to PNGV program goals.

FUEL PATHWAYS – In order to estimate upstream energy and emissions, a fuel-cycle path from primary energy recovery to fuel combustion in vehicles was specified for each fuel. For a given transportation fuel, a fuel cycle includes the following chain of processes: primary energy recovery; primary energy transportation and storage; fuel production; fuel transportation, storage, and distribution; and vehicular fuel combustion (Figure 1). Fuel-cycle activities that precede vehicular fuel combustion are usually referred to as upstream activities; vehicular fuel combustion is sometimes referred to as a downstream activity. Energy is consumed and emissions are generated during each of these activities. Emissions may be coincident with the activity or occur somewhat later, as in the case of fuel leakage and evaporation. The GREET model calculates fuel-cycle energy use and emissions by taking into account all these sources (Wang 1996).

Figure 3 shows fuel production pathways for each of the 11 different fuels considered in this analysis. The base case or benchmark fuel-cycle path was defined as petroleum to RFG for conventional vehicles. As shown in Figure 3, hydrogen was assumed to be produced from

either NG or solar energy and ethanol was assumed to be produced from either corn or biomass. Prior to 2020, all hydrogen was assumed to be produced from NG via steam reforming; beginning in 2020, an increasing share of new production was assumed to come from solar energy (via water electrolysis). Similarly, all ethanol production was assumed to be from corn until 2016, when production from cellulosic biomass was assumed to begin. Over time, as more new plants begin producing hydrogen from solar energy and ethanol from cellulosic biomass, these technologies' respective shares of total hydrogen and total ethanol production were assumed to rise steadily.

Energy consumption and emissions rates were calculated for a total of 14 fuel-cycle paths:

Petroleum to RFG – This path includes crude oil recovery in oil fields; crude oil transportation and storage; crude oil refining; and gasoline (i.e., RFG) transportation, storage and distribution. Among the upstream processes for this fuel cycle, crude oil refining consumes the most energy (with an energy efficiency of 83%, which is slightly below that of refining crude to conventional gasoline). As for GHG emissions, the venting of associated gas in oil fields is a significant source of CH_4 emissions.

Petroleum to RFD – This path includes crude oil recovery, transportation and storage; diesel production in crude refineries; and diesel transportation, storage and distribution. Again, the largest energy requirement for this cycle occurs at petroleum refineries (with an energy efficiency of 87%).

Petroleum to LPG – This path includes crude oil recovery, transportation and storage; LPG production in petroleum refineries; and LPG transportation, storage and distribution. Despite an energy efficiency of 93.5%, LPG production at petroleum refineries consumes the most energy; transportation of imported LPG consumes the next largest share. For this analysis, 40% of LPG supply was assumed to come via this pathway (i.e., from crude oil) under all scenarios. This is consistent with current U.S. production shares (EIA 1997a; EIA 1997b).

NG to CNG – On this path, natural gas is produced in, and processed near, NG fields and transported through pipelines to service stations where it is compressed to about 300 psi. NG pipelines are powered by NG-fueled engines and turbines. Of the stages making up this path, NG compression (by means of electric compressors at dispersed refueling facilities) has the lowest energy efficiency (95%).

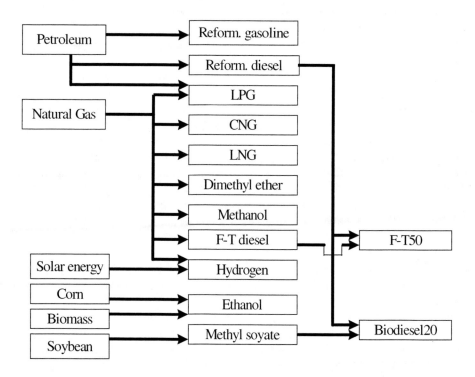

Figure 3. Fuel Pathways Considered in this Study

NG to LNG – On this path, natural gas is produced in and processed near NG fields, liquefied at LNG plants that are adjacent to NG processing plants, and then transported by rail and truck to LNG service stations. Of the stages making up this path, NG liquefaction uses the most energy (with an energy efficiency of 85%).

NG to LPG – For domestic supplies, this path includes natural gas recovery in NG fields, LPG production at NG processing plants near NG fields, and LPG transportation via rail or truck to LPG service stations. For imports, LPG is assumed to be transported to major U.S. ports via ocean tanker and then transported to LPG refueling stations via truck, rail and pipeline. Though high relative to other fuels, the efficiency of LPG production (96.5%) is the lowest of all the stages in the pathway. In this analysis, NG was assumed to account for 60% of LPG supply under all scenarios.

NG to DME – On this path, natural gas is recovered in and processed near NG fields. DME is then produced at plants that are adjacent to NG processing plants. DME production has the lowest energy efficiency (70%) of these upstream activities. For this analysis, all DME was assumed to be produced overseas from inexpensive NG, shipped by ocean tanker to main U.S. ports, and transported to service stations via truck, rail and pipeline.

NG to Methanol – The upstream stages of this path are similar to those of the NG-to-DME path except that methanol is produced instead of DME. For this analysis, all methanol was assumed to be imported, shipped to main U.S. ports via ocean tanker, and then transported to service stations via pipeline and truck. Methanol production

has the lowest energy efficiency (65%) of all the upstream activities in this path.

NG to F-T Diesel – On this path, natural gas is recovered in and processed near NG fields. At F-T plants that are adjacent to NG processing plants, F-T diesel is produced and blended with conventional diesel. F-T diesel blends are then transported to service stations in the same way as conventional diesel. Among the upstream activities of this cycle, F-T diesel production has the lowest energy efficiency (57%).

NG to H_2 – Hydrogen can be produced in centralized facilities (like it is now) or in decentralized facilities (like refueling stations using NG). The advantage of decentralized production is its avoidance of expensive H_2 distribution infrastructure. For this analysis, centralized production was assumed because the technology is proven and economies of scale can be realized. Although both liquid and gaseous H_2 can be used for H_2-powered fuel-cell vehicles (FCVs), gaseous H_2 was assumed for this analysis because liquefaction poses additional energy losses and emissions, and the transportation and storage of liquid H_2 can be expensive.

On this path, NG is recovered in and processed near NG fields, H_2 is produced at centralized plants adjacent to NG processing plants, and gaseous H_2 is transported to service stations via pipeline. In this analysis, H_2 was assumed to be compressed to about 6,000 psi at service stations. With an energy efficiency of 68%, H_2 production consumes the most energy of all the stages in this path. Because H_2 contains no carbon and no carbon sequestration was assumed in this study, the conversion of NG to H_2 produces considerable CO_2 emissions.

<u>Solar Energy to H₂</u> – Production of H_2 from solar energy via water electrolysis offers significant energy and environmental benefits, as well as the possibility of a practically unlimited energy source. In this study, H_2 was assumed to be produced in centralized facilities in such regions as the Southwestern United States where solar energy is abundant. Hydrogen was then assumed to be compressed moderately (to about 100 psi) and transported via pipeline to H_2 refueling stations. There, gaseous H_2 was assumed to be compressed to about 6,000 psi for use by H_2-powered FCVs. Electricity was assumed to be used for compressing H_2 and powering pipeline motors, and parameters typical of U.S. average electric generation were used to estimate emissions of criteria pollutants and GHGs from the electricity used. Note that the same assumptions were applied to H_2 transportation and compression for the above NG-to-H_2 path. Energy efficiencies for gaseous H_2 transportation via pipeline and H_2 compression at service stations were assumed to be 94% and 90%, respectively.

<u>Corn to Ethanol</u> – This path includes corn production and transportation; ethanol production; and ethanol transportation, storage, and distribution. GHG emissions from corn production come from fuels used for farming, harvesting, and corn drying, together with the amount from fertilizers and herbicides used during corn farming. Both wet- and dry-milling technology is currently used in the United States to produce ethanol. Wet-milling plants now account for about two-thirds of ethanol production capacity; dry-milling plants account for the remaining third. Because of tax incentives that are available in certain states and their generally lower capital requirement, most newer ethanol plants are small-scale, dry-milling plants. In this analysis, future corn-to-ethanol plant capacity was assumed to be evenly split between the two technologies (i.e., half wet-milling and the other half dry-milling). Note that this assumption implies that more dry-milling plants will be built in the future than wet-milling plants. For a detailed discussion of the technical assumptions regarding this cycle, see Wang et al. (1997).

<u>Biomass to Ethanol</u> – This path includes biomass production and transportation; ethanol production; and ethanol transportation, storage, and distribution. Biomass includes both woody and herbaceous feedstocks. In this analysis, the energy and emissions associated with biomass production were calculated in the same way as those for producing ethanol from corn. At cellulosic ethanol plants, the lignin portion of biomass was assumed to be burned to generate steam and electricity in co-generation systems. While combustion of biomass undoubtedly releases CO_2 emissions, this CO_2 came from the atmosphere through photosynthesis. Thus, CO_2 emissions from biomass combustion were treated as a transfer back into the atmosphere with a net effect of zero. For the same reason, CO_2 emissions from ethanol combustion by ethanol-powered vehicles were also assigned a net value of zero.

The electricity generated at ethanol plants was assumed to be exported to the electric power grid. Emissions credits for the generated electricity were calculated in GREET as a function of the amount of electricity generated and average emissions associated with electricity generation in U.S. electric utility systems.

<u>Soybeans to Biodiesel</u> – This path includes soybean farming; soybean transportation; soy oil extraction and transesterification; biodiesel blending; and biodiesel blend transportation, storage, and distribution. Among the upstream activities for this path, biodiesel production (including extraction and transesterification of soy oil) and soybean farming consume most of the energy and produce most of the emissions.

<u>Vehicular Emissions Improvements</u> – Emission standards are an important reason for considering alternative propulsion systems in the PNGV program. Approximately 40% of model year 1994 (MY94) passenger cars met the Phase 1 (commonly called Tier 1) emission requirements of the Clean Air Act Amendments; by MY96, all passenger cars were required to be in compliance with Tier 1 standards. These standards — for carbon monoxide (CO), oxides of nitrogen (NO_x), non-methane hydrocarbons (NMHC), and particulate matter smaller than 10 μm (PM_{10}) — are shown in Table 1. Note that the PM standard applies only to light-duty diesels. Tier 2 standards that require a further 50% reduction in emissions have been proposed but are not yet mandated. EPA is scheduled to rule on the need for these more stringent standards by December 1999 (EPA 1998a).

Table 1. Five-Yr 50,000-mi Emission Standards for Light-Duty Vehicles: 49 States and California (g/mi)

Standard	First MY	NMHC	NMOG	CO	NO$_X$	PM	HCHO (formaldehyde)
49 States (mandated)							
Tier 1	1994	0.25	N/A[a]	3.4	0.4	0.08[b]	N/A
Tier 2	2004[c]	0.125	N/A	1.7	0.2	0.08[b]	N/A
California (mandated)							
TLEV	1994	N/A	0.125	3.4	0.4	0.08[d]	0.015
LEV	1997	N/A	0.075	3.4	0.2	0.08[d]	0.015
ULEV	1997	N/A	0.040	1.7	0.2	0.04[d]	0.008
ZEV	2003	0[e]	0[e]	0[e]	0[e]	0[e]	0[e]
National LEV Program (voluntary)							
TLEV	2001	N/A	0.125	3.4	0.4	0.08[d]	0.015
LEV	2002	N/A	0.075	3.4	0.2	0.08[d]	0.015
ULEV	2003	N/A	0.040	1.7	0.2	0.04[d]	0.008

[a] Not applicable.
[b] Applies to all 49-state LDVs beginning in MY96.
[c] Need for these standards will be determined by EPA in 1999. They are not yet mandated. Definition of useful life increased to 10 yr or 100,000 mi.
[d] Applies to diesel vehicles only; standards are for 10 yr or 100,000 mi.
[e] Emissions from vehicle itself.

California has defined still stricter vehicle emission standards to be phased in over the next decade. The standards include formaldehyde and replace the non-methane hydrocarbon (NMHC) standard with a non-methane organic gas (NMOG) standard (which includes NMHC and several other organic gases). To meet a fleet-wide standard for NMOG, vehicle manufacturers must certify each of their vehicles in one of four emission categories: Transitional-Low-Emission Vehicles (TLEV), Low-Emission Vehicles (LEV), Ultra-Low-Emission Vehicles (ULEV), or Zero-Emission Vehicles (ZEV). A weighted average consisting of the emission standard for the category and the share of each manufacturer's California sales in that category will then be used to determine if manufacturers are meeting the fleetwide NMOG standard.

The California Air Resources Board recently adopted the so-called "LEV II" program (CARB 1998), which extends the already adopted TLEV, LEV, and ULEV standards to 120,000 mi; tightens PM emission standards to 0.04 g/mi for TLEV and 0.01 for LEV and ULEV (applicable at 120,000 mi); and adds a new vehicle category ñ super-ultra-low-emission vehicle (SULEV). Applicable for 120,000 mi, the proposed SULEV standards are 0.01 g/mi for NMOG, 1.0 for CO, 0.02 for NO_x, and 0.01 for PM.

Recently, EPA adopted a national low-emission-vehicle (NLEV) program to encourage the introduction of LEV types. Forty-five states and the District of Columbia will be covered under this program. The NLEV program is voluntary, and vehicle manufacturers can participate in lieu of complying with the individual requirements of any state except California.[4] The NLEV program begins in MY 2001 and is similar to the California LEV program with one major exception: Zero Emission Vehicles (ZEVs) are not required to be sold (EPA 1998b).

It is generally believed that 3X vehicles will be subject to Tier 2 standards for VOC, CO, and NO_x and the ULEV standard for PM. For this analysis, it was assumed that RFG-fueled SIDI engines will meet Tier 2 standards, but no further emissions reductions (e.g., LEV II standards) will occur. All other SIDI engines (fueled with methanol, ethanol, CNG, LNG, and LPG) were assumed to at least meet Tier 2 standards. If an alternative fuel offers inherently lower emissions than RFG, emission reductions were assumed for that fuel. Table 2 presents

---

[4] Twenty-three automakers that comprise nearly all of the U.S. LDV market have agreed to participate in the NLEV program. New York, Massachusetts, Maine, and Vermont have decided to pursue ZEV requirements (i.e., the California model) instead of participating in NLEV.

the emission reductions assumed for the five alternative SIDI fuels.

Recently, it has been proposed that CIDI 3X vehicles should be subject to a PM standard of 0.01 g/mi, which is equivalent to the PM emission rate of conventional gasoline SI vehicles. For this analysis, it was assumed that 3X vehicles with CIDI engines will at least meet Tier 2 standards for NO_x, CO, and VOC and current ULEV standards (i.e., 0.04 g/mi) for PM. As with SIDI alternative fuels, CIDI alternative fuels that offer inherently lower emissions were assumed to achieve further reductions relative to RFD. Table 3 presents the emission standards that the four CIDI fuels were assumed to meet. Note that the four fuels were assumed to produce no evaporative emissions since all have very low Reid vapor pressure (RVP). RFD, B20, and F-T50 were assumed to meet the current ULEV standard for PM. DME was assumed to meet the tighter LEV II ULEV standard (0.01 g/mi), which is equivalent to that of an SIDI engine.

It is generally believed that FCVs (using H_2, methanol, or RFG) will have substantially lower emissions than either SIDI or CIDI engines. Table 4 presents the emission assumptions used in this analysis. Note that entries are relative to Tier 2 RFG-fueled vehicles.

Table 2. Relative Emissions of Alternative SIDI Fuels

Pollutant	Percent Tier 2 RFG-Fueled SIDI Rate				
	Methanol	Ethanol	CNG	LNG	LPG
VOC (exhaust)	55	55	15	15	75
VOC (evaporative)	100	100	0	0	0
CO	60	60	40	40	60
NO_x	80	80	60	60	90
PM (exhaust)	10	10	1	1	1
PM (brake and tire)	100	100	100	100	100
CH_4	65	65	1000	1000	100
N_2O	100	100	100	100	100

Table 3. Emissions of Alternative CIDI Fuels

Pollutant	Emissions (g/mi), by Fuel			
	RFD[b]	DME	B20	F-T50
VOC (exhaust)	0.125	0.125	0.125	0.125
VOC (evaporative)	0	0	0	0
CO	1.7	1.7	1.7	1.7
NO_x	0.2	0.2	0.2	0.2
PM (exhaust)	0.04	0.01	0.04	0.04
CH_4[a]	0.008	0.008	0.008	0.008
N_2O[a]	0.005	0.005	0.005	0.005

[a] Based on GREET estimates for conventional diesel.
[b] Current California diesel has a sulfur content of about 150 ppm. RFD was assumed to have a sulfur content of 100 ppm in order to meet the 0.04 g/mi PM_{10} emission standard.

ENERGY AND EMISSIONS ESTIMATES

EMISSIONS OF CRITERIA POLLUTANTS – Figures 4-8 display percent changes in urban emissions of the five criteria pollutants for each of the propulsion system/fuel combinations examined. Each figure depicts results for a single pollutant as a series of curves showing annual percentage increases or decreases from the reference scenario forecast. Curves that are all but indistinguishable are combined to aid interpretation. Upstream and operational emissions are not shown separately because virtually all urban emissions are due to vehicle operation. Readers interested in further detail are urged to consult Appendixes A and B of Wang et al. which contain estimates of upstream, operational, and total emissions by propulsion system/fuel combination and scenario (1998).

Table 4. Relative Emissions of Alternative FCV Fuels

Pollutant	Percent of Tier 2 RFG-Fueled SIDI Rate		
	Hydrogen	Methanol[a] (Partial Oxidation)	Gasoline[a] (Partial Oxidation)
VOC (exhaust)	0	0.5	0.5
VOC (evaporative)	0	20[b]	50[b]
CO	0	1	1
NO_x	0	1	1
PM (exhaust)	0	0	0
PM (brake and tire)	100	100	100
CH_4[c]	0	0	0
N_2O[d]	0	0	0

[a] Based on Kumar (1997).
[b] Smaller tank size for 3X vehicles helps reduce evaporative emissions.
[c] Gasoline vehicle CH_4 emissions = 0.074 g/mi.
[d] Gasoline vehicle N_2O emissions = 0.005 g/mi.

Emissions estimates under both market share scenarios show similar results. However, the patterns are much more striking under the high-market-share scenario which, by definition, is a more extreme example of possible market penetration. Thus, the following discussion focuses on results from that scenario. Note that each technology/fuel alternative considered in the analysis was examined in the context of a scenario that contains a significant portion of conventional, as well as 3X, vehicles. Thus, emissions were computed for a combination of conventional and 3X technologies, and results are less striking than would be the case for 3X technologies alone.

Nitrogen Oxides (NO_x) – Figure 4 illustrates the impact of alternative 3X propulsion system/fuel combinations on urban NO_x emissions. Because it was assumed that the four CIDI fuels would meet equivalent Tier 2 emission standards, RFD, DME, F-T50, and B20 all fall within a narrow band and are essentially equivalent to RFG, MeOH, EtOH, LPG, and the gaseous-fueled alternatives. Methanol and gasoline fuel cells offer the largest reduction in urban NO_x emissions — 35% under the high-market-share scenario. Hydrogen fuel cells achieve somewhat lower NO_x reduction (approximately 32%) because of their relatively higher upstream emissions.

Carbon Monoxide (CO) – Figure 5 shows reductions in CO emissions under the high-market-share scenario. Again, reductions range up to about 35%, with fuel cells achieving the highest reductions and SIDI engines on any of six fuels achieving the lowest. Between these two clusters, however, the position of the other propulsion system/fuel alternatives differs markedly from NO_x results. Given the CI engine's proven record of relatively low CO emissions, it is not surprising that diesel-like fuels (RFD, DME, F-T50, B20) have the second-best CO reduction.

Volatile Organic Compounds (VOCs) – For VOC, reductions from reference scenario emissions range up to approximately 37% under the high-market-share scenario (see Figure 6). Hydrogen fuel cells are the clear leader from a VOC-reduction standpoint, with methanol fuel cells a close second and gasoline fuel cells third. CIDI engines on RFD, DME F-T50, or B20 and SIDI engines on LPG, CNG, or LNG achieve almost half the reduction of hydrogen fuel cells.

Sulfur Oxides (SO_x) – Unlike the other criteria pollutants, urban SO_x emissions are closely related to the volume of fuel used. Thus, relative to the reference scenario, all propulsion system/fuel alternatives reduce urban SO_x emissions because of their tripled fuel efficiency (Figure 7). Hydrogen fuel cells, LPG, CNG, ethanol, and DME achieve the biggest reductions, but urban SO_x represents a very small share (on the order of 13%) of the total SO_x attributable to light-duty vehicles. Most SO_x emissions come from upstream fuel processing, which tends to be outside urban areas.

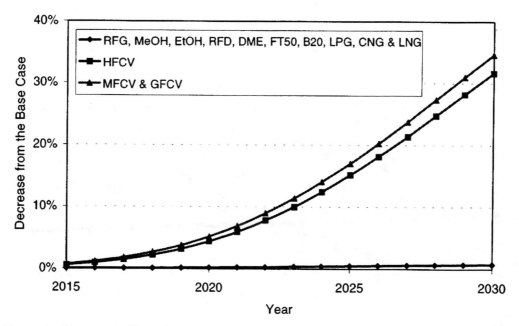

Figure 4. Changes in Fuel-Cycle Urban NO$_x$ Emissions by 3X Technology/Fuel Alternative

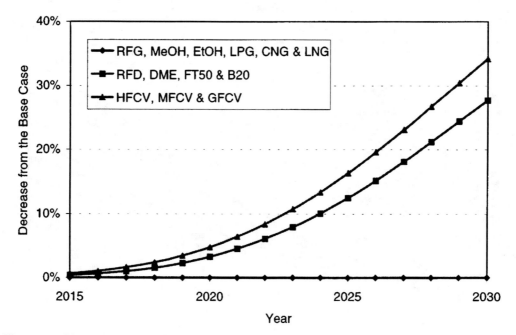

Figure 5. Changes in Fuel-Cycle Urban CO Emissions by 3X Technology/Fuel Alternative

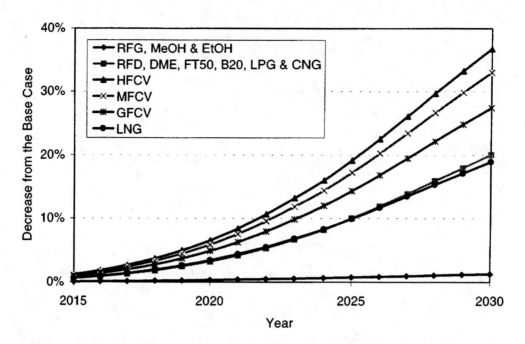

Figure 6. Changes in Fuel-Cycle Urban VOC Emissions by 3X Technology/Fuel Alternative

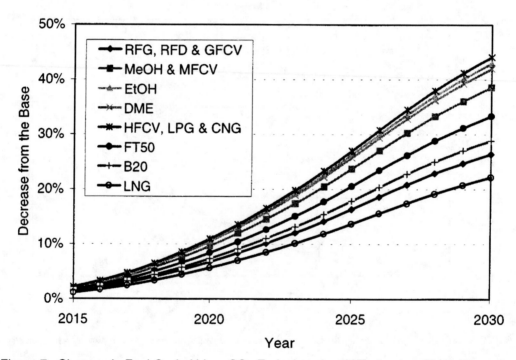

Figure 7. Changes in Fuel-Cycle Urban SO$_x$ Emissions by 3X Technology/Fuel Alternative

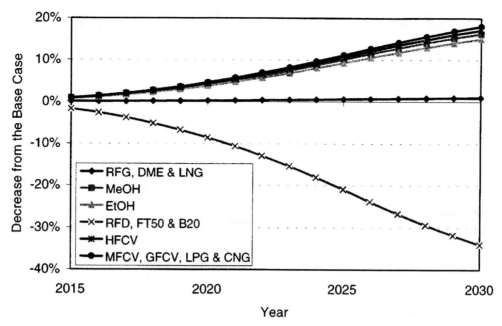

Figure 8. Changes in Fuel-Cycle Urban PM_{10} Emissions by 3X Technology/Fuel Alternative

Particulate Matter (PM_{10}) – Unlike total PM_{10} emissions, nearly half of which occur upstream, urban PM_{10} emissions are dominated by vehicle operations. Thus, with the exception of DME, diesel-like fuels increase PM_{10} emissions (Figure 8). Excluding RFG, DME, and LNG, which have little effect on PM, all the other alternatives decrease PM_{10} emissions by approximately 15% under the high-market-share scenario. Note that the increase for diesel-like fuels occurs despite the assumption of a "Tier 2 equivalent" exhaust emission standard of 0.04 g/mi (as compared with the current standard of 0.08 g/mi at 50,000 mi). DME, which produces virtually no particulate matter from fuel combustion, was assumed to achieve the proposed LEV II ULEV standard (including 0.01g/mi PM_{10}). Given that assumption, it comes as no surprise that DME and RFG have comparable urban PM_{10} emissions. For LNG, which produces similar urban PM emissions, upstream processes account for much of urban PM.

Note also that ethanol is not markedly different from the other SIDI fuel alternatives insofar as urban PM_{10} emissions are concerned. If total PM_{10} emissions had been shown in Figure 8, results would have appeared quite different. Ethanol accounts for the largest increase in total PM_{10} emissions, virtually all of which are due to agricultural processes. Thus, if produced in the quantities required under the high-market-share scenario, ethanol would generate considerable PM_{10} emissions but those emissions would be relatively benign from an urban perspective.

GREENHOUSE GAS EMISSIONS – Figure 9 displays changes in total greenhouse gas (GHG) emissions in the same format as that used for the criteria pollutant graphs. Note that because CO_2 comprises the bulk of GHGs and all propulsion system/fuel alternatives share the same fuel efficiency, emission reductions from non-renewable fuels are clustered. Under the high-market-share scenario, the range is from 23 to 45%. Chief among the low-GHG alternatives are ethanol-fueled SIDI engines and hydrogen fuel cells, both of which generate no CO_2 from vehicle operations. Hydrogen fuel-cell vehicles generate no CO_2 because no carbon is contained in the fuel. Ethanol-fueled SIDI engines are assumed to generate no CO_2 because the carbon in ethanol comes from carbon in the atmosphere via photosynthesis. When combined with the conventional vehicles (and their GHG emissions) in the high-market-share scenario, these low-GHG alternatives achieve overall reductions (from all light-duty vehicles, both 3X and conventional) of 46% (for ethanol) and 33% (for hydrogen).

Note also that shifts from current to advanced production technologies cause some GHG reduction curves to shift position relative to the others. Specifically, hydrogen shifts from a position at or near the bottom of the pack to second place by 2025. Ethanol, which is also assumed to shift to a more advanced production technology, has a change in slope, but it is less obvious relative to the other alternatives. After ethanol and hydrogen, the only two renewable fuels examined, LPG and the gaseous fuels, achieve the next-best reduction in GHGs. However, they are only marginally better than the other alternatives.

ENERGY CONSUMPTION – Figures 10-12 provide estimates of changes in total energy, fossil energy, and petroleum use for the high-market-share scenario relative to the reference scenario. Again, formats are identical to the above graphs.

Total Energy – As shown in Figure 10, total energy use by light-duty vehicles declines by 18-29% under the high-market-share scenario. By definition, all fuel/technology alternatives achieve 3X fuel economy. Thus, operational energy use declines by 27% in 2030 for all alternatives and the upstream energy requirements of the various fuels account for all the variation in total energy use among the alternatives.

Fossil Fuels – Reductions in fossil energy use by the 13 fuel/technology alternatives are shown in Figure 11. The ethanol- and hydrogen-fueled alternatives, both largely nonfossil fuels in 2030, achieve the largest reductions in fossil fuel use in that year (approximately 45% under the high-market-share scenario), followed by the biodiesel blend (B20) and RFD, LPG, and CNG, which achieve reductions of nearly 30%. The transition from fossil to nonfossil feedstocks is particularly evident in the hydrogen curve, as is a flattening out in all curves for the high-market-share scenario.

All (completely) fossil-fueled alternatives consume 11.1 quads of fossil fuels because of vehicle *operation* in 2030 under the high-market-share scenario vs. 10.8 quads for B20, 20% of which is nonfossil, and 8.2 quads for the nonfossil alternatives. Again, upstream energy use accounts for the variation in fossil energy use (for the entire fuel cycle) within the two groups of fossil- vs. nonfossil-fueled alternatives.

Petroleum – Several of the fuel/technology alternatives consume nonpetroleum fuels. To the extent that such fuels are derived from fossil sources (e.g., DME or methanol from natural gas), they offer little reduction in greenhouse gas emissions, despite potentially dramatic reductions in petroleum use. Figure 12 displays changes in petroleum use by technology/fuel alternative under the high-market-share scenario. Clearly, the alternatives cluster into three groups: largely petroleum fuels (i.e., RFG, RFD and B20), "part petroleum" fuels (i.e., F-T50 and LPG), and largely nonpetroleum fuels (i.e., hydrogen, methanol, ethanol, DME, CNG, and LNG). By 2030, the nonpetroleum alternatives achieve an approximately 45% reduction in total petroleum use under the high scenario relative to the reference scenario. The "part petroleum" alternatives (i.e., LPG and F-T50) achieve the next-best reduction — approximately 35% under that scenario.

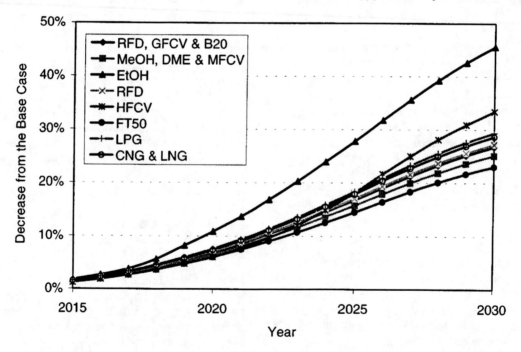

Figure 9. Changes in Fuel-Cycle GHG Emissions by 3X Technology/Fuel Alternative

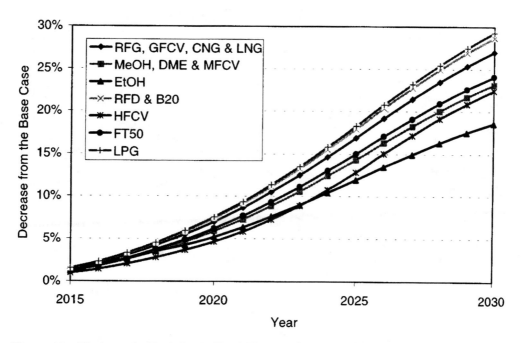

Figure 10. Changes in Fuel-Cycle Total Energy Use by 3X Technology/Fuel Alternative

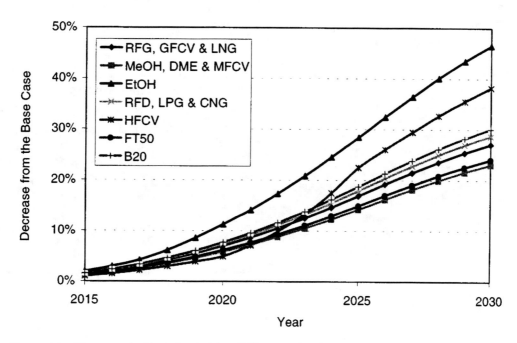

Figure 11. Changes in Fuel-Cycle Fossil Energy Use by 3X Technology/Fuel Alternative

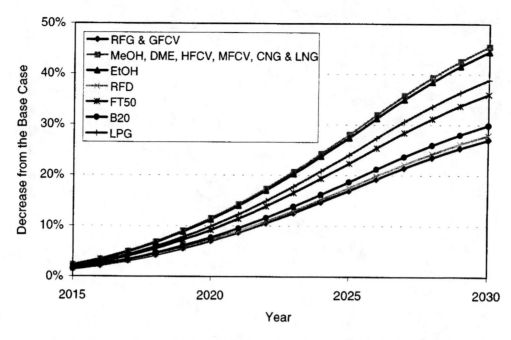

Figure 12. Changes in Fuel-Cycle Petroleum Use by 3X Technology/Fuel Alternative

CONCLUSIONS

In this study, 11 fuels (RFG, RFD, DME, methanol, ethanol, LPG, CNG, LNG, F-T50, B20, and hydrogen) that are potential candidates for use in 3X vehicles were evaluated in three power system applications (SIDI engine, CIDI engine, or fuel cell) for a total of 13 propulsion system/fuel combinations. Two scenarios depicting alternative levels of 3X market penetration were developed and used to estimate the fuel production and distribution infrastructure needed to satisfy the fuel demands of 3X vehicles and the fuel-cycle energy and emissions impacts of the 13 potential propulsion system/fuel combinations. Impacts were generated for each year from market introduction (2007 in the high-market-share scenario and 2013 in the low-market-share scenario) to 2030 for each of the propulsion system/fuel combinations.

Results indicate that energy and emissions impacts of 3X vehicles are highly dependent on market penetration and thus differ dramatically between the two scenarios examined in this study. Because impacts are relatively small under the low-market-share scenario, most of the discussion presented here focuses on the more significant results obtained for the high-market-share scenario. For all 3X propulsion system/fuel technologies considered, total energy and fossil fuel use by U.S. light-duty vehicles declines significantly relative to reference scenario estimates for 2030. Fuel savings occur as a result of fuel-efficiency improvements, which apply to all 3X technologies and reduce LDV energy use by more than 25%, as well as a result of fuel substitution, which applies to the nonpetroleum-fueled alternatives studied. Together, the two effects reduce LDV petroleum use in 2030 by as much as 45% relative to the reference scenario. GHG emissions follow a similar pattern. Total GHG emissions decline by 25-30% with most of the propulsion system/fuel alternatives. For those using renewable fuels (i.e., ethanol from biomass and hydrogen from solar energy), GHG emissions drop by 33% (hydrogen) and 45% (ethanol) relative to the reference scenario.

Among the five criteria pollutants, urban NO_x emissions decline slightly for 3X vehicles using CIDI and SIDI engines and drop substantially for fuel-cell vehicles (FCVs). Urban CO emissions decline for CIDI and FCV alternatives, while VOC emissions drop significantly for all alternatives except RFG-, MeOH- and EtOH-fueled SIDI engines. With the exception of CIDI engines using RFD, F-T50, or B20 (which increase urban PM_{10} emissions by over 30% in the high scenario), all propulsion system/fuel alternatives reduce urban PM_{10} emissions. Reductions are approximately 15ñ20% for fuel cells and methanol-, ethanol-, CNG-, or LPG-fueled SIDI engines (RFG- and LNG-fueled SIDI engines and DME-fueled CIDI engines have only very slight reductions).[5] Although urban SO_x emissions decline for all of the alternatives, SO_x emissions resulting from the use of LNG are higher than those resulting from the use of hydrogen, LPG, and CNG.

---

[5] Increased urban PM10 emissions from RFD, and to a lesser extent from F-T50 and B20, may also produce worse health effects because much of it is particulate matter of 2.5 μm or less ($PM_{2.5}$) which results in greater damage per unit than PM_{10} emissions from other fuels (e.g., ethanol) which tend to be in the 2.5–10 mm range.

The air quality implications of these emissions results should be interpreted cautiously. Changes in emissions of the five criteria pollutants do not necessarily translate into similar changes in air quality, simply because emissions from different fuels and upstream fuel-production activities occur in different locations and at different times and are dependent on atmospheric processes. Generally speaking, upstream emissions occur outside urban areas, while vehicular emissions occur within urban areas. Because of high population exposure (especially where mortality effects exist), emissions in urban areas generate far greater damage than those outside urban areas. That is why urban emissions have been estimated in this analysis. However, those estimates are based on broad, categorical data that may not be representative of all urban areas and that cannot take into account the effects of varying local climatic conditions. Further, the results reported here exclude the upstream emissions of criteria pollutants from methanol, DME, and LPG production which occurs in foreign countries.

So, which of the propulsion system/fuel combinations are best? If the aim is to reduce criteria pollutant and GHG emissions, the clear choice is renewables, especially hydrogen from solar energy. But is this a realistic choice? Cumulative capital needs were estimated as part of the analysis summarized here.[6] As expected, capital needs were found to vary by technology and scenario. Of particular interest, though, is that supplying the low-market-share scenario's fuel demand requires an incremental investment of less than $50 billion for all fuels except hydrogen, (which requires $130-150 billion). By contrast, production and distribution facilities with capacity sufficient for the high-market-share scenario require cumulative investments of approximately $50 billion for LNG, $90 billion for ethanol, $100 billion for methanol, $120-160 billion for CNG, $160 billion for DME, and $480-560 billion for hydrogen. Although these requirements are spread over many years, their sheer magnitude (and associated risk) could pose a serious challenge to the widespread introduction of 3X vehicles requiring new fuels. Clearly, hydrogen FCVs offer the largest energy and emissions benefits but with the greatest incremental capital cost. On the other hand, methanol and gasoline FCVs which offer somewhat less energy and emissions benefits, appear to have particularly promising benefits-to-costs ratios.

ACKNOWLEDGMENTS

This work was sponsored by the U.S. Department of Energy's Office of Advanced Automotive Technologies. The authors thank Pandit Patil and Ed Wall for their continued support; Margaret Singh and Kevin Stork for their analyses of fuel production and distribution costs; and Dan Sperling, Sujit Das, Tien Nguyen, and Dan Santini for their input to and review of this work.

REFERENCES

1. California Air Resources Board (CARB)

2. CARB, 1998, *Initial Statement of Reasons for Proposed Amendments to California Exhaust and Evaporative Emission Standards and Test Procedures for Passenger Cars, Light-Duty Trucks and Medium-Duty Trucks*, Staff Report, Mobile Sources Control Division, El Monte, Cal., Sept. 18.

3. EIA: Energy Information Administration.

4. EIA, 1996, *Annual Energy Outlook 1997 with Projections to 2015*, DOE/EIA-0383(97), U.S. Department of Energy, Washington, D.C., Dec.

5. EIA, 1997a, *Petroleum Supply Annual 1996, Vol. 1*, DOE/EIA-0340 (96)/1, U.S. Department of Energy, Washington, D.C., May.

6. EIA, 1997b, *Natural Gas Annual 1996*, DOE/EIA-0131 (96), U.S. Department of Energy, Washington, D.C., Sept.

7. EPA: U.S. Environmental Protection Agency

8. EPA, 1998a, *Tier 2 Report to Congress: Tier 2 Study*, Office of Air and Radiation, EPA420-R-98-008, July 31.

9. EPA, 1998b, "Control of Air Pollution from New Motor Vehicles and New Motor Vehicle Engines: State Commitments to National Low Emission Vehicle Program" [URL.http://www.epa.gov/fedrgstr/(as of July 15, 1998)]

10. FHWA: Federal Highway Administration.

11. FHWA, 1992, *1990 Nationwide Personal Transportation Survey*, public use tape, Washington, D.C.

12. Kumar, R., 1997, Argonne National Laboratory, personal communication.

13. Mintz, M., et al., 1994, *The IMPACTT Model: Structure and Technical Description*, ANL/ESD/TM-93, Argonne National Laboratory, Argonne, Ill.

14. NRC: National Research Council.

15. NRC, 1994, *Review of the Research Program of the Partnership for a New Generation of Vehicles*, National Academy Press, Washington, D.C.

16. Wang, M., et al., 1998, *Assessment of PNGV Fuels Infrastructure Phase 2 Report: Additional Capital Needs and Fuel-Cycle Energy and Emissions Impacts*, ANL/ESD-37, Center for Transportation Research, Argonne National Laboratory, Argonne, Ill., Aug.

17. Wang, M., et al., 1997, *Fuel-Cycle Fossil Energy Use and Greenhouse Gas Emissions of Fuel Ethanol Produced from U.S. Midwest Corn*, Center for Transportation Research, Argonne National Laboratory, prepared for Illinois Department of Commerce and Community Affairs, Springfield, Ill., Dec.

18. Wang, M., 1996, *GREET 1.0 — Transportation Fuel Cycles Model: Methodology and Use*, ANL/ESD-33, Argonne National Laboratory, Argonne, Ill.

---

[6] For a full discussion of the methodologies and assumptions of the cost analysis, see Wang et al. (1998).

2000-01-1541

SUV Powered by On-Board Generated H$_2$

**Steven C. Amendola, Phillip J. Petillo, Robert Lombardo,
Stefanie L. Sharp-Goldman, M. Saleem Janjua, Nicole C. Spencer,
Michael T. Kelly and Michael Binder**

Millennium Cell LLC

ABSTRACT

A Ford Explorer has been totally powered by internal combustion engine (ICE).fueled by hydrogen (H$_2$) gas. H$_2$ gas was generated on-board by a patented H$_2$ generation technology – a safe, ambient temperature chemical reaction with a water based solution. The novel feature of this system is that H$_2$ gas is safely generated on-board the vehicle as needed without resorting to bulky, pressurized cylinders. The entire H$_2$ generation system easily fits within the engine compartment and beneath the floor of the vehicle making this one of the few practical zero emission vehicles that has room for 5 passengers and cargo while offering competitive performance and driving range to current SUVs.

INTRODUCTION

Currently, the major energy supply for transportation applications is derived from combusting fossil fuels. This process simultaneously generates carbon dioxide, carbon monoxide, incompletely combusted hydrocarbons, and volatile organic compounds. The environmental impact of these pollutants, the realities of finite petroleum reserves, and the high cost of gasoline outside the U.S. are all factors in the search for alternative fuels.

Hydrogen powered vehicles are an attractive and environmentally desirable solution. H$_2$ can either be used directly as a fuel in internal combustion engines or electrochemically oxidized via Proton Exchange Membrane, PEM, fuel cells. In both cases, water and electrical energy are the primary products. H$_2$ powered vehicles can use advanced technologies to improve environmental quality and energy security, while providing the range, performance and utility of today's gasoline vehicles. H$_2$ is extremely energy rich on a weight basis but relatively poor (compared to gasoline) on a volumetric basis. Thus, large volumes of H$_2$ must be safely carried onboard a vehicle. This leads to the primary technical challenge in developing H$_2$ fueled vehicles i.e. how to generate and/or store adequate amounts of H$_2$ fuel to enable significant vehicle range.

H$_2$ can be stored/generated by various physical or chemical methods. These choices determine refueling time, cost, and infrastructure requirements, as well as energy efficiency, vehicle fuel economy, safety, and performance. Physical methods for storing H$_2$ include: storing H$_2$ as a low temperature liquid or as a compressed gas. Drawbacks of storing liquid H$_2$ include the substantial amount of electricity required for liquefying H$_2$, the difficulty in maintaining extremely low temperatures onboard a vehicle and considerable (>1.5 % /day) bleed-off and evaporation losses. High pressure cylinders are bulky, have limited storage capacity, and raise safety issues in event of collisions.

Chemical methods of storing H$_2$ include: reforming H$_2$ containing fuels (e.g. methanol or other hydrocarbons), storing H$_2$ in reversible hydrides, and reacting water or heat with chemicals to yield H$_2$ gas. Reforming requires significant heat and produces CO$_2$, a greenhouse gas. Storing H$_2$ in reversible hydrides is expensive, has very low H$_2$ storage efficiently by weight, and may require thermal energy to remove H$_2$ from the hydride. Producing H$_2$ via alkali metal hydride reactions is explosive and not easily controlled.

BOROHYDRIDE HYDROGEN GENERATOR

The portable borohydride hydrogen generator produces high-purity hydrogen at ambient temperature. This easily controllable system is particularly suitable for generating H$_2$ for either fuel cells or internal combustion engines. In our compact generator, a nonflammable, water-based, alkaline sodium borohydride (NaBH$_4$, tetrahydroborate) solution is used. NaBH$_4$, readily available commercially, is domestically produced from raw materials found in the US. When aqueous NaBH$_4$ solutions contact ruthenium, Ru, metal catalysts, these solutions hydrolyze to yield pure H$_2$ gas and water-soluble, sodium metaborate, NaBO$_2$.

$$NaBH_4 + 2 H_2O \longrightarrow 4 H_2 + NaBO_2$$
$$\text{catalyst} \qquad (1)$$

In addition to H_2, the other hydrolysis discharge product, sodium metaborate, is water soluble and environmentally innocuous since it is chemically similar to ingredients found in laundry detergents. Reaction [1] is totally inorganic (carbon-free) and produces no fuel cell poisons (i.e. virtually zero sulfur, CO or aromatics are produced). Reaction [1] is considerably safer, more efficient, and easily controllable than producing H_2 by reacting water with other reactive chemicals such as LiH or $LiAlH_4$. The heat generated by reaction [1], is only 60% of the heat produced by reacting other chemical hydrides with water.

To generate H_2, Ru catalyst supported on ion exchange resin beads, is simply allowed to contact $NaBH_4$ solution. This ensures rapid response to H_2 demand i.e. H_2 is generated only when Ru catalyst contacts $NaBH_4$ solution. One safety advantage is that high H_2 generation rates can be generated at close to ambient temperatures without requiring mechanical compression. In addition, since stabilized $NaBH_4$ solutions not contacting catalyst do not produce H_2, in the event of a spill, no H_2 will be generated. Since little free H_2 is stored when the system is not operating (H_2 is only generated as needed), concerns about onboard H_2 storage is reduced. These facts speak favorably about using $NaBH_4$ solutions to generate H_2 in vehicular applications. From an engineering perspective, $NaBH_4$ solutions have weight and volume advantages over other H_2 storage methods.

Another advantage of the borohydride H_2 generator is the safety of the reagents and the reaction. $NaBH_4$ solutions are stable, nonflammable and can be stored in plastic containers. No dangerous reactions occur when $NaBH_4$ solutions are exposed to water or air. Ru catalyst is equally as safe. Dispersing Ru on high surface area substrates decreases the amount of Ru required for the reaction to occur at reasonable rates and reduces the cost. The stability and safety of $NaBH_4$ solutions lend itself to rapid refueling. No other H_2 source presents easier protocols or logistics.

For the $NaBH_4$ generator, assuming 100% stoichiometric yield in reaction [1], each mole of $NaBH_4$ (37.8 g/mole) produces 4 moles of H_2. At 25°C, 4 moles H_2 occupies (4 moles H_2)(24.6 liters H_2/mole) = 98 liters. Therefore, one liter of 35 wt % $NaBH_4$ aqueous solution (containing 350 g/(37.8 g/mole) =9.3 moles $NaBH_4$) yields: (9.3 moles $NaBH_4$)(98 liters H_2/mole $NaBH_4$) ~ 910 liters H_2 = 74 grams H_2 /liter of 35 wt % $NaBH_4$ solution. This value exceeds current US Department of Energy Storage Goals of 70 g H_2/liter. Another way of expressing this storage efficiency is that the equivalent of 1 kg H_2 can be stored in only 14.2 kg of 35 wt % $NaBH_4$ solution (13.5 liters of solution). Thus, $NaBH_4$ solutions are a convenient, non pressurized method of storing H_2.

The following chart summarizes the amounts of H_2 that can be produced from $NaBH_4$ solutions. Ru catalyst weight and volumes for these reactions is negligible.

$NaBH_4 \rightarrow 4 H_2$
37.8g $\rightarrow$ 8 g H_2 or 98 liters at 25°C
1 liter of 35 wt % $NaBH_4$ solution $\rightarrow$ 910
liters H_2 = 74 g H_2

ENERGY CONTENT OF BOROHYDRIDE SOLUTIONS

As a benchmark, 5 kg H_2 (equivalent to 19 liters, 5 gallons of gasoline) is considered necessary for a general purpose vehicle since it provides a 320 km (200 mile) range in a 17 km/liter (40 mpg) conventional car, or a 640 km (400 mile) range in a 34 km/liter (80 mpg) series hybrid (or fuel cell) vehicle. Volumes required to store 5 kg H_2 in cryogenic containers, in pressurized tanks, and in $NaBH_4$ solutions can be compared. For liquid H_2 stored cryogenically, 5 kg H_2 requires ~110 liters. Storing 5 kg H_2 in pressurized vessel at 24.8 MPa (3600 psi) requires ~320 liters. Storing 5 kg H_2 in 35 wt % $NaBH_4$ solution requires 65 liters (18 gallons). We emphasize that although we are basing volumes solely on $NaBH_4$ solution volumes, this is not a bad estimate for what is expected for H_2 generation system. This is because $NaBH_4$ solution volumes occupy the bulk of H_2 generation system volume. Ru catalysts take up negligible weight/volume. Since $NaBH_4$ solutions are not pressurized, storage tanks can be constructed from lightweight plastic which do not add considerable weight and/or volume.

Storing 5 kg H_2 Required in	Volume
pressurized cylinders	320 liters (85 gal)
cryogenic containers	110 liters (29 gal)
$NaBH_4$ solutions	65 liters (18 gal)

MILEAGE EQUIVALENT WITH BOROHYDRIDE SOLUTIONS

A midsize vehicle requires ~ 0.3 kWh/mile. From the calculations above, 1 kWh = requires 738 liters H_2. To travel 1 mile (0.3 kWh), requires = 20 g H_2 which can be produced by 0.26 liters of a 35 wt % $NaBH_4$ solution. Therefore, the mileage equivalent of 35 wt % $NaBH_4$ solution is ~15.8 mpg.

VEHICLE DESCRIPTION

A 1991 Ford Explorer was converted to run as a ZEV in a series hybrid configuration. The original engine and transmission were removed and in its place, we installed a 30 HP water-cooled, gasoline engine that had been converted to operate on directly injected H_2 gas. This engine will provide power to recharge a 144 volt (12 X

12V) battery pack by running an electrical generator. The generator and batteries together will provide additional power for acceleration and hill-climbing. The car will be driven by a DC electric motor (180 HP peak) running from the battery pack. The motor, with a drive ratio of 3.73:1 will provide a direct drive at the rear wheels. We are also considering increasing vehicle performance by use of a 5.13:1 gear ratio. With this higher gear ratio, maximum vehicle speeds of 80 mph are expected.

The $NaBH_4$ H_2 generator rapidly produces H_2 as needed. A small mechanical pump meters $NaBH_4$ solution onto a Ru coated catalyst bed to generate H_2. The entire H_2 generation system easily fits in the engine compartment and beneath the vehicle floor. This practical, H_2 powered, zero emission vehicle will have ample room for 5 passengers and cargo while offering competitive performance and driving range to current gasoline powered sport utility vehicles.

CONCLUSIONS

A full-size SUV was totally powered by a H_2 gas fueled internal combustion engine. H_2 gas was produced on-board as needed, by a safe H_2 generator using an ambient temperature chemical reaction and a (mostly) water based $NaBH_4$ solution. Using $NaBH_4$ solutions as a H_2 source reduces the inherent safety concerns associated with long term H_2 storage. Since $NaBH_4$ solution simply contacts Ru to produce H_2, it can be used for numerous application where H_2 gas is used e.g. PEM fuel cells or as a direct fuel in H2 powered internal combustion engines. $NaBH_4$ generators can be quickly refueled by simply filling the reservoir with fresh solution. Ru catalysts are reusable.

ACKNOWLEDGEMENTS

We greatly benefited from numerous scientific discussions and enlightening technical suggestions from Mike Strizki of the NJ Department of Transportation who motivated us to initiate this study. Heartfelt kudos to Martin M. Pollak Esq. and Jerome I. Feldman Esq. of GP Strategies Inc. for their unfailing enthusiasm, encouragement, and optimism throughout this effort. We are most grateful to Weinroth-Andersen Inc. for providing generous financial support for this applied research program. Our research was partially supported by the State of New Jersey Commission on Science and Technology.

2000-01-2824

Hydrocarbon Emissions from a SI Engine Using Different Hydrogen Containing Gaseous Fuels

T. K. Jensen, J. Schramm, C. Søgaard and J. Ahrenfeldt
The Technical University of Denmark

ABSTRACT

Experiments have been conducted on a gas fueled spark ignition engine using natural gas and two hydrogen containing fuels. The hydrogen containing fuels are Reformulated Natural Gas (RNG) and a mixture of 50% (Vol.) natural gas and 50% (Vol.) producer gas. The producer gas is a synthetic gas with the same composition as a gas produced by gasification of biomass.

The hydrocarbon emission, measured as the percentage of hydrocarbons in the fuel, which passes unburned through the engine, was for the mixture of natural gas and producer gas up to 50% lower than the UHC emissions using natural gas as fuel. The UHC emission from the experiments using reformulated natural gas was 15% lower at lean conditions. Furthermore, both hydrogen-containing fuels have a leaner lean burn limit than natural gas.

The combustion processes from the experiments have been analyzed using a three-zone heat release model, which is taking the effect of crevices into account. The analysis showed that the two hydrogen containing fuels cause an increase in the turbulent flame speed resulting in faster combustion compared to natural gas. Even though both hydrogen containing fuels have a lower adiabatic flame temperature the increased turbulent flame speed causes a higher maximum temperature of the burned gas compared to natural gas. It is this increased maximum burned gas temperature which causes a higher NO_x emission from the hydrogen containing fuels.

INTRODUCTION

Methane, which is the main component in natural gas, contains a smaller fraction of carbon than other hydrocarbon fuels and, because of that, there is a lower emission of CO_2 from the combustion of methane. This is one of the reasons why utilization of natural gas has increased over the past years.

The regular shape of the methane molecule means that it is relatively stable. Furthermore, the laminar flame speed of methane is lower than for most other hydrocarbon fuels. This is a part of the explanation why the emission of unburned hydrocarbons is higher for methane than most other hydrocarbon fuels when it is used as fuel for combustion engines. This is most pronounced at lean burn condition. However, lean burn combustion is widely used because it allows a high thermal efficiency of the engine and low levels of NO_x emission without a catalyst [1].

Researchers have successfully enriched natural gas by addition of hydrogen which is very reactive compared to methane. When hydrogen is used as a supplementary fuel for a hydrocarbon fuel the higher reactivity of hydrogen is important in the beginning of the combustion where the flame kernel is developed [2]. The improved flame kernel development implies that the cyclic variations will be reduced [3]. Especially at lean combustion the cyclic variations are smaller and this is due to the broad flammability limits of hydrogen [4], [5]. Furthermore the laminar flame speed of hydrogen is higher than for hydrocarbon fuels [6], [7]. The broader flammability limits and the lower duration of the combustion caused by higher flame speed mean that addition of hydrogen to natural gas makes it possible to run the engine leaner with low levels of emission. The emission of CO, NO_x and UHC is reduced by increasing the air-fuel ratio until partial burning becomes predominant [8]. In this work combustion of two different hydrogen containing fuels will be examined and compared with natural gas. The first fuel is a steam reformulated natural gas and the other is a mixture of 50% (vol.) producer gas and 50% (vol.) natural gas. The fuels are described in further details later. The mixture of producer gas and natural gas has been examined earlier by Jensen et al. [9]. That investigation showed that addition of producer gas to natural gas reduced the emission of unburned fuel significantly.

Fuel description

One of the fuels examined is Reformulated Natural Gas, in the following abbreviated RNG. The RNG is produced by reaction between natural gas and water vapor at approximately 400-420°C on a nickel catalyst. The steam reforming process is described by Søgaard et al. [11].

The other fuel examined is a mixture of 50% natural gas and 50% producer gas (Vol.). This corresponds to 13% producer gas on energy basis. The producer gas is a synthetic gas with the same composition as a gas produced by thermal gasification of biomass. At The Technical University of Denmark a gasification plant is producing gas with following composition: 30-35% H_2, 15-20% CO, 15% CO_2, 1-2% CH_4 and the rest being N_2. The gasification process is described by Henriksen et al. [13].

These two fuels are compared to Danish natural gas. The composition of the three examined fuels is given in Table 1.

Composition	RNG [%]	NG [%]	PG + NG [%]
CH_4	70.9	87.6	44.4
C_2H_6	0	6.6	3.3
C_3H_8	0	3.0	1.5
C_4H_{10}	0	1.3	0.7
H_2	22.6	0	17.0
CO	0	0	9.6
CO_2	6.5	1.3	8.1
N_2	0	0.4	15.6

Table 1. Composition of the examined fuels.

The different composition of the examined fuels means that the fuels have different combustion properties. The lower heating value (LHV) and the stoichiometric air-fuel ratio are given in Table 2. The combination of lower heating value and stoichiometric air-fuel ratio for the three fuels means that approximately the same energy flow is supplied to the engine for the examined fuels if the intake pressure is the same. The largest difference is 3% at λ=1 and the difference decreases as the relative air-fuel ratio increases as shown in Figure 1. The adiabatic flame temperature calculated as constant volume combustion at an initial state corresponding to the state at the time of ignition is given in Table 2. Here it is shown that natural gas has the highest adiabatic flame temperature. However, the difference is small.

	RNG	NG	PG + NG
LHV / [MJ/m^3]	27.5	39.3	22.9
A/F$_{stoic}$	14.3	16.3	8.58
T_{ad}	2734 K	2743 K	2709 K

Table 2. The lower heating value, the stoichiometric air-fuel ration and adiabatic flame temperature of the examined fuels.

Another important property for the fuel is the specific heat capacity of its combustion products. This thermodynamic property is influencing both the level of the emissions and the brake efficiency. Despite the difference in composition of the three fuels, the composition of the products are rather similar. This means that the specific heat of the products also is similar at a given temperature and relative air-fuel ratio. Figure 2 shows the specific heat of the combustion products at 1500 K as a function the relative air-fuel ratio. It is shown that the specific heat of the products decreases as the relative air-fuel ratio increases. In reality this effect is even further pronounced since leaning the fuel-air mixture causes a decrease in the temperature leading to an even lower specific heat capacity at lean burn conditions.

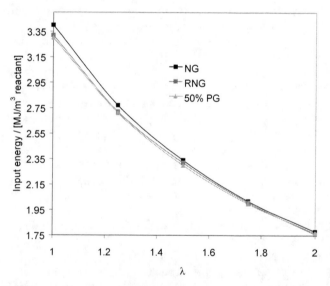

Figure 1. Amount of energy in the mixture of fuel and air for the three examined fuels.

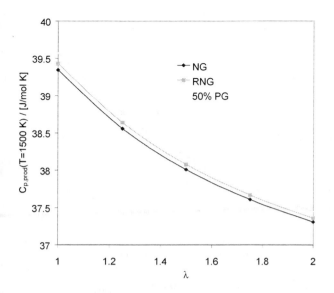

Figure 2. The mole specific heat capacity for the combustion products at 1500 K versus the excess of air for the three fuels examined in this work.

Experimental investigation

Engine experiments have been carried out using the three described fuels. The engine, which has been employed for the experiments described in this paper, is a naturally aspirated 1-cylinder, 4-stroke SI engine.

The main technical data of the engine is:

Bore	85 mm.
Stroke	85 mm.
Compression Ratio	12 : 1
Number of valves	2

The engine has a pancake shaped combustion chamber.

The volume of the crevices has been measured on the engine while disassembled. The volume of the crevices was estimated to be 1200 mm^3, corresponding to 0.25% of the displacement volume and 3% of the compression volume.

The experiments have been carried out with the relative air-fuel ratios varied from $\lambda \approx 1$ to the lean burn limit for the fuels examined. In all experiments ignition timing was adjusted to achieve the maximum brake torque and the engine speed was kept constant at 1500 rev/min. Furthermore, during all experiments, the engine was operated with wide open throttle. Of course, throttling affects the results of the experiments, but this effect is outside the scope of this paper.

In all experiments carried out, the concentration of HC, NO_x, CO_2, CO and O_2 was measured in the exhaust gas. Furthermore, the flow of fuel and air was measured and the pressure in the combustion chamber was measured for 50 consecutive combustion cycles.

Results

The brake efficiency measured for the examined fuels is depicted on Figure 3. Here it is seen that in general the hydrogen containing fuels result in a higher efficiency than natural gas. Furthermore it is seen that for $\lambda<1.5$ the highest efficiency is achieved by using RNG while at leaner combustion conditions the highest efficiency is achieved by using the mixture of producer as and natural gas.

In natural gas there is 1.14 mole C_1 as HC per mole fuel, in RNG there is 0.709 mole C_1 and in the mixture of natural gas and producer gas there is 0.639 mole C_1 per mole fuel. This difference in the concentration of HC in the fuels means that it isn't possible directly from the measured concentration of HC in the exhaust gas to assess to which extent the fuel has been oxidized either in the propagating flame or later by the post oxidation processes. Therefore it is chosen to compare the percentage of hydrocarbons in the fuel which passes unburned through the engine rather than the measured concentration of hydrocarbons in the exhaust gas. This implies the assumption, that what is measured in the exhaust gas is unburned fuel. However, this not correct since, for instance, C_2H_2 and C_2H_4 is produced during the combustion but these species are only produced in small amounts. Applying this assumption, the percentage of hydrocarbons in the fuel which passes unburned through the engine is determined and is shown in Figure 4. Here it is seen that the amount of hydrocarbons in the fuel, which remains unburned, is 15-50% lower for the mixture of natural gas and producer gas than for natural gas within the lean burn limit of natural gas. Furthermore, it is shown that a larger share of the fuel hydrocarbons remains unburned using RNG compared to natural gas at slightly lean combustion conditions and that the emission of unburned fuel HC is approximately 15% smaller for RNG compared to natural gas near the lean burn limit of natural gas.

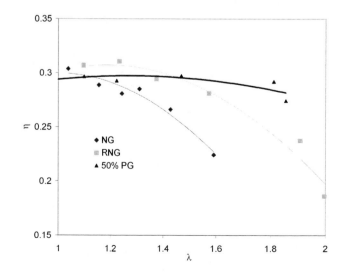

Figure 3. The brake efficiency depicted versus the excess of air for the examined fuels.

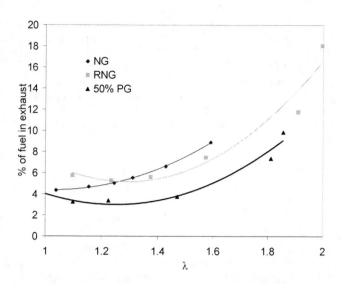

Figure 4. The percentage of hydrocarbons in the fuel that passes unburned through the engine.

In Figure 5 the specific emission of oxides of nitrogen is depicted versus the excess of air and in Figure 6 versus the specific UHC emission. Figure 5 shows that the specific NO_x emission is highest for the mixed fuel and lowest for natural gas. In Figure 6 it is shown that there is a trade off between the two types of emissions. This means that for none of the examined fuels it is possible to choose an operation condition where the emissions of both NO_x and UHC are low. The lowest NO_x emission is measured using RNG but at conditions where the UHC is unacceptably high. The lowest UHC is achieved by using a mixture of producer gas and natural gas at conditions causing high levels of NO_x. However, apparently the lowest level of the emission from these three fuels can be achieved by the mixture of natural gas and producer gas.

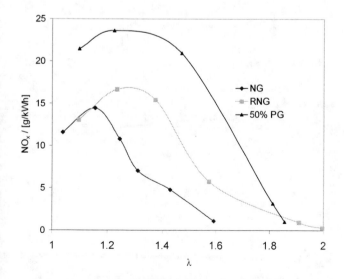

Figure 5. The specific NO_x emission for the examined fuels versus λ.

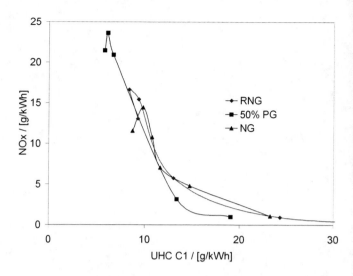

Figure 6. The specific NO_x emission versus the specific UHC emission for the examined fuels.

The specific emission of CO is depicted versus the excess of air in Figure 7 and in Figure 8 versus the specific emission of UHC. These two figures show that the emission of CO is high when the mixture of producer gas and natural gas is used. The relative high content of CO in the producer gas is probably the explanation for the higher levels of CO emission from the mixture of natural gas and producer gas. This means that it is expected that the measured CO concentration mainly is unburned fuel rather than a result of the combustion process as it was in the case when natural gas or RNG was used as fuel. Very low CO emissions have been measured from RNG. It is seen that the CO emission is reduced by approximately 80% by reformulating the natural gas.

Experiments have been carried out earlier using a mixture of natural gas and producer gas as fuel [9]. In [9] it was found that the emission of oxides of nitrogen was lower for the mixture of producer gas and natural gas than it for natural gas. Furthermore, the efficiency was higher for natural gas than for the mixed fuel in the experiments carried out earlier. For the experiments described in this paper, the opposite is the case. For the experiments described here, only at slightly lean combustion conditions the efficiency was higher for natural gas than for the mixed fuel. At leaner combustion conditions the highest efficiency is achieved by using the mixed fuel.

All experiments were carried out on the same engine but the shape of the combustion chamber is significantly different for the two different sets of experiments. In the first set of experiments the compression ratio was 10:1 and a bowl-type piston crown was used. During the experiments described in this paper the compression ratio was increased to 12:1 and a flat top piston crown was used.

The authors believe that is it the different shape of the combustion chamber which is reason why the results shown in this paper differ from what was achieved earlier. The bowl-shaped piston crown means that a higher turbulent intensity is present during the combustion compared to using a flat piston crown. Apparently, the combustion of the natural gas is improved relative to combustion of the mixed fuel by using the bowl-shaped piston crown instead of the flat-top piston crown. If this is the case, it explains the described differences in the two sets of experiments.

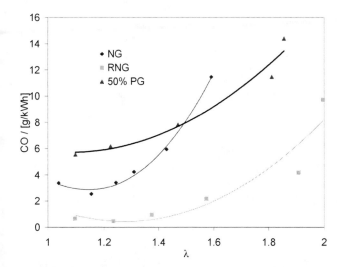

Figure 7. The specific CO emission for the examined fuels versus λ.

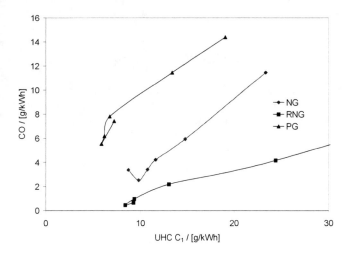

Figure 8. The specific emission of CO for the examined fuels versus the specific emission of UHC emission.

In-cylinder pressure analysis

The recorded pressure-time histories from the experiments have been analyzed by using a three-zone thermodynamic heat release model which is taking the effect of crevices into account. In the model the combustion chamber is divided into three zones consisting of respectively gas trapped in the crevices, unburned gas and burned gas.

The model is based on the following assumptions:

- All gases can be treated as ideal gases.

- The pressure in the combustion chamber is uniform at any instant.

- Both the crevices and the unburned gas zone, consist of a perfectly mixed mixture of fuel, air and residual gases.

- The burned gas zone consists of perfectly mixed and completely oxidized gases.

- The temperature in any of the three zones is uniform at any instant.

- Blowby does not occur.

- The crevices in the combustion chamber can be modeled as one crevice and the temperature of the gas discharged from the crevices is the same as the cylinder wall temperature.

- Heat losses from the burned and unburned gas zones can be calculated by using the model described by Woschni [12].

- Conduction of heat from one zone to another is negligible.

At the end of the combustion where the heat release rate is low and the amount of unburned gas is low, the validity of the two last assumptions is probably not very good. At this stage of the combustion, the heat loss from the combustion chamber is higher than the heat liberated by the combustion processes. Therefore, the limited accuracy of the empirical heat loss model will affect the results of the calculations to a higher extent then at earlier stages of combustion. At the end of the combustion process, when the volume of the unburned gas zone is small, the assumption of no heat conduction from one zone to another might also be rather rough. The model is described in details by Jensen and Schramm [13].

In Figure 9 the maximum pressure for the experiments carried out is depicted for the three fuels as a function of the relative air-fuel ratio. Here it is seen that the maximum pressure is highest for the mixture of natural gas and producer gas and lowest for the experiments using natural gas as fuel. Furthermore it is seen that the curves describing the maximum pressure for the hydrogen containing fuels are converging as the relative air-fuel mixture is increased. However, the measured maximum pressures depicted are rather scattered.

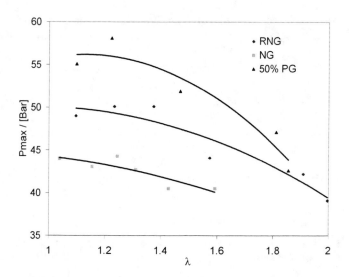

Figure 9. The maximum pressure during combustion of natural gas, RNG and a mixture of natural gas and producer gas.

In Figure 10 the maximum calculated temperature of the burned gas zone is depicted. Here it is shown that at all examined air-fuel ratios the maximum temperature of the burned zone is approximately 100 K higher for the mixture of natural gas and producer gas than for the measurements carried out using natural gas as fuel. Furthermore it is seen that near stoichiometric conditions the maximum temperature of the burned gas when using RNG as fuel is similar to those calculated when the fuel is natural gas. The curves showing the maximum burned gas temperature for the two hydrogen containing fuels are converging as the air-fuel mixture is getting leaner. It is this higher maximum temperature which is expected to cause the higher NO_x emission from the two hydrogen-containing fuels.

The turbulent flame propagation in the combustion chamber during the combustion has been analyzed. In order to do this a geometric routine has been implemented to the model. The geometric model is based on the assumption that the flame is propagating as a sphere from the spark plug and it calculates the area of the propagating flame front, which is in contact with the unburned gas zone and the radius of flame.

The turbulent flame speed, S_T, is calculated as

$$S_T = \mathbf{6} \cdot N \frac{M}{\rho_u \cdot A_f} \frac{dx_b}{d\theta}$$

where N is the engine speed (rev/min.), M is the total mass of gas in the combustion chamber, ρ_u is the density of the gas in the zone of unburned gas, A_f is the surface area of the propagating flame front, x_b is the mass burned fraction and θ is the piston position measured as crank angle degree. The turbulens will affect the shape of the propagating flame and the centre of flame might also be carried away from the spark plug

during the flame kernel development by convection. The effect of displacement of the flame centre on the flame speed has been analyzed by using five different values of the distance from spark plug to the cylinder wall as input to model. The five values were the measured distance, the measured distance ± 2.5 mm. and ± 5.0 mm. The effect of the displacement of the centre of the flame on the calculated flame speed is shown in Table 3.

displacement [mm.]	flame speed [m/s]
-5	6.17
-2.5	5.98
0	5.77
2.5	5.52
5	5.21

Table 3. The influence of the displacement of the flame centre on the calculated flame speed.

It is seen that displacement of the centre of the flame has some influence on the calculated flame speed. The flame speeds presented in this paper are calculated from pressure-time histories which all are average histories of 50 consecutive cycles. This means that effect cyclic variation is more or less eliminated. Since the ignition timing was almost the same at the same relative air-fuel ratio for the three examined fuels, it is expected that the convection is similar near the spark plug during the flame kernel development period. This does not mean that displacement of the centre of the flame does not occur, it means that, on average and at same relative air-fuel ratio for the three fuels, the flame kernel is expected to be moved in the same direction and therefore affect calculated flame speed in the same way independent of the fuel.

In the applied engine nothing special has been done in order to create swirl or turbulens in the intake mixture and since the spherical shape of the flame only is considerably distorted at high swirl [14] it is expected to be reasonable assume the flame front to be spherical. However, maybe the calculated flame speeds are not the exactly the same as the actual flame speeds, the authors are convinced that this simple method for calculating the flame speed is suitable for comparing flame speeds for different fuels. Especially when all the experiments has been carried on the same engine with the throttle in the same position as it is case for the experiments presented here. This turbulent flame speed has earlier been calculated in a similar way by Lancaster et al. [15].

The calculated turbulent flame speed at the time when 50% of the charged fuel is burned is depicted in Figure 11. It has been chosen to depict the flame speed at this instant because it is rather constant. On the figure it can be seen that for all three fuels an increment of the relative air-fuel ratio decreases the turbulent flame speed. At all values of λ, the turbulent flame speed is higher for the mixture of producer gas and natural gas than for natural gas. Near stoichiometric combustion condition the turbulent flame speed of RNG is comparable with the flame speed of natural gas. Near the lean burn limit for the two hydrogen-containing fuels, the flame speed is highest for RNG.

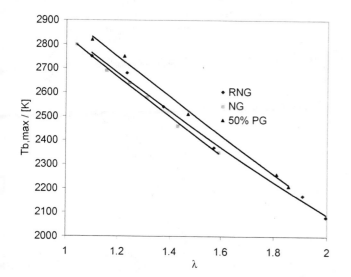

Figure 10. The maximum temperature of the burned zone during the combustion of natural gas and a mixture of natural gas and producer gas.

The combustion process can be divided into different stages. The first stage is a flame development period the second is a rapid burning period [15]. The flame development period is defined as the time it takes before 10% of the fuel is burned. On Figure 12 the duration of the flame development period, given as crank angle degrees, is shown for the three examined fuels. Here it is shown that the flame development period is approximately the same for all three fuels near stoichiometric combustion conditions. Furthermore, the flame development period for the two hydrogen containing fuels is almost the same for all values of λ and it is shown that it is shorter for the two hydrogen containing fuels that for natural gas at lean burn conditions

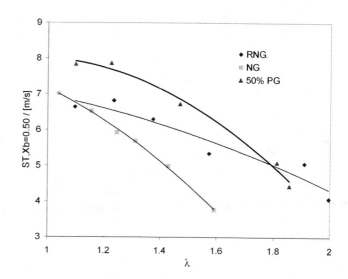

Figure 11. The turbulent flame speed for the three examined fuels versus the piston position at the instant where 50% of the fuel is burned.

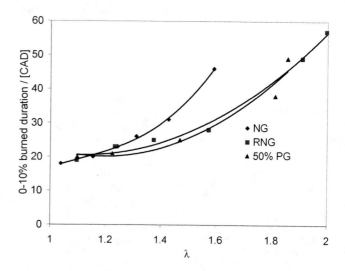

Figure 12. The time (measured as crank angle degrees) it takes to burn the first 10% of the fuel versus the excess of air for the natural gas and a mixture of natural gas and producer gas.

The cycle by cycle variation is determined as the coefficient of variation, COV, of the indicated mean effective pressure, IMEP. The COV is defined in [15] as

$$COV_{IMEP} = \frac{\sigma(IMEP)}{\overline{IMEP}}$$

where $\sigma(IMEP)$ is the standard deviation and $\overline{IMEP}$ is average IMEP of the 50 consecutive combustion cycles. The COV for the three fuels is given in Figure 13. In this figure it is shown that the cyclic variations is a little higher for mixture of producer gas and natural gas than for the other two fuels until partial burning is beginning to occur using natural gas as fuel. Only near the lean burn limit the COV is the same for the two hydrogen containing fuels. This means that the lower emission of unburned fuel from the mixture of natural gas and producer gas compared to RNG and natural gas can not be explained by smaller cyclic variations.

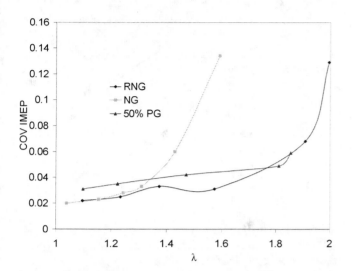

Figure 13. The coefficient of variation in the indicated mean pressure.

Comparison of the combustion at $\lambda=1.23$

In the following the combustion process is analyzed at the same relative air-fuel ratio, $\lambda=1.23$, for the three examined fuels in order to be able to compare the combustion process. The time of ignition for the different fuels was:

Natural gas	30° BTDC
RNG	35° BTDC
Natural gas + producer gas	35° BTDC

The pressure-time history for the three experiments is shown in Figure 14. It is seen that the pressure rise is fastest for the mixture of producer gas and natural gas and slowest for natural gas.

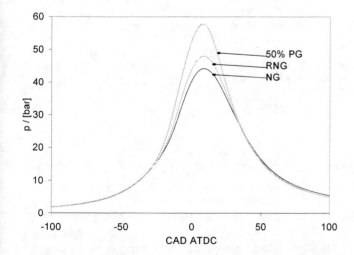

Figure 14. Pressure-time history for the three examined fuels. The shown pressures are average values of 50 consecutive combustion cycles. $\lambda=1.23$.

Figure 15 is showing the calculated mass burned fraction for three fuels. It is seen that the combustion starts earlier for the mixture of producer gas and natural gas than for RNG despite the fact that the ignition timing is the same. As one could expect from the pressure curves, the steepest mass burned fraction curve is achieved using the mixture of producer gas and natural gas as fuel. It is also seen that at the end of the combustion the two other fuels are burning faster.

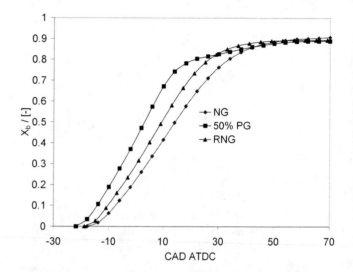

Figure 15. Mass burned fraction. $\lambda=1.23$.

In order to examine this unexpected shape of the mass burned fraction curve for the mixed fuel, the individual combustion cycles were analyzed. In Figure 16 and Figure 17 the calculated volume of the burned and the unburned gas zones, (V_b) and (V_u), the total volume of the combustion chamber (V), the mass fraction of fuel in the crevices (x_{cr}) and the mass burned fraction (x_b) are shown for one individual combustion cycle for respectively natural gas and for a mixture of producer gas and natural gas. The data shown in Figure 16 and Figure 17 is not for representative cycles but for cycles where the combustion is faster than the average cycles. It was necessary to analyze the fast combustion cycles in order to find an explanation for the unexpected shape of the mass burned fraction curve for the mixed fuel. In Figure 16 it is shown that around 10 crank angle degrees after top dead centre the mass burned fraction curve for the mixed fuel is almost flat and after a little while it start to rise again. It is also seen that the volume of the unburned gas zone is steadily decreasing and approaching a level close to zero. These two events happen at the same instant corresponding to, the only place where unburned fuel is present is in the crevices. After a 4-5 crank angle degrees, when the pressure in the chamber is decreasing due to the expansion, the gas in the crevices is flowing back into the chamber. When this happens the mass burned fraction is increasing again. This means, that the second rise of the x_b-curve is caused by oxidation of fuel discharged from the crevices.

After the volume of the unburned gas has reached a minimum, it apparently starts to raise again. This can be explained by unburned gas is flowing faster of out of the crevices than it is oxidized. In reality, the flow from the crevices and back into the cylinder is, to some extent, restricted but there is no flow restrictions implemented in the crevice model. This means that the gas is flowing slower out of the crevices than predicted by the model. Therefore, it is not possible tell whether the volume of unburned gas is increasing again after a minimum or not.

From Figure 17 is seen that for natural gas used as fuel the volume of unburned gas is more slowly decreasing and does reach a minimum at the end of the combustion. Therefore a two-step combustion doesn't occur for natural gas. It doesn't occur for RNG either.

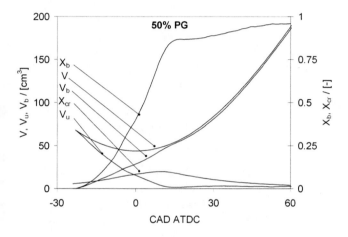

Figure 16. The total volume, the volume of the burned and unburned gas, the mass fraction of gas burned and the fraction of gas in the crevices for one individual.

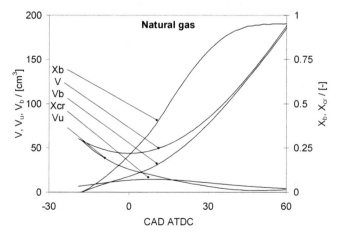

Figure 17. The total volume, the volume of the burned and unburned gas, the mass fraction of gas burned and the fraction of gas in the crevices for one individual.

In Figure 15 it is also shown that at a certain piston position at the later part of the combustion a larger of

fraction of the fuel is burned when RNG is used as fuel compared to using the mixed fuel. This might appear to be a contradiction to what has just been concluded, namely that it is lack of fuel in the combustion chamber which is causing the unexpected shape of the x_b-curve of the mixed fuel. But the reason way this is possible is, that using RNG as fuel causes a lower pressure raise relative to the mixed fuel and therefore a smaller share of the fuel is trapped in the crevices.

The calculated temperature of the burned gas and unburned gas is shown in Figure 18 for the three fuels. It is here seen that the highest temperature of the burned gas is achieved with the mixed fuel but in the later part of the expansion stroke the temperature is higher for the other two fuels than for the mixed fuel. The calculated temperature of the unburned zone decreases drastically at around 15 degrees after top dead centre. This occurs when the gas trapped in the crevices starts to flow back into the combustion chamber. In reality this temperature drop will not be as pronounced as indicated by the applied model. In the model it is assumed that heat transport between the different zones is negligible but at the end of the combustion process when the mixture of producer gas and natural gas is used as fuel the amount of unburned is so small that it is not reasonably to neglect heat transfer. However, this is not expected to have an influence on the other results.

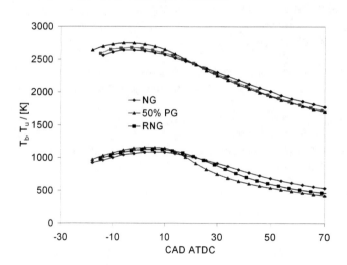

Figure 18. The calculated temperature of the burned and the unburned gas. $\lambda=1.23$.

The turbulent flame speed during the combustion versus the mass burned fraction is shown in Figure 19. During the main part of the combustion the turbulent flame speed is almost constant for all three fuels but the flame speed decreases earlier for the slower burning natural gas than for the two hydrogen containing fuels. At the end of the combustion of the mixture of producer gas and natural gas the flame speed apparently is getting constant for a short period. This is not the case. It is more likely that it is the effect of fuel and air flowing out of the crevices and subsequently oxidized which is causing this odd behavior of the turbulent flame speed.

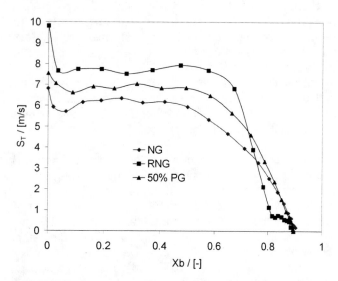

Figure 19. The turbulent flame speed for the three examined fuels versus the mass burned fraction. λ=1.23.

CONCLUSION

Experiments have been conducted on a naturally aspirated gas fueled spark ignition engine using natural gas and two hydrogen containing fuels, namely steam Reformulated Natural Gas, RNG, and a mixture of 50% (Vol.) natural gas and 50% (Vol.) producer gas. The experiments have shown that:

- the lean burn limit is extended when using the two hydrogen containing fuels compared to natural gas.

- the fraction fuel which passes unburned through the engine is reduced by 50% for the mixture of producer gas and natural gas and by 15% at lean burn condition using RNG.

- the NO_x emission is higher for the two hydrogen containing fuels than for natural gas and it is not possible to choose a fuel-air ratio where both UHC and NO_x emission are low.

- the CO emission is reduced by 80% by reformulating the natural gas. For the mixture of producer gas and natural gas the CO emission is higher than for natural gas at all examined air-fuel ratios. This is probably due to the content of CO in the mixed fuel.

The pressure-time history was recorded and have been used for analyzing the combustion process. This analysis showed that

- if partial burn effects are not being predominant the cyclic variation is highest for the mixed fuel at all examined air-fuel ratios. Therefore it is not the effect

on cyclic variation which causes the low emission of UHC when the mixed fuel is applied.

- the flame development period is reduced at lean burn condition using the two hydrogen containing fuels compared to using natural gas.

- the turbulent flame speed is highest for the mixed fuel and lowest for the natural gas.

- at all examined air-fuel ratios the maximum temperature of the burned gas zone was approximately 100K higher for the mixed fuel than for natural gas. This is the reason for high amount of NO_x produced during combustion of the mixed fuel.

REFERENCES

[1]   Germane, Geoff J. et al., "Lean combustion in spark-ignited internal combustion engines -a review", SAE paper 831694. 1983.

[2]   De Risi, A et al. "A study of H_2, CH_4, C_2H_6 Mixing and Combustion in a Direct-Injection Stratified-Charge Engine" SAE paper 971710. 1997

[3]   Jamel, Y et al. "On-Board Generation of Hydrogen-Rich Gaseous Fuel- A Review", Int. J. Hydrogen Energy, Vol. 19 pp.557-572. 1994.

[4]   Meyers, D. P., "The Hybrid Rich-Burn / Lean-Burn Engine", Journal of Engineering for Gas Turbines and Power, Vol. 119, January 1997.

[5]   Smith, Jack A., et al. "The Hybrid Rich-Burn / Lean-Burn Engine, Part 2" ASME paper ICE-Vol. 27-4. Book No. G1011D. 1996.

[6]   Apostolescu, Nicolae et al. "A Study of Hydrogen-Enriched Gasoline in a Spark Ignition Engine", SAE paper 960603. 1996.

[7]   Swain, M. R. et al." The effect of Hydrogen Addition on Natural Gas Engine Operation", SAE paper 932775. 1993.

[8]   Bell, Stuart R. et al. "Extension of the Lean Operating Limit for Natural Gas Fueling of a Spark Ignited Engine Using Hydrogen Blending", Combust. Sci. and Tech. Vol. 123 pp.23-48. 1997.

[9]   Jensen, T. K. et al. "Unburned Hydrocarbon Emissions from SI Engines

Using Gaseous Fuels", SAE paper 1999-01-0571. 1999.

[10] Søgaard, C "Reduction of UHC Emissions From Natural Gas Fired SI Engines - Production and Application of Reformulated Natural Gas", SAE paper 00FL-621. 2000.

[11] Henriksen, U. et al. "Gasification of Straw in a Two-stage 50 kW Gasifier", 8th European Conference on Biomass for Energy, Environment, Agriculture and Industry". Vienna. 1994.

[12] Woschni, G. "A Universally Applicable Equation for the Instantaneous Heat Transfer Coefficient in the Internal Combustion Engine", SAE paper 670931. 1967.

[13] Jensen, T. K. et al. "A Three-Zone Heat Release Model for Analysis Combustion in SI Engine. –Effects of Crevices and Cyclic Variations on UHC Emissions" SAE paper 00FL-623. 2000

[14] Witze, P.O. "The Effect of Spark Location in a Variable-Swirl Engine" SAE paper 820044, SAE Trans., vol. 91. 1982.

[15] Lancaster, D. R. et al. "Measurement and Analysis of Engine Pressure Data" SAE paper 750026, SAE Trans., vol. 84. 1975.

[16] Heywood, John .B. "Internal Combustion Fundamentals". McGraw-Hill, Inc. 1988

2001-01-0252

Combustion Characteristics of H2-CO-CO2 Mixture in an IC Engine

Toshio Shudo and Yasuo Nakajima
Musashi Institute of Technology

Koichiro Tsuga
JATCO Transtechnology Co.

ABSTRACT

Reformed fuel from hydrocarbons or alcohol mainly consists of hydrogen, carbon monoxide and carbon dioxide. The composition of the reformed fuel can be varied to some extent with a combination of a thermal decomposition reaction and a water gas shift reaction. Methanol is known to decompose at a relatively low temperature. An application of the methanol reforming system to an internal combustion engine enables an exhaust heat recovery to increase the heating value of the reformed fuel. This research analyzed characteristics of combustion, exhaust emissions and cooling loss in an internal combustion engine fueled with several composition of model gases for methanol reformed fuels which consist of hydrogen, carbon monoxide and carbon dioxide. Experiments were made with both a bottom view type optical access single cylinder research engine and a constant volume combustion chamber.

The thermal decomposition reaction produces 1mol of carbon monoxide and 2mol of hydrogen from 1mol of methanol, and the water gas shift reaction produces the same amount of hydrogen and carbon dioxide from carbon monoxide and water. Therefore, an increase in hydrogen means an increase in carbon dioxide, an inert gas with large heat capacity, and a decrease in the heating value of the fuel. This research cleared a balance of the combustion promotion by the increased hydrogen and the demerit by the increased heat capacity of mixture due to the increased carbon dioxide. The highest overall thermal efficiency including the exhaust heat recovery is obtained in the combustion of the methanol-reformed fuel with just the thermal decomposition reaction, which consists of 33% of hydrogen and 67% of carbon monoxide.

INTRODUCTION

Hydrocarbons and alcohol can be reformed into a flammable gas consists of mainly hydrogen and carbon monoxide. Internal combustion engines fueled with the reformed gas are supposed to have high overall thermal efficiency by an exhaust heat recovery. Especially the methanol decomposition reaction proceeds at relatively low temperature, and there has been some experimental reports on internal combustion engines fueled with methanol reformed fuel [1-2].

Methanol can be decomposed into hydrogen and carbon monoxide. The carbon monoxide reacts with steam to produce hydrogen and carbon dioxide in water gas shift reaction. Therefore, it is possible to vary the composition of the methanol-reformed gas with a combination of the thermal decomposition reaction and the water gas shift reaction. The ignitability of carbon monoxide is highly promoted by a small amount of hydrogen addition [3]. The heating value and the carbon dioxide content of the reformed gas changes with the hydrogen content. Because the change in the gas composition in the fuel influences combustion characteristics and thermal efficiency of internal combustion engines, a fuel composition with the maximum heating value does not always bring the highest thermal efficiency. Therefore, it is important to analyze the optimum composition of the methanol-reformed fuel for the higher thermal efficiency in internal combustion engines. The carbon dioxide in the fuel also influences NOx exhaust emission. The fuel composition should be optimized for both the thermal efficiency and the exhaust emission.

Hydrogen has a higher burning velocity and a shorter

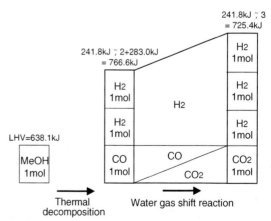

Figure 1. Reformation of methanol

Table 1 Composition of fuel tested in this research

(a) H$_2$-CO mixture

	Case 1	Case 2	Case 3	Case 4
H$_2$	0.67	0.80	0.90	1.00
CO	0.33	0.20	0.10	0.00
H$_2$/(H$_2$+CO)	0.67	0.80	0.90	1.00
CO$_2$	0.00	0.00	0.00	0.00
LHV (MJ/Nm^3)	11.40	11.17	10.98	10.80

(b) H$_2$-CO-CO$_2$ mixture (methanol reformed fuel)

	Case 1'	Case 2'	Case 3'	Case 4'
H$_2$	0.67	0.70	0.73	0.75
CO	0.33	0.18	0.08	0.00
H$_2$/(H$_2$+CO)	0.67	0.80	0.90	1.00
CO$_2$	0.00	0.12	0.19	0.25
LHV (MJ/Nm^3)	11.40	9.83	8.90	8.10

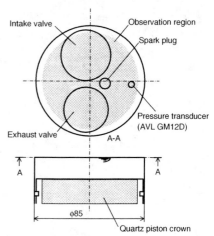

Figure 2. Combustion chamber and observation region

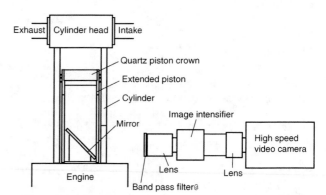

Figure 3. Bottom view engine and optical system in this research

quenching distance [4] and sometimes causes a higher cooling loss to the combustion chamber wall as compared with hydrocarbon fuels [5-6]. Because a hydrocarbon combustion with a hydrogen premixing also has a high cooling loss [7-8], it is important to analyze the cooling loss in the combustion of the reformed gas including hydrogen. This research experimentally analyzed the combustion and the exhaust emission in an internal combustion engine fueled with several compositions of H$_2$-CO-CO$_2$ mixture. The basic characteristics of combustion and the cooling loss of the fuel were analyzed in a constant volume combustion chamber.

METHANOL REFORMING AND TESTED FUEL

REFORMING REACTION

1mol of methanol can be reformed into 2mol of hydrogen and 1mol of carbon monoxide with the thermal decomposition reaction (1) on an adequate catalyst at the temperature of around 600K.

$$CH_3OH \rightarrow CO + 2H_2 \quad \text{(endothermic)} \qquad (1)$$

The endothermic reaction turns 1mol of methanol with LHV of 638.1kJ into the reformed fuel with total LHV of 766.6kJ as shown in Figure 1. The heating value of the reformed fuel is increased by 20.1% based on the liquid methanol. The carbon monoxide in the reformed gas can be turned into 1mol of hydrogen and 1mol of carbon dioxide with the following water gas shift reaction.

$$CO + H_2O \rightarrow CO_2 + H_2 \quad \text{(exothermic)} \qquad (2)$$

The following endothermic reaction is an overall reaction to produce the largest amount of hydrogen from methanol.

$$CH_3OH + H_2O \rightarrow CO_2 + 3H_2 \quad \text{(endothermic)} \qquad (3)$$

However, because of the exothermic reaction (2) the increase in LHV in the reaction (3) is just 13.7% based on liquid methanol. The reaction (2) can vary the composition of H$_2$-CO-CO$_2$ in the reformed fuel. For instance, PEFC can work on only the reformed fuel with reaction (3) to avoid the poisoning of catalyst metal on the electrode by the carbon monoxide in fuel. In addition, the working temperature of

the PEFC is not enough for the methanol-decomposition reaction, and the methanol must be partially oxidized in the reforming process to provide the heat for the reaction. Therefore, the reformed fuel in a methanol-reform type PEFC has smaller heating value compared to the methanol. Contrary, internal combustion engines can use the carbon monoxide as a fuel, and the increase in heating value in the fuel reforming is ideally expected up to 20.1% with the exhaust heat recovery.

On the other hand, combustion characteristics of hydrogen are largely different from those of carbon monoxide. A change in the reformed fuel composition varies not only the heating value but also the combustion characteristics of the reformed fuel, and both the heating value and the combustion characteristics influence on the thermal efficiency of internal combustion engines. Because of the change in the combustion characteristics, a fuel composition with the maximum heating value does not always bring the highest thermal efficiency. Therefore, it is important to analyze the optimum fuel composition for thermal efficiency.

TESTED FUEL

The model fuels for the reformed gases in this research are shown in Table 1. Proportions of hydrogen and carbon monoxide in the fuels are 67%:33%, 80:20, 90:10, and 100:0. The fuel with 67% of hydrogen and 33% of carbon monoxide corresponds to the thermally decomposed methanol. The four cases of fuels in Table 1(a) exclude carbon dioxide, and the other four cases of fuels in Table

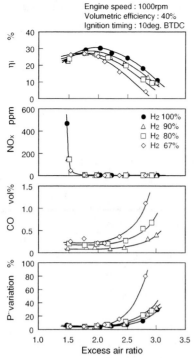

Figure 4. Influence of excess air ratio on combustion and exhaust emissions

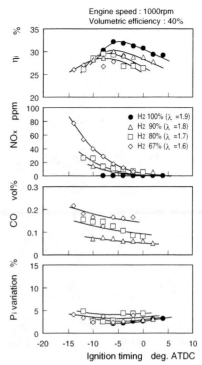

Figure 6. Influence of ignition timing on combustion and exhaust emissions

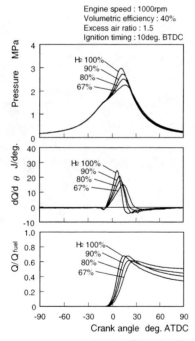

Figure 5. Pressure, apparent rate of heat release rate, and normalized cumulative apparent heat release

1(b) include carbon dioxide produced in the reaction (2).

EXPERIMENTAL APARATUS

The engine tested in this research was a four-stroke cycle single cylinder spark-ignition engine with a bore of 85mm, a stroke of 88 mm and a compression ratio of 13 as shown in Figure 2. Fuel gas flow was measured with a mass flow meter (Oval MSFLO-OVAL), and continuously supplied into the intake manifold. In-cylinder pressure was measured

with a piezo-electric type pressure transducer (AVL GM12D) installed in the cylinder head. For each experimental condition, 100cycles of pressure data were averaged and used to analyze indicated thermal efficiency, apparent rate of heat release, combustion variation and others. The combustion variation was evaluated with the ratio of standard deviation of IMEP to cycle-averaged IMEP. NOx concentration in the exhaust gas was measured with a chemi-luminescence detector, and carbon monoxide concentration was with a non-dispersed infrared detector. For some experimental cases, combustion flame was observed through a transparent silica piston crown from bottom side. The flame images were intensified with an image intensifier (Hamamatsu photonics C4273mod) and recorded with a memory type high-speed video camera (Photoron FASTCAM ultima) as shown in Figure 3.

RESULTS AND DISCUSSIONS

COMBUSTION OF H_2-CO MIXTURE IN AN ENGINE

Influence of excess air ratio

Figure 4 shows the influence of excess air ratio on combustion and exhaust emissions. Tested fuels are mixtures of hydrogen and carbon monoxide shown in Table 1(a). Hydrogen contents in the mixtures are 67, 80, 90, and 100%. The engine speed was set at 1000rpm, the ignition timing was at 10degree BTDC, and the volumetric efficiency was 40% including fuel gas. A decrease in hydrogen content makes indicated thermal efficiency lower in lean mixture conditions because of the lowered burning velocity. Therefore, fuel with low hydrogen content has its best thermal efficiency at the smaller excess air ratio. CO exhaust emission increases with the decrease in hydrogen content of the fuels, and the increase is significant in leaner mixture conditions. Cycle by cycle combustion variation

Engine speed : 1000rpm
Volumetric efficiency : 40 %
Excess air ratio : 1.5
Ignition timing :10deg BTDC

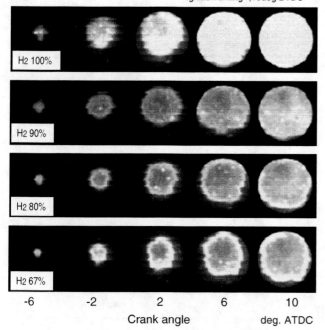

Engine speed : 1000rpm
Volumetric efficiency : 40 %
Excess air ratio: 1.5
Ignition timing : 10deg BTDC

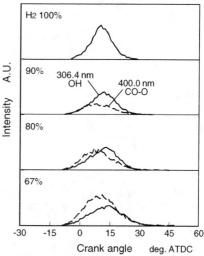

Figure 9. OH radical emission and CO-O recombination reaction emission in combustion flame

Figure 7. Flame images for different fuel compositions

Engine speed : 1000rpm
Volumetric efficiency : 40%
Excess air ratio : 1.5
Ignition timing : 10deg BTDC

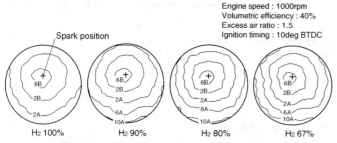

Figure 8. Time-series flame front trace

H2 67%, CO33 %
Engine speed : 1000rpm
Volumetric efficiency : 40 %
Excess air ratio : 1.5
Ignition timing : 10deg.BTDC

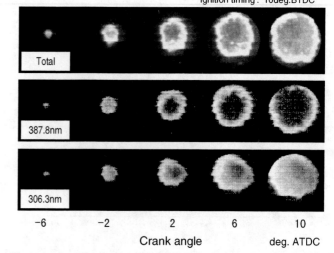

Figure 10. Distribution of OH radical emission and CO-O recombination reaction emission

tends to increase with excess air ratio, and the variation is significant in combustion of a low hydrogen fuel at a lean mixture condition. The thermal efficiency has a peak at excess air ratio of around 1.7 to 2.0 in the four cases of fuels, and the combustion variation is quite small with those mixture conditions.

Figure 5 shows the in-cylinder pressure, the apparent rate of heat release, and the cumulative heat release at an excess air ratio of 1.5. The ignition timing was set at 10degree BTDC. The cumulative heat release Q/Q_{fuel} is a cumulative heat release Q normalized with a heating value of the fuel supplied per cycle Q_{fuel}. The cumulative heat release Q calculated with the in-cylinder pressure is influenced by the cooling loss and described with the cumulative real heat release Q_B and the cumulative cooling loss Q_C as follows.

$$Q = Q_B - Q_C \qquad (4)$$

Because the cumulative real heat release Q_B is a product of the supplied fuel heat Q_{fuel} and the combustion efficiency η_u, the normalized cumulative heat release corresponds to a function of combustion efficiency η_u and the cooling loss ratio ϕ_w.

$$Q/Q_{fuel} = \eta_u \cdot (1 - \phi_w) , \quad \phi_w = Q_C/Q_B \qquad (5)$$

The apparent rate of heat release $dQ/d\theta$ and the normalized cumulative heat release Q/Q_{fuel} are influenced by both the cooling loss and the combustion efficiency during the combustion period. However, a negative rate of apparent heat release and the normalized cumulative heat release Q/Q_{fuel} after the end of combustion is influenced by almost only the cooling loss. Therefore, the cooling loss characteristics can be evaluated with the negative rate of apparent heat release and the normalized cumulative heat release Q/Q_{fuel} after the combustion period. The figure shows that a decrease in hydrogen content elongates combustion period due to a decreased burning velocity of fuel, but simultaneously decreases the negative rate of apparent heat release. It means a decrease in the cooling loss. The decreased cooling loss maintains indicated thermal efficiency, in spite of the decreased degree of

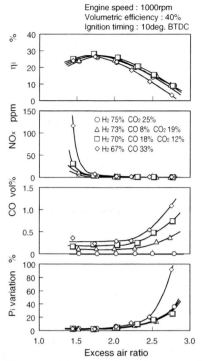

Engine speed : 1000rpm
Volumetric efficiency : 40%
Ignition timing : 10deg. BTDC

Figure 11. Influence of excess air ratio on combustion and exhaust emissions

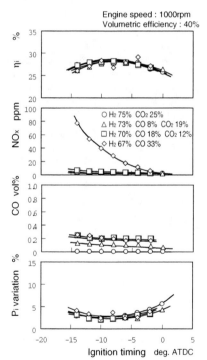

Engine speed : 1000rpm
Volumetric efficiency : 40%

Figure 12. Influence of ignition timing on combustion and exhaust emissions

constant volume due to the longer combustion period.

Influence of ignition timing

Figure 6 shows the influence of the ignition timing on combustion and exhaust emissions. The excess air ratio was set at the optimum for indicated thermal efficiency for each case of fuel compositions. The optimum ignition timing for the thermal efficiency MBT tends to advance with a

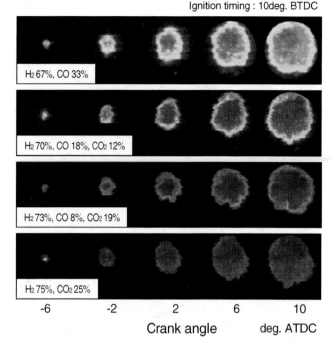

Engine speed : 1000rpm
Volumetric efficiency : 40%
Excess air ratio : 1.5
Ignition timing : 10deg. BTDC

Crank angle   deg. ATDC

Figure 13. Flame images for different fuel compositions

decrease in hydrogen content of fuel. The spark advance makes NOx and CO exhaust emissions higher, and a low hydrogen fuel has a higher NOx exhaust emission because of a smaller excess air ratio. The influence of the ignition timing on the combustion variation is quite small in the tested conditions in this research.

Observation of combustion flame

Figure 7 shows the combustion flames of the four cases of fuels. Figure 8 shows time-series trace of flame front. The excess air ratio was set at 1.5, and the ignition timing was at 10degree BTDC. The flame propagation velocity decreases with the decease in hydrogen content of fuel.
Figure 9 shows the results by a monochromator. The observation field of the monochromator is almost all of the combustion chamber. The wavelength 306.4nm corresponds to a radical emission of OH, and the another wavelength 400nm represents emission of CO-O recombination reaction. With decease in the hydrogen content of fuel, the intensity of both the emissions decrease and a ratio of CO-O recombination emission to OH radical emission increases. The CO-O recombination emission has a similar trend to the apparent rate of heat release in Figure 5. On the other hand, the peak of the OH radical emission is later than that of the CO-O recombination emission, and the OH radical emission has a similar trend to the in-cylinder gas mean temperature.
Figure 10 shows the flame images recorded through two kinds of band-pass filters with center wavelengths of 306.3nm and 387.8nm respectively. The wavelength 306.3nm was chosen for imaging OH radical emission distribution in the combustion chamber, and the 387.8nm is for CO-O recombination emission distribution. Because these two series of images are recorded in different cycle with the same operating condition, these are slightly

101

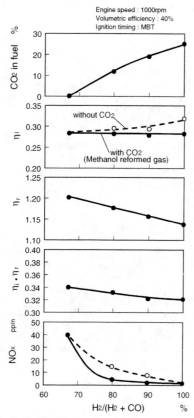

Engine speed : 1000rpm
Volumetric efficiency : 40%
Ignition timing : MBT

CO_2 in fuel %

without CO_2

η_i

with CO_2
(Methanol reformed gas)

η_r

$\eta_i \cdot \eta_r$

NO_x ppm

$H_2/(H_2 + CO)$ %

Figure 14. Overall thermal efficiency based on liquid methanol

influenced with a cycle by cycle variation of combustion. In the results, CO-O recombination emission exists only in the flame front, and OH radical emission appears not only in but also behind the flame front. The CO-O recombination reaction occurs only in the burning region, and OH radical exists even in the burnt gas region at high temperature.

COMBUSTION OF H_2-CO-CO_2 MIXTURE IN AN ENGINE

Influence of excess air ratio

Figure 11 shows the influence of excess air ratio on combustion of H_2-CO-CO_2 mixture. The spark ignition timing was set at 10degree BTDC. The maximum indicated thermal efficiencies with the four cases of fuels are similar values. It is because of the increase in carbon dioxide with the large heat capacity by the increased hydrogen. The combustion promotion by hydrogen and the combustion suppression by carbon dioxide balanced. NOx emission tends to increase with the hydrogen content at excess air ratio of less than 2.0, but NOx emission is almost zero in the larger excess air ratio than 2.0. The increase in NOx emission is because of a higher combustion temperature by the increased carbon monoxide with a larger heating value than hydrogen and by the decreased carbon dioxide. The CO exhaust emission and the combustion variation tend to increase with a decrease in hydrogen content especially at the leaner mixture conditions.

Influence of ignition timing

Figure 12 shows the influence of the ignition timing on

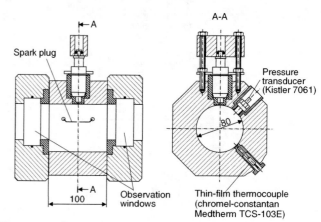

Figure 15. Constant volume combustion chamber tested in this research

combustion and exhaust emissions. The NOx exhaust emission is higher in a fuel without carbon dioxide because of the higher combustion temperature. All of the four fuels have the similar values of the indicated thermal efficiency and the combustion variation. The CO exhaust emission increases with the carbon monoxide content in fuel.

Observation of combustion flame

Figure 13 shows the combustion flame of the four cases of the methanol-reformed fuel. The excess air ratio was set at 1.5, and the spark ignition timing was at 10 degree BTDC. The figure shows that the fuel composition hardly influence the flame propagation velocity. It is supposedly caused by the balance of the combustion promotion by hydrogen and the combustion suppression by carbon dioxide. It is also a reason for the small influence of the fuel composition on the indicated thermal efficiency shown in Figure 12. Figure 13 shows a high intensity of CO-O recombination reaction emission in the flame front in carbon monoxide rich fuel. It can be attributed to the same reason for the results in Figure 9.

OVERALL THERMAL EFFICIENCY

Figure 14 shows the overall thermal efficiency based on the liquid methanol, versus the hydrogen ratio in the flammable content of the fuel. The η_r, a determined degree of fuel heat increase, represents a ratio of heating value of reformed fuel to the heating value of the liquid methanol. Therefore, a product of the degree of fuel heat increase in the fuel reforming and indicated thermal efficiency of the engine $\eta_i \cdot \eta_r$ shows the overall thermal efficiency based on the liquid methanol. The η_r was ideally calculated by taking just reactions (1) and (2) into consideration. Results of indicated thermal efficiency and NOx exhaust emission by the fuels excluding carbon dioxide are also shown for a comparison. The excess air ratio was set at the optimum for indicated thermal efficiency for each composition of fuels, and spark ignition timing is at the optimum MBT. Although the fuel without carbon monoxide brings the highest indicated thermal efficiency of the four cases of fuels excluding carbon dioxide, the fuel composition hardly influence the indicated thermal efficiency in the fuels including carbon dioxide. On the other hand, the degree of fuel heat increase is the maximum at the fuel with 67% of hydrogen. Therefore, an internal combustion engine fueled

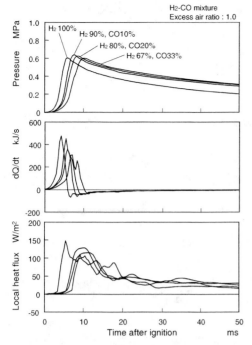

Figure 16. Combustion and instantaneous heat flux for different fuel compositions (H_2-CO mixture)

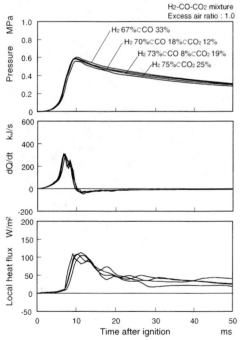

Figure 17. Combustion and instantaneous heat flux for different fuel compositions (H_2-CO-CO_2 mixture)

with the methanol-reformed fuel has the maximum overall thermal efficiency with the fuel produced in the thermal decomposition reaction shown in reaction (1). However, from the viewpoint of NOx reduction, it is better to use a fuel including carbon dioxide. A fuel composition with a small amount of carbon dioxide may be the optimum for both the high thermal efficiency and the low NOx emission in an internal combustion engine fueled with the methanol-reformed fuel.

RESULTS IN CONSTANT VOLUME COMBUSTION CHAMBER

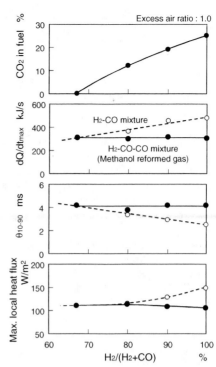

Figure 18. Influence of fuel composition on combustion characterisitics and maximum instantaneous heat flux

Fundamental combustion characteristics of the reformed fuels were studied in a constant volume combustion chamber shown in Figure 15. The chamber was made of stainless steel and has capacity of $640 cm^3$. Combustion pressure was measured with a piezo-electric type pressure transducer （Kistler 7061）and used to calculate apparent rate of heat release. Combustion chamber wall temperature was measured with a thin-film type thermo-couple （Medtherm TCS-103E, chromel-constantan）, and used to analyze instantaneous heat flux to combustion chamber wall. The instantaneous heat flux was calculated with assumption of one-dimensional heat conduction in the wall [9]. Mixture was prepared in a mixing chamber with a stirrer. Excess air ratio of the mixture was controlled with partial pressure of each gas. The preset mixture was induced into the constant volume combustion chamber and ignited with a spark plug 60s after the induction to eliminate the influence of the gas flow on the combustion. Initial pressure in the chamber is atmospheric at all the conditions.

Figure 16 and 17 show the results in the constant volume combustion chamber. Fuels are H_2-CO mixtures in Figure 16 and H_2-CO-CO2 mixtures i.e. methanol reformed fuels in Figure 17. Excess air ratio was set at 1.0 for all the cases. Figure 18 shows the influences of the fuel composition on combustion and cooling loss to combustion chamber wall. In combustion of H_2-CO mixtures, the combustion period is shortened and the cooling loss increases with the increase in hydrogen content in the fuel. On the other hand, combustion of H_2-CO-CO_2 mixtures has small change in the combustion period and the cooling loss regardless the hydrogen content in the fuels. It is due to the increase in CO_2 with the increased hydrogen. The promotion of combustion by hydrogen and the suppression by CO_2 with the large heat capacity are supposedly balanced. Those are supposedly reasons for the small change in the indicated thermal efficiency in the internal combustion engine despite the change in composition of methanol-reformed fuel shown in Figure 14.

CONCLUSION

This research analyzed characteristics of combustion, emissions and cooling loss in an SI internal combustion engine fueled with several composition of H_2-CO-CO_2 mixtures as model gases for methanol-reformed fuels. The results in this research can be summarized as follows.

1. A decrease in hydrogen deteriorates the thermal efficiency of an internal combustion engine fueled with H_2-CO mixture.
2. The composition of H_2-CO-CO_2 mixture reformed from methanol has a small influence on the thermal efficiency of an internal combustion engine.
3. Because the hydrogen increases with carbon dioxide in methanol reforming, the carbon dioxide suppresses the combustion promotion by the hydrogen.
4. A change in the composition of methanol-reformed fuel hardly influences the combustion period and the cooling loss.
5. The maximum overall thermal efficiency is obtained in the thermally decomposed methanol, because of the largest heating value and the small change in the engine efficiency for the fuel composition.

ACKNOWLEDGMENTS

Authors appreciate Mr. Takuya Nagano of Yamaha Motor Co., a former student in Musashi Institute of Technology, for his helps in experiments of this research.

REFERENCES

1. Hirota, T., et al., "Study of Methanol-Reformed Gas Engine", Transaction of JSAE, No.20, (in Japanese) (1980).
2. Numata, et al., "Research and Development of a Methanol Reformed Gas Engine for Hybrid Electric Vehicles", Proceedings of JSAE, No.72-99, (in Japanese with English summary) (1999).
3. Tanford, C., et al., "Equilibrium Atom and Free Radical Concentrations in Carbon Monoxide Flames and Correlation with Burning Velocities", Journal of Chemical Physics, Vol.15, No.7, (1947).
4. Lewis, B., et al., "Combustion, Flames and Explosion of Gases", Academic Press, (1961).
5. Shudo, T., et al., "Analysis of Thermal Efficiency in a Hydrogen Premixed Spark Ignition Engine", Proceedings of ASME, AES-Vol.39, (1999).
6. Shudo, T., et al., "Analysis of Degree of Constant Volume and Cooling Loss in an SI Engine Fuelled with Hydrogen", International Journal of Engine Research, No.4, (2001).
7. Shudo, T., et al., "Analysis of Direct Injection Spark Ignition Combustion in Hydrogen Lean Mixture", Proceedings of FISITA World Automotive Congress, F2000A085, (2000).
8. Shudo, T., et al., "Combustion Promotion and Cooling Loss in Hydrogen Premixing to Methane DI Stratified Charge Combustion", Transaction of JSAE, Vol.32, No.1, (in Japanese with English summary) (2001).
9. Isshiki, S., et al., "Heat Transfer by Combustion in Closed Chamber", Transaction of JSME, Vol.39, No.328, (in Japanese) (1973).
10. Shudo, T., et al., "Combustion and Emissions in a Methane DI Stratified Charge Engine with Hydrogen Premixing", Elsevier JSAE Review, Vol.21, No.1, (2000).
11. Shudo, T., et al., "Characteristics of an Internal Combustion Engine Fueled with Hydrogen-Carbon Monoxide Mixture", Transaction of JSAE, Vol.31, No.4, (in Japanese with English summary) (2000).

NOMENCLATURE

IMEP:	indicated mean effective pressure
MBT:	optimum spark ignition timing for indicated thermal efficiency
LHV:	lower heating value of fuel
λ:	excess air ratio
Q_{fuel}:	heat of fuel supplied in a cycle
Q:	cumulative apparent heat release
Q/Q_{fuel}:	normalized cumulative apparent heat release
θ:	crank angle
$dQ/d\theta$:	apparent rate of heat release in the IC engine
Q_B:	cumulative real heat release
Q_C:	cumulative cooling loss
ϕ_w:	cooling loss ratio, Q_C/Q_B
η_u:	combustion efficiency
η_i:	indicated thermal efficiency
η_r:	degree of fuel heat increase in fuel reforming
dQ/dt:	apparent rate of heat release in the constant volume combustion chamber
θ_{10-90}:	combustion period from 10 to 90% of mass burnt fraction

CONTACT

Toshio Shudo, Ph.D.
Assistant Professor
Department of Energy Science and Engineering
Musashi Institute of Technology
Tamazutsumi 1, Setagaya-ward, Tokyo, Japan
Phone: +81 3 3703 3111, extension 3509
Facsimile: +81 3 5707 2127
e-mail: shudo@herc.musashi-tech.ac.jp
URL: http://www.herc.musashi-tech.ac.jp

2002-01-2196

Performance and Fuel Consumption Estimation of a Hydrogen Enriched Gasoline Engine at Part-Load Operation

G. Fontana, E. Galloni, E. Jannelli and M. Minutillo
Department of Industrial Engineering University of Cassino

ABSTRACT

Hydrogen and gasoline can be burned together in internal combustion engines in a wide range of mixtures. In fact, the addition of small hydrogen quantities increases the flame speed at all gasoline equivalence ratios, so the engine operation at very lean air-gasoline mixtures is possible. In this paper, the performance of a spark-ignition engine, fuelled by hydrogen enriched gasoline, has been evaluated by using a numerical model. A hybrid combustion model for a dual fuel, according to two one-step overall reactions, has been implemented in the KIVA-3V code. The indicated mean pressure and the fuel consumption have been evaluated at part load operating points of a S.I. engine designed for gasoline fuelling. In particular, the possibility of operating at wide-open throttle, varying the equivalence ratio of air-gasoline mixture at fixed quantities of the supplemented hydrogen, has been studied.

INTRODUCTION

In recent years, the interest in the use of hydrogen, as an alternative fuel for spark-ignition engines, has grown according to energy crises and pollution problems. Comparing the properties of hydrogen and gasoline, it is possible to underline the possibilities, for hydrogen-fueled engines, of operating with very lean (or ultra-lean) mixtures [1, 2], obtaining interesting fuel economy and emission reductions. However, today the concept of hydrogen enriched gasoline, as fuel for internal combustion engines, has a greater interest than pure hydrogen powered engines because it involves fewer modifications to the engines and their fueling systems. In fact, hydrogen and gasoline can be burned together in a wide range of air-fuel mixtures, providing such good performances, as high thermal efficiency and reduced pollutant emissions. Problems associated to the use of hydrogen are fuel production and storage. Generally, hydrogen can be produced by electrolysis of water, using electricity obtained from fossil fuel (conventional power plants) and from renewable energy sources, as well. Hydrogen can be stored in the vehicle as a compressed gas, a cryogenic liquid or as gas dissolved in metal

hydrides. However a great amount of hydrogen stored on board implies an increase of the vehicle weight and aerodynamics limitations. Many other problems derive, for example, from the absence of distribution systems. For these reasons, the on-board hydrogen generation (steam reforming and partial oxidation) starting from liquid fuels, such as gasoline, is an optimal solution in order to introduce hydrogen in internal combustion engines market [3]. However, if hydrogen is extracted from gasoline, it is necessary to evaluate the optimum level of hydrogen production, from an economic point of view, according to the efficiency of the hydrogen generator. In the present work a numerical investigation has been conducted to foresee the performance of a small, multi-valve, spark ignition engine operating at part-load with hydrogen as additional fuel to gasoline. The investigation has been carried out considering different equivalence ratios of the hydrogen-gasoline-air mixtures, and in particular varying both the hydrogen-air equivalence ratio (ϕ_h) and the gasoline-air equivalence ratio (ϕ_g). The possibility of controlling load by using ultra-lean mixtures coupled with the unthrottled operation has been evaluated. The idea is that a small hydrogen amount can permit leaning the gasoline-air mixture in order to vary the engine load varying the fuel flow rate, alike in a diesel engine, with an evident gain in terms of pumping loss. The transition from high to low torque, at a fixed speed, varying both ϕ_h and ϕ_g, has been studied. Furthermore, considering the possibility of using a hydrogen generator on-board, it has been calculated the remarkable variation of fuel consumption when operating with hydrogen-gasoline mixtures. The hydrogen utilization in an internal combustion engine, allowing a high air overabundance (driving at wide-open-throttle with lean mixtures) without a decrease of engine performance, is a sound solution aimed to the pollutant emissions reduction [4, 5].

THE MODEL

The purpose of the present work is to study the effect of hydrogen enriched gasoline on the performance, emissions and fuel consumption of a small spark-ignition engine. A numerical investigation has been conducted using a computational model, suitably developed in order to foresee the combustion process of dual fuel mixtures. The computational model, based on the KIVA3-V code [6, 7] was previously tested by simulating the behavior of a gasoline engine (FIAT FIRE 1242 16V) [8]. The code has been modified to simulate the behavior of the engine fueled with gasoline-hydrogen mixtures. Twelve chemical species are considered: gasoline, H_2, O_2, N_2, CO_2, H_2O, H, O, N, OH, CO and NO. The Zeldovich kinetic model has been used to predict the thermal NO formation, while four equilibrium reactions have been used to evaluate the chemical dissociation of CO_2, H_2O, N_2, O_2 [9]. The combustion model assumes a single step reaction mechanism for the hydrogen oxidation and gasoline oxidation. The global reaction rates ω_r are calculated using the hybrid model [10, 11]:

$$\omega_r = \omega_m \qquad \text{For} \quad \omega_m \geq \omega_k$$

$$\frac{1}{\omega_r} = \frac{\tau}{\omega_m} + \frac{1-\tau}{\omega_k} \qquad \text{For} \quad \omega_m < \omega_k \qquad (1)$$

where ω_m is calculated according to the Magnussen [12] equation and ω_k is calculated according to the Arrhenius equation, so both turbulent and chemical kinetic aspects are accounted for. τ is an empirical coefficient that must be tuned in order to well reproduce the cylinder pressure curve. The combustion model is able to simulate the oxidation of a hydrogen-enriched fuel by means of a serial mechanism. Hydrogen and gasoline have different oxidation rates, so the hydrogen oxidation is considered at first and then that of gasoline. Hydrogen burning accelerates the gasoline combustion as shown in literature [13, 14, 15], this feature has been evidenced by the model results, but the chemical effects due to the active radicals released by the hydrogen oxidation have been neglected. The combustion rates are calculated using the hybrid model (equation 1); for gasoline the values summarized in table 1, to calculate kinetic and turbulent rates, are used; for hydrogen, the rate in the Arrhenius expression (table 2) is taken from

Westbrook [16], while the eddy break-up rate in the Magnussen model is calculated assuming the same gasoline constants. Finally, the parameter τ, that is a characteristic engine coefficient, is assumed equal to the value 0.43. In fact, in a previous paper [8], it has been found that, assuming for τ an average value equal to 0.43, the numerical model is able to predict the whole mapping of the FIAT engine fueled with gasoline.

RESULTS AND DISCUSSIONS

The generated mesh has 10832 cells reproducing the whole domain. Since the exhaust-suction stroke is not considered, all the computations start 148° before top dead center (BTDC), when the intake valves close. At the same time all the scalar variables are considered to be uniform, while an initial tumble motion is considered. Due to lack of experimental data available for the hydrogen-fueled engine, the above-mentioned initial conditions are derived from the engine firing with gasoline. The air mass flow is assumed to be that measured at 2500 rpm at full load operation (wide-open throttle). In each case studied, the reduction of the volumetric efficiency, due to the lower hydrogen density, has been considered. The studied cases, shown in table 3, refer to the engine operating points: 2500 rpm, variable torque and variable spark timing. All the points are engine part-load operating points; at 2500 rpm, the engine full-load torque is greater than 100 Nm. For spark timing, the reference engine values have been assumed. In the gasoline-fueled engine (reference engine), the variation of torque is obtained, of course, controlling the engine load, keeping about stoichiometric equivalence ratios. The hydrogen supplementation increases the combustion speed and extends the equivalence ratio lean-limit of the mixture reducing the misfiring. Thus, operation at ultra-lean air-gasoline mixtures is possible and the engine load can be controlled by varying the gasoline flow-rate at wide-open throttle, limiting the pumping loss. The lowest hydrogen-air equivalence ratio has been assumed according to the minimum value for which a hydrogen-fueled engine has been run in the experimental tests conducted by Varde and Frame [17] and Gentili [18].

Arrhenius Model				Eddy-Breakup Model	
$C_f = 46 \cdot 10^{10}$	$E_a = 3 \cdot 10^4$ cal/mole	$\alpha = 0.25$	$\beta = 1.5$	$A = 18$	$B = 0.5$

Table 1 - Characteristic constants for gasoline oxidation rate

Arrhenius Model				Eddy-Breakup Model	
$C_f = 1.8 \cdot 10^{13}$	$T_a = 17614$	$\alpha = 1$	$\beta = 1$	$A = 18$	$B = 0.5$

Table 2- Characteristic constants for hydrogen oxidation rate

So the hydrogen addition can guarantee a regular running with many advantages in terms of emission levels and fuel consumption reduction.

Calculated Torque	Hydrogen equivalence ratio ϕ_h	Gasoline equivalence ratio ϕ_g	hydrogen/ gasoline [% in mass]
15 Nm	0.2	0.1	92 %
	0.24	0.05	201 %
	0.3	0.0	-
30 Nm	0.2	0.18	47 %
	0.24	0.14	75 %
	0.3	0.08	159 %
45 Nm	0.2	0.27	33 %
	0.24	0.23	47 %
	0.3	0.20	66 %
60 Nm	0.2	0.43	21 %
	0.24	0.34	28 %
	0.3	0.33	40 %
75 Nm	0.2	0.59	15 %
	0.24	0.54	20 %
	0.3	0.48	27 %

Table 3 – The cases studied

In table 3, the hydrogen-air equivalence ratio ϕ_h, the gasoline-air equivalence ratio ϕ_g and the corresponding engine torque values are reported. It is possible to notice that, at low torque values, the engine operates with pure hydrogen or with more hydrogen than gasoline. In figure 1, a particular engine map, where hydrogen and gasoline equivalence ratios change together, is shown. It can be marked that the hydrogen injection, contributing to a stable flame development, allows the engine to run with very lean mixtures at low load while, firing with pure gasoline, the engine operates with slightly rich mixtures.

Figure 2 identifies the comparative NOx exhaust emissions of the engine operating with straight gasoline and with several hydrogen-gasoline mixtures, according to the five torque values examined.

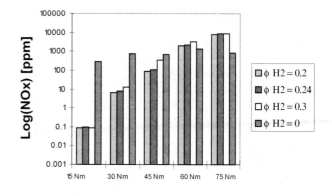

Figure 2 – Comparison of NOx emissions

Because of the relationship between NOx formation and temperature inside the cylinder, an increase of NOx concentration characterizes mixtures with high amount of hydrogen (faster burning rate). However, in Figure 2, a less NOx concentration for hydrogen-gasoline-air mixtures than for gasoline-air mixtures can be noted, except in correspondence of high torque values (60 Nm and 75 Nm). The reason of a low concentration is in the air overabundance (driving at WOT and at lean mixtures) that limits the peak temperatures. Obviously, it is possible to mark (Figure 3) a decrease of exhaust carbon dioxide, according to low gasoline equivalence ratios (ϕ_g). However it is important to underline that this reduction of CO_2 emission, as shown in figure 3, is effectively obtained whether hydrogen is produced from a non-fossil energy source.

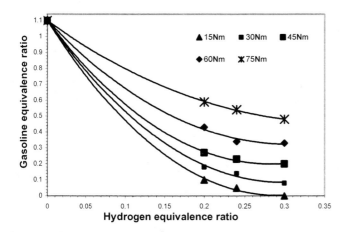

Figure 1 – Torque variation at 2500 rpm

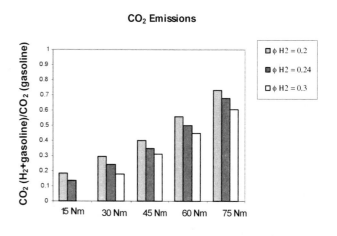

Figure 3 – CO_2 emissions according to gasoline reference engine

Higher CO_2 levels are produced in the case of an on-board gasoline reforming generation. Furthermore, in the present work, the fuel consumption of the hydrogen-gasoline engine has been compared to the consumption of the gasoline engine (reference engine). The possibility of on-board hydrogen generation has been evaluated and three efficiency values of the hydrogen generator have been considered. The results of this analysis are shown in tables 4, 5, and 6. The hydrogen generator efficiency, used in the above tables, is defined as the ratio of the produced hydrogen mass to the consumed gasoline mass. It is possible to notice that in all of the cases analyzed, the utilization of hydrogen as gasoline additional fuel permits reducing the global gasoline consumption (engine + hydrogen production). The results, shown in the tables, have been obtained comparing the measured gasoline mass flows, relatively to different torque values at 2500 rpm (reference engine), and the calculated gasoline mass flows, considering the several gasoline-hydrogen mixtures in the same operating conditions (torque, air mass flow, spark timing, and engine speed).

Torque	*Reduction of gasoline consumption [%]*		
	Efficiency of hydrogen generator from gasoline		
	0.4	*0.6*	*0.8*
15 Nm	12 %	33 %	43 %
30 Nm	19 %	34 %	41 %
45 Nm	25 %	36 %	42 %
60 Nm	17 %	27 %	31 %
75 Nm	12 %	20 %	24 %

Table 4 – Variation of fuel consumptions, $\phi_h = 0.2$

Torque	*Reduction of gasoline consumption [%]*		
	Efficiency of hydrogen generator from gasoline		
	0.4	*0.6*	*0.8*
15 Nm	14 %	38 %	50 %
30 Nm	21 %	38 %	47 %
45 Nm	26 %	40 %	46 %
60 Nm	19 %	30 %	36 %
75 Nm	14 %	23 %	28 %

Table 5 – Variation of fuel consumptions, $\phi_h = 0.24$

Torque	*Reduction of gasoline consumption [%]*		
	Efficiency of hydrogen generator from gasoline		
	0.4	*0.6*	*0.8*
15 Nm	12 %	41 %	56 %
30 Nm	21 %	42 %	52 %
45 Nm	22 %	38 %	46 %
60 Nm	19 %	32 %	39 %
75 Nm	15 %	27 %	32 %

Table 6 – Variation of fuel consumptions, $\phi_h = 0.3$

CONCLUSION

In this paper a numerical investigation has been conducted to foresee performances, exhaust emissions and fuel consumption of a small, multi-valve, spark ignition engine fueling with hydrogen enriched gasoline. The investigations have been carried out considering several equivalence ratios of both hydrogen-air mixture and gasoline-air mixture, to the aim of obtaining different torque values at 2500 rpm, but at wide-open throttle in each case. In fact, the transition from high to low torque, at the given speed, has been studied varying both ϕ_h and ϕ_g. The computational analysis has marked the possibility of operating with high air overabundance (lean or ultra-lean mixtures) without a performance decrease, but with great advantages on pollutant emissions and fuel consumption. The results have shown a reduction of NO_x and CO_2 emission. The utilization of hydrogen as gasoline additional fuel permits reducing the global gasoline consumption (engine + hydrogen production). These results have been obtained by comparing the gasoline consumption, measured on the reference engine, to the gasoline consumption calculated considering both gasoline injection and hydrogen production, by varying the efficiency of hydrogen generator (on-board reactor).

REFERENCES

1. Mathur H.B., Das L.M., "Performance characteristics of a Hydrogen Fueled SI Engine using Timed Manifold Injection", Int. J. Hydrogen Energy, vol 16, pp. 115-127, 1991.
2. H.R. Ricardo, "Further Note on Fuel Research", Proc. Automobile Engr (Lond.) 18 (1923).
3. Y. Jamal, M.L. Wyszynski, "On-board Generation of Hydrogen-Rich Gaseous Fuels- A Review", Int. J. Hydrogen Energy, vol. 19, No.7, pp. 557-572, 1994
4. Apostolescu N., Chiriac R., "A Study of Combustion of Hydrogen-Enriched Gasoline in a Spark Ignition Engine", SAE Paper, 960603, 1996.

5. Petkov T., Veziroglu T.N., Sheffield J.W., "An Outlook of Hydrogen as an Automotive Fuel", Int. J. Hydrogen Energy, vol. 14, No.7, pp. 449-474, 1989.

6. Amsden, P. J. O'Rourke, and T. D. Butler, "KIVA-II: A Computer Program for Chemically Reactive Flows with Sprays", Los Alamos National Laboratory report LA-11560-MS (May 1989)

7. Amsden, "KIVA-3V: A Block-Structured KIVA Program for Engines with Vertical or Canted Valves", Los Alamos National Laboratory report LA-13313-MS (July 1997)

8. G. Fontana, E. Galloni, E. Jannelli, "Multidimensional Modeling of Premixed Turbulent Combustion: Models and Experimental Test of a Small Spark-Ignition Engine", ASME Spring Conference, Philadelphia, 2001.

9. K. Meintjes, A.P. Morgan, "Equilibrium Equations for a Model of Combustion", Report no. GMR-4361, General Motors Research Laboratories, Warren, Mich., 1983

10. G. Fontana, E. Galloni, E. Jannelli, "Numerical simulation of a Four Cylinder, 16 V, Spark-Ignition Engine", Proceedings of 32nd ISATA, pp.169-176, Vienna, 1999.

11. G. Fontana, E. Galloni, E. Jannelli, M. Minutillo, "Numerical Modelling of a Small, Multivalve, Spark-igniton Engine: Predicted and Measured Performance at Part-Load Operation, 6° Convegno Internazionale Automobili e Motori High-Tech", Modena, 2000.

12. F. Magnussen, B. H. Hjertager, "On Mathematical Modeling of Turbulent Combustion with Special emphasis on Soot Formation and Combustion", 16th Symposium (International) on Combustion, The Combustion Institute, 1977.

13. J.B. Heywood, "Internal Combustion Engine Fundamentals", Mc Graw-Hill, 1988.

14. B.E. Milton and J.C. Keck, "Laminar Burning Velocities in Stoichiometric Hydrogen and Hydrogen-Hydrocarbon Gas Mixtures", Combustion and Flame, 58, 1984, pp. 13-22.

15. C.K.Westbrook and F.L. Dryer, "Chemical Kinetic Modeling of Hydrocarbon Combustion", Progr. Energy Combust. Science, Vol. 10, 1984. pp. 1-57.

16. Marinov, N., Westbrook, C.K. and Pitz, W.J., "Detailed and Global Chemical Kinetics Model for Hydrogen" in Transport Phenomena in Combustion, Volume 1 (S. H. Chan, edited), Talyor and Francis, Washington, DC, 1996. (UCRL-JC-120677.)

17. Varde, K.S., Frame, G.M., "A Study of Combustion and Engine Performance Using Electronic Hydrogen Fuel Injection", Int. J. Hydrogen Energy, vol 9, No.4, pp. 327-332, 1984.

18. Gentili R., "Lean Air-Fuel Mixtures Supplemented with Hydrogen for S.I. Engines: a Possible Way to Reduce Specific Fuel Consumption?", Int. J. Hydrogen Energy, vol. 10, No.7/8, pp. 491-495, 1985.

19. Das L.M., "Hydrogen-Oxygen Reaction Mechanism and its Implication to Hydrogen Engine Combustion", Int. J. Hydrogen Energy, vol. 21, No.8, pp. 703-715, 1996

20. Mathur H.B., Khajuria P.R., "A Computer Simulation of Hydrogen Fueled Spark Ignition Engine" Int. J. Hydrogen Energy, vole 11, No.6, pp. 409-417, 1986.

21. May H., Gwinner D., "Possibilities of Improving Exhaust Emissions and Energy Consumption in Mixed Hydrogen-Gasoline Operation", Int. J. Hydrogen Energy, vol. 8, No.2, pp. 121-129, 1982.

22. Swain M.R., Schade G.J., Swain M., "Design and Testing of a Dedicated Hydrogen-Fueled Engine", SAE Paper, 961077, 1996.

23. Li Jing-Ding, Lu Ying-Oing, Du Tian-Shen, "Improvement on the Combustion of a Hydrogen Fueled Engine", Int. J. Hydrogen Energy, vole 11, No.10, pp. 661-665, 1986.

24. G. Yu, C.K. Law and C.K. Wu, "Laminar Flame Speeds of Hydrocarbon + Air Mixtures with Hydrogen Addition", Combustion and Flame, 63, 1986, pp. 339-347.

25. J. I. Ramos, "Internal Combustion Engine Modeling", Hemisphere, 1989

26. Z. Han, R. D. Reitz, "Turbulence Modeling of Internal Combustion Engines using RNG k-ε Model", Combustion Science and Technology, 1995

CONTACT

fontana@unicas.it

galloni@unicas.it

jannelli@unicas.it

minutillo@unicas.it

1999-01-0619

A Numerical Study of a Free Piston IC Engine Operating on Homogeneous Charge Compression Ignition Combustion

S. Scott Goldsborough and Peter Van Blarigan
Sandia National Laboratories

ABSTRACT

A free piston, internal combustion (IC) engine, operating at high compression ratio (~30:1) and low equivalence ratio (ϕ~0.35), and utilizing homogeneous charge compression ignition combustion, has been proposed by Sandia National Laboratories as a means of significantly improving the IC engine's cycle thermal efficiency and exhaust emissions. A zero-dimensional, thermodynamic model with detailed chemical kinetics, and empirical scavenging, heat transfer, and friction component models has been used to analyze the steady-state operating characteristics of this engine. The cycle simulations using hydrogen as the fuel, have indicated the critical factors affecting the engine's performance, and suggest the limits of improvement possible relative to conventional IC engine technologies.

INTRODUCTION

In an effort to improve the fuel economy and exhaust emissions of the internal combustion (IC) engine, Sandia National Laboratories has been developing a free piston, IC engine-generator that will have application to both transportation and stationary power systems (1). The goal is to maximize the thermal efficiency at a particular operating point, while minimizing the emissions released. Fuel cell type performance is desired. The principle of operation is to ignite and burn lean homogeneous charge mixtures (equivalence ratio ~ 0.35) at high compression ratio (~30:1) in a near constant-volume combustion process. (Equivalence ratio (ϕ) is defined here as the ratio of the actual fuel-to-air ratio to the stoichiometric ratio.) This operation should improve the indicated thermal efficiency relative to conventional stoichiometric spark-ignition operation, while reducing the peak cylinder temperatures to a level where essentially no oxides of nitrogen (NO_x) are formed. Regulated emissions such as unburned hydrocarbons (HC) and carbon monoxide (CO) are expected to be controlled using oxidation catalyst technologies.

The Sandia approach utilizes a free piston / double-ended cylinder arrangement where a linear alternator is integrated directly into the cylinder's center section. Homogeneous charge compression ignition (HCCI) combustion at alternating cylinder ends is used to drive permanent magnets fixed to the piston, back and forth through the alternator's coils. The alternator serves to generate useful electrical power, and to control the piston's motion by dynamically varying the rate of electrical generation. Engine startup is also achieved using the alternator. Charging of the engine's cylinders is accomplished using a two-stroke cycle process. The proposed free piston device is illustrated in Figure 1.

BACKGROUND

The use of HCCI combustion as an alternative to spark-ignition and compression-ignition combustion in internal combustion engines has been investigated for some time (2-7). In this process, premixed fuel-air charges are compression heated to the point of autoignition. Numerous ignition points throughout the mixture ensure rapid combustion (3), while low equivalence ratios can be utilized, since no flame propagation is required (4-6). The operating compression ratio can also be increased, since higher temperatures are required to autoignite lean mixtures (6,7). Low NO_x emissions generally result (5).

Free piston geometries as an alternative to crankshaft-driven geometries have been incorporated into some unique IC engine applications. Examples include gasifiers for single stage turbines (8), as well as hydraulic pumps (9,10). The free piston configuration results in a variable compression ratio engine with reduced mechanical friction, and thus the capability of operation on a multitude of fuels (8). NO_x emissions can be reduced due to the unique free piston dynamics (10).

Recent studies of HCCI combustion using a free piston have shown that very rapid combustion (near constant-volume) is possible with certain fuels, and that ideal Otto cycle performance can be closely approached (1,11).

The issue of burn duration as a limitation to indicated thermal efficiency improvements with increased compression ratios (12) is practically eliminated in this combustion scheme. Further, NO_x emissions can be controlled by sufficient dilution of the fuel/air mix with excess air.

The purpose of the present study is to investigate the operating characteristics of the free piston engine-generator design using a full cycle engine model. A zero-dimensional, thermodynamic approach was used, where the potential of the engine cycle in terms of the thermal efficiency and emissions performance was examined through various cycle simulations. Only steady-state operation was considered. The critical factors affecting the engine's performance, and the limits of improvement possible relative to conventional IC engine performance were studied.

The remainder of this paper is organized as follows. The details of the engine model are presented next, followed by a discussion of the simulated cases and the simulation results. The implications of the model results are then discussed, and the conclusions of this study summarized.

MODEL DESCRIPTION

A complete description of the model used in this study can be found in ref. (13). However, a summary of the model is presented here as well.

The free piston engine model consists of two main parts: one which calculates the thermodynamics of the engine cycle, and another which describes the motion of the free piston. A detailed analysis of the changing piston dynamics was necessary since the free piston's motion could not be prescribed a priori. The thermodynamic analysis includes the compression, combustion, expansion and gas exchange parts of the cycle, and accounts for heat exchange between the cylinder gases and walls of the combustion chamber. The free piston analysis couples the calculated cylinder pressure differential across the piston with a description of the piston ring - wall friction, and the electromagnetic forces generated by the linear alternator. The details of the linear alternator component, however, are not considered.

THERMODYNAMICS – The analysis of the in-cylinder gas cycle is based on a zero-dimensional, thermodynamic approach where fluid dynamic and spatial effects are not considered. Here the state of the in-cylinder gases is determined by application of the energy conservation equation to the cylinder control-volume. This analysis is identical for both sides of the engine, and can be expressed as,

$$\frac{dU}{dt} = \frac{dQ}{dt} - P\frac{d\forall}{dt} + \sum_i \dot{H}_i - \sum_e \dot{H}_e \qquad (1)$$

A graphical representation of this analysis is presented in Figure 2.

Throughout the majority of the engine model (through the compression, combustion, and expansion phases of the cycle) the cylinder charge is assumed to exist as a homogeneous medium, uniform in temperature and composition. During the gas exchange process however, the cylinder gases are considered to exist as two distinct regions: one a zone of burned combustion gases, and another a zone of pure fresh charge. The two zones are assumed to remain immiscible, but within each zone the mixture is considered homogeneous. These gas zones mix instantaneously at the end of the scavenging process.

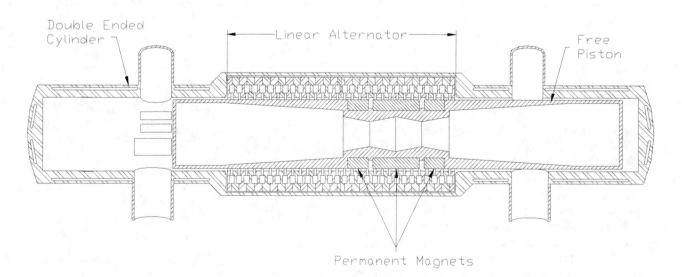

Figure 1. Free piston engine.

Compression, Combustion and Expansion Phases – Within the free piston engine model, the HCCI combustion process is modeled as an entirely chemical phenomenon, where the initiation and propagation of combustion are assumed to be completely dependent on kinetic mechanisms. No spark ignition or injection timing are considered, nor are any flame propagation or mixing-type phenomena. This approximation was considered reasonable due to the nature of the HCCI process, where the fuel and air are initially premixed (assumed to a good degree in this application), and combustion is initiated by compression heating alone.

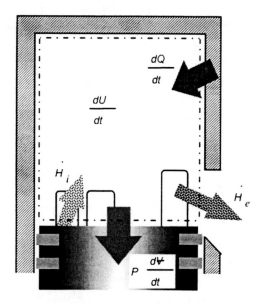

Figure 2. In-cylinder control volume.

Because of this approximation, the equations were written such that chemical reactions could proceed at any time during the compression and expansion strokes. Equation (1) follows as,

$$\frac{d}{dt}\left(\forall \sum_j c_j u_j^* \right) = \frac{dQ}{dt} - P\frac{d\forall}{dt} \tag{2}$$

Here the u_j^*'s are determined based on c_p and ΔH^o data from the JANAF tables. In this formulation the gas temperature is implied by the u_j^*'s.

The species concentrations are governed by mass conservation, where this is written on a molar basis as

$$\frac{d}{dt}\left(c_j \forall \right) = \sum_n \Re_{j,n} \tag{3}$$

During this part of the cycle it is assumed that no mass enters or leaves the cylinder control volume; thus, no leakage term or crevice volume is taken into account. The $\Re_{j,n}$'s are determined using the law of mass action, where this is applied to all of the reactions affecting the concentration of species j. The rate constants for the individual reactions are calculated assuming modified Arrhenius kinetics,

$$k_n = A_n T^{m_n} \exp\left(\frac{-E_{a,n}}{R_u T}\right) \tag{4}$$

Because the HCCI process is modeled as a set of chemical reactions, the kinetic mechanism employed has a large influence on the timing of combustion, as well as the extent to which it is completed, and pollutant formation. A detailed kinetic mechanism was employed in the model in hopes of achieving a greater degree of accuracy relative to a reduced mechanism.

The rate of heat transfer to and from the cylinder gases is determined assuming uniform, quasi-steady behavior, where the overall heat transfer coefficient is calculated using Woschni's correlation (14). Data from HCCI experiments using hydrogen in a rapid compression-expansion machine (RCEM) were used to adjust the heat transfer constants for this correlation (13).

The piston work is determined based on the cylinder pressure, calculated assuming an ideal gas, and the rate of volume change, calculated based on the piston dynamics. A pancake geometry is assumed for the combustion chamber.

Gas Exchange Phase – The gas exchange, or scavenging part of the cycle is modeled using quasi-steady intake and exhaust flow models, and an empirical correlation for the exhaust gas purity. The exhaust gas purity is defined as the mass fraction of fresh charge in the exhaust stream during the process. The use of an empirical, or semi-empirical correlation was important in quantifying the two-stroke cycle behavior for this zero-dimensional model, since the burned gas / fresh charge interaction could not be specified otherwise. A zonal formulation is used in this part of the engine model.

The energy and mass conservation equations are rewritten as

$$\left[\frac{d}{dt}\sum_j n_j u_j^* \right]_{b,f} = \left[\frac{dQ}{dt} - P\frac{d\forall}{dt} + \sum_i \dot{H}_i - \sum_e \dot{H}_e \right]_{b,f} \tag{5}$$

and

$$\left[\frac{dn_j}{dt} \right]_{b,f} = \left[\sum_i x_{j,i} \dot{N}_i - \sum_e x_{j,e} \dot{N}_e \right]_{b,f} \tag{6}$$

where these are applied separately to the burned gas and fresh charge zones. Here the inflow and outflow of each specie are calculated throughout the scavenging process. It is assumed that no chemical reactions occur, and that the zones do not exchange mass or heat with one another.

The enthalpy flows into and out of each zone are determined based on the temperature, composition and total flow rates of the incoming and exhausting streams. The molar flow rates are calculated assuming quasi-steady compressible flow behavior. The flow rates are given as

$$\dot{N}_{i,e} = C_D A_{eff} \left(\frac{P_o^2}{R_u T_o MW} \right)^{1/2} PR^{1/\gamma} \left[\frac{2\gamma}{(\gamma-1)} \left[1-(PR)^{(\gamma-1)/\gamma} \right] \right]^{1/2} \Bigg|_{i,e} \quad (7)$$

when the flow is unchoked, and

$$\dot{N}_{i,e} = C_D A_{eff} \left(\frac{P_o^2}{R_u T_o MW} \right)^{1/2} \gamma^{1/2} \left[PR_c \right]^{(\gamma+1)/2\gamma} \Bigg|_{i,e} \quad (8)$$

when the flow is choked. The discharge coefficients are determined using Annand's empirical correlation (15).

The effective flow areas for each zone are calculated based on the total available intake and exhaust port areas, as determined by the instantaneous piston position, and prescribed ratios for the fresh charge to burned gas flows. For simplicity, only forward flow is considered, where it is assumed that some kind of reed valve can be used.

The fresh charge to burned gas flow ratios are determined as follows. For flow through the intake port, it is assumed that all of the incoming mass goes directly into the fresh charge zone. Thus,

$$\frac{\dot{N}_{i,b}}{\dot{N}_{i,f}} = 0; \quad \dot{N}_{i,f} = \dot{N}_i \quad (9)$$

so that

$$(A_{eff})_{i,f} = A_i ; \qquad (A_{eff})_{i,b} = 0 \quad (10)$$

For flow through the exhaust port, both the fresh charge and burned gases are allowed to exit the cylinder. Short-circuiting is accounted for in this way.

The ratio of the exhausting gas flows is prescribed according to Sher and Harari's empirical correlation for the exhaust gas purity (16). (This is often referred to as the 'S'-shaped correlation.) The exhaust gas purity is calculated here as,

$$\xi = H_{sc}^{\left[\frac{1}{s_2} - \frac{s_1}{s_2}(1-s_2) \right] \frac{\Lambda_{de}}{\Lambda_{de,max}}} \quad (11)$$

where the scavenging parameters, s_1 and s_2, have been adjusted to fit data from a small loop scavenged engine (17). The exhaust flow rate ratio follows as,

$$\frac{\dot{N}_{e,f}}{\dot{N}_{e,b}} = \frac{\frac{\partial}{\partial t} \left[N_e \, \xi \, \frac{MW_e}{MW_f} \right]}{\frac{\partial}{\partial t} \left[N_e (1-\xi) \frac{MW_e}{MW_b} \right]} \quad (12)$$

Heat transfer to and from the fresh charge and burned gas zones is determined as before, using Woschni's correlation (14). Here the zonal surface area is assumed to be proportional to the volume occupied by each zone. The piston work is determined by calculating the cylinder pressure, assumed uniform across the zones, and the rates of change of the zonal volumes. The rates of zonal volume change are determined based on the total volume change, and the volumetric ratio change, calculated assuming ideal gases.

PISTON DYNAMICS – The piston dynamics are determined by analyzing the forces acting on the free piston. In the engine-generator these include the front and back cylinder pressures, the piston ring - wall friction, and the alternator-induced, electromagnetic force. These forces are summarized in Figure 3, where a free body diagram of the piston is illustrated.

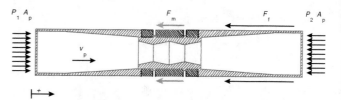

Figure 3. Free body diagram of the free piston.

Quite different from the conventional crankshaft-driven piston motion, the free piston's acceleration is determined as,

$$a_p = \frac{(P_1 - P_2)A_p - F_{frict} - F_{mag}}{m_p} \quad (13)$$

The frictional force is modeled here by including both static and viscous components,

$$F_{frict} = sign(v_p)\left[f_1 + f_2 |v_p| \right] \quad (14)$$

where the model constants, f_1 and f_2, have been correlated to match experimental data from the RCEM (1). The electromagnetic force is modeled as

$$F_{mag} = M v_p \quad (15)$$

where the alternator coil strength is assumed to be constant throughout the cycle, but is adjusted for each simulated operating condition.

SOLUTION TECHNIQUE – The full cycle thermodynamic analysis is achieved by integrating with respect to time, the energy and mass conservation equations through the compression and expansion strokes of the piston. Only one oscillation is required since this is a two-stroke cycle engine. The piston motion is determined by integrating with respect to time the acceleration equation. During the compression, combustion and expansion phases the chemical kinetics software HCT (18), developed by Lawrence Livermore National Laboratory for one-dimensional unsteady problems, is used to calculate the thermodynamic processes. A separate finite difference code is used during the scavenging phase.

An elaborate initialization scheme was used to start the free piston motion and to initialize the engine variables. This was required due to the characteristics of the modeled engine: the free piston motion is tightly coupled to the in-cylinder thermodynamics, the start of combustion is not controlled by any timing parameter, and the two-stroke cycle scavenging process, which influences the rest of the engine cycle, is dependent on the free piston motion and the cylinder blowdown conditions. The bulk of the computational time (approximately 75 cycles) is dedicated to running the initialization code so that consistent (cycle-to-cycle) cylinder pressure and piston position profiles are achieved. In the initialization routine combustion is modeled as a constant-volume process at top dead center (TDC), or at a specified (guessed) autoignition temperature. The HCT software is accessed only after the initialization scheme is complete. This initialization procedure resulted in similar piston position profiles for the non-HCT cycle, and the cycle in which HCT was employed, using hydrogen as the fuel.

In addition to the initialization scheme, an iteration process was employed, where engine parameters such as the port sizing, alternator coil strength, and equivalence ratio, are adjusted from run to run, so that the desired simulation conditions (e.g. ϕ, TDC combustion, etc.) could be achieved. A complete discussion of the initialization procedure and the code's operation can be found in ref. (13).

MODEL SIMULATIONS

Various cycle simulations were conducted in order to investigate the characteristics of the engine-generator, and to determine the engine's thermal efficiency and pollutant emissions potential. The studies detailed here include: variations in intake equivalence ratio and scavenging efficiency, control of the scavenged temperature, control of the compression ratio, and a comparison of the cycle performance utilizing the free piston motion relative to the cycle performance utilizing a crankshaft-driven piston. The scavenging efficiency (η_{sc}) is defined here as the mass fraction of fresh charge in the cylinder at port closure. The scavenged temperature (T_{scav}) is the mass averaged temperature of the cylinder charge at port closure.

The cycle simulations were run on a P166 PC where typical execution times were approximately 4 minutes. Hydrogen was chosen as the fuel for this study due to the extensive knowledge available regarding its combustion kinetics, and its potential as a renewable fuel source for this engine. In addition, extensive experimental data from the RCEM was available for this fuel. Further, reduced computational times relative to other fuels, and the fact that NO_x is the sole pollutant were also factors. The kinetic mechanism used for this fuel/air mixture can be found in ref. (19).

ENGINE DIMENSIONS – The engine dimensions and baseline operating conditions used for the simulations are listed in Table 1. Figures 4a and b illustrate the dimensions for reference.

For this investigation the engine was assumed to be piston ported where the effective bore-to-compression stroke ratio was 1:2. This was chosen in order to minimize the surface area-to-volume ratio near TDC, and to reduce the possibility of cylinder head - piston contact during high compression ratio operation. The location of the ports and the duration of port opening were critical to the scavenging process, however, these could only be specified based on the relative free piston motion, not by any crankshaft timing, or other means.

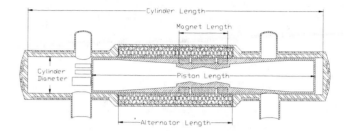

(a) Free piston engine dimensions.

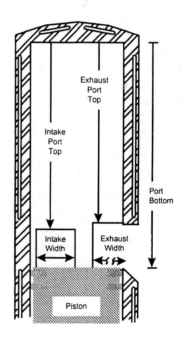

(b) Free piston engine dimensions.

Figure 4.

The total port openings are given in Table 1 where the ports were assumed to be rectangular in shape and each represented by a single discharge coefficient. The port dimensions listed in Table 1 determined the maximum available port area during the gas exchange process and were used to fix the scavenging efficiency for this zero-dimensional model, as opposed to changes in the intake pressure or piston speed, as is common for simulations of crankshaft-driven engines. For this study, the exhaust port width ranged from 0.31 to 8.25 cm, while the width of the intake port remained constant. The wide range of exhaust port widths was necessary in order to investigate the range of scavenging efficiencies desired at the particular operating conditions.

Table 1. Simulated free piston engine specifications.

Cylinder length	65.00 cm
Cylinder bore	7.00 cm
Piston length	41.8 - 47.2 cm
Piston mass	2700 g
Typical stroke	16.40 cm
Exhaust port top	14.00 cm
Intake port top	14.15 cm
Exhaust port width	0.31 - 8.25 cm
Intake port width	8.00 cm
Exhaust port bottom	17.25 cm
Intake port bottom	17.25 cm
Alternator length	29.50 cm
Magnet length	10.25 cm
Intake temperature	300 K
Intake pressure	1.50 bar
Exhaust pressure	1.00 bar
Wall temperature	700 K

The piston length was adjusted through this study so that a wide range of compression ratios (3:1 up to 38:1) could be accommodated. This was required due to the double ended function of the piston.

MODEL RESULTS

Validation of the free piston engine model proved difficult due to the limited amount of experimental data available for the engine-generator. However, each of the component models (e.g., chemical kinetics, scavenging, etc.) has been validated previously. Where possible in this study, the results from the model simulations are compared to data that is currently available.

The typical operation of the free piston engine as predicted by the full cycle model is presented first. Figures 5 and 6 show the piston dynamics where typical position vs. time, and velocity vs. position curves are plotted. Illustrated in these figures are the differences between the free piston motion and the motion of a crankshaft-driven piston of the same stroke and piston frequency. (For this comparison it was assumed that the ratio of connecting rod length to the crank radius, for the crankshaft-driven piston, is 2.5:1.)

It can be seen in Figure 5 that the motion of these two pistons is very different. Evident in this figure is the fact that the free piston spends less time at TDC relative to the crankshaft-driven piston. This is also illustrated in Figure 6 where it is seen that the free piston accelerates and decelerates more quickly at the ends of the stroke. (These curves are similar to the free piston results presented in refs. (9) and (10).) The inherent characteristic of shorter residence time at TDC for the free piston could be attractive in terms of heat transfer losses and NO_x formation since shorter time at higher temperature is desirable.

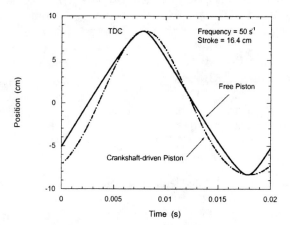

Figure 5. Piston position vs. time.

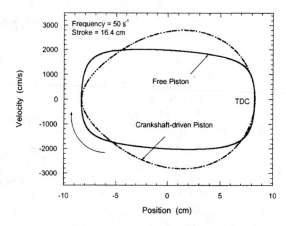

Figure 6. Piston velocity vs. piston position.

Figure 7 plots a typical pressure vs. time curve for the hydrogen-fueled engine. The pressure plotted here is non-dimensionalized using the scavenged pressure (P_{scav}), the pressure at port closure. Figure 8 plots a log pressure vs. log volume curve for the free piston engine cycle. The volume is non-dimensionalized using the engine's trapped volume (V_t). It can be seen in these figures that the engine model predicts very rapid combustion for the autoignition process, almost constant-volume. The combustion duration, as determined from the rate of pressure increase, is on the order of 10-50 µs. This calculation agrees well with the results obtained from the RCEM experiments (1).

Here it is worth noting that the free piston may prove to be much better suited than a crankshaft-driven configuration to the high rates of combustion seen in this engine application. This is due to the lack of mechanical structures in the one-moving-part engine.

VARIATIONS IN EQUIVALENCE RATIO AND SCAVENGING EFFICIENCY – To determine the range of operation and performance for this engine, simulations were conducted at intake equivalence ratios of 0.15, 0.30, 0.45 and 0.60, and at scavenging efficiencies of 0.5, 0.7 and 0.9. Such a wide range of operating conditions seemed possible for the free piston engine-generator operating on HCCI combustion.

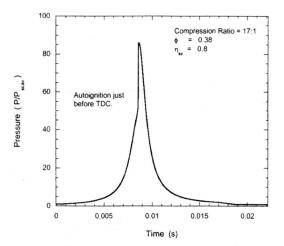

Figure 7. Pressure vs. time.

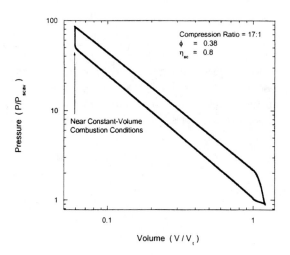

Figure 8. Log pressure vs. log volume.

It should be noted here that the free piston engine is designed to operate over a range of compression ratios where optimal steady-state operation occurs when the piston compresses the cylinder gases just to the point of autoignition. Additional compression after combustion is not usually required. The operating compression ratio of the engine then, is governed by its operating conditions (ϕ, η_{sc}, intake temperature, etc.), as adjusted to provide the amount of compression heating required for autoignition.

For these runs the extent of piston compression was controlled by adjusting the alternator coil strength so that combustion would occur at, or just before TDC. Some variation was allowed in this constraint due to the convergence criteria of the engine code. Iteration to the precise

operating parameters proved to be very time consuming and was not considered necessary for this study. The following discussion therefore is based on simulations where the operating compression ratio was limited to +15% greater than the compression ratio required for TDC autoignition.

Figures 9 and 10 illustrate the variations in operating compression ratio with equivalence ratio and scavenging efficiency, respectively. As can be seen, increased equivalence ratios lower the achievable compression ratio, while increased scavenging efficiencies raise the operating compression ratio. These characteristics are due mainly to the effect that ϕ and η_{sc} have on the temperature of the cylinder gases at the start of compression. This is described as follows. Increased equivalence ratios lead to higher combustion temperatures which result in more heat per unit mass at the end of the cycle. For a fixed scavenging efficiency this leads to higher temperatures at the end of the scavenging process. Higher scavenged temperatures decrease the amount of compression heating required for autoignition. (The effect of equivalence ratio on the autoignition compression ratio is secondary in this case to the scavenged temperature effect.)

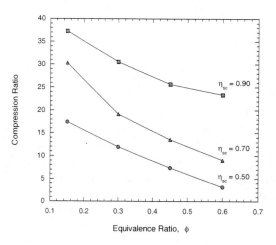

Figure 9. Compression ratio vs. equivalence ratio.

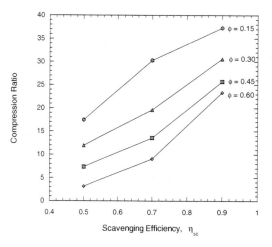

Figure 10. Compression ratio vs. scavenging efficiency.

117

On the other hand, increased scavenging efficiencies result in a greater fraction of fresh charge at the start of compression, and thus a lower thermal mass. A greater percentage of the hot combustion gases is removed from the cylinder with more scavenging, and thus the compression heating required for autoignition is increased. Although increased scavenging efficiencies lead to higher charge energy densities, due to the fact that there is less dilution by residual gases, the dominant effect is the reduction in the scavenged temperature.

These points are clearly illustrated in Figure 11 where the engine's operating compression ratio is plotted versus the scavenged temperature. The scatter present in this plot is thought to be due to the over-compression of some of the runs, as a result of the code's convergence criteria.

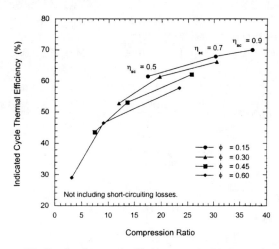

Figure 12. Cycle thermal efficiency vs. compression ratio.

As a practical matter, extremely low equivalence ratio operation will prove impractical due to frictional losses in the engine. Additionally, short-circuiting losses will tend to degrade the engine's cycle thermal efficiency for highly scavenged operation, as more fresh gas is required to displace the burned charge more completely. These points will have to be investigated in the future, however the following plot is offered to illustrate the extent to which short-circuiting can negate the improvements in cycle thermal efficiency gained by increased scavenging efficiencies.

In Figure 13 the engine's overall efficiency is plotted versus its operating compression ratio. Here the overall efficiency is defined as the ratio of the engine's indicated power output, neglecting pumping work, to the fuel energy flow through the engine.

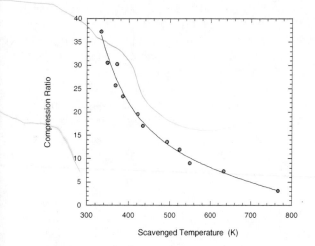

Figure 11. Compression ratio vs. scavenged temperature.

Figure 12 shows the variation in indicated cycle thermal efficiency with operating compression ratio. Here the indicated cycle thermal efficiency is defined as the ratio of the engine's work output, as determined by integrating the pressure-volume curve, to the fuel energy trapped in the cylinder at port closure. This efficiency does not account for frictional losses, pumping work, or fuel lost through short-circuiting (where unburned fuel escapes through the exhaust port). The short-circuiting losses will be discussed shortly, however. As expected, the engine's indicated cycle thermal efficiency increases with increased operating compression ratio.

It appears then that both the equivalence ratio and scavenging efficiency significantly affect the achievable thermal efficiencies of this engine through the scavenged temperature. Improvements in the gas mixture's specific heat ratio due to decreased equivalence ratios and improved scavenging also contribute to thermal efficiency increases, however, this effect is of secondary importance here (where relative changes of only a few percent (for this range of equivalence ratio and scavenging efficiency) are generally seen).

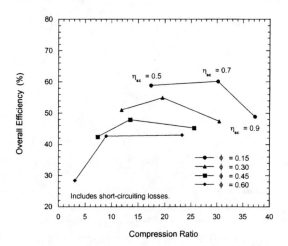

Figure 13. Overall efficiency vs. compression ratio.

In examining this plot it should be recalled that the empirical model used to describe the scavenging process (Eq. (11)) tends to predict significant increases in fuel short-circuiting for highly scavenged systems. (This problem could be addressed in the engine design through fuel injection techniques or stratified scavenging. However, the potential for these solutions was not investigated in this study.)

The pollutant emissions performance of the free piston engine is presented next. Here only NO_x emissions are given since hydrogen was used as the fuel. (HC and CO emissions from operation with hydrocarbon fuels are expected to be controllable using oxidation catalyst technologies.)

The NO_x performance is characterized in this study based on the dilution ratio, as opposed to the equivalence ratio. This is due to the fact that two-stroke cycle scavenging affects the amount of fuel energy trapped in the cylinder at port closure (and therefore the resulting fuel to air plus residual gas ratio), and this may be quite different than the intake equivalence ratio. The dilution ratio is defined here as the ratio of the actual fuel to air plus residual gases ratio to the stoichiometric fuel to air ratio.

Figure 14 illustrates how the NO_x concentrations increase with increasing dilution ratio. In these results the NO_x levels include NO, NO_2 and N_2O emissions, though NO was the predominant specie. The increases in NO_x with dilution ratio are due mainly to the increased combustion temperatures that result from higher charge energy densities. The scatter in the simulated points in this figure is thought to be primarily a function of over-compression of the cylinder charge after autoignition, as discussed previously.

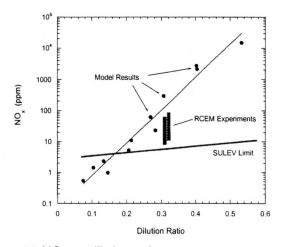

Figure 14. NO_x vs. dilution ratio.

Also shown in Figure 14 are selected results from RCEM experiments using hydrogen (1). The range in the experimental data is due to both over- and under-compression of the cylinder charge. As can be seen, the simulation results tend to over-predict the NO_x concentrations, though the increasing trend with dilution ratio seems to match the data. The discrepancies between the calculations and the data could be due to the model's uniform charge assumption, and the inadequacy of the heat

transfer model to accurately represent the RCEM heat transfer characteristics.

Finally, shown in Figure 14 is the California Air Resources Board's (CARB) proposed Super Ultra Low Emissions Vehicle (SULEV) NO_x standard (20). (This NO_x limit assumes a vehicle efficiency of 60 miles per gallon.) The SULEV standard is considered the goal to be met by the free piston design concept. From Figure 14 it is clear that there is a limit of operation for compliance, without exhaust gas aftertreatment.

The following studies were conducted at a dilution ratio of 0.2 since this appeared to be within the range of compliance of the SULEV standard, extrapolating from the experimental data. Actual operation could probably be greater than 0.2, considering the discrepancies between the model and the experimental results.

CONTROL OF THE SCAVENGED TEMPERATURE – Figure 11 illustrated the detrimental effect that high scavenged temperatures have on the operating compression ratio of this engine. The corresponding decrease in cycle thermal efficiency was presented in Figure 12. In order to reduce the scavenged temperatures high scavenging efficiencies could be used, however increased short-circuiting, as predicted by the empirical scavenging model, would probably negate any improvements.

This is illustrated in Figure 15 where the cycle thermal efficiency and overall efficiency are plotted versus the scavenging efficiency. The short-circuiting ratio, defined here as the mass fraction of fresh charge that blows through the cylinder before port closure, is included for reference. This ratio provides an indication of the amount of fresh gas that must be sacrificed to remove increasing amounts of burned charge from the cylinder.

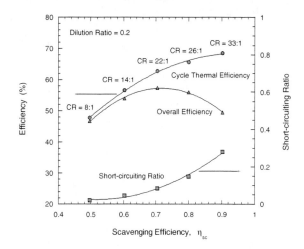

Figure 15. Cycle thermal efficiency, overall efficiency and short-circuiting ratio vs. scavenging efficiency.

The use of low scavenging efficiencies, along with sufficient control of the scavenged temperature by reduced intake temperatures, in order to improve the overall efficiency of the engine was investigated. Several scavenging efficiencies were simulated, where the cylinder charge was diluted by varying fractions of residual gases and intake air. Figure 16 illustrates the variation in cycle thermal efficiency and overall efficiency with scavenging efficiency. Here the scavenged temperature was maintained at 340 K. The operating compression ratio in these runs was 35:1.

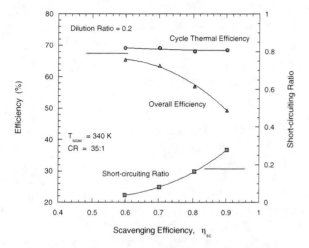

Figure 16. Cycle thermal efficiency, overall efficiency and short-circuiting ratio vs. scavenging efficiency.

It can be seen in this figure that the cycle thermal efficiency remains relatively constant for the various scavenging efficiencies investigated. Thus, it appears that operation on high fractions of residual charge (near 40%) could yield equivalent thermal efficiencies as low equivalence ratio operation (for the same overall dilution ratio), as long as high compression ratios can be reached. The result of this is that, thermal control of the cylinder charge, in conjunction with reduced scavenging, could be used to increase the cycle thermal efficiency, while reducing short-circuiting from the engine.

COMPRESSION RATIO CONTROL – The free piston design of this engine offers a reduction in the number of moving parts and reduced mechanical friction. However, this is at the cost of sophisticated electronic control. The absence of mechanical linkages necessitates the ability of the linear alternator to moderate the free piston dynamics through each and every stroke, such that the required compression ratio is achieved.

As can be inferred from the scatter in Figure 14, variations in charge over-compression can affect the NO_x emissions of this engine. To determine the extent to which this is a critical parameter, the engine code was used to investigate the over- and under-compression of the cylinder charge and the subsequent effects on the engine's cycle thermal efficiency and NO_x emissions. The range of piston motion was controlled by adjusting the alternator coil strength. A constant intake temperature of 300 K was used, and the engine was simulated at a scavenging efficiency of 0.8. Dilution ratios of 0.2 and 0.3 were used.

Figures 17 and 18 illustrate the effects of variations in the piston motion. The cycle thermal efficiency and NO_x emissions are plotted versus operating compression ratio, where this represents the compression ratio achieved by the engine, though in these figures this can be greater than that required for TDC autoignition. Figure 17 shows little variation in the cycle thermal efficiency over the range investigated, as long as the piston motion leads to proper autoignition and complete combustion. (Autoignition and incomplete combustion could occur with insufficient compression.) This seems to indicate that heat loss near TDC is not greatly affected by small changes in the piston's dynamics, even though significantly increased pressures result. (The inclusion of a blowby loss term near TDC could affect the thermal efficiency when the cylinder gases are over-compressed, however, this was not investigated.) This finding of little change in heat loss is supported by the experimental results presented in ref. (1).

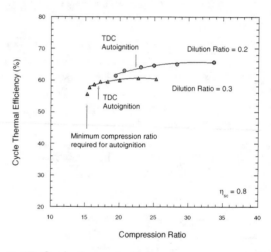

Figure 17. Cycle thermal efficiency vs. compression ratio.

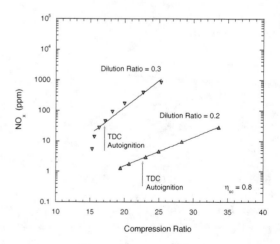

Figure 18. NO_x vs. compression ratio.

On the other hand, however, NO_x emissions were found to be very sensitive to the range of over- or under-compression, increasing in orders of magnitude for variations in the piston motion of only 1-2% of the entire piston stroke (though this represents a significant range of compression energy). This result appears to be attributable to the higher temperatures reached with increased compression.

VARIATIONS IN THE FREE PISTON DYNAMICS – The free piston dynamics are governed to a large extent by the mass of the piston. The degree to which the piston mass affects the motion of the piston and the resulting cycle thermal efficiency, and NO_x emissions performance was studied. In this analysis the piston mass was doubled from 2700 to 5400 g. This change in piston mass had a great affect on the frequency of piston oscillation (49 Hz → 38 Hz) as well as the required alternator coil strength for stable operation. However, the residence time at TDC (the time that the in-cylinder gases spend at high compression (e.g., above a compression ratio of 15:1 for an operating compression ratio of 25:1)) was only slightly altered (e.g., from 1.15 ms to 1.6 ms). In terms of the heat transfer mechanisms and the NO_x kinetics, this relative change was small. The result of this was that the thermal efficiency and NO_x emissions were unchanged between these runs.

The extent of benefit that the free piston geometry provides relative to a crankshaft-driven piston configuration, in terms of improved cycle thermal efficiency and NO_x emissions was also analyzed. This was done by comparing the results of simulations for the different piston configurations. Two cases were studied, one where a simulated crankshaft-driven piston maintained the same oscillating frequency as the free piston, and another where the crankshaft-driven piston was limited in its maximum velocity to that of the free piston's maximum velocity. The stroke in all cases was the same (9.40 cm). A connecting rod length to crank radius ratio of 2.5:1 was assumed.

Figures 5 and 6 illustrated the free and crankshaft-driven piston's motion where the oscillating frequency is 50 s^{-1}. Figures 19 and 20 compare the piston dynamics where the maximum piston velocity is limited to 2000 cm/s. Again, position vs. time and velocity vs. position curves are offered for comparison.

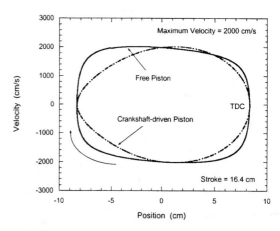

Figure 20. Piston velocity vs. piston position.

As noted earlier, the free piston motion is very different from the crankshaft-driven piston motion. These differences are even more pronounced for the case where the piston's maximum velocity is restricted. (This maximum velocity constraint is generally important in conventional crankshaft-driven engines, and thus the comparison in the simulated results for this condition is expected to be more important.)

The results from these comparative simulations indicated that there was little variation in the cycle thermal efficiency and NO_x emissions performance for the two pistons oscillating at the same frequency. However, the crankshaft-driven piston that was limited in its maximum velocity affected the cycle thermal efficiency (-10%) by increasing the heat losses at TDC, and the NO_x emissions (+20%) by holding the combustion gases at the high temperatures for a longer period of time.

DISCUSSION

The free piston engine model has been used to investigate the engine-generator's steady-state operating characteristics over a wide range of operating conditions. A detailed investigation of the engine's thermodynamic and NO_x emissions performance, where hydrogen was used as the fuel, has been performed. The results of the cycle simulations, however, must be considered with great care. Limitations of the model include the zero-dimensional characterization of the engine processes, as well as the applicability of the detailed chemical kinetics model to describe the combustion process, and the use of the empirical heat transfer, scavenging, and friction models employed.

However, many of these simplifications seem reasonable. The engine is expected to operate (at least through most of the cycle) on a well-mixed basis, and combustion is achieved through homogeneous charge compression ignition. Heat transfer has been shown to be small in the free piston configuration (1), and the friction model has been used to successfully replicate experimental data (13).

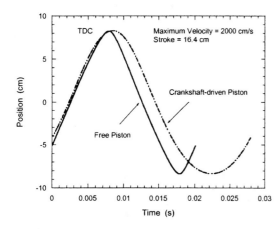

Figure 19. Piston position vs. time.

The engine simulation results are compared to the performance of conventional spark-ignition and compression-ignition IC engines next. Achieving almost constant-volume combustion, and capable of operation on very lean mixtures, while utilizing very high compression ratios, the free piston engine concept presented here seems to meet the requirements for optimized Otto cycle operation.

Based on the model's simplifications, the simulations indicate that this engine is capable of achieving indicated ideal cycle thermal efficiencies near 65% (not including short-circuiting losses) (56% confirmed through RCEM experiments (1)), while complying with the proposed SULEV NO_x emissions standards. This is a substantial gain in cycle thermal efficiency relative to modified Diesel and spark-ignition engines operating on hydrogen, where indicated efficiencies of 40% and 43%, respectively, have been reported (21,20). Further, the possibility of stringent NO_x emissions compliance without the need for exhaust gas after-treatment seems to be an inherent capability of the free piston engine operating on very lean mixtures (11). With the advantage of high compression ratio while eliminating the need for a massive or mechanically complex design, the free piston engine poses as a significant advancement in IC engine technology.

CONCLUSIONS

The steady-state operation of a free piston internal combustion engine operating on homogeneous charge compression ignition combustion has been analyzed using a zero-dimensional, thermodynamic model with detailed chemical kinetics, and empirical scavenging, heat transfer, and friction component models. The results of this analysis indicate that:

1. The operating compression ratio in this engine is variable and particularly dependent on the operating conditions of the engine (e.g. equivalence ratio, scavenging efficiency, intake temperature, etc.). It seems to be possible to control the operating compression ratio, and thus maximize the achievable thermal efficiency, by adjusting the engine's input parameters.

2. The HCCI combustion process is calculated to be very rapid, approaching almost constant-volume. This result agrees well with the experimental results from (1). This feature of the combustion system would essentially eliminate the issue of burn duration as a restriction to achieving high efficiency due to high compression ratios.

3. The NO_x emissions levels are predicted to be significantly reduced over conventional IC engine technologies since very low equivalence ratio homogeneous-charge combustion is possible. With the reduced engine friction for this one moving part configuration, such low equivalence ratio operation may be an attractive option. The possibility of meeting stringent NO_x emissions standards without the need for exhaust gas after-treatment may enable the use of oxidation catalysts for HC and CO emissions from hydrocarbon fueled operation.

4. The free piston engine's successful operation will greatly depend on the scavenging process. As suggested through the cycle analyses, the two-stroke cycle gas exchange process can significantly affect the engine's cycle efficiency by limiting the achievable operating compression ratio, as well as by short-circuiting significant amounts of the unburned fuel through the exhaust port. Losses introduced by the scavenging system for this two-stroke cycle engine must be controlled so that the advances achieved through the engine's unique free piston design are not negated. This must be addressed in the engine's further development.

5. The piston dynamics will be critical in ensuring complete combustion through the HCCI process, and in minimizing NO_x emissions. Variations in the piston motion near TDC could lead to poor combustion (under-compression) or significant NO_x emissions (over-compression).

ACKNOWLEDGEMENTS

The authors gratefully acknowledge the assistance of Dr. Charles Mitchell of Colorado State University and Dr. William Pitz of Lawrence Livermore National Laboratory for their help in the development and implementation of the free piston engine code. Dr. Pitz provided a great deal of guidance with the HCT software and the hydrogen combustion kinetics. Lila Chase of Lawrence Livermore National Laboratory supplied a copy of HCT, and provided technical assistance with this code.

The work presented here is from a thesis submitted to the Academic Faculty of Colorado State University in partial fulfillment of the requirements for the degree of Master of Science, and was supported by the U.S. Department of Energy, Office of Solar Thermal, Biomass Power and Hydrogen Technologies, and the Laboratory Directed Research and Development Program at Sandia National Laboratories.

REFERENCES

1. Van Blarigan, P., Paradiso, N. and Goldsborough, S., "Homogeneous Charge Compression Ignition with a Free Piston: A New Approach to Ideal Otto Cycle Performance," SAE Paper 982484, 1998.

2. Alperstein, M., Swim, W. B. and Schweitzer, P. H., "Fumigation Kills Smoke - Improves Diesel Performance," SAE Transactions, **66**, 574-588, 1958.

3. Onishi, S., Jo, S. H., Shoda, K., Jo, P. D. and Kato, S., "Active Thermo-Atmospheric Combustion (ATAC) - A New Combustion Process for Internal Combustion Engines," SAE Paper 790501, 1979.

4. Najt, P. M. and Foster, D. E., "Compression-Ignited Homogeneous Charge Combustion," SAE Paper 830264, 1983.

5. Thring, R. H., "Homogeneous-Charge Compression-Ignition Engines," SAE Paper 892068, 1989.

6. Christensen, M., Johansson, B. and Einewall, P., "Homogeneous Charge Compression Ignition (HCCI) Using Isooctane, Ethanol and Natural Gas - A Comparison with Spark Ignition Operation," SAE Paper 972874, 1997.

7. Christensen, M., Johansson, B., Amneus, P. and Mauss, F., "Supercharged Homogeneous Charge Compression Ignition," SAE Paper 980787, 1998.

8. Klotsch, P., "Ford Free-Piston Engine Development," SAE Paper 590045, 1959.

9. Baruah, P. C., "A Free-Piston Engine Hydraulic Pump for an Automotive Propulsion System," SAE Paper 880658, 1988.

10. Somhorst, J. H. E. and Achten, P. A. J., "The Combustion Process in a DI Diesel Hydraulic Free Piston Engine," SAE 960032, 1996.

11. Van Blarigan, P., "Advanced Hydrogen Fueled Internal Combustion Engines," Energy & Fuels, **12**, 72-77, 1998.

12. Caris, D. F. and Nelson, E. E., "A New Look at High Compression Engines," SAE Transactions, **67**, 112-124, 1959.

13. Goldsborough, S. S., "A Numerical Investigation of a Two-Stroke Cycle, Hydrogen-Fueled, Free Piston Internal Combustion Engine," M.S. Thesis, Colorado State University, Summer 1998.

14. Woschni, G., "A Universally Applicable Equation for the Instantaneous Heat Transfer Coefficient in the Internal Combustion Engine," SAE Paper 670931, 1967.

15. Annand, W. J. D., "Compressible Flow Through Square-Edged Orifices: An Empirical Approximation for Computer Calculations," International Journal of Mechanical Engineering Science, **8**, 448-449, 1966.

16. Sher, E. and Harari, R., "A Simple and Realistic Model for the Scavenging Process in a Crankcase-Scavenged Two-Stroke Cycle Engine," Proceedings of the Institution of Mechanical Engineers, **205**, 129-137, 1991.

17. Blair, G. P., "The Correlation of Theory and Experiment for Scavenging Flow in Two-Stroke Cycle Engines," SAE Paper 881265, 1988.

18. Lund, C. M., "HCT: A General Computer Program for Calculating Time-Dependent Phenomena Involving One-Dimensional Hydrodynamics, Transport, and Detailed Chemical Kinetics," Lawrence Livermore National Laboratory Report UCRL-52504, rev. 1995.

19. Marinov, N. M., Westbrook, C. K. and Pitz, W. J., "Detailed and Global Chemical Kinetics Model for Hydrogen," in Transport Phenomena in Combustion, Chan, S. H., ed., Taylor and Francis: Washington, DC, **1**, 118-141, 1996.

20. Van Blarigan, P., "Development of a Hydrogen Fueled Internal Combustion Engine Designed for Single Speed/Power Operation," SAE Paper 961690, 1996.

21. Homan, H. S., Reynolds, R. K., DeBoar, P. C. T. and McLean, W. J., "Hydrogen-Fueled Diesel Engine without Timed Ignition," International Journal of Hydrogen Energy, **4**, 315-325.

NOMENCLATURE

A	preexponential constant
A_{eff}	effective port area
A_p	piston crown cross sectional area
a_p	piston acceleration
c	concentration
C_D	discharge coefficient
c_p	specific heat at constant pressure
E_a	activation energy
F_{frict}	frictional force
F_{mag}	electromagnetic force
f_1, f_2	friction constants
ΔH°	heat of formation at temperature, T°
$\dot{H}$	enthalpy flow
k	rate constant
M	alternator coil strength
m	non-Arrhenius exponent
m_p	piston mass
MW	molecular weight
N_e	moles in exhaust stream over one time step
$\dot{N}$	total molar flow rate
n	moles of specific specie
P	pressure
P_o	stagnation pressure
PR	pressure ratio, P/P_o
PR_c	critical pressure ratio, $\left(\dfrac{2}{\gamma+1}\right)^{\gamma/(\gamma-1)}$
Q	heat transfer
$\mathfrak{R}$	rate of specie production
R_u	universal gas constant
s_1, s_2	scavenging parameters
T	temperature

T_o	stagnation temperature
U	internal energy
u^*	molar specific internal energy
$\not{V}$	volume
v_p	piston velocity
x	molar fraction
$\dfrac{d}{dt}$	full derivative with respect to time
$\dfrac{\partial}{\partial t}$	partial derivative with respect to time
γ	ratio of specific heats
H_{sc}	instantaneous scavenging efficiency
$\dot{\Lambda}_{de}$	instantaneous rate of change in delivery ratio
$\dot{\Lambda}_{de,max}$	maximum rate of change in delivery ratio
ξ	exhaust gas purity

Subscripts

b , f	burned gas zone, fresh charge zone
i , e	incoming, exiting
j	specific specie
n	specific reaction

2002-01-0240

Ford P2000 Hydrogen Engine Design and Vehicle Development Program

William F. Stockhausen, Robert J. Natkin, Daniel M. Kabat, Lowell Reams, Xiaoguo Tang, Siamak Hashemi, Steven J. Szwabowski and Vance P. Zanardelli
Ford Motor Co.

ABSTRACT

In late 1997 Ford Motor Company Scientific Research Laboratory started the project to design and develop a practical, low-cost hydrogen fueled internal combustion engine (H_2ICE) vehicle. This type of vehicle could serve as an interim step to drive the development of the hydrogen infrastructure before the widespread use of fuel cell vehicles. This paper will discuss the design and development approach and results for a dedicated engine optimized for operation on hydrogen, the unique and custom instrumentation necessary when working with hydrogen, the engine dynamometer development program, the unique triple-redundant vehicle safety system, and the final implementation into the Ford P2000 experimental vehicle.

INTRODUCTION

Hydrogen has been referred to as the ultimate fuel. It is the most plentiful element in the universe, and on a basis of carbon atoms per fuel molecule, it is at the cleanest end of the fuel spectrum with none. This makes hydrogen the fuel of most promise because it has the potential of producing only water when reacted with the oxygen from air. When produced renewably it has the potential of eliminating all carbon based emissions from the fuel production/use cycle. All of this is fine in principle, but tuning the technology to accomplish this objective is a challenge.

Ford Motor Company has previously published reports on the Fuel Cell Vehicle [1]. Other papers [2][3] concentrated on higher-level issues such as infrastructure and storage challenges, the best applications for specific alternative fuels (including electricity), and wells-to-wheels CO_2 emissions. Much work has been done over the last several decades [4 - 10] about converting ICEs to hydrogen; however, there is a paucity of information on attempts to optimize an ICE and complete vehicle system for hydrogen and the

resulting data reporting the efficiency, FTP cycle emissions and other pertinent technical details.

This report and its accompanying reports [11,12,13] summarize the Ford Motor Company Scientific Research Laboratory's effort to design, develop and implement a hydrogen powered internal combustion engine (H_2ICE) vehicle. The objective of the program, which started in late 1997, was to conduct the basic engine related research necessary to implement a safe, practical, low-cost, dedicated H_2ICE vehicle that delivered carry-over function to today's gasoline powered vehicles, minimized engine-out emissions, and improved fuel efficiency.

ENGINE DESIGN

Early in the program, a thorough literature search was conducted to obtain information that would allow the initial engine design to deliver maximum efficiency and output. Hydrogen is a fuel with unique properties such as extreme lean flammability (4%), which can be used to advantage for lean combustion extension, and high octane value which can be used to advantage with increased compression ratio (CR). Over a fairly wide range of fuel/air ratios it also has low ignition energy requirement, which leads to pre-ignition tendencies [11]. This information, coupled with subsequent experience, resulted in gasoline engine upgrades to optimize for the use of hydrogen fuel in the following four areas:

- Oil control - for the minimization of hot oil/oil ash residue remaining in the combustion chamber from the previously fired cycle, as well as for the reduction of carbon-based emissions.
- Cooling – preventing combustion chamber hot spots.
- Ignition – prevention of premature ignition due to residual voltage in the spark gap.
- Port fuel injectors (PFI) - upgraded for the dryness of hydrogen fuel.

The engine selected for the program was the Ford 2.0 liter Zetec DOHC, 16-valve, four-cylinder engine with an 84.8mm bore and 88mm stroke. The production engine is produced both in Europe and Mexico for the Ford Focus. The Zetec engine uses an aluminum cylinder head with a cast iron block. The European Zetec cylinder block was chosen because of its centralized water pump configuration and its capability to incorporate piston-cooling oil squirters, which testing later determined were not needed.

OIL CONTROL

Because the source of engine-out carbon-based emissions (and possibly preignition as well) in the H2ICE is the consumption of lubricating oil, special efforts were taken in the engine design to minimize oil consumption. Changes made to the engine design for reduced oil consumption and also hot spots were:

- Deck plate boring and honing of the cylinder block improved bore cylindricity by a factor of four relative to production levels. The cylinder block final bore size was 85.215 ± .005 mm diameter which was 0.4mm oversize from production allowing use of available oversize piston rings. A special boring and honing procedure was used to give a final peak-honed surface finish of 0.2 – 0.3μm Ra for quick ring seating and reduced oil volume retention. Studs with nuts and washers were used for fastening the cylinder head to the block to achieve the most consistent torque/clamping loads. A special lubricant and nut torque sequence was developed to give a clamping load of 54,000 Newtons using the Zetec multi-layer steel head gasket.
- High silicon hypereutectic cast aluminum pistons for improved strength and low thermal expansion. The piston skirt diameter was held to 85.185 ± .005 mm.
- Tight piston skirt to wall clearance (0 - 10μm) using a 15μm per side solid lubricant skirt coating
- Minimum end gaps for piston rings (0.25 mm-top, 0.5 mm-2nd, 0.4 mm-oil).
- Low leakage valve stem seals, less than .001 gm/hr [14].

COOLING AND CYLINDER HEAD DESIGN

A special high-output cylinder head was employed that was developed in an earlier project and deemed beneficial for hydrogen use because it provided the following features:

- Improved intake and exhaust port flow for increased volumetric efficiency. It was anticipated that with the fast burn rate of hydrogen at richer fuel/air ratios, the need for high charge motion was reduced [11].
- Cross-flow coolant flow path that improved cooling around the spark plug and exhaust seats [14].
- A compact combustion chamber design that gave the capability of high CR. A CR as high as 18:1 could be obtained using a 5mm pop-up piston dome and machining 1.0 mm from the cylinder head face. The first build was at 14.5:1, which turned out to be a good balance for heat losses for this size cylinder [15].
- Two exhaust valves on a four valve engine helped to reduce valve temperature versus a single larger valve. Sodium cooled exhaust valves were added for further temperature reduction [8][14].
- Valve seat inserts were upgraded to cast tool steel and a hard surface coating was applied to the valve seat area to minimize valve seat recession. This was based on Ford's previous experience with dry gaseous fuels such as CNG [16]

IGNITION UPGRADES

Figure 1 shows the three types of spark plug tips tested. Platinum was not used since platinum was thought to be a possible ignition catalyst for hydrogen [17]. The side gap design had low thermal mass, delivered the best results with efficient electrical energy delivery to the gap, and had stable initial flame kernel development (0-2% burn rate). The production Zetec double-fire coil was replaced with prototype separate coil-on-plug units used on an Aston Martin V12 engine. The coil secondary was grounded to prevent any electrical charge residual in the electrodes between firings [18].

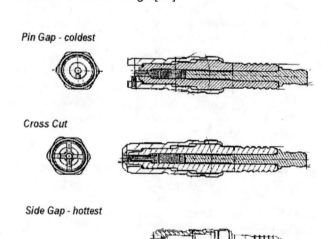

Pin Gap - coldest

Cross Cut

Side Gap - hottest

Figure 1: Hydrogen engine spark plugs

OTHER HARDWARE CHANGES

- The base gasoline piston pin and connecting rod were upgraded for 90 bar peak cylinder pressure. This was done after early dynamometer testing with hydrogen fuel indicated significantly higher in-cylinder pressure when pre-ignition was encountered. The piston pin size was increased from ø20.6 mm to ø23 mm using shortened "Burgess-Norton" 5.0L V8 piston pins. The connecting rod was changed to a H-beam 4140 steel billet rod from "Arrow Precision", an after-market racing vendor in England. The Mahle pistons required machining revisions for the larger piston pins. The crankshaft was considered adequate for 90 bar.

- Several piston ring land failures occurred in early dynamometer testing that were caused by severely high (>130 bar) cylinder pressure from pre-ignition. The initial piston design with a top land of 6 mm was then modified by moving the top compression ring up by 2 mm and thus increasing the second ring land width by 2 mm. It was calculated that the 2 mm thicker 2nd ring land reduced the stress by 47%. It was recognized that this change would not eliminate failure at these high pressures, but could provide enough improvement in fatigue strength to allow engine-mapping sweeps at maximum load.

- A separate fuel injector adapter housing spacer that bolted to the cylinder head was used for evaluation of different fuel injectors and also had the capability of two injectors per cylinder. Figure 2 shows the fuel injector adapter housing with two injectors installed per cylinder.

- Addition of a charge motion control valve (CMCV) upstream of the intake port for increased charge motion if needed at extreme lean conditions such as idle. Figure 2 shows the CMCV unit with a 1/4 area opening in the valve plate.

- For the capability of mostly unthrottled, lean operations in the vehicle an electronic throttle was added.
  A brushless alternator and spark arrested marine starter were added for the vehicle application. The brushless alternator was a water-cooled unit that required fabrication of a special housing. The cabin heater coolant circuit was routed to the alternator ahead of the heater.

FUEL INJECTOR DEVELOPMENT

After reviewing the literature and balancing the advantages/disadvantages of the four basic types of known hydrogen fueling systems (carburetor, throttle body-injection, intake port fuel injection (PFI) and direct in-cylinder injection (DI) it was decided that PFI offered the most advantage for excellent cylinder-to-cylinder fuel distribution, minimum fuel charge in the intake manifold (to minimize the effects of any intake backflash), and low cost and complexity. Initially, the fuel injector spacer was designed to accommodate two of the production-type compressed natural gas (CNG) injectors [16] per cylinder. The purpose of double injectors was to be able to inject stoichiometric mixtures within the duration of the intake stroke at high speeds (6000 rpm) which reportedly mitigated pre-ignition problems [7]. After just thirty hours of engine dynamometer testing with the CNG injector, pintle sticking/seizure problems were encountered that required the initiation of a parallel injector durability development program [19].

While electronic, timed PFI injectors have been developed for gasoline, alcohol, and CNG none have been developed for hydrogen. Due to the very low density of hydrogen much larger volumes of fuel must be injected per stroke, and of the aforementioned fuels hydrogen has the least lubricity (some suspended oil aerosol from CNG industrial compressors aids CNG injector durability whereas hydrogen compressors are typically more tightly sealed).

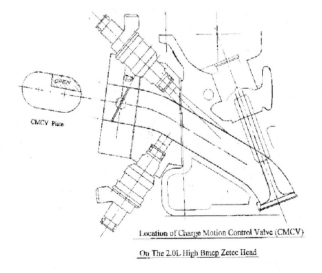

Figure 2: Dual injector and charge motion control valve (CMCV)

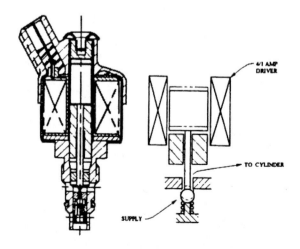

Figure 3: Candidate injector 1

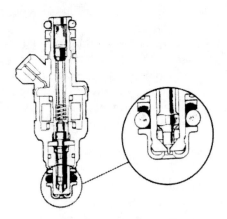

Figure 4: Candidate injector 2

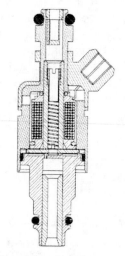

Figure 5: Candidate injector 3

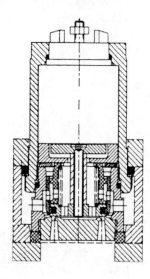

Figure 6: Candidate injector 4

Because there was no known previous hydrogen PFI injector durability work, Ford and the University of California – Riverside, College of Engineering, College of Environmental Research and Technology (UC-R, CE-

CERT) partnered in evaluating injectors from four manufacturers, some of which were treated with hard surface finishes and solid-film lubricant coatings. A hydrogen-fed, computer-controlled injector durability test stand was set up at UC-R, CE-CERT to operate four injectors simultaneously on a 24 hour/day, 7 day/week schedule. Pulse widths and frequency cycles were specified to simulate actual engine operating conditions. Pressure transducers and flow meters were employed to measure flows as well as shot-to-shot repeatability. Testing proceeded for more than 800 hours (64 million cycles - equivalent to about 72,000 km of vehicle travel) or until failure.

Figures 3, 4, 5, and 6 show cross sections of the candidate injectors, several versions of which completed the test requirements (candidate #3 passed without modifications - see [19] for more details of durability tests).

ENGINE DYNAMOMETER INSTRUMENTATION AND FUEL SUPPLY

Instrumentation and safety systems as well as engine combustion evaluation procedures had to be modified from conventional gasoline procedures or derived specifically for hydrogen.

INSTRUMENTATION AND FUEL SUPPLY

Because hydrogen produces no fuel-related carbon emissions, unique and custom instrumentation (and suitable data reduction algorithms) had to be investigated and tested on the engine to assure accurate and repeatable measurements of:

- Mass measured gaseous hydrogen fuel flow (coriolis type metering system with pneumatically actuated solenoid valves for range selection and nitrogen purge functions, constructed to meet very rigid Class 1, Division 1, Group B fire safety standards),
- Fuel/air (F/A) ratio (mass flow measured, lab calibrated UEGO sensor, intake and exhaust hygrometer measurements, moisture-corrected O_2 emission analyzer),
- Unburned exhaust hydrogen (mass spectrometer tuned for hydrogen), and
- Calculations based on total exhaust chemistry including water, oxygen, nitrogen, NOx and unburned hydrogen.

Initially the engine was fueled by commercial type hydrogen cylinders that required frequent replacement. In August 1999, a liquid and high-pressure gaseous hydrogen vehicle refueling station was installed in the

Research and Engineering center - the first North American, OEM car manufacturer installation. A stainless steel supply line was fed to the engine dynamometer test cells from the station alleviating the time consumed with bottle replacement.

Efforts to quantify preignition on the engine dynamometer resulted in the development of an automatic, peak cylinder pressure threshold triggered, fuel shutdown algorithm that prevented the destruction of engine hardware. Safety was of paramount importance throughout the project with much effort being put forth to assure a safe development program both in the dynamometer and vehicle. Operational safety of the dynamometer was enhanced by the usage of a hydrogen-flame sensing, infrared video camera targeted on the engine during operation. More details of the test cell instrumentation and fuel supply are in reference [12].

ENGINE DYNAMOMETER DEVELOPMENT

The engine dynamometer development program identified many important combustion characteristics of hydrogen including:

- Hydrogen as a fuel for a H_2ICE has unique properties such as extremely low carbon related emissions, very broad flammability limits with a fast burn rate, and high octane number that can all be used to improve the emissions and fuel economy of IC engines.

- Pre-ignition (unintentional ignition of the fresh charge after the intake valve closes) imposes a limit on maximum torque output that is primarily a function of maximum equivalence ratio (Φ) capability, the magnitude of which depends on CR, spark timing, charge density and engine speed (Φ = fuel/air observed / fuel/air stoichiometric).

- Relative to gasoline, pre-ignition reduced torque output on the 14.5:1 CR, 2.0L Zetec engine by about 35% at low and mid speeds, and, because pre-ignition becomes more severe at higher speeds, reduced peak power by about 50%.

- The hydrogen engine will run unthrottled as lean as Φ = 0.12; however, best specific fuel consumption occurs at Φ = 0.25 so some throttling would be advantageous.

- NO_x emissions are primarily a function of equivalence ratio with the concentration less than 10ppm at $\Phi < 0.38$ and less than 100ppm at $\Phi < 0.5$. The concentration increases dramatically at $\Phi > 0.5$.

- For the 2.0L Zetec the indicated thermal efficiency peaks at about 52% and the brake thermal efficiency peaks at about 38%.

- Because hydrogen has a very broad flammability and burn rate range the H_2ICE engine can be controlled similar to today's production gasoline engine where fuel/air ratio is constant and desired torque is a function of both fuel and air flow, and also similar to a diesel where air is unthrottled and desired torque is proportional to fuel flow.

The results obtained represent the most complete multi-cylinder, production-feasible, H_2ICE data set known. Refer to [11] for more complete details.

VEHICLE SYSTEMS DEVELOPMENT

DESCRIPTION OF THE VEHICLE HARDWARE

The vehicle used for the hydrogen internal combustion engine demonstration project was the Ford P2000 vehicle, which is an all-aluminum vehicle developed for the PNGV program. As a platform it was also used for the corporate fuel cell and hybrid work and so provided a convenient means of comparing fuel economy and performance. The vehicle is similar to a Ford Contour but stretched to create the interior space of a Taurus at an approximately 1000 lb. weight saving. Figure 7 shows a photograph of the vehicle. The vehicle uses the production Zetec intake air and exhaust system except the catalyst brick was removed from the catalyst can. Due to unthrottled operation an electric vacuum pump with controller similar to the pump used on the Ford Electric Ranger Truck was packaged under the left front fender to ensure power brake functionality during unthrottled operation.

Figure 7: Ford P2000 Hydrogen IC engine vehicle

POSITIVE CRANKCASE VENTILATION (PCV) SYSTEM FOR WOT OPERATION

With the possibility of unthrottled engine operation resulting in high levels of manifold absolute pressure (MAP) similar to a diesel engine, it was necessary to revise the PCV system (Figure 8). A coalescing style air/oil separator was selected to minimize hydrocarbon emissions from blowby oil carry-over. The inlet side of the air/oil separator was connected to the outlet of the normal Zetec side-of-block baffle oil separator. Oil collected on the coalescing filter was gravity drained back to the oil pan. A one-way check valve in the drain line prevents oil from being pulling out of the engine oil pan in the event of a plugged coalescing filter. The outlet side of the coalescing filter is connected to the throat of a venturi placed upstream of the air cleaner. The venturi provides the necessary pressure drop for positive flow through the crankcase and the separator. A connection from the air cleaner to the cylinder head valve cover provides makeup air to the engine.

It was estimated from previous PCV work that the maximum blow-by flow for a 2.0L engine could be from 0.034 cubic meters/minute (new) to 0.056 cubic meters/minute (high mileage). Figure 8 shows a schematic of the PCV venturi system with an alternate flow path under the throttle plate for part-throttle operation.

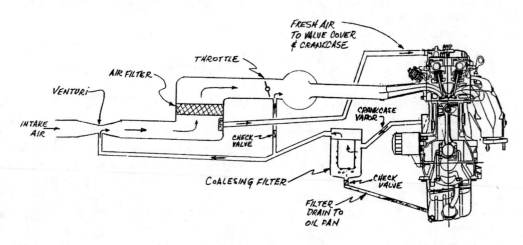

Figure 8: PCV system

HYDROGEN FUEL SYSTEM

Obtaining sufficient on-board mass storage of hydrogen with volume comparable to gasoline is a major challenge. Metal hydride systems tend to be heavy and require dynamic control for hydrogen release. Liquid cryogenic systems have been tried [20] and demonstrated but are not seen as a long term solution due to the additional energy requirement for liquifaction. It was decided for this program to utilize high-pressure gaseous tank storage as the easiest to use with minimum handling complexity. The vehicle employed two carbon fiber reinforced aluminum tanks with a total volume of 87 liters, which at 250 bar (3600 psi) held 1.5 kg of fuel. Conventionally available pressure capability has since increased to 345 bar (5000 psi), which is a 39% improvement in density. On-board storage was not a prime objective of the program but a parallel effort was on-going to stay abreast of developing technologies and to encourage outside efforts to develop higher density storage systems. Additional details of the on-board fuel storage and supply system are in a related paper [13].

HYDROGEN TRIPLE-REDUNDANT SAFETY SYSTEM

Safety was of paramount importance throughout the project with an unprecedented effort put forth to assure a safe development program both in the dynamometer and vehicle. Operational safety of the dynamometer was enhanced by the usage of a hydrogen flame sensing infrared video camera as was mentioned earlier.

For the vehicle a triple-redundant safety system was developed, which includes under hood, trunk and passenger compartments. This system consists of active and passive low-pressure ventilation, as well as a hydrogen sensing network with four sensors distributed throughout the vehicle. Figure 9 shows the locations of the vehicle hydrogen ventilation system components and the safety sensors. The system also includes an instrument panel display and automated emergency-operating sunroof. All vehicular electric motors were specified to be brushless or sealed to eliminate potential ignition sources. This vehicle safety and ventilation system, albeit in excess of production requirements, is the most advanced of its kind known.

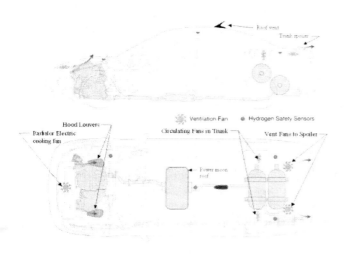

Figure 9: Vehicle safety and ventilation system

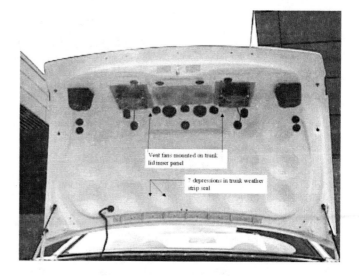

Figure 11: Trunk exhaust fans

Passive and active ventilation systems function under hood, in the passenger cabin, and in the trunk. The passive system includes two Jaguar XK8 hood louvers located above the engine as shown in Figure 10. Also included in the passive system are several small depressions in the front part of the trunk weather strip seal allowing natural venting of the trunk (Figure 11).

The active system includes four small brushless ventilation fans in the trunk (Figures 11, 12), a power operated moon roof (Figure 13), and the engine radiator cooling fan. Two of the trunk fans provide mixing circulation (Figure 12) and two fans vent the trunk through an Escort ZX2 trunk lid mounted hollow spoiler with holes drilled on the underside of the spoiler (Figure 14).

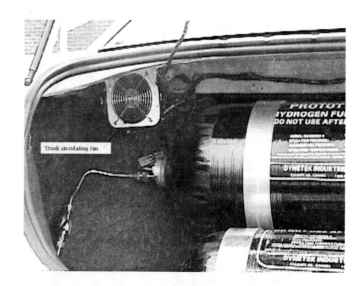

Figure12: Vehicle trunk

Figure 10: Hood louvers for passive ventilation

Figure 13: Vehicle's moon roof

Figure 14: Vehicle rear spoiler

Also as part of the safety system, four hydrogen sensors are employed - two located in the trunk, one in the passenger compartment incorporated into the interior dome light (Figure 15), and one under the hood. The lower flammability limit (LFL) of hydrogen is 40,000 ppm or 4% (properties of hydrogen and vehicle considerations are reviewed in [21][22]). The hydrogen sensor safety system is capable of programming three levels of hydrogen concentration warning. However, for the vehicle, only the first level set at 15% of LFL or 6,000 ppm was used. This first level alerts the driver, activates the ventilation system, closes the fuel solenoids and shuts down the engine. The four sensor monitors were packaged under the driver seat.

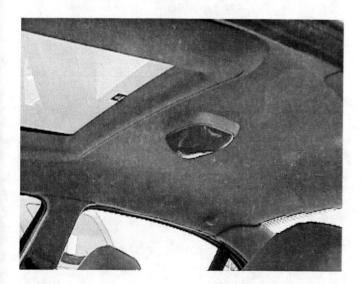

Figure 15: Vehicle interior hydrogen detector location

VEHICLE ENGINE CONTROLS

An innovative vehicle engine controls system had to be formulated for the hydrogen vehicle to optimize fuel economy and emissions [11][13][23]. Two stages of control strategy development were undertaken. The second strategy incorporated a constant F/A ratio approach except for enrichment for heavy load operation. Creative methods were incorporated to avoid pre-ignition and backflash on transients and for safe engine startup and shutdown. Due to time constraints in the project a final unthrottled strategy was not fully developed. This final strategy was an unthrottled, lean strategy where fuel flow as a function of torque demand was devised. This diesel-like strategy would still have some throttling at idle for best fuel efficiency [13].

The vehicle has undergone FTP fuel economy and emissions testing and, with the second strategy and limited calibration development, has shown a metro-cycle fuel economy improvement of 14% relative to gasoline (on an energy equivalent basis). Due to the stringent oil control and lean operation, feedgas exhaust emission of 0.0076/0.008/0.74/1.4 grams/mile, for HC/CO/NOx/CO_2 respectively, were less than SULEV requirements for all emissions except NOx. Further strategy and calibration development and incorporation of an exhaust after treatment system should achieve SULEV for NOx as well.

Acceleration performance for 0-96 km/hr (0 - 60 mph) was 16.8 seconds. At equal performance to gasoline the metro-cycle economy improvement would be reduced to about 11%.

CONCLUSIONS

In keeping with its proactive position in the area of alternative fuels, Ford Motor Company has designed and developed the first North American OEM, production feasible, hydrogen-fueled ICE vehicle. This vehicle, the H_2ICE P2000, incorporates a custom hydrogen fuel supply system, an integrated triple-redundant safety system, and a dedicated, optimized engine for the use of hydrogen fuel. For a near-term approach to near zero regulated as well as CO_2 emissions, the P2000 demonstrates a practical, low-cost, lightweight, compact-package with significant improvement in fuel efficiency. It could serve as a stepping-stone in driving the establishment of a hydrogen infrastructure, which would be required for widespread introduction of fuel cell vehicles.

Future development efforts could comprise additional control strategy refinement on the P2000 that should reduce engine out NOx emissions. Other recommended future work include:

- Boosting the air charge for torque and power recovery [17, 23] should be investigated to not only improve H_2ICE power density, but also to deliver further efficiency improvement, hopefully without increasing NO$_x$ emissions.

- Depending on the control strategy used and the power/weight ratio of the H₂ICE vehicle, an EGR strategy and/or an exhaust after treatment system may be required to meet future stringent NOx emission standards.

ACKNOWLEDGMENTS

The authors wish to acknowledge Rod Tabaczynski and Michael Pulick for their support and encouragement during the duration of this project. We also wish to acknowledge the following hydrogen experts that have been helpful in getting started: Dr. Walter Peschka, Dr. Furuhama and staff at Musashi Institute, Dr. Michael Swain at University of Miami, and Frank Lynch at Hydrogen Components, Inc. We would also like to acknowledge the help of our partners Joseph Norbeck and Jim Heffel at University of California-Riverside.

REFERENCES

1. Adams, J. A., et al, The Development of Ford's P2000 Fuel Cell Vehicle, SAE 2000-01-1061.
2. Kukkonen, C., Hydrogen as an Alternative Automotive Fuel, SAE 810349.
3. Kukkonen, C. & Shelef, M., Hydrogen as an Alternative Automotive Fuel: 1993 Update, SAE 940766.
4. Norbeck, J. M., et al, Hydrogen Fuel for Surface Transportation, SAE, Inc. 1996.
5. Furuhama, S., et al, Development of a Liquid Hydrogen Car, Int. J. Hydrogen Energy, Vol. 3, pp61, 1978.
6. Oehmichen, M., Wasserstoff als Motortreibmittel, VDI Heft 68, pp.1-30, 1942.
7. Das, L. M., Exhaust Emission Characterization of Hydrogen-operated Engine System: Nature of Pollutants and Their Control Techniques, Int. J. Hydrogen Energy, Vol. 16, No. 11, pp. 765-775, 1991.
8. Peschka, W. & Escher, W., Germany's Contribution to the Demonstrated Technical Feasibility of the Liquid-Hydrogen Fueled Passenger Automobile, SAE 931812.
9. Swain, M. R., et al, Considerations in the Design of an Inexpensive Hydrogen-Fueled Engine, SAE 881630.
10. Mamiya, K. & Yagi, T., Emission Characteristics of a Hydrogen Fueled Vehicle, Presentation from Mazda Motor Corporation to SAE, Washington, D.C., May, 1993.
11. Tang, X., et al, Ford P2000 Hydrogen Engine Dynamometer Development, SAE 2002-01-0242.
12. Natkin, R. J., et al, Ford Hydrogen Engine Laboratory Testing Facility, SAE 2002-01-0241.
13. Szwabowski, S. J., et al, Ford's Hydrogen IC Engine Powered Vehicle, SAE 2002-01-0243.
14. Swain, M. R., et al, Design and Testing of a Dedicated Hydrogen-Fueled Engine, SAE 961077.
15. Muranaka, S., et al, Factors Limiting the Improvement in Thermal Efficiency of S.I. Engine at Higher CR, SAE 870548.
16. Lapetz, J., et al, Ford's 1996 Crown Victoria Dedicated Natural Gas Vehicle, SAE 952743.
17. Furuhama, S. & Fukuma, T., High Output Power Hydrogen Engine with High Pressure Fuel Injection, Hot Surface Ignition and Turbo-charging, Int. J. Hydrogen Energy, Vol. 11, No. 6, pp. 399-407, 1986.
18. Kondo, T., et al, A Study on the Mechanism of Backfire in External Mixture Formation Hydrogen Engines, SAE 971704.
19. Kabat, D. M. & Heffel, J. W., Durability Implications of Neat Hydrogen under Sonic Flow Conditions on Pulse-Width Modulated Injectors, Int. J. Hydrogen Energy, 2001
20. Wolf, J. F., et al, BMW's Energy Strategy – Promoting the Technical and Political Implementation, SAE 2000-01-1324.
21. Hansel, J. G., et al, Safety Considerations in the Design of Hydrogen-Powered Vehicles, Int. J. Hydrogen Energy, Vol. 18, No. 9, pp.783-790, 1993.
22. DeLuchi, M., Hydrogen Vehicles: An Evaluation of Fuel Storage, Performance, Safety, Environmental Impacts, and Cost, Int. J. Hydrogen Energy, Vol. 14, pp. 81-130, 1989.
23. Withalm, G. & Gelse, W., The Mercedes-Benz Hydrogen Engine for Application in a Fleet Vehicle, CERI World Hydrogen Energy Conference, Vienna.

2002-01-0373

Application of Advanced Simulation Methods and Their Combination with Experiments to Modeling of Hydrogen Fueled Engine Emission Potentials

Miloš Polášek, Jan Macek, Michal Takáts and Oldřich Vítek
Czech Technical University in Prague, Josef Bozek Research Center

ABSTRACT

The paper deals with an application of advanced simulation methods to modeling of hydrogen fueled engines. Two models have been applied – 0-D algorithm and CFD. The 0-D model has been based on GT-Power code. The CFD model has been based on Advanced Multizone Eulerian Model representing general method of finite volume. The influence of main engine parameters, e.g. air excess, spark timing, compression ratio, on NO_x formation and engine efficiency has been investigated. Both models have been calibrated with experimental data. Examples of results and comparison with experiments are shown. The means of reducing NO_x formation are discussed.

INTRODUCTION

As reserves of fossil fuels have decreased, alternative fuels as well as alternative power sources have become more and more important. It is necessary to find such fuel or power source, which is able to meet all current and future requirements concerning friendliness to the environment and reasonable overall net efficiency of energy transformation. Recently, a lot of work has been done on development of fuel cells (FC). The fuel cells represent an alternative power source using hydrogen as the primary source of energy. Unfortunately, the time when the fuel cells will be fully prepared and competitive with current internal combustion (IC) engines seems to be rather far-off. Moreover, specific energy referenced to weight (or volume) and costs are still inconvenient compared to those of IC engines. Therefore, it is necessary to fill the gap until they are widely applicable and to accelerate infrastructure building in the meantime.

The fueling of classical IC engines with hydrogen represents the other possibility of hydrogen utilization as the alternative renewable fuel. It is a very convenient fuel for reciprocating engines, which is able to fulfill all future requirements concerning emission formation and engine efficiency. It has strong advantages since it can be used in the classical IC engine without considerable change in the basic design concept. Nevertheless, it seems that the hydrogen engine should be completely re-designed concerning fuel/air mixing, combustion, exhaust gas aftertreatment and turbo/super charging if the potentials are to be fully utilized. The problems of fuel storage and availability, i.e. necessary infrastructure, are not discussed in the paper. The change of fuel via ICE → hydrogen ICE → hydrogen FC seems to be competitive to the currently used ICE → reformer methanol/gasoline FC → hydrogen FC. In the latter pattern many advantages of FC, namely high efficiency and no CO_2 emission, are lost due to reforming of significant carbon content fuels such as methanol or hydrocarbons with unused chemical energy of carbon oxidation.

This paper is focused on the application of numerical simulation tools of different levels to modeling of in-cylinder phenomena of dedicated hydrogen fueled IC engines and comparing them with experiments.

PROBLEM DESCRIPTION AND MAIN AIMS

The main topic of the work is the general simulation of thermodynamic cycle of a hydrogen engine and detailed description of in-cylinder phenomena. The influence of the main engine parameters, i.e. air excess, spark timing, etc., on engine efficiency and NO_x formation has been investigated. In the case of the hydrogen engine, combustion products are composed of a high amount of water vapor, oxygen if lean operation is assumed and

nitrogen oxides. This is the reason why strong emphasis has been put on the simulation capability of engine emission characteristics in models. In connection with this, the means of reducing NO$_x$ formation and autoignition have to be proposed and proved. In addition, lean operation limits have to be taken into consideration.

Most numerical experiments have been performed on 1-cylinder experimental engine with the variable compression ratio. This engine has been redesigned for operating with hydrogen fuel. Hydrogen is injected directly into the cylinder. Base engine parameters are shown in table 1. Integral parameters of the engine operating prevailingly with extremely lean mixture and in-cylinder pressure have been measured – [1]. All experiments have been done at the Technical University of Liberec.

bore	82.5 mm
stroke	114.3 mm
compression ratio	10, 13
engine speed	910 r.p.m.
aspiration	natural
mixture formation	direct injection, electromagnetic valve

Table 1: Main parameters of the investigated engine

The main aim of the presented work is the employment of advanced simulation tools to modeling of main engine parameters and NO$_x$ formation of a hydrogen fueled engine. Possibilities of NO$_x$ formation reduction have been proposed and proved. The simulations are aimed at providing a useful tool to help find compromise between specific power, engine efficiency, NO$_x$ production and autoignition. Two kinds of models of different levels have been employed to meet the aim successfully – a 0-D model and a CFD based multidimensional algorithm. In the case of the multidimensional method, formulation of appropriate CFD algorithm, sub-models and treatment of specific problems with numerical integration, i.e. stiffness, have been considered as another important aim of the work presented. Both models have been combined with experimental data not only to calibrate the achieved results but also to modify the sub-models. All computations and model formulations have been performed with the possibility of future application to simulation of a full-scale engine and extrapolation of results to other cases of interests where lack of experimental data can be found. The CFD algorithm should provide us with capabilities of modeling some specific features of hydrogen fueled engines, e.g. mixture formation, influence of cylinder wall temperature on autoignition, etc.

From the point of view of the application of multidimensional algorithms to simulation of in-cylinder phenomena of hydrogen fueled IC engines, the model has to treat correctly the following main features:

- IC engine specific Mach and Reynolds numbers. Phenomena of ICE are characterized by low Mach number and also low Reynolds number but with very strong influence of gas compressibility and significant influence of turbulence excited by special means.
- Moveable boundaries of a computational domain due to the piston movement. It causes changes of aspect ratio of the computational domain in a range typically from 1:1 up to 1:20 during a piston stroke.
- Combustion phenomena and its interaction with turbulence. The combustion model represents the crucial one of the whole algorithm.

All features summarized above call for formulation of proper physical models but it is necessary to take into account also properties of numerical methods used and possible problems with numerical integration, e.g. stiffness of the equation set, etc.

All results simulations in the paper assume the engine operating with homogeneous mixture in spite of the direct injection applied. To fulfill this assumption, the early injection timing, i.e. the injection during intake stroke or early stage of compression, was used in experiments to assure enough time for mixture formation.

THEORETICAL BACKGROUND

0-D/1-D SIMULATIONS - The 0-D simulations have been performed employing the GT-Power code – [2]. All simulations have been done on a 1-cylinder engine – see table 1. This model has mainly been used to simulate the influence of the main engine parameters on NO$_x$ emission production.

The engine model has been built to describe the experimental setup as precisely as possible. Combustion has been modeled in the standard way using a two-zone combustion model with prescribed rate-of-heat release. The main reason for this has been to verify the possibility of using standard Wiebe's law in connection with measured combustion duration for simulation of in-cylinder phenomena of hydrogen IC engines. This is important from the point of view of extending the use of the algorithm to simulation of full-scale engine and extrapolation of the results to other cases of interest where detailed experimental data is not available. NO$_x$ formation has been solved employing the extended Zeldovich reaction mechanism. An important problem in the case of the hydrogen engine and, in general, hydrogen combustion simulations is a correct description of mixture properties because of the high concentration of water vapor in combustion products, i.e. exhaust gases. It calls for involvement of suitable state equation of real gas. Nevertheless, perfect gas has been assumed in all simulations up to now in rate-of-heat-release evaluations and inverse simulations. The errors are somewhat compensated by this approach because the same assumptions and mixture models neglecting influence of real gas properties and water vapor are used in both the experimental data evaluation and simulations.

CFD SIMULATIONS – The CFD simulations are based on Advanced Multizone Eulerian Model (AMEM) developed by the authors – [3]. It represents general method of finite volumes taking into account moveable boundaries. This feature is very important and necessary in the case of in-cylinder phenomena modeling of ICE since the computational domain, i.e. combustion chamber, has moving boundaries as well as changing geometry due to piston and valve movement.

Basic conservation equations – AMEM is based on integral form of budgeting equation taking into account moveable boundaries. This equation is deduced in general form for an open thermodynamic system communicating with its environment.

$$\frac{d}{dt}\int_{V_j} s\,\rho\,dV = - \int_{A_j=\partial V_j} s\,\rho\left((\vec{w}_F-\vec{w}_G).\vec{n}\right)\delta A + \int_{A_j}(\vec{j}_s\cdot\vec{n})dA + P_{\phi,j} - D_{\phi,j} \quad (1)$$

$$F = \int_{V_j} s\,\rho\,dV \quad (2)$$

A bulk quantity F on the left-hand side in (1) means mass of specie, energy, etc. The s is a mass-specific, by flow convected density of quantity F. The equation (1) considers diffusion fluxes j_s and both production and dissipation terms P, D. The Important feature is involvement of relative velocity on a moveable boundary in convective term defined according to fluid (material) velocity w_F and grid velocity w_G.

Conservation equation for mass, momentum and energy – The specific budgeting equation in a form matching the finite volume approach are deduced from (1). Generally, each finite volume (FV) of a computed domain communicates with neighboring ones via boundary interfaces – fig. 1. Therefore, the fluxes through the boundary interfaces are either given by the neighboring finite volumes or they have to be calculated from interface boundary conditions. The scheme in fig. 1 demonstrates an example of finite volume mesh for cylindrical coordinates.

The chosen system of coordinates is important for vector quantity conservation. Momentum budget may be treated in its vector components but the transformation of substantional derivative depends on the system of coordinates. It is simple in the case of Cartesian coordinates only. However, the use of Cartesian coordinates does not create any limitation concerning geometry because arbitrarily chosen finite volumes described by the orientation of surface elements (unit vector n) are assumed. The use of other rectangular systems of coordinates is mentioned and demonstrated in an example in [4].

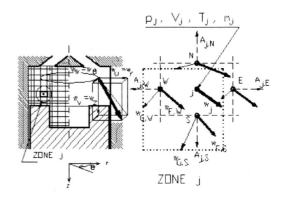

Figure 1: **Finite Volumes (Zones) of AMEM**

The conservation equations for a bulk quantity F in (2) (mass of a specie, energy, etc.) in a FV of volume V create a system that may be written in a general column vector form.

The left-hand side column vector of unknown time and space functions is composed of specie masses m_i (the sum of them being the bulk mass m), components of momentum $m.w_i$ in directions of Cartesian coordinates α, β, γ. Total energy includes specific internal energy u, bulk flow (average) kinetic energy $K=w^2/2$ and its turbulent part k. Hence, the total specific energy is $e=u+w^2/2+k$ and the total energy is $m.e$.

The variables of turbulence models and other closure ones are stated usually in the form of conservation equations. Then they can be easily included in the vector F (e.g., mass specific values k and ε of k-ε two equation model has been used as an example). The vector F is written in line, i.e., transposed form:

$$F^T = [m_1,....,m_s,m.w_\alpha,m.w_\beta,m.w_\gamma,m.e,m.k,m.\varepsilon] \quad (3)$$

$$m = \sum_i^s m_i \qquad (4)$$

$$\rho = \frac{m}{V} = \sum_i^s c_i \qquad (5)$$

The bulk mass is given by the sum of specie masses, the bulk density is given by the sum of specie mass concentrations c_i. The time derivative of F must include both the changes of mass and mass-specific quantities (concentrations, velocity components, specific total energy, etc.). Thus, mass derivatives occur in each differential equation but it does not cause problems in the solving algorithm because mass derivatives themselves do not depend on other unknown derivatives. It will be clear after the whole system is described.

The densities of convective fluxes in the first integral term of the right-hand side are:

$$S^T = [c_1, ..., c_s, \rho.w_\alpha, \rho.w_\beta, ..., \rho.e, \rho.k, \rho.\varepsilon] \qquad (6)$$

Vectors of diffusion flux densities are presented here with some terms for example only; α, β, etc., are unit Cartesian system coordinate vectors. All transport coefficients (mass diffusion of a specie i into a mixture D_i; thermal conductivity λ; dynamic viscosity μ) are used as effective ones, i.e., as sums of molecular and turbulent values. All of them are generally variable and convected by the flow. Components of the viscous stress tensor may be calculated, e.g., according to Stokes hypothesis with effective viscosity for k-ε model; diffusion of enthalpy (Dufour phenomenon - it might be important especially during combustion) is added to heat conductivity. Stresses cause forces on the FV surfaces that are calculated for arbitrarily oriented surface according to:

$$\vec{j}^T = [D_i\vec{\nabla}c_1, ..., \begin{Bmatrix} (-p+\tau_{\alpha\alpha})\vec{\alpha} \\ \tau_{\beta\alpha}\vec{\beta} \\ \tau_{\gamma\alpha}\vec{\gamma} \end{Bmatrix}, \begin{Bmatrix} \tau_{\alpha\beta} \\ (-p+\tau_{\beta\beta})\vec{\beta} \\ \tau_{\gamma\beta}\vec{\gamma} \end{Bmatrix}, \begin{Bmatrix} \tau_{\alpha\gamma} \\ \tau_{\beta\gamma}\vec{\beta} \\ (-p+\tau_{\gamma\gamma})\vec{\gamma} \end{Bmatrix}, \qquad (7)$$

$$\lambda\vec{\nabla}T + \sum_i^s h_i.D_i\vec{\nabla}c_i, difk, dif\varepsilon, ...]$$

where first index of tangential stresses denotes direction of the resulting force and second one denotes the outward normal to the surface. The symbol *dif* means specific terms of turbulence model concerning diffusion of transported quantities, which is not present here in all details.

Closure formulae - It is well known that the system of conservation equations is not sufficient for the solution. Thermic state variables - temperature T and pressure p

as well as specific volume (or density ρ) coupled by thermic and caloric state equations are required to amend conservation equations themselves and for boundary conditions. Temperature T may be determined from the caloric state equation

$$T = T(u, \frac{1}{\rho}), \text{ where } u = e - \frac{w^2}{2} - k \qquad (8)$$

where u is calculated from e subtracting known specific kinetic energies. For real gases, iterative procedure calculates temperature dependence of mean specific heat capacity and the density correction. Regression substitutions in (8) may be used to accelerate this procedure. Then pressure p is found from a thermic state equation at a known bulk density in the computed FV.

The iterative procedure solving thermic state variables is very useful in case of employment of real gas state equation inside AMEM. The use of common Bennedict-Webb-Rubin state equation has been preliminary tested. It is important just in the simulations considering hydrogen combustion because of the high concentration of water vapor in combustion products. However, all results presented in the paper results from computations using state equation of perfect gas.

Combustion modeling – The combustion model represents the crucial one of the whole algorithm. It is necessary to develop suitable combustion models, which are able to qualitatively treat hydrogen flame correctly in both cases of premixed combustion and diffusion combustion. It is also necessary to describe the influence of turbulence on combustion. In the work presented, two combustion models have been applied – a simple semi-empirical approach similar to the PDF one and a complex combustion model based on detailed description of reaction mechanism.

The semi-empirical combustion model represents the simplest way to describe combustion of premixed mixture. The model uses two parameters characterizing rate-of-heat-release and flame front propagation velocity. Fuel is simplified to one chemical specie only. Therefore, the rate-of-heat-release represents the main reaction coordinate and it defines the first model parameter. In the case of multidimensional approach, the spatial distribution of the flame has to be defined. To fulfil this, the shape of flame front is prescribed. Therefore, flame front velocity propagation remains the last unknown to be defined. It represents the second model parameter closing the equation set. The parameters of the model can be chosen according to experimental data or empirical assumption. It is possible to use an algebraic formula giving laminar velocity of flame propagation corrected to turbulence.

It is important to stress that this semi-empirical model does not restrict any other calculation of chemical kinetics and it provides us with the possibility of combining it with the solution of NO_x formation according to extended Zeldovich mechanism. The model splits unstable feedback between stiff chemistry of combustion, formation dependent on sometime unreliable model of turbulence and heat-release source term in the total energy equation.

The complex combustion model based on detailed description of reaction mechanism has been applied to modeling of hydrogen combustion in the next stages of the code development. Chemical kinetics have been applied to both slow reactions and fast ones. Reaction rates have been solved according to Arrhenius law in an automated way using the solver CHEMKIN – [5]. The involvement of chemical kinetics causes considerable problems with the numerical integration of the obtained equation set, i.e. laws of conservation, due to stiffness of source terms. Proposed measures of overcoming it are described in the section dealing with the numerical solution.

The reaction mechanism of hydrogen combustion has been described using the scheme published in [6]. The suitability of the mechanism for the simulation of hydrogen combustion in IC engines has to be proved. It was tested by the authors on simulation of laminar flame speed and ignition delays in a shock tube showing good agreement with the experimental results. Nevertheless, all tests have been done at lower temperature and pressure compared to in cylinder conditions of an engine. The reaction mechanism can give unrealistic data concerning ignition delays, etc., if the state is changed too much.

Comparing both models, the complex one is capable of a detailed solution of chemistry, influence of flow and temperature field on combustion, autoignition, etc. The model can be used for the modeling of premixed mixture combustion as well as diffusion flame. Unfortunately, the costs for its application are very high due to demands on computational time. The two-parameter model can be advantageously applied for verification of the whole algorithm.

Regarding the simulation of autoignition, it is important to employ it in some manner in computations especially if turbocharging and high compression ratio are assumed. It calls for formulation of an effective algorithm which can be based on the fact that detailed modeling of autoignition is not needed and only a 'autoignition identification' is sufficient in most cases. Therefore, the two-parameter model (or even two-zone model applied in the 0-D simulations) can be combined with the solution of chemical kinetics. However, the solution of chemical kinetics and rate-of-heat-release has to be split to eliminate high demands on computational time. Using this procedure, the rate-of-heat-release is solved according to a simple model such as Wiebe's law and composition changes are corrected solving chemical kinetics equations. The algorithm works as the marker of autoignition only. In most cases where autoignition occurs, the numerical algorithm becomes unstable but the indication of autoignition is the main point of such a simulation and a further solution is not so important.

Boundary conditions – A brief remark on the involvement of heat transfer via cooled walls into AMEM is given below. The heat transfer via cooled walls represents a very important phenomenon of ICE modeling. It has a significant effect on the thermodynamic cycle and its efficiency, autoignition invocation, etc. In the cases where boundary layer is dominant the standard turbulence models with extrapolation to the wall using some form of wall functions are not suitable – [7].

In presented computations, a semi-empirical Eichelberg equation solving heat transfer coefficients according to mean parameters in a cylinder – (9) - has been used for the estimation of heat fluxes to cooled walls.

$$\alpha_T = 7.8 \cdot 10^{-3} \cdot c_s^{0.3333} \cdot p_{mean}^{0,5} \cdot T_{mean}^{0,5} \qquad (9)$$

where α_T denotes heat transfer coefficients and c_s is mean piston velocity, p_{mean} and T_{mean} denote mean in-cylinder pressure and temperature. The mean piston velocity represents a scale for the in-cylinder charge motion. Owing to the mean in-cylinder parameters used, this procedure enables estimation of only the mean heat transfer coefficients which are not valid locally at each finite volume. However, if DNS is not used, this approach is able to estimate heat fluxes of a coarse mesh.

Model construction and numerical solution – The model calls for the employment of a suitable numerical scheme and a method of numerical integration. In this work, convective terms have been approximated using up-wind method of 1st order of accuracy in space. The second order method, i.e. central differencing, has been tested as well. Diffusion terms have been approximated using central differences. The obtained equation set has been numerically integrated by the multi-stage Runge-Kutta method of either 2nd or 4th order of accuracy in time.

As has been already stated, the involvement of the complex combustion model based on chemical kinetics brings about significant problems with numerical integration due to the stiffness of chemistry pertinent source terms in laws of conservation. To overcome it, the numerical solution has been split into three stages:

- 1st stage – flow phenomena solution.
- 2nd stage – solution of combustion pertinent source terms. Mixture composition changes and rate-of-heat-release due to chemical reactions have been solved using special procedure VODE – [8].
- 3rd stage – linking of flow phenomena solution with the chemistry one. Common dependencies from the previous stages are updated.

Regarding the computational mesh used in the computations, very simple geometry of the modeled cylinder enabling the use of an orthogonal grid of finite volumes has been considered. The computational grid has been recalculated in each time step to correspond with the instantaneous piston position.

DESCRIPTION OF MODELED CASES

All 0-D simulations have been performed on a 1-cylinder engine. The main engine parameters are shown in table 1. As stated above, the model layout has been proposed to simulate experimental setup. Influence of spark timing, air excess and compression ratio on properties of the engine cycle has been investigated. In the case of computations concerning air excess and spark timing influence, parameters of combustion have been fitted to the experimentally obtained data. In accordance with the aims, the main stress has been put on the modeling of NO_x formation. The possibility of the high compression ratio use in combination with extremely retarded spark advance has been solved as well. It has been taken into consideration as the efficient measure reducing NO_x formation together with a reasonably high engine efficiency and helping to avoid autoignition.

The CFD computations have been performed on a cylinder with flat piston geometry. The model is not limited to the simple geometry at all. It has been used to suppress influence of complicated numerics. In spite of the simple geometry, the code represents a comprehensive model considering moveable boundaries, combustion, heat transfer via cooled walls, heat conductivity of gas and diffusion. Perfect gas has been assumed but with temperature and mixture dependent specific heat capacities.

DISSCUSSION OF RESULTS

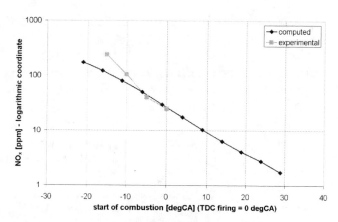

Figure 2: Influence of start of combustion on NO_x production. Air excess of 2, compression ratio of 10, 910 r.p.m., homogeneous charge combustion. Comparison of computed and measured data.

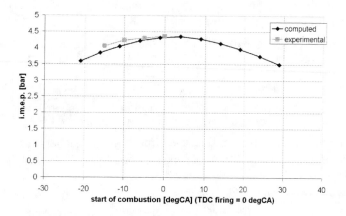

Figure 3: Influence of start of combustion on i.m.e.p. Air excess of 2, compression ratio of 10, 910 r.p.m., homogeneous charge combustion. Comparison of computed and measured data.

0-D SIMULATIONS OF 1-CYLINDER EXPERIMENTAL ENGINE – The influence of spark timing, respectively start-of-combustion, is shown in figs. 2, 3 and 4. All simulations have been performed at constant air excess of 2. Start-of-combustion angle has a very strong impact on NO_x formation due to the decreasing in-cylinder temperatures if retarding spark advance timing – see fig. 2. Figs. 3 and 4 show typical dependencies of indicated mean effective pressure and indicated specific fuel consumption on the start-of-combustion angle. These dependencies clearly demonstrate that optimal spark timing is very close to TDC because of fast combustion, i.e. short combustion duration.

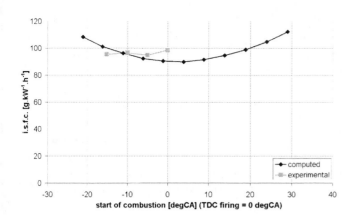

Figure 4: Influence of start of combustion on i.s.f.c. Air excess of 2, compression ratio of 10, 910 r.p.m., homogeneous charge combustion. Comparison of computed and measured data.

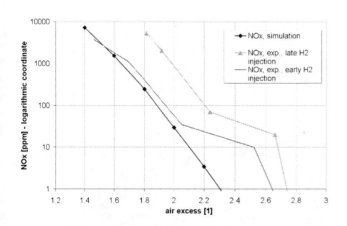

Figure 5: Influence of air excess on NO$_x$ production. Spark advance of 5 degCA BTDC, compression ratio of 10, 910 r.p.m., homogeneous charge combustion (*early injection timing*). Experimental results of stratified mixture combustion are also plotted (*late injection timing*). Comparison of computed and measured data.

The influence of air excess at constant spark timing of 5 deg CA BTDC is shown in figs. 5, 6, 7, 8 and 9. In fig. 5, the comparison of computed and measured NO$_x$ concentration in exhaust gases is presented. Two cases have been examined experimentally – premixed mixture combustion and stratified mixture one. Fuel is injected directly into the cylinder in both cases. However, injection timing is completely different. The curve designated 'early injection' represents premixed mixture combustion because fuel is injected into cylinder during intake stroke so that a nearly homogeneous mixture could be created. In the second case, fuel is injected into the cylinder during compression stroke close to spark timing angle – late injection. Only the early injection case has been considered in simulations but, in principle, it is also

possible to model the latter one using corrected rate-of-heat-release patterns inside the GT-Power model. Results show very strong dependence of NO$_x$ formation on air excess. Higher concentrations of NO$_x$ for stratified combustion have been observed due to faster combustion accompanied by higher in-cylinder temperatures. The predicted NO$_x$ emissions on a specific scale are shown in table 2. The specific NO$_x$ emissions are referenced to the indicated engine power.

air excess [1]	1.4	1.6	1.8	2
NO$_x$ [g.kW^{-1}.h^{-1}]	46.7	11.25	2.02	0.28

Table 2: Specific NO$_x$ emissions referenced to indicated engine power

Comparison of computed and measured indicated mean effective pressure is shown in fig. 6. The naturally aspirated engine has been considered in all simulations. Thus, low indicated mean effective pressure has been achieved considering lean mixture operation. The comparison of predicted and measured maximal in-cylinder pressure demonstrates good agreement – fig. 7. In fig. 8, the influence of air excess on indicated specific fuel consumption is presented. A difference between computed and measured data has appeared. The predicted indicated specific fuel consumption has been optimistic compared to the measured one. It can be corrected by further tuning of model parameters (heat transfer coefficients, etc.).

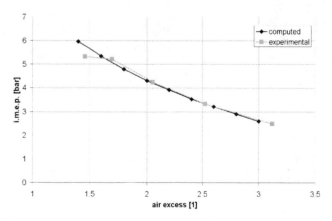

Figure 6: Influence of air excess on i.m.e.p. Spark advance of 5 degCA BTDC, compression ratio of 10, 910 r.p.m, homogeneous charge combustion. Comparison of computed and measured data.

Computed maximal in-cylinder temperature and combustion duration used in simulations are presented in fig. 9. The combustion duration has been chosen according to the experimental data. An important feature of hydrogen is that combustion is fast enough to achieve high engine efficiency even if a very lean mixture is used. The combustion is faster than in methane or gasoline engine and it does not require complicated measures to increase combustion rate. Ignition limits of hydrogen cover a wider range of air excess as well.

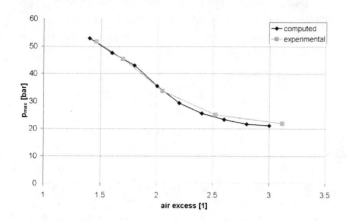

Figure 7: Influence of air excess on maximum firing pressure. Spark advance of 5 degCA BTDC, compression ratio of 10, 910 r.p.m., homogeneous charge combustion. Comparison of computed and measured data.

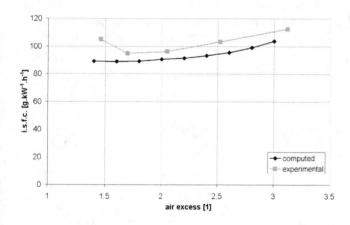

Figure 8: Influence of air excess on i.s.f.c. Spark advance of 5 degCA BTDC, compression ratio of 10, 910 r.p.m., homogeneous charge combustion. Comparison of computed and measured data.

The influence of compression ratio used is shown in figs. 10 and 11. The simulations have considered constant air excess of 2. The compression ratio increase to 13 from original value of 10 has been modeled. Expected NO_x concentration increase with higher compression ratio used has been observed – see fig. 10. Nevertheless, NO_x production remains acceptable if extremely retarded spark advance is considered and it is possible to take advantage of better engine efficiency. It is demonstrated in fig. 11 where the indicated specific fuel consumption is

shown. The extremely retarded spark timing is useful also from the point of view of autoignition avoidance. During combustion, unburned mixture is compressed and preheated by propagating flame which could cause autoignition invocation. However, it is compensated by increasing cylinder volume if spark timing is shifted to the expansion stroke. The simulations considering only two compression ratios are not sufficient to find an optimum from the point of view of compromise between fuel consumption, i.e. engine efficiency, and NO_x production. Thus, it will be necessary to perform a more detailed study.

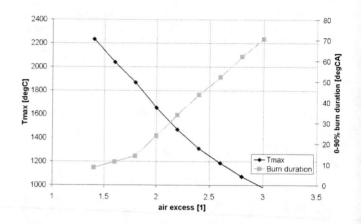

Figure 9: Influence of air excess on maximum in-cylinder temperature and combustion duration used in the simulations. Spark advance of 5 degCA BTDC, compression ratio of 10, 910 r.p.m., homogeneous charge combustion.

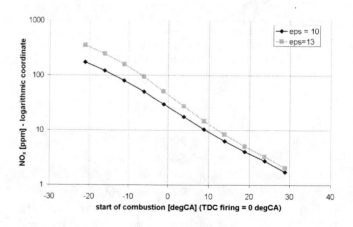

Figure 10: Influence of compression ratio on NO_x production. Variable spark timing. Compression ratios of 10 and 13, 910 r.p.m., homogeneous charge combustion.

CFD SIMULATIONS - The results of the CFD computations presented in the paper demonstrate capabilities of those algorithms. Unfortunately, the results are not fully practically applicable yet. Hydrogen self-ignition has been modeled - up to now - to test the algorithm. The results are presented in figs. 12 – 16. Only results from the complex combustion model are described. The algebraic two-parameter combustion model has been employed to test the algorithm in the first stage and the main stress has been put on the chemical kinetics involvement during the work elaboration.

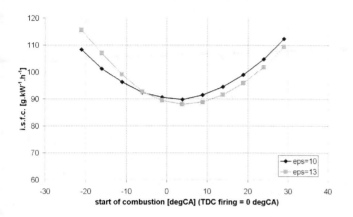

Figure 11: Influence of compression ratio on i.s.f.c. Variable spark timing. Compression ratios of 10 and 13, 910 r.p.m., homogeneous charge combustion.

In figs. 12, 13 and 14, time pattern of mass fraction of species taking place in the reaction mechanism used is described. Air excess of 4 and initial temperature of 1500 K have been used in these simulations. The graphs show very fast kinetics to be solved. The effects are amplified by the high initial temperature used but possible problems of numerical solution are quite obvious even if the flame front propagation will be modeled. The velocity of flame front propagation should not be a problem but local reaction rates will be very near to the presented rates in the vicinity of flame front. It causes stiffness of equation set.

The main advantage of CFD models is documented in figs. 15 and 16. In fig. 15, isolines of mass fraction of hydrogen are described. Axis are oriented so that position x=0 corresponds with cylinder axis and y=0 corresponds with piston top. These pictures clearly show influence of temperature at early stage of combustion. The mixture burns very intensively in sub-volumes with the highest local temperature but combustion is considerably slower near to the cooled wall of the cylinder. Isolines of mass fraction of nitrogen oxide after end of combustion show the capability of such models of the qualitatively correct description of strong temperature dependence of NO_x kinetics.

As has been already stated, the presented results are not practically applicable yet, but they show the future prospects of the algorithm. It is able to model autoignition phenomena, to connect detailed solution of temperature field with combustion and it can be used to predict the influence of hot spots on combustion chamber walls, etc.

The AMEM has been also amended for a simple ignition model so that flame front propagation could be modeled. Unfortunately, problems in linkage of both chemical kinetics solution and flow phenomena solution have appeared. To overcome these problems, two modifications of the code have been proposed. The first measure represents minor change of the algorithm in which treatment and linkage of kinetic source terms with other equations has been slightly rearranged. The second one is based on the considerable change of combustion model in the algorithm. Reactions are divided into two groups – fast reactions and slow ones. The fast reactions are solved using chemical equilibrium equations whereas the slow ones, typically NO formation, are solved using Arrhenius law. The procedure could be combined with prescribed flame front propagation. This formulation has strong disadvantage in necessity of solution of differential equation and strongly nonlinear algebraic chemical equilibrium equations.

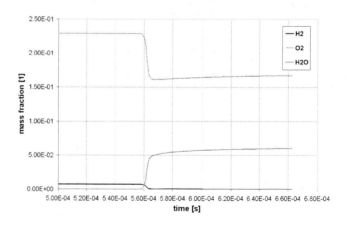

Figure 12: CFD simulations of hydrogen self-ignition. Main products.

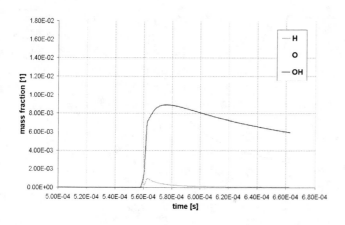

Figure 13: CFD simulations of hydrogen self-ignition.

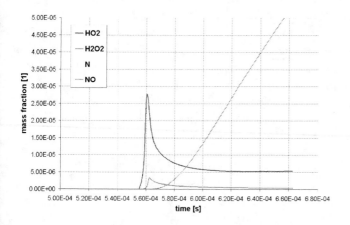

Figure 14: CFD simulations of hydrogen self-ignition.

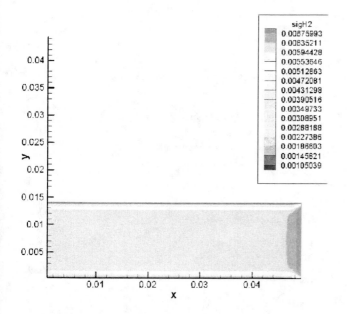

Figure 15: H_2 mass fraction after self-ignition.

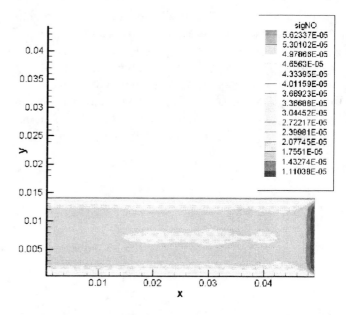

Figure 16: NO mass fraction after H_2 self-ignition. H_2 combustion finished, equilibrium of NO is not reached.

For completeness, examples of velocity fields for a flat piston geometry and a piston bowl are presented in figs. 17, 18 and 19. In fig. 17, the example of velocity field in a cylinder with a flat piston is shown. The influence of viscosity on velocity field is clearly demonstrated by eddies in piston/cylinder wall contact corner. They are generated due to the moving piston because the boundary layer is packed and rolled during compression stroke. Figs. 18 and 19 show the computed velocity field in a cylinder considering piston bowl. As stated above, the simple geometry with a flat piston has been used in computations involving chemical kinetics. The results employing more general geometry show future prospects of the algorithm. The more complicated combustion chamber geometry represents one of the possible measures of avoiding autoignition. It is caused by the large eddies created during compression stroke, i.e. toroidal ones located in piston bowl in presented case, serve as the source of turbulence at the end of the compression which has a direct influence on in-cylinder transport phenomena. Thus, it has, together with the direct effects of flow field, a strong impact on combustion, autoignition, etc. However, it is important to emphasize that the velocity fields considering the simple piston bowl geometry are presented here to demonstrate capabilities of the model used only. The combustion chamber has to have a different shape to be used in a real engine, e.g. without sharp edges to eliminate overheating, etc.

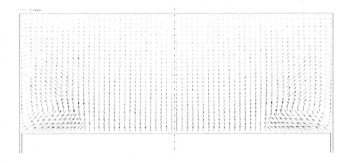

Figure 17: Example of velocity field. Flat piston geometry. Computation without combustion, dynamic viscosity of 1e-3 Pa.s. Piston position at 10 deg. CA BTDC.

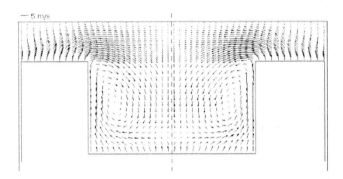

Figure 18: Example of velocity field. Piston bowl geometry. Dynamic viscosity of 1e-3 Pa.s. Swirl number at TDC of 0. Piston position at 5 deg. CA BTDC.

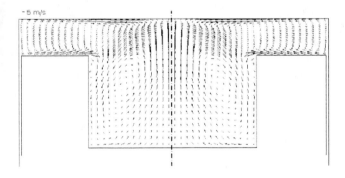

Figure 19: Example of velocity field. Piston bowl geometry. Dynamic viscosity of 1e-3 Pa.s. Swirl number at TDC of 2.33. Piston position at 5 deg. CA BTDC.

CONCLUSION

Conclusions of the work presented can be summarized as follows:

- The 0-D model is ready for practical application to modeling of NO_x formation of hydrogen engines. Further fitting to experimental data can be done. The model gives good qualitative and quantitative agreement with experimental data. The model is ready to help in finding a compromise between specific power/engine efficiency/NO_x production/autoignition.
- It is possible to considerably reduce NO_x formation using a very lean mixture. The main advantage of hydrogen combustion is the reasonably high combustion velocity even if extremely lean mixture is used. It does not restrict achievable engine efficiency.
- The combination of a high compression ratio, very lean mixture and extremely retarded spark advance provide a strong advantage from the point of view of NO_x formation and engine efficiency. It is also worthwhile due to autoignition avoidance. Combustion occurs during the expansion stroke when the increasing cylinder volume compensates the compressing and preheating of unburned mixture due to propagating flame.
- A dedicated engine using high compression ratio, and direct injection is necessary to utilize all the potential provided by hydrogen.
- The results of 0-D simulations are promising from the point of view of the model application to full-scale turbocharged engine.
- CFD model provides very high potential concerning modeling of in-cylinder phenomena of IC engines. It is very computationally time-demanding. Unfortunately, the practical usability of such a model is still very restricted.
- The multidimensional models can be preferably applied to modeling of fuel injection, autoignition, etc.
- Difficulties with stiffness have to be treated if chemical kinetics is involved. Using special numerical procedure for the stiff problem integration does not completely eliminate the high demands on computational time.

ACKNOWLEDGMENTS

This work has been subsidized by grants of the Czech National Grant Agency No. 101/98/K001, 101/97/K053 and 101/99/D014. This help has been gratefully appreciated.

REFERENCES

1. Takats, M.: Hydrogen Combustion in Cylinder of Reciprocating Engine – Thermodynamic Analysis of Indicator Diagram. In: Report No. Z99-02. Prague: CTU Prague, Dept. of Aerospace and Automotive Eng. 1999. (In Czech)
2. GT-Power, User's manual – Vers. 5.1. Gamma Technologies Inc., 601 Oakmont lane, Suite 220, Westmont, IL, USA
3. Macek, J. - Polášek, M.: Advanced Eulerian Multizone Model - Versatile Tool in Moveable Boundary Problem Modeling. In: ISCFD'99 - All Contributions, CD-ROM. Vol. 1. Bremen: ZARM Universität Bremen. 1999. p. 1-20.
4. Macek,J.-Steiner,T.: Advanced Multizone Multidimensional Models of Engine Thermoaerodynamics. Proc.of 21st CIMAC Congress Interlaken, London 1995, pap. D10, pp 1-18
5. Kee,R.J. – Rupley,F.M. – Meeks,E. – Miller,J.A.: CHEMKIN III: A Fortran Chemical Package for the Analysis of Gas-Phase Chemical and Plasma Kinetics. Report of Sandia National Laboratories SAND96-8216. Sandia National Laboratories, Livermore, CA 94551-0969. 1996
6. Marinov, N., Westbrook, C.K. and Pitz, W.J.,"Detailed and Global Chemical Kinetics Model for Hydrogen" in ransport Phenomena in Combustion, Volume 1 (S. H. Chan, edited), Talyor and Francis, Washington, DC, 1996
7. MACEK, J. - POLÁŠEK, M.: AMEM - Multigrid-in-Time Approach for Engines. In: MOTORSYMPO '99. Vol. 1. Praha: CVUT. 1999. p. 1-10. ISBN 80-01-01985-3
8. Netlib Public Library. www.netlib.org.

CONTACT

Miloš Polášek
Czech Technical University in Prague
Josef Bozek Research Center
Technicka 4
CZ – 166 07 Prague
Czech Republic
e-mail: polasekm@fsid.cvut.cz

ADDITIONAL SOURCES

All measurement used in the paper have been performed in laboratories at the Technical University of Liberec, Hálkova 6, CZ – 461 17 Liberec, Czech Republic

DEFINITIONS, ACRONYMS, ABBREVIATIONS

AMEM	Advanced Multizone Eulerian Model
BTDC	Before Top Dead Center
CA	Crank Angle
CFD	Computational Fluid Dynamics
FC	Fuel Cell
FV	Finite Volume
IC engine	Internal Combustion Engine
TDC	Top Dead Center

Electronic Fuel Injection for Hydrogen Fueled Internal Combustion Engines

James W. Heffel, Michael N. McClanahan
and Joseph M. Norbeck
University of California, Riverside, CE-CERT

ABSTRACT

This paper describes an evaluation of a series of commercially available natural gas fuel injectors, originally designed for heavy-duty diesel application, for use with hydrogen fuel in an electronic fuel-injected internal combustion engine. Results show that sonic flow, pulse-width-modulated electronic gaseous fuel injectors provide accurate and stable metering of hydrogen gas at fuel pressures between 25 and 200 psig. A linear flow rate of hydrogen was observed with a low standard deviation error during pulse width modulation. Plots of flow rate of hydrogen (mg/injection) versus pulse width (PW) are presented for inlet pressures from 25 to 200 psig for selected injectors. In addition, injector response tests were conducted and found to have time delays (time it takes the injector to open) between 2.6 ms and 2.3 ms at 25 psig inlet pressure. Time-delay times increased linearly between 4.0 ms and 3.0 ms at 200 psig. In addition, leak tests were performed on all injectors and the measured leak rates were found to be insignificant. The precise metering of fuel is critical for the performance of hydrogen powered engines. The performance of the injectors evaluated was within the necessary tolerances for hydrogen applications with internal combustion engines.

INTRODUCTION

In recent years, an increased concern regarding future availability of fossil fuels and global climate change has provided a renewed interest in hydrogen as a transportation fuel. The advantages of hydrogen are well known, having been considered as a fuel for internal combustion engines for more than 100 years [1-5]. There are significant technical barriers that have limited the practical implementation of hydrogen as a transportation fuel. One of the primary problems is premature ignition. Backfire conditions can develop if the premature ignition occurs near the fuel intake valve and the resultant flame travels back into the induction system.

Under most operating conditions, controlling the engine at a lean air/fuel ratio decreases flame speed and com-

bustion chamber temperature, and reduces oxides of nitrogen emissions [6-11, 27]. The low energy density of hydrogen, the non-linear relationship between air/fuel ratio and emissions of oxides of nitrogen, and the pre-ignition problem require precise fuel metering beyond that required for other fuels.

There have been numerous attempts to solve the fuel-metering problem for hydrogen; some examples include high-pressure (3000psig) fuel injection, modifications to carburetion, and direct manifold injection [3-23]. None have been successful due to numerous technical barriers. A recent review of the advantages and disadvantages of these technologies is presented in Norbeck, et al [24] and is beyond the scope of this paper.

Electronic fuel injection for gasoline fueled engines using solenoid-actuated injectors to provide fuel into the throttle body or into the intake manifold near the intake valve, has been developed over the past three decades in response to the need to reduce emissions and improve fuel economy. Microprocessor-based systems using feedback sensors and electronically controlled actuators allow precise fuel metering, cylinder distribution, cold start enrichment closely controlled to the temperature of the engine and air, deceleration and overspeed cutoff, and protection from driver abuses such as flooding. Their use has led to significant reductions in engine out emissions and increased fuel economy.

Fuel injectors designed to operate with gasoline cannot be used for hydrogen since much larger volumes of fuel must be injected. This is due to the extremely low energy density of hydrogen compared with gasoline. For example, at a stoichiometric mixture, hydrogen will occupy approximately 30% of the cylinder volume. Special electronic fuel injectors have been designed for natural gas, but to date, no injectors have been manufactured for hydrogen applications.

This report presents the evaluation of several different sizes of commercially available solenoid-actuated injectors that were developed for use with heavy duty, large displacement engines, operating on natural gas. BKM,

INC., in San Diego, CA. [25, 26] manufactured all the injectors evaluated.

EVALUATION CRITERIA

An initial evaluation of the hydrogen mass flow requirements and an estimate of the maximum injector pulse width were made for two engine configurations operating on hydrogen. In this evaluation it was assumed that the fuel pressure would be between 20 and 200 psia, and the air/fuel ratio, expressed in terms of equivalence ratio, would be between 0.4 and 1.0.

The first application considered was a variable speed engine (VSE) for a conventional vehicle with the following specifications:

- Eight cylinders, naturally aspirated
- Approx. displacement of 350 cu. in. (5.7 liters)
- Maximum engine speed of 5,000 rpm

The second application considered was for a constant speed engine (CSE) that could be used to drive an Auxiliary Power Unit (APU) for a hybrid electric vehicle with the following specifications:

- Four cylinders, naturally aspirated
- Approx. displacement of 136 cu. in. (2.2 liters)
- Optimum APU speed of 3,500 rpm (±10%)

MASS OF HYDROGEN REQUIRED – An estimate of the hydrogen mass needed for each injection was done assuming that the VSE and CSE engines operated at 100% volumetric efficiency at wide open throttle with an equivalence ratio of 0.4 and volumetric cylinder displacements of 5.7 l and 2.20 l, respectively. With these operating characteristics, the mass of hydrogen was estimated to be 8.4 mg/injection for the VSE, and 6.9 mg/injection for the CSE.

INJECTOR PULSE WIDTH ESTIMATES – Injection of the hydrogen late in the air intake cycle enables the incoming air (no fuel) to cool hot spots that may have formed on the valves or in the combustion chamber, thus minimizing the chances of the fuel igniting prematurely (preignition). Because injection time is inversely proportional to engine speed, this is not a difficult requirement for low engine speeds. However, for high engine speeds this requirement becomes very critical.

The injector pulse width (the amount of time an injector is energized during the intake cycle) is based on starting hydrogen injection at 35° ATDC of the exhaust gas stroke and terminating injection at BDC of the intake stroke. This results in a hydrogen induction duration of about 145° (180° minus 35°). This parameter was changed (optimized) during engine dynamometer testing, but it served as an initial estimate.

Maximum Injector Pulse Width for the VSE – Using 5,000 rpm as the top engine speed for the VSE, the crankshaft will complete one revolution in 12 ms (1 min/5000 rev x 60 s/min). Of the 360° of crank rotation during one revolution, 145° is used for hydrogen induction. This leaves 4.83 ms (12 ms x 145°/360°) available for hydrogen induction at maximum engine speed.

Maximum Injector Pulse Width for the CSE – Using 3,500 rpm as the top engine speed for the constant speed engine, the crankshaft will complete one revolution in 17 ms (1 min/3500 rev x 60 s/min). Of the 360° of crank rotation during one revolution, 145° is used for hydrogen induction. This leaves 6.90 ms (17 ms x 145°/360°) available for hydrogen induction at maximum engine speed.

Summary of the Injector Mass/Flow Requirements – It was estimated that the VSE fuel injector needs to be able to supply 8.4 mg of hydrogen in a time span of 4.83 ms, and the CSE fuel injector needs to be able to supply 6.9 mg of hydrogen in a time span of 6.90 ms.

FUEL INJECTOR SELECTION – Based on the above information it was determined that four models of Servojet pulse-width-modulated electronic natural gas injectors, (SP-010, SP-014, SP-021, and SP-051) could possibly meet the fuel metering specifications needed for the two engine applications. The technical details of these injectors have been published elsewhere. [25]

TEST SET-UP

A layout of the Injector Test Bench is shown in **Figure 1**. The test bench has two hydrogen flow meters plumbed in series. One flow meter is manufactured by Porter and the other by the Micromotion. The Porter meter was used to measure flows under 35 standard cubic feet per minute (scfm). For flows between 35 and 150 scfm, the Micromotion was used. The accuracy of the Porter and Micromotion meters are 0.5% and 0.1% of full scale, respectively.

Hydrogen from the hydrogen storage system is regulated to the test pressure at Pressure Regulator 1 (PR1) prior to entering the flow meters. From the flow meters, the hydrogen enters the H_2 inlet manifold where the pressure and temperature of the hydrogen are measured and recorded. The inlet manifold distributes the hydrogen to four injectors, which pass the fuel through to the exhaust manifold. The hydrogen is exhausted outside the building via a vent stack.

To operate the injectors, driver circuitry was developed and incorporated into the test bench. The system is capable of firing one to four injectors simultaneously or sequentially to characterize hydrogen flow under a range of possible engine operating conditions.

A function generator and pulse-width modulator provide the input signal to the injector drive circuitry. The function generator is used to control the base frequency representing the engine rpm.

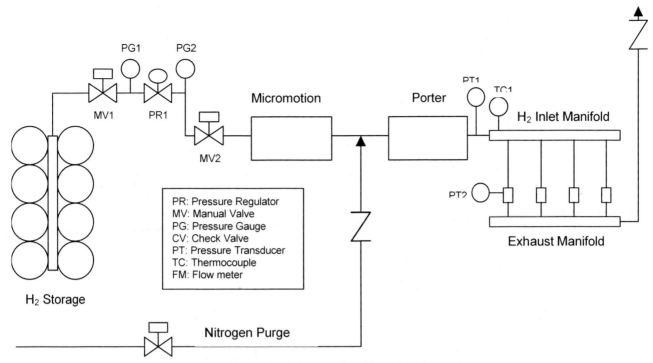

Figure 1. Injector Flow Test Bench

The pulse-width modulator controls the time during a cycle that the injector is energized (duty cycle). This modulator can control pulse widths from 5% to 95% duty cycle. By using the modulator in conjunction with a signal generator, frequency and duty cycle can be controlled within the entire range of the test matrix. Injector firing frequency and pulse-width duration are displayed on a Dual Beam Oscilloscope.

FUEL INJECTOR LEAKAGE – Leakage rates for the four injector models were measured to evaluate the possibility of pre-ignition due to hydrogen leakage past the injectors into the engine block. The four injector models were tested at pressures of 50, 100, 150, and 200 psig. Pressurized hydrogen gas from the K-bottle rack was piped to the injector assembly. At the outlet port of the injector assembly a plastic tube was attached. This plastic tube was then connected to a Humonics Optiflow 520 Digital Flowmeter. The hydrogen leakage flow rate in cc/min was recorded at 30-second intervals for a total of 30 minutes.

EXPERIMENTAL RESULTS

INJECTOR DYNAMIC FLOW CHARACTERISTICS – Due to the high speed at which the injectors open and close and the low flow rate for each opening, it is impossible to measure the flow rate per injection directly. To determine the mass of hydrogen per injection the following experimental protocol was used:

Each injector flow test was repeated 4 times at 4, 6, 8, 10, 12, 14, and 16 ms. pulse widths. A 50% duty cycle was chosen so the hydrogen flow rate for each test would remain fairly constant for each PW (even though the

injector is open 4 times longer for a PW of 16 ms versus a PW of 4 ms, for a duty cycle of 50% it is also closed 4 times longer.)

The following test data were measured and recorded at a rate of approximately 10 samples per second:

- Time (sec)
- Hydrogen flow rate (lb./min)
- Manifold rail pressure (psia)
- Hydrogen temperature (°C)

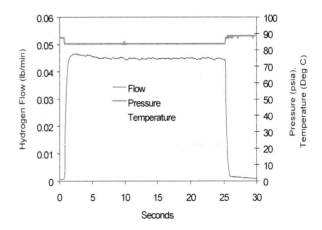

Figure 2. A typical plot showing the relationship of hydrogen flow, temperature, and pressure during an injector 50% duty-cycle test at 75 psig.

A typical test consisted of adjusting the PW modulator to the proper PW (as displayed on the Tektronix 5112 Dual Beam Oscilloscope), starting the data acquisition system, initiating injection firing, terminating injection firing after 25 seconds, and stopping data acquisition at 30 seconds. This was repeated four times for each PW. Each test yielded approximately 300 data points. A typical plot of this data is shown in **Figure 2**. Data from each test were then analyzed.

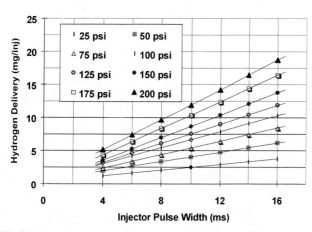

Figure 3.   Figure 3. H2 Mass Flow Rate at various pressures versus PW (ms) of injector data for SP-021.

To eliminate the perturbations in the flow rate due to the initiation and termination of flow at the beginning and end of each test, only 10 to 20 seconds (test duration) of the steady-state flow during the middle of the test were used. The average flow rate during this period was multiplied by the test duration to yield the total mass of hydrogen that passed through the injector. Knowing the test duration and the frequency at which the injectors are being operated, the number of injections can be calculated. This value is then divided into the total mass to yield the flow rate (mg/inj).

This protocol was repeated for pressures of 25, 50, 75, 100, 125, 150, 175 and 200 psig for injectors SP-014 and SP-021. For injector SP-051, pressures of 50, 100, 150, and 200 psig were used. It was found that the SP-010 injector would not open at 200 psig and performed poorly at 150 psig. The SP-010 was therefore only tested at 50 and 100 psig. A summary of the flow rate, standard deviation in flow rate and inlet pressure relative to PW is listed in the Appendix for the SP-014, SP-021, SP-051, and SP-010 injectors, respectively. The injectors were evaluated for sufficient mass flow at the critical pulse widths based on the VSE and CSE rpm requirements. For the maximum pulse width of 4.83 ms (8.4 mg/inj required for the VSE at 5000 rpm) only the SP-010 operating at a rail pressure of 100 psig provides enough hydrogen mass. However, for the maximum pulse width

of 6.90 ms (6.9 mg/inj required for the CSE at 3500 rpm) the SP-010, 014, 021, and 051 provides sufficient mass at 50, 125, 175, and 200 psig respectively. **Tables 1 and 2** present this data for both engine scenarios operating at their maximum rpm with an equivalence ratio of 0.4.

Table 1.   Hydrogen Flow at a Pulse Width of 4.83 ms (requirement for a Variable Speed Engine).

Inlet Pressure	SP-014	SP-021	SP-051	SP-010
(psig)	(mg/inj)	(mg/inj)	(mg/inj)	(mg/inj)
25	1.38	1.39	-	-
50	3.10	2.25	1.81	7.49
75	4.36	2.78	-	-
100	4.91	3.49	3.08	8.58
125	5.13	3.78	-	-
150	5.68	4.23	3.88	-
175	6.41	5.18	-	-
200	7.23	6.09	4.82	-

Table 2.   Hydrogen Flow at a Pulse Width of 6.90 ms (requirement for a Constant Speed Engine).

Inlet Pressure	SP-014	SP-021	SP-051	SP-010
(psig)	(mg/inj)	(mg/inj)	(mg/inj)	(mg/inj)
25	1.82	1.82	-	-
50	4.08	2.97	2.45	10.24
75	5.65	3.82	-	-
100	6.53	4.76	4.18	13.83
125	7.12	5.28	-	-
150	7.93	6.00	5.51	-
175	8.99	7.26	-	-
200	10.17	8.43	6.94	-

The flow rate curves at various pressures and pulse widths derived from these tests are shown for the SP-021 injector in **Figure 3**.

INJECTOR DYNAMIC RESPONSE CHARACTERISTICS – Response time of a solenoid depends on the physical size of the solenoid, power input, drive circuitry, mass and travel of the armature, and load. The opening time of the injector, which is the time required for the armature to reach its fully opened position after initiation of the drive circuit pulse input, depends on the gas supply pressure. The closing time also depends on supply pressure.

Since it is more important to know when flow actually starts than when the injector is physically open, it was decided to measure the dynamic pressure pulse at the outlet of the injector instead of trying to measure position of the injector ball poppet. To accomplish this measurement, a piezoelectric pressure transducer Model 112A (PT3 in Figure 1) was installed in the manifold block to measure the outlet pressure of injector #1.

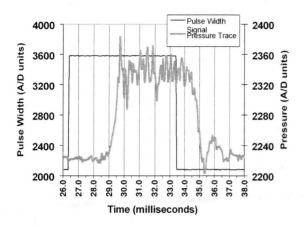

Figure 4. Pressure Rise and Pulse Width Signal vs. Time

During data collection, the rail pressure and the PW remained constant for each test. Data were recorded by feeding both signals from the oscilloscope into an Analog-to-Digital converter board interfaced with a Pentium computer. Data were recorded at a 30 kHz sample rate.

A typical plot of pressure rise and PW voltage versus time is shown in **Figure 4**. The time delay (the time it takes for the pressure to reach 50% of its final pressure after the PW signal is sent) is the start of the square wave to the midpoint of the first large peak of the pressure signal. For each pressure and PW, three to five cycles (injection firings) were measured and recorded.

Plots of the pressure and pulse width profiles for each of these runs were made by overlaying the pressure traces relative to injection firing (initial rise in the PW profile). Because the amplitude for the PW and pressure profiles is not required for the calculations to determine the time delays, the computer units were not converted to standard engineering units.

The results of these tests for the SP-014 and SP-021 injectors are shown in **Figure 5**. Here it is seen that the SP-021 injector opens faster than the SP-014 injector does. This is mainly due to the larger poppet ball (more mass) and the larger orifice (more surface area for the pressurize gas to force the valve closed) of the SP-014 injector.

INJECTOR LEAKAGE CHARACTERISTICS – The leakage flow rates for the four BKM injector models were quite low. The flow rates were so slow it was determined that the escaped hydrogen gas was insignificant.

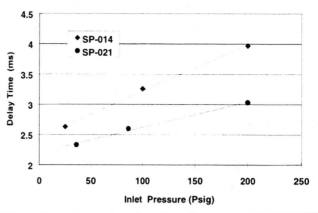

Figure 5. Injector Opening Time Delay vs. Inlet Pressure

Of all the injectors, the SP-051 injector displayed the highest leakage rate at 200 psig (approximately 250 cc/min). It was also observed that each time an injector was run for a brief period of time, the leakage rates decreased. This is most likely due to the poppet ball inside the injector becoming better seated after each run. A plot displaying the leakage rates for all four injectors is presented in **Figure 6**.

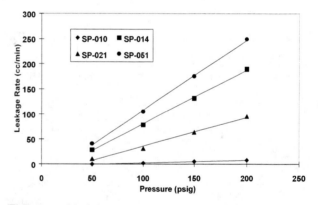

Figure 6. Hydrogen Injector Leakage Rates

CONCLUSIONS AND RECOMMENDATIONS

Our results indicate that the sonic flow, pulse-width-modulated electronic injectors for gaseous fuels used in this project provided precise, stable and reliable metering of hydrogen gas. This is evident in the linear flow and the low standard deviation error during pulse-width-modulation.

Mass flow rates (mg of hydrogen/injection) for each injector are within the requirements of 4.83 ms (variable speed engine) and 6.90 ms (constant speed engine), respectively. Our results show that the SP-010 is capable of meeting the requirement of 8.4 mg of hydrogen per injection for the variable speed engine. For the constant speed engine our data showed that all the injectors are capable of meeting the design requirement of 6.9 mg of hydrogen per injection.

Although the performance and durability of these injectors were excellent for these limited tests, the question of long term durability using hydrogen has not been fully addressed. It is recommended that further testing, preferably on an operating engine, be conducted to address this issue.

ACKNOWLEDGMENTS

CE-CERT would like to thank the South Coast Air Quality Management District for sponsoring this project, BKM, Inc. for providing the injectors and technical data, and Bourns Instruments for providing pressure sensors. CE-CERT would also like to thank Tim Fleenor for instrumentation and test set up; and student researchers Russ Gauthier, Ryan Gross, Forrest Jehlik, and Anthony Rossi for assisting in the experiments.

REFERENCES

1. Cecil, W., "On the application of hydrogen gas to produce a moving power in machinery; with a description of an engine which is moved by the pressure of the atmosphere, upon a vacuum caused by explosions of hydrogen and atmospheric air," *Trans. Cambridge Philos. Soc.*, Vol. 1, p. 217, 1822.
2. Van Vorst, W.D. and Woolley, R.L., "Hydrogen-fueled surface transportion," in Hydrogen: Its Technology and Implications, Edited by K. E. Cox and D. K. Williamson Jr., CRC Press, Boca Raton, FL, p. 1979.
3. Das, L.M., "Hydrogen Engines: A View of the Past and a Look into the Future," *Int. J. Hydrogen Energy*, Vol. 15, p. 425, 1990.
4. Ricardo, H.R., "Further Note on fuel research I.," *Proc. Inst. of Automobile Eng. of London*, Vol. 5, p. 327, 1924.
5. Oehmichen, M., "Wasserstoff als Motortreib-mittel," in Verein Deutsche Ingenieur, Deutsche Kraft-fahrtforschung, Verlag GmbH, Berlin, 1942.
6. Homan, H.S., De Boer, P.C.T. and McLean, W.J., "The effect of fuel injection on NOx emissions and undesirable combustion for hydrogen-fuelled piston engines," *Int. J. Hydrogen Energy*, Vol. 8, p. 131, 1983.
7. King, R.O. and Rand, M., "The oxidation, decomposition, ignition, and detonation of fuel vapours and gases. XXVII. The hydrogen engine," *Canadian J. Technol.*, Vol. 33, p. 445, 1955.
8. King, R.O., Wallace, W.A. and Mahapatra, B., "The oxidation, ignition and detonation of fuel vapours and gases-V. The hydrogen engine and detonation of the end gas by the ignition effect of carbon nuclei formed by pyrolysis of lubricating oil vapor," *Canadian J. Technol.*, Vol. 34, p. 1957.
9. Swain, M.R. and Adt, R.R., "The Hydrogen-Air Fueled Automobile," in 7th Intersociety Energy Conversion Engineering Conference, American Chemical Society, Washington, D.C., p. 1382, 1972.
10. Lynch, F.E., "Parallel induction: a simple fuel control method for hydrogen engines," *Int. J. Hydrogen Energy*, Vol. 8, p. 721, 1983.
11. MacCarley, C.A., "Electronic Fuel Injection Techniques for Hydrogen Fueled I.C. Engines," M.S. Thesis in Engineering, University of California, Los Angeles, 1978.
12. MacCarley, C.A. and Van Vorst, W.D., "Electronic fuel injection techniques for hydrogen powered I.C. engines," *Int. J. Hydrogen Energy*, Vol. 5, p. 179, 1980.
13. Mathur, H.B. and Das, L.M., "Performance Characteristics of a Hydrogen Fuelled S.I. Engine Using Timed Manifold Injection," *Int. J. Hydrogen Energy*, Vol. 16, p. 115, 1991.
14. Das, L.M., "Fuel Induction Techniques for a Hydrogen Operated Engine," *Int. J. Hydrogen Energy*, Vol. 15, p. 833, 1990.
15. Das, L.M., "Studies on timed manifold injection in hydrogen operated spark ignition engine: performance, combustion and exhaust emission characteristics," Ph.D. Thesis, Indian Institute of Technology, New Delhi, India, 1987.
16. Heffel, J. W. and Norbeck, J. M., "Preliminary Evaluation of UC Riverside's Hydrogen Truck", presented at the Sixth Annual National Hydrogen Meeting, Alexandria, VA, 1995.
17. Furuhama, S. and Fukuma, T., "High Output Power Hydrogen Engine with High Pressure Fuel Injection, Hot Surface Ignition and Turbo-Charging," in Hydrogen Energy Progress V, Edited by T. N. Veziroglu and J. B. Taylor, Pergamon Press, Elmsford, NY, p. 1493, 1984.
18. Furuhama, S., "Hydrogen Engine Systems for Land Vehicles," in Hydrogen Energy Progress VII, Edited by T. N. Veziroglu and A. N. Protsenko, Pergamon Press, Elmsford, NY, p. 1841, 1988.
19. Varde, K.S. and Frame, G.A., "Development of a High Pressure Hydrogen Injection for SI Engine and Results of Engine Behavior," in Hydrogen Energy Progress V, Edited by T. N. Veziroglu and J. B. Taylor, Pergamon Press, Elmsford, NY, p. 1505, 1984.
20. Krepec, T., Tebelis, T. and Kwok, C., "Fuel Control Systems for Hydrogen-Fueled Automotive Combustion Engines-- A Prognosis," *Int. J. Hydrogen Energy*, Vol. 9, p. 109, 1984.
21. Krepec, T., Giannacopoulos, T. and Miele, D., "New Electronically Controlled Hydrogen-Gas Injector Development and Testing," in Hydrogen Energy Progress VI, Edited by T. N. Veziroglu, N. Getoff and P. Weinzierl, Pergamon Press, Elmsford, NY, p. 1087, 1986.
22. Krepec, T., Giannacopoulos, T. and Miele, D., "New electronically controlled hydrogen-gas injector development and testing," *Int. J. Hydrogen Energy*, Vol. 12, p. 855, 1987.
23. Glasson, N. and Green, R., "High Pressure Hydrogen Injection," in Hydrogen Energy Progress IX, Edited by T. N. Veziroglu and C. D. -J. Pottier, International Association for Hydrogen Energy, Coral Gables, FL, p. 1285, 1992.
24. Norbeck, J.M., Heffel, J. W., Durbin, D. T., Tabbara, T., Bowden, J. M., and Montano, M. C., (1996). *Hydrogen Fuel for Surface Transportation*, Society of Automotive Engineers, Inc., Warrendale, PA.
25. Barkhimer, R.L., and Wong, H. (1995). Application of Digital, Pulse-Width-Modulated Sonic Flow Injectors for Gaseous Fuels, SAE Technical Paper 951912, also in *Gaseous-Fuel Engine Technology*, SAE Technical Book SP-1104.
26. Barkhimer, R.L.; Beck, N. J.; and Weseloh, W.E. (1983). Development of a durable, reliable and fast responding solenoid valve, SAE Technical Paper 831326.
27. Smith, J.R. (1994). Optimized Hydrogen Piston Engines, SAE Technical Paper 94C020, also in *Leading Change: The Transportation Electronic Revolution*; Proceedings of the 1994 International Congress on Transportation Electronics, P-283.

APPENDIX

SP-014 Injector Summary

Pulse Width (ms)	Flow rate (mg/injection)						517_14A	
	Run 1	Run 2	Run 3	Run 4	AVG Q	ST DEV	% DEV	AVG PSIA
4	1.22	1.22	1.22	1.22	1.22	0.0021	0.17%	35
6	1.54	1.54	1.55	1.55	1.54	0.0061	0.39%	36
8	2.11	2.10	2.11	2.10	2.11	0.0020	0.10%	36
10	2.55	2.55	2.56	2.55	2.55	0.0041	0.16%	36
12	2.90	2.89	2.89	2.86	2.88	0.0162	0.56%	36
14	3.32	3.32	3.31	3.31	3.32	0.0067	0.20%	35
16	3.71	3.72	3.72	3.73	3.72	0.0088	0.24%	35

Pulse Width (ms)	Flow rate (mg/injection)						511_14B	
	Run 1	Run 2	Run 3	Run 4	AVG Q	ST DEV	% DEV	AVG PSIA
4	2.73	2.75	2.76	2.77	2.75	0.0169	0.61%	58
6	3.37	3.23	3.45	3.46	3.38	0.1082	3.20%	58
8	4.41	4.50	4.65	4.61	4.54	0.1066	2.35%	59
10	5.61	5.69	5.78	5.74	5.71	0.0716	1.25%	59
12	6.49	6.49	6.53	6.51	6.51	0.0156	0.24%	60
14	7.38	7.32	7.32	7.43	7.36	0.0555	0.75%	60
16	8.11	8.29	8.17	8.34	8.23	0.1043	1.27%	61

Pulse Width (ms)	Flow rate (mg/injection)						510_14C	
	Run 1	Run 2	Run 3	Run 4	AVG Q	ST DEV	% DEV	AVG PSIA
4	3.86	3.88	3.85	3.87	3.86	0.0114	0.29%	83
6	5.10	5.07	5.10	5.11	5.09	0.0163	0.32%	84
8	6.14	6.10	6.22	6.22	6.17	0.0603	0.98%	83
10	7.64	7.71	7.71	7.74	7.70	0.0428	0.56%	83
12	8.56	8.69	8.66	8.67	8.64	0.0584	0.68%	84
14	9.96	10.02	9.99	10.01	9.99	0.0296	0.30%	84
16	11.27	11.37	11.31	11.37	11.33	0.0512	0.45%	84

Pulse Width (ms)	Flow rate (mg/injection)						604_14D	
	Run 1	Run 2	Run 3	Run 4	AVG Q	ST DEV	% DEV	AVG PSIA
4	4.61	4.56	4.55	4.47	4.55	0.0594	1.31%	108
6	5.42	5.41	5.40	5.41	5.41	0.0066	0.12%	109
8	7.46	7.52	7.55	7.57	7.52	0.0489	0.65%	110
10	9.12	9.06	9.05	9.05	9.07	0.0334	0.37%	110
12	10.53	10.56	10.50	10.51	10.53	0.0234	0.22%	111
14	12.09	12.05	12.05	12.04	12.06	0.0235	0.20%	110
16	13.61	13.57	13.68	13.71	13.64	0.0675	0.49%	110

Pulse Width (ms)	Flow rate (mg/injection)						521_14E	
	Run 1	Run 2	Run 3	Run 4	AVG Q	ST DEV	% DEV	AVG PSIA
4	4.40	4.37	4.33	4.32	4.35	0.0351	0.81%	134
6	6.40	6.41	6.41	6.41	6.41	0.0038	0.06%	133
8	8.08	8.08	8.06	8.05	8.07	0.0133	0.17%	134
10	9.91	9.96	9.97	10.06	9.97	0.0600	0.60%	134
12	12.02	12.02	12.03	12.02	12.02	0.0041	0.03%	133
14	13.90	13.94	13.96	13.97	13.94	0.0317	0.23%	133
16	15.96	15.98	15.94	15.94	15.96	0.0191	0.12%	134

Pulse Width (ms)	Flow rate (mg/injection)						521_14F	
	Run 1	Run 2	Run 3	Run 4	AVG Q	ST DEV	% DEV	AVG PSIA
4	5.05	5.05	5.03	5.03	5.04	0.0117	0.23%	157
6	6.63	6.63	6.62	#REF!	6.63	0.0043	0.07%	158
8	9.02	9.03	9.01	8.82	8.97	0.0994	1.11%	158
10	11.37	11.43	11.32	11.36	11.37	0.0480	0.42%	158
12	13.56	13.55	13.53	13.54	13.54	0.0120	0.09%	158
14	16.16	16.12	16.06	16.16	16.12	0.0445	0.28%	158
16	19.13	19.12	19.11	19.07	19.11	0.0281	0.15%	159

Pulse Width (ms)	Flow rate (mg/injection)						531_14G	
	Run 1	Run 2	Run 3	Run 4	AVG Q	ST DEV	% DEV	AVG PSIA
4	5.94	5.92	5.93	5.96	5.94	0.0178	0.30%	184
6	7.56	7.56	7.56	7.53	7.55	0.0122	0.16%	184
8	10.09	10.11	10.12	10.09	10.10	0.0112	0.11%	184
10	12.56	12.64	12.56	12.58	12.58	0.0378	0.30%	185
12	15.52	15.50	15.36	15.30	15.42	0.1065	0.69%	185
14	18.20	18.17	18.12	18.10	18.15	0.0473	0.26%	184
16	20.76	20.70	20.68	20.48	20.66	0.1190	0.58%	183

Pulse Width (ms)	Flow rate (mg/injection)						531_14H	
	Run 1	Run 2	Run 3	Run 4	AVG Q	ST DEV	% DEV	AVG PSIA
4	6.04	5.99	6.00	5.96	6.00	0.0334	0.56%	208
6	7.67	7.69	7.68	7.69	7.68	0.0053	0.07%	209
8	10.61	10.62	10.64	10.63	10.63	0.0138	0.13%	210
10	13.54	13.51	13.53	13.55	13.53	0.0157	0.12%	210

SP-021 Injector Summary

	Flow rate (mg/injection)						517_21A	
Pulse Width (ms)	Run 1	Run 2	Run 3	Run 4	AVG Q	ST DEV	% DEV	AVG PSIA
4	1.23	1.23	1.23	1.23	1.23	0.0023	0.19%	36
6	1.54	1.55	1.55	1.55	1.54	0.0049	0.32%	36
8	2.11	2.11	2.10	2.11	2.11	0.0049	0.23%	36
10	2.54	2.55	2.56	2.56	2.55	0.0095	0.37%	36
12	2.87	2.90	2.90	2.90	2.89	0.0145	0.50%	36
14	3.32	3.34	3.34	3.32	3.33	0.0105	0.31%	36
16	3.70	3.72	3.72	3.72	3.71	0.0093	0.25%	35

	Flow rate (mg/injection)						511_21B	
Pulse Width (ms)	Run 1	Run 2	Run 3	Run 4	AVG Q	ST DEV	% DEV	AVG PSIA
4	2.02	2.01	2.03	2.03	2.02	0.0080	0.40%	59
6	2.62	2.64	2.64	2.63	2.63	0.0077	0.29%	60
8	3.28	3.30	3.30	3.31	3.30	0.0106	0.32%	60
10	3.99	4.00	4.00	4.02	4.01	0.0132	0.33%	60
12	4.73	4.78	4.73	4.73	4.74	0.0251	0.53%	61
14	5.42	5.47	5.48	5.47	5.46	0.0277	0.51%	61
16	6.18	6.23	6.24	6.18	6.21	0.0327	0.53%	61

	Flow rate (mg/injection)						510_21C	
Pulse Width (ms)	Run 1	Run 2	Run 3	Run 4	AVG Q	ST DEV	% DEV	AVG PSIA
4	2.72	2.68	2.67	2.67	2.69	0.0206	0.77%	84
6	3.19	3.20	3.19	3.19	3.19	0.0038	0.12%	85
8	4.20	4.21	4.22	4.20	4.21	0.0065	0.15%	85
10	5.28	5.30	5.30	5.27	5.29	0.0135	0.26%	84
12	6.35	6.35	6.36	6.38	6.36	0.0159	0.25%	85
14	7.45	7.47	7.48	7.51	7.48	0.0231	0.31%	85
16	8.36	8.46	8.48	8.47	8.44	0.0547	0.65%	85

	Flow rate (mg/injection)						524_21D	
Pulse Width (ms)	Run 1	Run 2	Run 3	Run 4	AVG Q	ST DEV	% DEV	AVG PSIA
4	2.88	2.88	2.88	2.86	2.87	0.0065	0.23%	111
6	4.26	4.27	4.27	4.28	4.24	0.0145	0.34%	112
8	5.44	5.46	5.45	5.45	5.45	0.0034	0.06%	112
10	6.81	6.83	6.83	6.83	6.67	0.1291	1.94%	111
12	7.79	7.77	7.79	7.77	7.79	0.0138	0.18%	111
14	9.19	9.14	9.17	9.13	9.20	0.0162	0.18%	111
16	10.30	10.28	10.28	10.28	10.26	0.0107	0.10%	111

	Flow rate (mg/injection)						521_21E	
Pulse Width (ms)	Run 1	Run 2	Run 3	Run 4	AVG Q	ST DEV	% DEV	AVG PSIA
4	3.49	3.47	3.45	3.45	3.46	0.0196	0.57%	135
6	4.43	4.44	4.42	4.43	4.43	0.0093	0.21%	134
8	5.96	6.02	6.01	5.97	5.99	0.0274	0.46%	134
10	7.56	7.46	7.44	7.45	7.48	0.0553	0.74%	135
12	9.01	9.02	9.03	8.99	9.01	0.0185	0.21%	134
14	10.40	10.43	10.42	10.42	10.41	0.0136	0.13%	134
16	11.95	11.97	11.99	11.98	11.97	0.0138	0.12%	134

	Flow rate (mg/injection)						521_21F	
Pulse Width (ms)	Run 1	Run 2	Run 3	Run 4	AVG Q	ST DEV	% DEV	AVG PSIA
4	3.99	3.97	3.97	3.94	3.97	0.0233	0.59%	158
6	5.07	5.03	5.03	5.03	5.04	0.0189	0.38%	159
8	6.81	6.80	6.80	6.79	6.80	0.0098	0.14%	159
10	8.44	8.44	8.44	8.42	8.43	0.0066	0.08%	158
12	10.20	10.21	10.21	10.19	10.20	0.0089	0.09%	159
14	12.15	12.20	12.19	12.22	12.19	0.0288	0.24%	159
16	13.82	13.90	13.93	13.96	13.90	0.0606	0.44%	159

	Flow rate (mg/injection)						531_21G	
Pulse Width (ms)	Run 1	Run 2	Run 3	Run 4	AVG Q	ST DEV	% DEV	AVG PSIA
4	4.10	4.04	3.99	3.98	4.02	0.0549	1.36%	187
6	5.75	5.78	5.77	5.77	5.77	0.0138	0.24%	185
8	7.50	7.44	7.42	7.42	7.45	0.0366	0.49%	185
10	9.41	9.41	9.40	9.38	9.40	0.0129	0.14%	186
12	11.72	11.72	11.71	11.71	11.71	0.0054	0.05%	185
14	13.80	13.79	13.76	13.77	13.78	0.0197	0.14%	185
16	15.77	15.68	15.55	15.96	15.74	0.1748	1.11%	183

	Flow rate (mg/injection)						611_21H	
Pulse Width (ms)	Run 1	Run 2	Run 3	Run 4	AVG Q	ST DEV	% DEV	AVG PSIA
4	4.28	4.31	4.32	4.34	4.31	0.0219	0.51%	208
6	5.90	5.91	5.89	5.88	5.89	0.0092	0.16%	209
8	8.26	8.28	8.28	8.22	8.26	0.0270	0.33%	209
10	10.35	10.34	10.39	10.35	10.36	0.0202	0.19%	209

SP-051 Injector Summary

Pulse Width (ms)	Run 1	Run 2	Run 3	Run 4	AVG Q	ST DEV	807_51B % DEV	AVG PSIA
			Flow rate (mg/injection)					
4	1.56	1.55	1.54	1.54	1.55	0.0094	0.0061	62
6	2.21	2.21	2.20	2.20	2.21	0.0031	0.0014	62
8	2.79	2.79	2.78	2.79	2.79	0.0058	0.0021	61
10	3.36	3.37	3.36	3.37	3.37	0.0084	0.0025	61
12	4.01	4.01	4.02	4.02	4.01	0.0083	0.0021	61
14	4.62	4.63	4.63	4.63	4.62	0.0061	0.0013	61
16	5.24	5.25	5.26	5.26	5.25	0.0080	0.0015	61

Pulse Width (ms)	Run 1	Run 2	Run 3	Run 4	AVG Q	ST DEV	807_51D % DEV	AVG PSIA
			Flow rate (mg/injection)					
4	2.91	2.91	2.89	2.88	2.90	0.0152	0.0053	110
6	3.50	3.52	3.53	3.51	3.52	0.0100	0.0028	110
8	4.68	4.66	4.68	4.68	4.67	0.0076	0.0016	110
10	5.77	5.78	5.78	5.78	5.78	0.0041	0.0007	110
12	6.85	6.86	6.86	6.84	6.86	0.0094	0.0014	110
14	7.95	7.99	7.99		7.98	0.0227	0.0028	110
16		9.12	9.14	9.14	9.13	0.0127	0.0014	111

Pulse Width (ms)	Run 1	Run 2	Run 3	Run 4	AVG Q	ST DEV	807_51F % DEV	AVG PSIA
			Flow rate (mg/injection)					
4	3.46	3.45	3.46	3.48	3.46	0.0137	0.0039	159
6	4.69	4.70	4.76	4.70	4.71	0.0314	0.0067	159
8	6.20	6.23	6.26	6.26	6.24	0.0248	0.0040	159
10	7.81	7.83	7.84	7.84	7.83	0.0157	0.0020	159
12	9.47	9.54	9.53		9.51	0.0396	0.0042	159
14	11.02	11.06	11.09	11.07	11.06	0.0296	0.0027	159
16	12.78	12.79	12.81	12.83	12.80	0.0216	0.0017	159

Pulse Width (ms)	Run 1	Run 2	Run 3	Run 4	AVG Q	ST DEV	807_51H % DEV	AVG PSIA
			Flow rate (mg/injection)					
4	4.49	4.45	4.44	4.48	4.46	0.0247	0.0055	210
6	5.88	5.83	5.79	5.83	5.83	0.0383	0.0066	211
8	7.74	7.76	7.76	7.75	7.75	0.0094	0.0012	211
10	9.88	9.87	9.89	9.87	9.88	0.0120	0.0012	210
12	12.14	12.13	12.16	12.13	12.14	0.0145	0.0012	210
14	14.33	14.34	14.36	14.34	14.34	0.0127	0.0009	210
16	16.46	16.47	16.44	16.46	16.46	0.0108	0.0007	210

SP-010 Injector Summary

Pulse Width (ms)	Run 1	Run 2	Run 3	Run 4	AVG Q	ST DEV	918_10B % DEV	AVG PSIA
			Flow rate (mg/injection)					
4	6.69	6.21	6.06	6.04	6.25	0.3017	0.0298	59
6	8.82	8.79	8.78	8.79	8.80	0.0175	0.0119	59
8	11.93	11.95	11.94	11.95	11.94	0.0094	0.0125	59
10	14.68	14.73	14.64	14.64	14.67	0.0418	0.0125	60
12	17.21	17.20	17.13	17.12	17.17	0.0460	0.0131	59
14	19.63	19.63	19.64	19.68	19.65	0.0205	0.0126	59
16	22.08	22.14	22.09	22.10	22.10	0.0231	0.0120	60

Pulse Width (ms)	Run 1	Run 2	Run 3	Run 4	AVG Q	ST DEV	918_10D % DEV	AVG PSIA
			Flow rate (mg/injection)					
4	7.70	7.59	7.41	7.44	7.53	0.1363	0.0462	110
6	12.22	12.27	12.14	12.17	12.20	0.0557	0.0295	110
8	18.33	18.43	18.40	18.33	18.37	0.0492	0.0196	109
10	23.53	23.43	23.38	23.36	23.42	0.0746	0.0199	109
12	28.16	28.12	28.04	28.04	28.09	0.0624	0.0186	109
14	32.65	32.61	32.50	32.50	32.56	0.0756	0.0197	108
16	37.18	37.20	37.24	37.22	37.21	0.0259	0.0183	108

981922

Turbocharged Hydrogen Fueled Vehicle Using Constant Volume Injection (CVI)

**James W. Heffel, Michael N. McClanahan
and Joseph M. Norbeck**
University of California, Riverside, CE-CERT

Frank Lynch
Hydrogen Components, Inc.

ABSTRACT

A University of California, Riverside (UCR) 1992 Ford Ranger truck was converted to operate on hydrogen which is produced from water electrolysis at the UCR College of Engineering-Center for Environmental Research and Technology (CE-CERT) Solar Hydrogen Research Facility (SHRF).

The Ford Ranger's 2.3L engine was modified to operate as a lean-burn, hydrogen fuel internal combustion (IC) engine, using a Constant Volume Injection (CVI) system with closed-loop control and exhaust oxygen feedback. The vehicle had excellent starting, idle, and shut-down operation; a range in excess of 161km (100 miles); and initially operated with virtually no preignition problems typical of hydrogen fuel engines. At speeds above 64 km/h (40 mph), the vehicle exhibited performance characteristics similar to comparable gasoline-powered vehicles, although further improvements are needed at lower speeds. Tailpipe emissions for carbon monoxide, carbon dioxide, and hydrocarbons were found to be negligible and attributed to oil seeping by the piston rings. The emission rate for oxides of nitrogen (NO_x), as measured by the Federal Test Procedure (FTP-77), was 0.37 grams per mile. Evaluation of the vehicle by both on-road and chassis dynamometer testing shows that appropriate adjustment of the fuel injection system is critical to the overall performance, driveability, and NO_x emission rates. Operation of the engine at an equivalence ratio (actual air/fuel ratio divided by the stoichiometric air/fuel ratio) phi below 0.5 is critical for low NO_x emission, but this may result in less than optimal driveability given the vehicle's weight and engine displacement.

INTRODUCTION

In March, 1992, the College of Engineering-Center for Environmental research and Technology (CE-CERT) at UC Riverside was funded by the South Coast Air Quality Management District (SCAQMD) to develop and evaluate a prototype renewable transportation system based on hydrogen. The various aspects of this project addressed the renewable generation of hydrogen from water using solar energy; the development of a hydrogen-vehicle refueling station; and the development and evaluation of a prototype hydrogen-powered vehicle.

This paper describes the development and evaluation of the hydrogen-powered 1992 Ford Ranger truck with a turbocharged 2.3L engine which was converted to operate on hydrogen as part of the program.

BACKGROUND

Hydrogen has been considered as a fuel for internal combustion engines for more than 100 years. In 1820, Cecil was the first to recommend the use of hydrogen as a fuel for powering engines [1].

One of the primary problems that have been encountered in the development of operational hydrogen engines is premature ignition. Premature ignition occurs when the cylinder charge is ignited before the ignition by the spark plug. It results in an inefficient, rough running engine. A flashback in the intake manifold can also develop if the premature ignition occurs near the fuel intake valve and the resultant flame travels back into the induction system. Premature ignition is a much greater problem in hydrogen-fueled engines than in gasoline-fueled engines, due to hydrogen's low minimum ignition energy and wide flammability range.

Many attempts have been made to eliminate or minimize the preignition problem by means of the fuel delivery system. The four general fuel delivery systems that have been investigated over the years are: carburetion, inlet manifold injection, inlet port injection, and direct cylinder injection. The first three techniques involve forming the fuel-air mixture during the intake stroke, either through the carburetor, in the intake manifold or an inlet port.

Direct cylinder injection is more technologically sophisticated and involves forming the fuel-air mixture inside the combustion cylinder after the air intake valve has closed. A brief review of some of the developments of different fuel delivery systems is given here.

The simplest method of delivering fuel to a hydrogen engine is via a carburetor. This system is similar to that used for gasoline. The advantage of this system is that it is fairly simple and does not require a sophisticated high-pressure injector. The disadvantage of this technique is that engines with carburetors are more susceptible to irregular combustion due to preignition and backfire problems. Ricardo performed some of the first rigorous experiments in the 1920s using a single-cylinder engine with a carburetor [2, 3, 4, 5]. He experienced severe preignition and flashback at equivalence ratios greater than unity. King et al. also experienced preignition problems when they used carburetors with hydrogen engines [2, 3, 4, 6, 7]. To eliminate these problems, they made several modifications to the engine, such as using cold spark plugs and an aged sodium-filled exhaust valve.

Troubled by the continual preignition problems, a number of researchers began exploring alternatives to carburetion. A simple solution is a technique called inlet manifold injection. This technique involves injecting fuel directly into the intake manifold rather than drawing fuel through the carburetor. Typically, the timing of the fuel injection is controlled so the hydrogen is not injected into the manifold until after the beginning of the intake stroke, at a point where conditions are much less severe and the probability for premature ignition is reduced. The air, which is injected separately at the beginning of the intake stroke, dilutes the hot residual gases and cools off any hot spots. A similar technique is inlet port injection. With this system, fuel and air are injected from separate ports into the combustion chamber during the intake stroke and, as such, are not premixed in the intake manifold. Both of these methods can be used successfully to overcome preignition problems.

In the 1970s and 1980s, Swain and Adt at the University of Miami carried out experiments using a method known as the Hydrogen Induction Technique, in which hydrogen was supplied separately from the air through a flow tube close to the intake valve [8]. Their results demonstrated that inlet injection techniques could be used successfully to suppress preignition.

Lynch et al. also investigated a technique called parallel induction [9]. The basic concept is similar to inlet port injection in that fuel and air enter the combustion chamber during the intake stroke, but are not premixed in the intake manifold. The engines used in Lynch's tests also incorporated turbochargers with aftercooling to increase the power output of the engine. The parallel induction scheme led to the development of the Constant Volume Injection (CVI) system that is utilized on the hydrogen truck discussed in this paper.

In the 1970s, MacCarley and Van Vorst extensively studied the prospects of using different fuel delivery systems to control backfire problems using a converted U.S. Postal Service Jeep [10,11]. They found that the performance of the port injection system was more favorable than that of the direct injection system since it was not subject to any problems related to poor fuel-air mixing. Their research showed that the conventional gasoline injectors were much too small for the required hydrogen flow. They designed, fabricated and tested several prototype injectors. By use of a capacitor discharge driver for the coil they were able to achieve very fast opening and closing, as low as 1 ms. The most advanced injector was able to inject 240 cc hydrogen at 207 kPa (30 psig) in a 5 ms pulse.

Timed manifold injection has been advocated by researchers at the Indian Institute of Technology in New Delhi, India [12, 13, 14]. These researchers found that engines using a timed manifold induction system enjoyed some of the benefits of both compression ignition and spark ignition engines and operated essentially free of combustion problems. These engines had power outputs comparable to that of a spark-ignition engine, while achieving diesel-like "quality" control and thermal efficiency. Studies were also done with low-pressure, direct cylinder injection, but it was found that this technique was more susceptible to incomplete combustion.

As hydrogen engines continued to progress in sophistication, direct injection into the combustion cylinder during the compression stroke became more commonly used. Utilizing this technique, premature ignition during the intake stroke and backfire are completely avoided since the intake valve is closed when the fuel is injected. The power output can also be increased with this method to be 20% more than that of a gasoline engine and 42% more than that of a hydrogen mixture formed using a carburetor. The development of more modern, higher-pressure injectors has been more difficult, as problems associated with the high operating pressure and the precise control of the injection parameters necessary under these conditions must first be overcome. The valve must travel considerably faster than a low-pressure injector valve in order to obtain shorter injection times. The leak rate also must be controlled so it is below the minimum amount needed to cause ignition before the main injection. This can be an especially difficult problem since the hydrogen molecule is small and has a high rate of diffusivity, enabling it to escape from even the smallest leaks. Finally, mechanical parts of the injector itself must be able to withstand severe thermal conditions and pressures as high as 10 MPa [14].

Many other groups have contributed to the development of high-pressure hydrogen injection to its present level of technology. Included among these are researchers in the United States [4, 15], and Canada [16, 17, 18]. A more complete review is given in Norbeck et al. [19].

VEHICLE AND ENGINE MODIFICATIONS

The vehicle used for the conversion was a stock 1992 Ford Ranger truck. A truck was used because it allowed for easy installation of the hydrogen storage vessels and was capable of handling the additional weight. In addition, it is expected that this vehicle will be used in the future as an on-road hydrogen storage test-bed, so ease of installation and removal of various types of hydrogen storage systems was also important. The fuel storage system installed in the truck-bed facilitates this requirement quite effectively. This converted vehicle is identified in this paper as UCR1.

VEHICLE DESCRIPTION BEFORE CONVERSION – The vehicle came equipped with a four-speed automatic transmission, a 4.11:1 rear differential, and power steering (no air conditioning). The stock engine is a 2.3L, four cylinder engine with a cross flow head design (intake ports located on the opposite side of the exhaust ports). This head design simplified the modifications required to add the CVI and exhaust system to the engine by not having to install both systems on the same side of the engine.

The specifications of the stock engine supplied with this vehicle are as follows:

- Naturally aspirated
- 9:1 compression ratio
- Dual spark plug distributorless ignition
- Multiport electronic fuel injection
- Exhaust gas oxygen sensor (EGO)
- Exhaust gas recirculation (EGR)
- Three-way catalyst

ENGINE MODIFICATIONS – All the gasoline-related components were removed from the engine so the new hardware (turbocharger, intercooler, exhaust and air induction system, etc.) could be sized for the engine compartment. The engine was then removed from the vehicle and modified as follows:

Casting irregularities (potential "hot spots") in the intake and exhaust ports and combustion chamber were removed. As the result of smoothing out the combustion chamber, the compression ratio of the engine reduced from the original 9:1 to 8:1.

The intake and exhaust ports were opened up, increasing the flow capacity of the heads by 30%.

An Engle TCS210H camshaft with a 0.495-inch lift and 214° duration was installed.

4. The stock dual spark plug distributorless ignition system was replaced with an Engine Electronics Compu-fire distributorless ignition system with fixed timing .

5. Champion G-81 spark plugs (10 mm) and shroud chambers in the spark plug ports on the intake side of the cylinder head were installed **(Figure 1)**. The shroud chambers and colder rated spark plugs were used to minimize the chance that the spark plug would cause preignition due to a hot electrode.

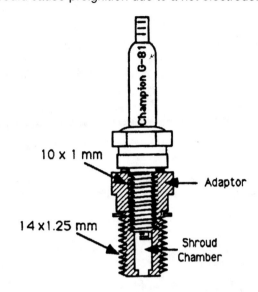

Figure 1. Spark plug adapter with shroud chamber.

6. The redundant spark plugs on the exhaust side of the cylinder head were removed and replaced with "blank" plugs.

7. A special high volume crankcase ventilation system was installed to purge unburned air-fuel mixtures from the crankcase.

8. A pressure relief valve on the rocker arm cover was installed to prevent engine damage in the event of ignition of gases in the crankcase.

9. A Turbonetics Model T-2 turbocharger with 0.48 a/f ratio turbine and integral wastegate (set at 12 psi) was installed.

10. An air-to-air intercooler was installed.

11. A Hydrogen Consultant, Inc. (HCI) Constant Volume Injection (CVI) system for delivering timed aliquots of hydrogen to the ports in phase with the intake stroke of the engine (i.e., sequential multi-port injection) was installed.

12. An NGK exhaust gas oxygen sensor—suitable for operation at ultra-lean conditions—was installed.

13. A CVI control module designed by HCI was installed. (The original engine control module (ECM) remains in the vehicle, but is used only to provide control of the automatic transmission).

14. New air induction and exhaust systems were installed.

VEHICLE MODIFICATIONS – Modifications to the truck include the following:

1. The gasoline tank was removed.

2. Three hydrogen storage vessels were installed in the bed of the truck. Storage vessels were provided by Comdyne Industries and have the following specifications:

   a. Aluminum liner with a full fiberglass wrapping.

   b. DOT approved for CNG, with DOT exemption for use with hydrogen.

   c. Maximum working pressure of 25 MPa (3,600 psi).

   d. Each vessel has its own thermal/pressure relief valve and isolation valve.

   e. Two of the three vessels have a "water volume" of 136 L (4.8 cubic feet) for a hydrogen storage capacity of 940 scf (2.18 kg of hydrogen at 25 MPa).

   f. The third vessel has a "water volume" of 45.3 L (1.6 cubic feet) for a hydrogen storage capacity of 313 scf (.73 kg of hydrogen at 25 MPa).

   g. Total hydrogen storage capacity is 2193 scf (5.1 kg of hydrogen at MPa). Total "water volume" is 317.3 Lr (11.2 cubic feet).

   h. Total energy storage capacity of 616,000 Btu (lower heating value), or a gasoline equivalent of approximately 19.3 L (5.1 gallons).

3. A hydrogen ancillary system was installed.

4. The rear differential gear ratio was changed from 4.11:1 to 4.56:1.

5. The stall speed of the torque converter was increased from 600 rpm to 1400 rpm.

CONSTANT VOLUME INJECTION SYSTEM – The fundamental principle of operation of a constant volume fuel injection system is the ideal gas pressure-volume-temperature relationship:

$$PV = nRT$$

where:

P is the absolute pressure

V is the volume

n is the number of moles of gas

R is the universal gas constant

T is the absolute temperature.

A CVI system consists of chambers that are charged and discharged by poppet valves in phase with the operating cycle of an internal combustion engine. CVI is a form of sequential multi-port fuel injection. **Figures 2a, b, and c** illustrate the CVI process for a 1-cylinder gaseous fueled engine. A multi-cylinder CVI system would have a chamber and a set of valves to serve each engine cylinder. **Figure 3** shows a cross-section of the CVI unit.

In **Figure 2a**, a chamber with fixed volume, V_{CVI}, is pressurized to P_{fuel} at approximately constant temperature, T. In **Figure 2b**, the charging valve is closed and the cham-

ber stores a measured charge of fuel gas quantified by the pressure and temperature of the gas, and by the volume of the chamber:

$$n_i = (V_{CVI}/RT) \times P_{fuel}$$

In **Figure 2c**, the discharge valve opens and releases the stored fuel gas into the intake port of an engine. The discharge valve opens in phase with the intake stroke so fuel and air flow into the engine cylinder together. At the end of an injection event, the pressure in the CVI chamber is equal to the manifold air pressure (MAP). The quantity of residual fuel gas in the chamber is

$$n_f = (V_{CVI}/RT) \times MAP$$

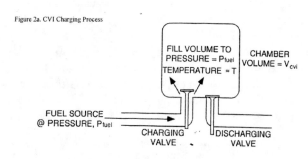

Figure 2a. CVI Charging Process

FILL VOLUME TO PRESSURE = P_{fuel} TEMPERATURE ≈ T
CHAMBER VOLUME = V_{cvi}
FUEL SOURCE @ PRESSURE, P_{fuel}
CHARGING VALVE   DISCHARGING VALVE

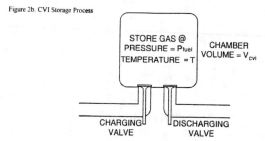

Figure 2b. CVI Storage Process

STORE GAS @ PRESSURE = P_{fuel} TEMPERATURE ≈ T
CHAMBER VOLUME = V_{cvi}
CHARGING VALVE   DISCHARGING VALVE

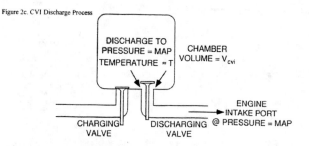

Figure 2c. CVI Discharge Process

DISCHARGE TO PRESSURE = MAP TEMPERATURE ≈ T
CHAMBER VOLUME = V_{cvi}
ENGINE INTAKE PORT
CHARGING VALVE   DISCHARGING VALVE @ PRESSURE = MAP

Figure 2. CVI processes.

The difference between the fuel gas inventory in the chamber before and after the injection event is the amount that is delivered to the intake port,

$$\Delta n_{fuel} = n_i - n_f = (V_{cvi}/RT)(P_{fuel} - MAP) \qquad \text{(Eq. 1)}$$

By maintaining a fixed ratio between P_{fuel} and MAP, i.e.,

$$P_{fuel} = kMAP,$$

Equation (1) may be rewritten as,

$$\Delta n_{fuel} = (V_{cvi}/RT)(k\,MAP - MAP)$$

And rearranged as,

$$\Delta n_{fuel} = (V_{cvi}/RT)(k-1)\,MAP \qquad \text{(Eq. 2)}$$

The air/fuel charge in an engine cylinder also may be characterized, to a good approximation, by pressure, temperature and volume, as illustrated in **Figure 4**.

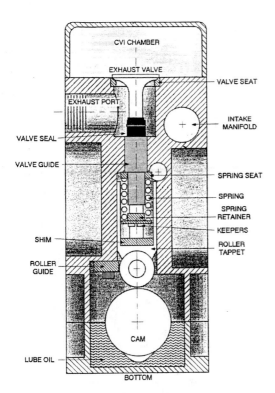

Figure 3. Exhaust valve cross-section of a CVI fuel injection system. Most of the parts are off-the-shelf automotive components that are already in production.

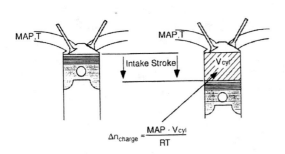

Figure 4. The number of moles air/fuel charge drawn into an engine cylinder is approximately proportional to swept cylinder volume and manifold pressure, and inversely proportional to manifold air temperature.

$$\Delta n_{charge} \approx \frac{MAP \cdot V_{cyl}}{RT} \qquad \text{(Eq. 3)}$$

The mole fraction of fuel in the mixture is, to a first order approximation, equation (2) divided by equation (3):

$$\frac{\Delta n_{fuel}}{\Delta n_{charge}} = \frac{(V_{cvi}/R)(k-1)MAP}{(V_{cyl}/R)MAP} = (k-1)\frac{V_{cvi}}{V_{cyl}} \qquad \text{(Eq. 4)}$$

The first order operating principles of CVI show that a constant air/fuel ratio may be maintained by holding k (the ratio of fuel pressure to manifold air pressure) constant. This requires a pressure ratio regulator. A typical value of k in CVI system is 2.5 (phi of 0.4). Naturally aspirated engines operated near sea level have maximum

MAP near 100 kPa (14 psia). This corresponds to a maximum CVI fuel supply pressure of 250 kPa (36 psia). In turbocharged engines, CVI fuel supply pressure up to 700 kPa (100 psia).

The volume of the CVI chamber is directly dependent on the engine cylinder volume and with the desired mole fraction of fuel gas in the intake mixture.

The simplified, first order considerations discussed above show that fuel and air flow in a CVI-equipped engine are directly related to pressure and volumes. Variations in air and fuel temperatures also affect flow. So do variations in the volumetric efficiencies of the engine and CVI. These effects are small, relative to the first order operating principles, yet significant.

To handle these real-world variations, CVI electronics adjust the value of k (equation 4) applied by an electronic fuel pressure regulator. This is an analog equivalent to adaptive learning. During a transient to a new operating condition, the old value of k is the first "guess" for the new condition. Since the old condition k was adjusted for temperature effects, this "learned" information is carried forward. **Figure 5** is a sketch of a complete CVI control system. If k is 2.5, based on idealized calculations, the feedback electronics may revise it to, 2.4 or 2.6, as necessary to account for small errors in the first order assumptions. The full range of CVI's controller is 2 < k < 3.

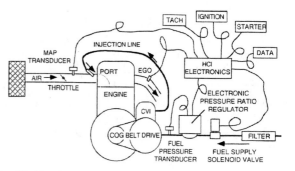

Figure 5. The CVI control system employs feedforward control, based on a "learned" value of fuel/air pressure ratio, k. k is continually revised by exhaust oxygen feedback. The electronics also switch the ignition and fuel solenoid valve during engine starting and stopping.

SYSTEM OPERATION

Hydrogen is stored on-board the vehicle at 3,600 psi in lightweight composite pressure vessels. The high-pressure hydrogen from the storage vessels enters the first of two pressure regulators where the pressure is reduce to about 100 psi. The hydrogen then enters the second regulator (a Buzmatic electronically controlled ratio pressure regulator) which controls the flow of hydrogen to the CVI unit. The flow rate of hydrogen is determined by a computer control module which receives inputs from the MAP and exhaust gas oxygen (EGO) sensors. Hydrogen from the Buzmatic regulator enters the CVI unit, which controls the injection of the hydrogen to each of the cylinders. The combusted hydrogen (steam) enters the exhaust manifold, where an EGO sensor determines whether the air/fuel ratio is too rich or too lean. Depending on the signal form the EGO sensor, the computer control module will send a signal to the Buzmatic regulator to reduce or increase the flow of hydrogen to the CVI unit.

EVALUATION PROCESS

CE-CERT's evaluation of UCR1 consisted of four specific testing sequences briefly described below.

VEHICLE RANGE AND DRIVEABILITY – The two specific objectives of on-road evaluation were to assess the initial driveability of the vehicle as compared with conventional gasoline-powered vehicles, and to obtain an initial estimate of the range of the vehicle with typical on-road operation. The first evaluation was to focus on the vehicle's acceleration, braking, handling and response during typical on-road driving. The second was to verify that the vehicle has an overall range of at least 100 miles without refueling.

EMISSION TESTING – The vehicle was tested at the Emissions Laboratory of the American Automobile Association of California in Los Angeles using a baseline FTP-77 emission test on a chassis dynamometer. Of specific concern were the NO_x emissions for the various phases of the FTP, although carbon monoxide, carbon dioxide, and hydrocarbon emissions also were measured.

CHASSIS DYNAMOMETER EVALUATION OF ENGINE OPERATION – The vehicle was driven on CE-CERT's Clayton water-brake chassis dynamometer to evaluate the vehicle's specific fuel consumption, power output, torque, engine temperature, speed, and air/fuel ratio at a series of steady-state engine conditions. The purpose of these tests was to provide a careful evaluation of the engine and fuel delivery system as compared to gasoline technology.

ENDURANCE TESTING – It was intended to accumulate a minimum of 16,100 km (10,000 miles) on the vehicle, and to evaluate the vehicle's engine and suspension system as a function of mileage. The current mileage on the vehicle is slightly over 2,253 km (1,400 miles), as it has been mostly used for static displays and our research. Thus, the endurance evaluation is on-going.

RESULTS

VEHICLE DRIVEABILITY AND RANGE – **Table 1** provides a summary of the overall performance characteristics of the vehicle. A detailed description of the evaluation follows.

Table 1.   Summary of UCR1 Performance Characteristics

Gross Weight	1701 kg (3,750 lbs)
Fuel Capacity	2,193 scf
Quarter Mile Time	24.97 sec.
Quarter Mile Speed	92 km/h (57 mph)
Vehicle Range (city driving)	187 km (116 miles)
Approximate Top Speed	145 km/h (90 mph)
Gasoline Equivalent Fuel Economy	9.6 L/km (22.7 mpg)
NO_x Emission (from FTP 77)	0.37 g/mile
Max. Power (@4,000rpm)*	48kW (64HP)
Optimal Efficiency (@ 3,500 rpm)*	21.5%
Specific Fuel Consumption*	.229lb/HP-hr

* Measured at the wheels of the vehicle

Engine starting, idling and shutdown – Engine starting, idling, and shutdown were very good. During engine startup there is a slight delay, as expected, due to a delay programmed into the CVI controller to restrain firing the spark plugs until the engine has turned a few revolutions. This delay allows any residual gas that may have accumulated in the combustion chambers to be vented. After a few turns of the engine, the controller commands the ignition system to fire the spark plugs and engine start-up occurs immediately. During the first 20 to 30 seconds of engine idle, the engine operates in an open-loop (high idle) configuration, which allows the exhaust gas oxygen (EGO) sensor to warm up. After the EGO sensor warms up, the control system returns to a closed-loop configuration (normal idle). For engine shutdown, the controller sends a signal which closes the fuel inlet solenoid valve, but continues to fire the spark plugs until the engine runs out of fuel, at which point the engine stops. This occurs within one second of shutoff. This procedure is used to ensure that no hydrogen is left in the combustion chamber. These systems continue to operate as intended and in a satisfactory manner.

Vehicle range and maximum speed – An estimate of the range was obtained by driving the vehicle over a prescribed route within the Los Angeles basin. The route began in the vicinity of the Los Angeles Zoo and culminated in Diamond Bar, near the offices of the SCAQMD. The vehicle used 5.1kg (11.2 lbs) of hydrogen in 187 km (116 miles) before running out of fuel. The fuel use is equivalent to 19.3 L (5.1 gallons) of gasoline. Thus, the city driving fuel economy was estimated to be 9.6 L/km (22.7mpg). This is excellent fuel economy considering the vehicle weighs 1701 kg (3750 pounds). Since the vehicle exceeded the desired 161 km (100 miles) range, no other evaluation of range was done. Based on on-road tests, the top speed of UCR1 is approximately 145 km/h (90 mph), similar to that of a comparable gasoline vehicle.

Acceleration and braking – The initial driveability evaluation showed low acceleration at low speeds when compared with a comparable gasoline-powered vehicle. This is as expected, and is the result of two primary factors.

First, the vehicle has considerably more weight than the original vehicle, due to the addition of the fuel storage tanks, roll bar, CVI and turbocharger systems. The gross weight increased 32.5% from the stock vehicle weight of 2830 pounds to 3750 pounds. Second, experience showed that ultra lean operation of the engine would reduce NO_x. The intent was to attempt to keep the equivalence ratio (Φ) below 0.5. However, lean operation directly limits the fuel flow to the engine and, at times, significantly affects the vehicle's driveability, particularly at low speeds.

In an attempt to compensate for the low acceleration and driveability, several modifications were made. The torque converter was modified to allow for higher engine rpm at low speeds. It was observed quantitatively that this improved the low-speed operation of the vehicle considerably.

The turbocharger had been added to compensate for the lean operation and the low volumetric energy density of hydrogen. Unfortunately, the turbocharger provides little benefit at low engine speeds. As a result, the vehicle's driveability (acceleration) in the city suffers significantly. Highway driveability, however, is more comparable to the gasoline version of the truck since the engine is able to utilize the assistance of the turbo boost.

Figure 6 shows a comparison of the velocity of UCR1 and a comparable 2.3L Ranger truck operating on gasoline (GAS). These data were obtained with a data acquisition system. The penalty of the added weight and of the lean operation of the vehicle is clearly evident. For example, after 10 seconds, the hydrogen vehicle's velocity was below 48 km/hr (30 mph), whereas the gasoline vehicle reached a velocity in excess of 80 km/hr (50 mph). Over a quarter-mile, GAS obtained a speed of 113 km/hr (70 mph) in 19.7 seconds compared with 92 km/hr (57 mph) in 25.0 seconds for the hydrogen truck. The data in Figure 6 also illustrate preignition problems with the CVI fuel system. Preignition begins to occur approximately after 17 seconds, and is characterized by the onset of variability (roughness) in the velocity curve. Also observe that the GAS has a manual transmissions (indicated by the shifts in the acceleration curves) as compared to the modified automatic transmission of UCR1.

At speeds above approximately 64 km/hr (40 mph), the vehicle's acceleration and overall driveability were comparable to those of the gasoline-powered vehicle, although occasional engine backfires (preignition) occurred. We have found that initially the vehicle had infrequent occurrences of preignition, but that the incidence and severity increased with mileage accumulation. Typically, preignition occurs at full throttle and at high engine rpm. We believe that oil contamination from the CVI manifold and from the engine sump is the primary source of the preignition. Considerable oil residue is evident in the top of the CVI unit and inside the four hoses which connect the CVI output to the inlet port area.

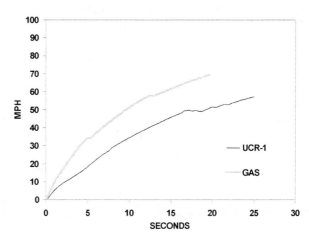

Figure 6. Acceleration of UCR1 compared with the stock gasoline Ford Ranger truck (GAS).

EMISSIONS – Immediately following the completion of the vehicle and engine modifications, and prior to any attempts to optimize the engine performance, a Federal Test Procedure (FTP) emissions test was performed at the American Automobile Association (AAA) of Southern California's chassis dynamometer facility in Los Angeles, California. The overall NO_x emission rate is 0.37 g/mile (the average of 0.395 g/mile, 0.423 g/mile, and 0.262 g/mile for bag 1, bag 2, and bag 3, respectively). This value is higher than expected but within the California Air Resources Board's (CARB's) TLEV emission standard for NO_x of 0.40 g/mile for gasoline-powered vehicles. The CO, CO_2, and HC emission rates (from oil leakage passing the piston rings) were respectively 0.00 g/mile, 3.03 g/mile, and 0.01 g/mi. It should be mentioned that these values were for an "un-optimized" vehicle, with no catalytic converter.

The very low CO and HC numbers were expected. Although the NO_x emissions meet CARB's LEV requirement, they are still about a factor of ten higher than what was expected for a hydrogen-fueled engine operating ultra-lean. For instance, **Figure 7**, taken directly from the published measurements of Das [5], illustrates how NO_x is dependent on equivalence ratio, and suggests the importance of operating an engine at Φ less than 0.5.

Chassis Dynamometer Testing – The performance tests and the FTP measurements discussed above highlighted the vehicle's limitations and capabilities, but also led to questions regarding the nature and origin of the preignition and higher than expected NO_x emissions. In an attempt to gain insight into these questions, a series of chassis dynamometer tests was conducted on the vehicle using CE-CERT's Clayton Industries 150 kW (200 HP) waterbrake dynamometer Model C-796 and portable emission bench.

Steady-state Tests – A series of carefully controlled steady-state and wide-open-throttle (WOT) tests was conducted on the vehicle at engine speeds from 2000 to 4500 rpm (500 rpm intervals). Mass air flow, hydrogen mass flow, NO$_x$ emissions, calculated equivalence ratio phi, and the vehicle exhaust gas oxygen signal as a function of engine speed are shown in **Figure 8** (please note that the NO$_x$ data has been divided by 100 in order for it fit on the graph). As expected, mass air flow and hydrogen mass flow increase with increasing engine rpm. The equivalence ratio phi decreased with increased engine rpm but was well above 0.5. Highest NO$_x$ concentrations were observed at low rpm and full throttle, but generally decreased with increased rpm.

Average horsepower, calculated specific fuel consumption, and efficiency are shown in **Figure 9**. Engine efficiency was relatively constant with rpm and remained at, or slightly below, 20%. The intercooler and exhaust gas temperatures and average horsepower are shown in **Figure 10**. Notice the correlation between exhaust gas temperature and horsepower. Note also that the intercooler on this vehicle is very effective at reducing the inlet air temperature from the turbocharger which is at approximately 200 degrees F to near ambient temperature.

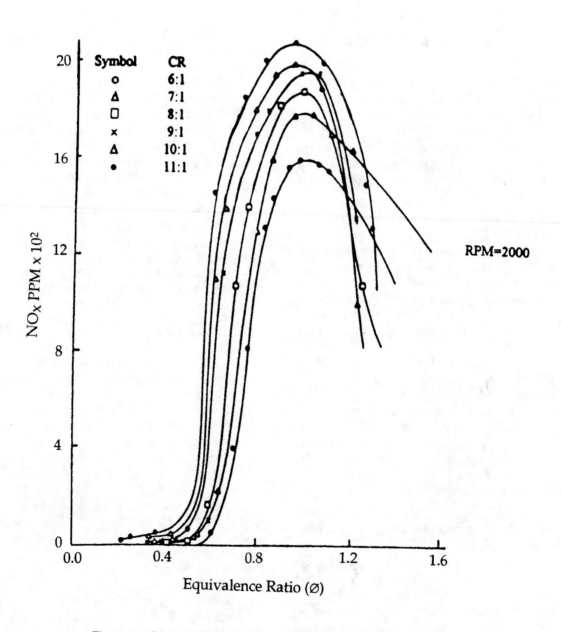

Figure 7. Plot of NO$_x$ Output vs. Equivalence Ratio (Φ) [5]

<u>Transient Tests</u> – Transient tests were performed on the dynamometer to evaluate the response time of the CVI control system. The results of these tests are shown in **Figure 11**. With the vehicle on the chassis dynamometer, the driver suddenly "stepped on," then quickly released, the accelerator, giving time between for the engine to come up to power. This provides "step" changes to the air box throttle plate, from closed to full open and then to closed. **Figure 11** illustrates the throttle opening/closing sequence. At opening there is a sudden increase of mass air flow, followed by a slow increase of hydrogen flow and power.

fuel flow. The fuel flow is based on an input from the manifold air pressure (MAP) sensor which is interpreted by the CVI control system. **Figure 12** illustrates that when the MAF quickly ceases, the control signal from the CVI control system lags by about a tenth of a second, and the hydrogen pressure only starts to respond after 0.1 second. Thus, the exhaust gas oxygen signal has a large "spike," showing a momentary large enrichment of the mixture. Although NO_x emissions were not measured during these tests, it was found that following the end of each steady-state test (throttle closed) there were considerable "spikes" in NO_x concentrations. An example of this is shown in **Figure 13**. We believe these "spikes" are the result of the sudden fuel enrichment caused by the closing of the throttle plate. These NO_x spikes along with the engine operating at a richer mixture than Φ of 0.5, are believed to be the major reason for the higher than expected NO_x emission during the FTP test.

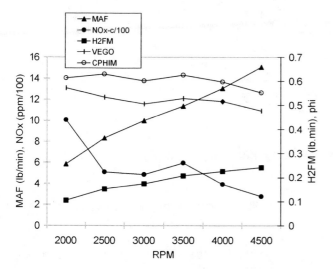

Figure 8. NO_x, mass air flow (MAF) and hydrogen mass flow (H2FM), exhaust gas oxygen signal (VEGO), and phi (CPHIM) calculated from mass flows.

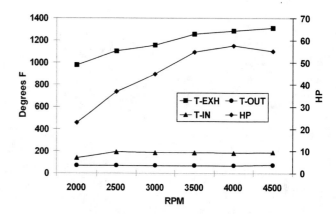

Figure 10. Temperature of exhaust gas (T-EXH), intercooler inlet (T-IN), intercooler outlet (T-OUT), and power (HP) vs. RPM

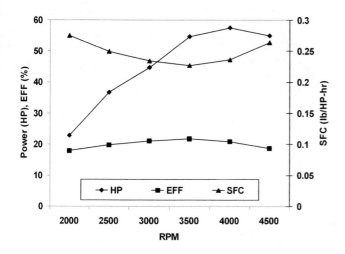

Figure 9. Power (HP), Specific Fuel Consumption (SFC), and Efficiency (EFF) vs. RPM (Summary Data)

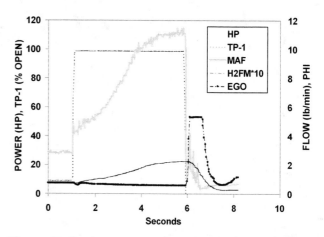

Figure 11. Engine response to a sudden throttle open, then closed. Power and throttle position are shown on the left scale. Mass air flow (MAF), hydrogen mass flow (H2FM), and exhaust gas oxygen sensor (EGO) are shown on the right scale.

On this scale, the closure occurs at time 0.10 seconds to 0.16 seconds. Unlike carbureted engines, where the accelerator pedal controls both the fuel flow (accelerator pump) and the air flow, this system directly controls the air flow (air throttle plate) but only indirectly controls the

The closure is illustrated in expanded scale in **Figure 12**.

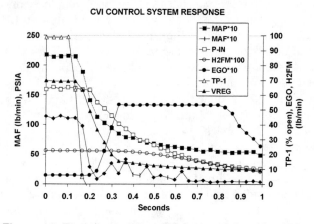

Figure 12. Throttle position, EGO signal (in phi units), VREG and hydrogen mass flow on the right scale. MAF, MAP and hydrogen pressure at CVI inlet are on the left scale.

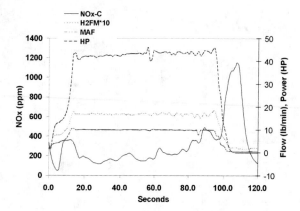

Figure 13. Power (HP), mass air flow (MAF), NO$_x$, and hydrogen mass flow (H2FM) vs. time (3000 RPM).

HARDWARE PROBLEMS OBSERVED DURING EVALUATION

TURBO BOOST BYPASS VALVE – There are two methods typically used to install a turbocharger on an engine: "draw through" and "blow through" configurations. In the "draw through" configuration, the turbocharger is installed between the intake manifold and the fuel metering system. In this configuration the fuel metering system is never exposed to pressures above ambient. In the "blow through" configuration, the fuel metering system is installed between the turbocharger and the intake manifold and is, therefore, subjected to the turbo boost pressure. This configuration is used on the hydrogen vehicle. For this type of configuration, a device called a turbo boost bypass valve (TBBV) is required between the outlet of the turbocharger and the butterfly valve of the air metering system. The purpose of this device is to relieve (vent) the boost pressure in the line when the butterfly is closed.

During our evaluation of the vehicle, we observed that the original TBBV (Audi 034145710A) would fail in the open

position. This would cause the boost pressure to be vented prior to its entering the engine. Failure of this device did not inflict any damage on the engine, but did cause a reduction in power due to the elimination of the turbo-boost. The original TBBV was replaced with a Turbotonics 30270 and no subsequent problems have occurred.

NOISE FROM HYDROGEN PRESSURE REGULATOR – A pressure regulator, located between the hydrogen storage vessels and Buzmatic (hydrogen flow control) regulator, is used to reduce the high-pressure hydrogen from the gas storage cylinder to approximately 690kPa (100 psi). This regulator is a diaphragm-type and produces an audible howling noise during medium to hard acceleration. The regulator was removed and tested to verify proper operation. During these tests, no noise was noticed. Thus, the howling sound detected when it is in the vehicle is probably due to the feedback from the Buzmatic regulator located in close proximity to the regulator's outlet.

SUMMARY OF EVALUATION

While some technical difficulties have been identified, the overall emission and performance characteristics of the vehicle met all the program goals. In some instances the vehicle performed beyond expectations. The following is a summary of the important findings of our evaluation.

- Operation of a hydrogen fuel internal combustion engine using a constant volume injection (CVI) system with closed-loop control and exhaust oxygen feedback has been demonstrated. Overall, the system operates as expected, although some suggestions for improvements of the system are given below.

- The vehicle has a range in excess of 161km (100 miles) between refueling stops, and it can be operated at moderate loads without appreciable preignition problems.

- The top speed is approximately 145 km/hr (90 mph), which is similar to the standard 2.3L Ranger truck. However, acceleration is less than the gasoline Ford Ranger, due in part to an increase in vehicle weight as a result of the conversion, the automatic transmission and the response time (lag) of the turbocharger. This is the one area that requires significant improvement.

- After several hundred miles of operation, oil is now being passed from the CVI block through the connections to the intake ports. This appears to be a major cause of preignition, and adversely affects the performance of the vehicle at high torque and rpm.

- The CVI system was supposed to be tuned to operate at an equivalence ratio (Φ) of less than 0.5. Our experience suggests Φ is over 0.5, and excursions over 0.6 are indicated by the EGO sensor signal.

- CO, CO_2 and HC emissions, attributed to oil seeping by the piston rings, were found to be negligible. Likely, as a consequence of the CVI operating at an equivalence ratio above 0.5, NO_x emission levels are higher than expected. The FTP emission rate for the vehicle for NO_x was 0.37 g/mile.
- Dynamic response of the CVI unit contains some anomalous behavior, likely due both to slow response of the electromechanical components of the CVI system and limitations of the analog electronics used in the CVI controller. Our experience indicates that preferred operation of the vehicle is based on proportionate response rather than sudden actions.
- Turbo boost at low speeds and low power levels is a limiting factor for overall driveability.

RECOMMENDATIONS FOR FURTHER RESEARCH AND TECHNICAL IMPROVEMENT – The following are some recommendations for continued research and evaluation of the vehicle and for technical improvements for future hydrogen vehicles.

- The vehicle's city and highway fuel economy needs to be better understood, and requires a modified EPA and MPG test procedure.
- The performance characteristics and NO_x emissions need to be measured as a function of equivalence ratio.
- The current ignition system needs to be modified or replaced so the timing can be advanced during engine operation. A timing map for the engine under various operating conditions (i.e., A/F ratios engine speed, load, etc.) needs to be developed. In addition, using the dual spark capability of this engine design should be reinvestigated.
- The pressure reduction regulator needs to be relocated from the engine compartment to the bed of the truck where the hydrogen storage vessels are located. This would reduce the length of high-pressure fuel line required.
- A high-pressure, electrically actuated solenoid valve at the outlet of the hydrogen storage tanks' manifold needs to be installed. This valve will be used as a redundant shutoff valve in the event of a hydrogen leak and will be configured to close when it loses power (fail-safe position).
- A thermocouple (TC) should be installed in one of the hydrogen storage vessels. This TC would be used to measure the hydrogen temperature increase during refueling. The signal from the TC would also be used in conjunction with the output signal for the hydrogen storage pressure transducer to determine the mass of hydrogen on-board the vehicle (i.e., fuel level gauge).
- The compression ratio of the engine needs to be incrementally increased from 8:1 to the highest compression ratio possible prior to the onset of knock. This will increase the engine efficiency.
- A microprocessor-based controller for the CVI system using mass air flow as one of the input parameters needs to be developed and installed.
- A means of preventing oil flow from the CVI system to the engine needs to be investigated.

ACKNOWLEDGMENTS

CE-CERT would like to thank the South Coast Air Quality Management District for sponsoring this project, and its partners Hydrogen Components, Incorporated and Advanced Machine Dynamics for achieving significant results. CE-CERT would also like to thank the following organizations for their assistance in the project: Praxair, The Electrolyser Corporation, Burke-Porter, Bourns Incorporated, Comdyne, Advantage Suspension Works, and the Automobile Club of Southern California.

REFERENCES

1. Cecil, W. "On the application of hydrogen gas to produce a moving power in machinery; with a description of an engine which is moved by the pressure of the atmosphere, upon a vacuum caused by explosions of hydrogen and atmospheric air," *Trans. Cambridge Philos. Soc.*, Vol. 1, p. 217, 1822.
2. Van Vorst, W.D. and Woolley, R.L., "Hydrogen-fueled surface transportation," in <u>Hydrogen: Its Technology and Implications</u>, Edited by K. E. Cox and D. K. Williamson Jr., CRC Press, Boca Raton, FL, p. 1979.
3. Das, L.M., "Hydrogen engines: A view of the past and a look into the future," *Int. J. Hydrogen Energy*, Vol. 15, p. 425, 1990.
4. Homan, H.S., De Boer, P.C.T. and McLean, W.J., "The effect of fuel injection on NO_x emissions and undesirable combustion for hydrogen-fuelled piston engines," *Int. J. Hydrogen Energy*, Vol. 8, p. 131, 1983.
5. Ricardo, H.R., "Further note on fuel research I.," *Proc. Inst. of Automob. Eng. of London*, Vol. 5, p. 327, 1924.
6. King, R.O. and Rand, M., "The oxidation, decomposition, ignition, and detonation of fuel vapours and gases. XXVII. The hydrogen engine.," *Canadian J. Technol.*, Vol. 33, p. 445, 1955.
7. King, R.O., Wallace, W.A. and Mahapatra, B., "The oxidation, ignition and detonation of fuel vapours and gases-V. The hydrogen engine and detonation of the end gas by the ignition effect of carbon nuclei formed by pyrolysis of lubricating oil vapor," *Canadian J. Technol.*, Vol. 34, p. 1957.
8. Swain, M.R. and Adt, R.R., "The hydrogen-air fueled automobile," in <u>7th Intersociety Energy Conversion Engineering Conference</u>, American Chemical Society, Washington, D.C., p. 1382, 1972.
9. Lynch, F.E., "Parallel induction: a simple fuel control method for hydrogen engines," *Int. J. Hydrogen Energy*, Vol. 8, p. 721, 1983.
10. MacCarley, C.A., "Electronic fuel injection techniques for hydrogen fueled I.C. engines," M.S. Thesis in Engineering, University of California, Los Angeles, 1978.
11. MacCarley, C.A. and Van Vorst, W.D., "Electronic fuel injection techniques for hydrogen powered I.C. engines," *Int. J. Hydrogen Energy*, Vol. 5, p. 179, 1980.
12. Mathur, H.B. and Das, L.M., "Performance characteristics of a hydrogen fuelled S.I. engine using timed manifold injection," *Int. J. Hydrogen Energy*, Vol. 16, p. 115, 1991.

13. Das, L.M., "Fuel induction techniques for a hydrogen operated engine," *Int. J. Hydrogen Energy*, Vol. 15, p. 833, 1990.

14. Das, L.M., "Studies on timed manifold injection in hydrogen operated spark ignition engine: performance, combustion and exhaust emission characteristics," Ph.D. Thesis, Indian Institute of Technology, New Dehli, India, 1987.

15. Varde, K.S. and Frame, G.A., "Development of a high pressure hydrogen injection for SI engine and results of engine behavior," in Hydrogen Energy Progress V, Edited by T. N. Veziroglu and J. B. Taylor, Pergamon Press, Elmsford, NY, p. 1505, 1984.

16. Krepec, T., Tebelis, T. and Kwok, C., "Fuel control systems for hydrogen-fueled automotive combustion engines -- a prognosis," *Int. J. Hydrogen Energy*, Vol. 9, p. 109, 1984.

17. Krepec, T., Giannacopoulos, T. and Miele, D., "New electronically controlled hydrogen-gas injector development and testing," in Hydrogen Energy Progress VI, Edited by T. N. Veziroglu, N. Getoff and P. Weinzierl, Pergamon Press, Elmsford, NY, p. 1087, 1986.

18. Krepec, T., Giannacopoulos, T. and Miele, D., "New electronically controlled hydrogen-gas injector development and testing," *Int. J. Hydrogen Energy*, Vol. 12, p. 855, 1987.

19. Norbeck, J.M., Heffel, J. W., Durbin, D. T., Tabbara, T., Bowden, J. M., and Montano, M. C., (1996). *Hydrogen Fuel for Surface Transportation*, Society of Automotive Engineers, Inc., Warrendale, PA.

DIESEL ENGINES

1999-01-2521

Acceptability of Premixed Hydrogen in Hydrogen Diesel Engine

Hiroyuki Nagaki
NEDO Industrial Technology Researcher

Hirohide Furutani, Sanyo Takahashi
Mechanical Engineering Laboratory, AIST, MITI

ABSTRACT

The Acceptability of boiloff hydrogen to an Inert Gas Circulating Hydrogen Diesel System, providing a high thermal efficiency, zero nitrogen oxides and carbon dioxide emissions, is discussed. To simulate a reciprocating engine cycle, a rapid compression-expansion machine is used. The machine brings fundamental data, such as hydrogen jet penetration injected in high pressure chamber and combustion characteristics. The results show an acceptable amount of hydrogen premixed to intake mixture without major negative effects. They suggest that most of boiloff hydrogen, inevitable in the facility where liquid hydrogen is used, could be supplied to the intake mixture as part of the fuel of the hydrogen diesel engine, saving a pumping loss to compress it up to an injection pressure.

INTRODUCTION

Hydrogen has been researched for years as an alternative fuel for internal combustion engines because of its advantages of the simple and clean reaction, the high energy density, the wide ignition range and the high burning velocity[1][2]. Application of hydrogen to diesel engines for cogeneration systems is expected to bring a high efficiency, a wide range of applicability and zero emissions[3][4]. When transferring and storing it is better to keep hydrogen in the liquid state due to its higher density and less leakage than in the gaseous state. It is, however, difficult to keep it below 20K, the boiling temperature of hydrogen, over the whole facility from storage tanks through energy conversion units. In spite of the efforts made to reduce boiloff hydrogen at consumer sites as low as possible, its levels are estimated to reach around 15 percent of all the hydrogen supplied to a site, so that any methods to utilize boiloff hydrogen have to be developed for the effective use of hydrogen. Among them, hydrogen-combustion turbine generators have been expected to be the only resort so far. However, it is estimated that around 10 percent of the power output from the generators are consumed to compress boiloff hydrogen to supply as part of the fuel[5]. This loss can be increased

in hydrogen diesel engines because of its higher injection pressure and spoils the benefit of the high efficiency inherent in the diesel engine. The present study investigates the effects of premixed hydrogen to the oxygen-argon mixture on the combustion characteristics using a rapid compression-expansion machine(RCEM)[6] and the hydrogen jet penetration processes into a high pressure chamber. Furthermore, the acceptability of boiloff hydrogen to be premixed to the intake mixture of a hydrogen diesel engine is discussed.

EXPERIMENTAL INVESTIGATION

The experimental apparatus used for these investigations, shown in Figure 1, is composed of an RCEM and several units of a hydrogen injection, a mixture supply, a data acquisition, a shadowgraph optics and a exhaust gas analysis. Hydrogen-oxygen-argon mixture is introduced to a combustion chamber and rapidly compressed to a high pressure and temperature where highly pressurized hydrogen is then injected. Premixed hydrogen levels are set to the excess oxygen ratios of 10, 15, 20 and infinity, or no hydrogen premixed, whereas the ratio of oxygen to argon is constant at 1:4. The amount of hydrogen injected

Table 1 RCEM Specifications

Type	Electro-hydraulic-control
Bore, mm	100
Maximum Stroke, mm	120
Maximum compression ratio	10
Operation speed, rpm(Hz)	1000(16.7)
Allowed in-cylinder pressure, MPa	15
Chamber temperature, deg. C	80
Piston driving force, kN	150

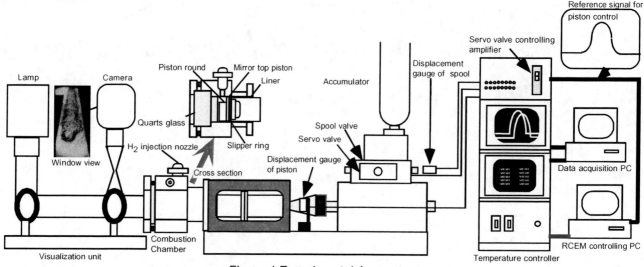

Figure 1 Experimental Apparatus

is set so that the total excess oxygen ratio becomes 2. The injection timing of hydrogen is set at the top dead center where in-cylinder pressure reaches its maximum. Mixing and combustion processes are observed from in-cylinder pressure, piston stroke, needle lift, framing photographs and exhaust gas analysis as well as other required signals.

EXPERIMENTAL SETUP

The experimental setup allows a variety of operations under different conditions. The RCEM can simulate single compression and expansion strokes of a practical reciprocating engine. The combustion chamber has a bore of 100mm and the maximum stroke is 120mm. The electro-hydraulic- controlled piston provides variable compression ratios up to around 10. The piston speed corresponds to 1000rpm of a modeled engine. The temperatures of cylinder liner and piston are controlled to 80 degrees Celsius. Charged pressures are 0.1, 0.2 and 0.3MPa for hydrogen jet observations and 0.1MPa for combustion tests. The combustion chamber, shown in Figure 2, is made up of a mirror top piston, a piston round, a cylinder liner and a quartz head-cum-window enabling optical access to the chamber for the shadowgraphy. Processes of injection and ignition are observed with an image converting mercury lamp through the quarts window to the mirror top piston. The reflection is then led to the camera via a half mirror before the window. On the upper side of the camera at frame rates of 6000fps and 12000fps and an exposure time of 40μs. A light of 1000W is supplied from a combustion chamber an original electro-hydraulic-

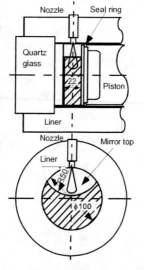

Figure 2 Combustion Chamber

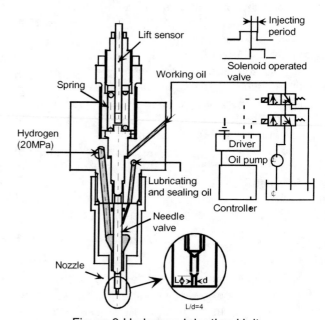

Figure 3 Hydrogen Injection Unit

controlled injection nozzle for pressurized hydrogen, shown in Figure 3, is mounted. This type of injection nozzle allows single-shot injection, tight sealing and flexible injection timing and period. Hydrogen is compressed by a piston compressor and buffered in a small tank before the nozzle of diameter of 0.7mm. The hydrogen injection pressure are 10, 15 and 20Mpa for hydrogen penetration analysis and 10Mpa for combustion tests. After combustion the exhaust gas is analyzed with a gas chromatograph attached to the exhaust line to obtain concentrations of residual hydrogen, oxygen and nitrogen.

The main measuring data include in-cylinder pressure, piston displacement giving the in-cylinder volume and temperature, and needle lift, as shown in Figure4, as well as the charged pressure before compression, hydrogen injection pressure, the amount of hydrogen injected, the temperatures of cylinder liner and piston, the photographs of hydrogen jet, mixing and self-ignition images and exhaust gas composition. Based on these data, obtained

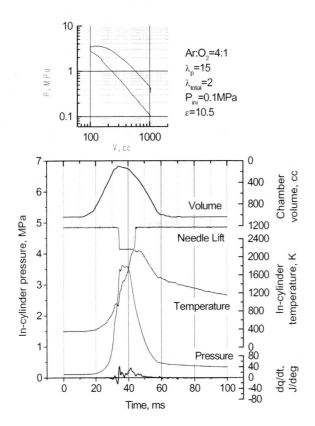

Figure 4 Measuring Data

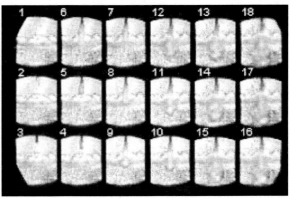

Figure 5 Typical Injection Process

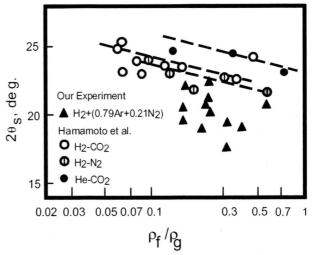

Figure 6 Hydrogen Jet Angle

are the hydrogen jet angle, the hydrogen jet penetration history, the self-ignition condition, the ignition delay, the indicated efficiency and the combustion efficiency.

RESULTS AND DISCUSSION

HYDROGEN PENETRATION AND JET ANGLE

It is important to understand the behavior of the jet of hydrogen injected into a high pressure and temperature atmosphere because of the following reasons; to investigate the self-ignition phenomenon, to find out the appropriate ignition timing and emplacement of ignition device in the case that an ignition support is required and to design the combustion chamber. Of those theoretical and experimental studies for designing diesel engine combustion chambers, Waguri et al.[7], based on a momentum theory, express the penetration of fuel spray in diesel engines with non-dimensional parameters, and Hamamoto et al.[8] confirm that the momentum theory is applicable to the transient gas jet where the penetration depth, X_s, is expressed in the following equation.

$$\frac{X_s^2}{d_n \cdot U_0 \cdot t} = \frac{\sqrt{\alpha \cdot \rho_f / \rho_g}}{\tan(\theta_s)}$$

where d_n is the nozzle diameter, U is the injected gas velocity at nozzle outlet, t is the time from injection start, α is the discharge coefficient of nozzle, ρ_f is the density of injected gas at nozzle outlet, ρ_g is the density of surrounding gas and θ_s is the average half cone angle of jet. In the above equation a dimensional analysis shows that the jet angle, θ_s, is expressed as the ratio of ρ_f to ρ_g. This means that penetration depth increases in proportion to the square root of t and in inverse proportion to θ_s.

In the present investigations hydrogen is injected into nitrogen-argon surrounding gas, imitative of oxygen-argon mixture, rapidly compressed by the RCEM. The operations are conducted at the charged pressures of 0.1, 0.2 and 0.3MPa, the compression ratios of around10 and the injection pressures of 10, 15 and 20MPa.

Figure 5 shows a typical hydrogen penetration process injected at the top dead center. Numbers on the photograph indicate the shooting order. The jet angles are measured from the photographs at 1ms after injection start because they are almost constant until the jets reach the opposite side of the chamber wall except just after injection. Figure 6 shows the comparison of the measured jet angles with those reported by Hamamoto et al. Although the present data are more scattered, they are very close each other. This means that the hydrogen jet angle is also expressed by the ratio of the density of injected gas at nozzle outlet, θ_f, to that of surrounding gas, ρ_g.

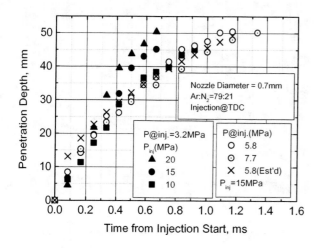

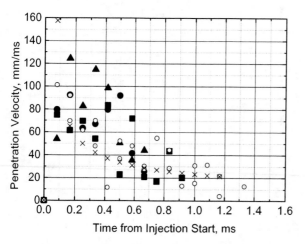

Figure 7 Hydrogen Jet Penetration History

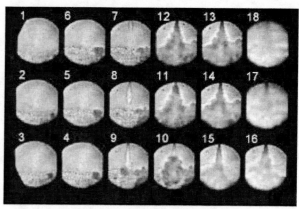

Figure 8 Typical Ignition Process

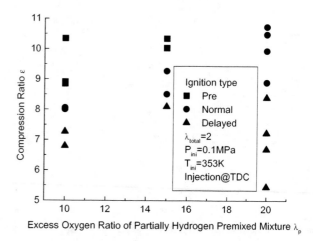

Figure 9 Dependency of Self-ignition Types on Compression Ratio and Ecess Oxygen Ratio of Intake Mixture

Figure 7 shows the growth of hydrogen jet penetration and its velocity. Despite that the equation indicates the penetration depth increases in proportion to the square root of time, in the measured results it is, on the contrary, in proportion to time until 0.4ms and such a relationship can be observed after 0.4ms. This tendency can be also reported for the sprays of diesel fuel and other gases[9]. One of the reasons for the difference in the penetration depths between the experimental results and the theoretical ones could be the increase in the hydrogen flow quantity until the needle valve lifts enough to allow full open of the nozzle outlet. Figure 7 includes the calculated penetration depth where the measured average half cone jet angle of 10degree has been inputted. Regardless of many assumptions in the equation, both results agree well. It makes possible to predict the depth of hydrogen jet penetration and the distribution of hydrogen concentration in chamber, provided that the average half cone jet angle has been given.

SELF-IGNITION AND IGNITION DELAY

Figure 8 shows a typical injection and ignition process with a hydrogen injection into the hydrogen-premixture in the excess oxygen ration of 15. There can be observed three kinds of ignition starts. One is the pre-ignition start in hydrogen premixed area in the first shot, another is the

self-ignition start of the injected hydrogen in the 9th shot, and the other is the time that the in-cylinder pressure starts rising in the 13th shot. Defined here as the ignition start is the last one since it governs the rest of the combustion process.

Figure 9 shows that the self-ignition types depend on the compression ratio, ε, and the excess oxygen ration of the intake mixture, λ_p. Self-ignition is here divided into three types; 1)pre-ignition, the ignition before injection, 2)normal ignition, with ignition delay up to 1ms from injection and 3)ignition delay, with that over 1ms. In a series of experiments prior to this study, it has been confirmed that self-ignition of hydrogen premixed mixture without hydrogen injection happens statistically near 850K and higher temperature and premixed hydrogen concentration lead to more reliable self-ignition. This is the reason that the pre-ignition as shown in Figure 9 can happen even in the compression ratios where the normal ignition is observed in the higher excess oxygen ratios. In the case of λ_p of 10, the compression ratios above 9 cause the pre-ignitions and they have to be lowered down to 8 for the normal ignition. This results in the lower thermal efficiency. It can be said that the intake mixture of λ_p of 10 is too rich to obtain the normal ignition. On the contrary, the mixture

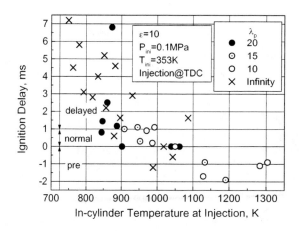

Figure 10 Ignition Delay

of λ_p of 15 allows the normal ignition just below the compression ratio of 10. It suggests that a limit of λ_p for normal ignition at the compression ratio of 10 is near 15. The higher temperatures bring the shorter ignition delays, as shown in Figure 10, although the data are scattered in a relatively wide range. The ignition delays around 1000K, at the top dead center in the compression ratio of 10, with and without partial premixing of hydrogen gather near zero. This means the hydrogen premixing does not elongate nor shorten the ignition delay around 1000K. These results suggest that an acceptable level of hydrogen premixing for ignition characteristics is near the λ_p of 15.

INDICATED THERMAL EFFICIENCY

Figure 11 illustrates the dependency of the indicated thermal efficiency, η_i, on λ_p excluding the pre-ignition data. The indicated thermal efficiency is a measure of how efficiently an internal combustion engine converts the heat that it receives to work excluding the mechanical loss, enabling to compare the different engine versions. Although the level of indicated mean effective pressures

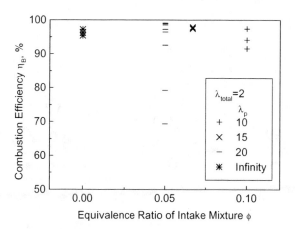

Figure 12 Effect of Premixed Hydrogen on Combustion Efficiency

are different between premixed and no premixed conditions due to their different charged pressures, the indicated thermal efficiencies lie between 45% and 55%. This means hydrogen premixing does not have a significant influence on the indicated efficiency. This range is also reported on a direct-injection hydrogen diesel engine[10]. It suggests that hydrogen diesel engine has higher indicated thermal efficiency as compared to the diesel-fueled air-working-fluid compression-ignition engine.

COMBUSTION EFFICIENCY

Figure 12 illustrates the dependency of the combustion efficiency, η_B, on λ_p excluding the pre-ignition data. The combustion efficiency is here a ratio of hydrogen consumed in a combustion to that premixed and injected to the chamber. Higher combustion efficiencies lead to lower fuel consumption. As shown in Figure 12, the combustion efficiencies gather around 95%, except the two data lower than 80% whose ignition delays are much longer among the hydrogen premixed conditions in the low compression ratios. This suggests that partially premixing of hydrogen to the intake mixture does not result in lowering the fuel consumption, as long as normal ignition is maintained.

ADVENTAGES OF HYDROGEN PREMIXING

From the considerations above the excess oxygen ratio of intake mixture of 15 could represent the maximum amount of hydrogen premixed to the intake mixture. Assuming the total excess oxygen ratio of 2 and the boiloff rate of 15 percent, the partially premixing of hydrogen enables to utilize 87 percent of the boiloff hydrogen. The pumping loss for the rest of boiloff hydrogen up to injection pressure could become significantly less than that for all the boiloff hydrogen without the premixing. This results in the higher thermal efficiency of the system because of the higher output power and the less liquid hydrogen supply to the engine.

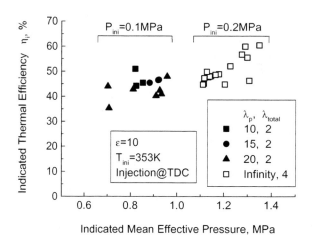

Figure 11 Effect of Premexed Hydrogen on Indicated Thermal Efficiency

CONCLUSION

The developed equipment composed of an RCEM, a high pressure hydrogen injection system, a visualization optics and a combustion product analyzer for the investigations of premixed hydrogen into an intake mixture as part of the fuel for an inert gas circulating hydrogen diesel system has brought a variety of knowledge on the hydrogen jet penetration and the combustion characteristics. The depth of hydrogen jet penetration into a high pressure combustion chamber can be estimated by a theoretical model for diesel spray, provided that the average half cone jet angle has been given. Because higher temperature and premixed hydrogen concentration lead to higher ignitability, pre-ignition likely happens in the excess oxygen ratio of under 15 in the compression ratio of 10. The premixed hydrogen into the oxygen-argon mixture does not have negative effects on the indicated thermal efficiency and the fuel consumption under the normal ignition conditions. These indicate the acceptable level of premixed hydrogen for a hydrogen diesel engine is in the excess oxygen ratio of above 15. It suggests that 87 percent of the boiloff hydrogen from hydrogen diesel cogeneration system can be supplied to the intake mixture as part of the fuel resulting in saving of the pumping loss, otherwise required to compress it up to the injection pressure, and longer fuel supply at the consumer site.

ACKNOWLEDGMENTS

This work was supported by New Energy and Industrial Technology Development Organization(NEDO) of Japan under the contract project of World Energy Network(WE-NET).

REFERENCES

1. S. Furuhama and K. Yamane, "Combustion Characteristics of Hydrogen-fueled Spark Ignition Engine," (in Japanese), J. SAE Jpn, No.6, 1973, p12-19.
2. G. A Karim and M. E. Taylor, "Hydrogen as a Fuel and the Feasibility of a Hydrogen-Oxygen Engine," SAE Paper 730089, 1973.
3. M. Chiba, H. Arai and K. Fukuda, "International and National Program and Project 'Hydrogen Energy Technology Development in Japan: New Sunshine Program,'" Proc. the 11th World Hydrogen Energy Conference, 1996, p.13
4. H. Ishida and Y. Tosa, "Fundamental Study on the Inert Gas Circulating Hydrogen Diesel System," (in Japanese), Proc. Thermodynamics, JSME, 1997, p30-31.
5. New Energy and Industrial Technology Development Organization, "International Clean Energy Network Using Hydrogen Conversion(WE-NET)," Annual Summary Report on Results, 1997.
6. S. Kobori and T. Kamimoto, "Development of a Rapid Compression-Expansion Machine," (in Japanese), J. JSME, Vol.62, No.593, 1996, p.392-397.
7. Y. Waguri, M. Fujii, T. Amitani and R. Tsuneya, "Studies on the penetraion of Fuel Spray of Diesel Engine," (in Japanese), J. JSME, Vol.24, No.156, 1958, p.820-826.
8. Y. Hamamoto and A. Sakane, " Study on Penetration of Transient Gas Jet," (in Japanese), J. JSME, Vol.53, No.496, 1987, p.3810-3813.
9. H. Hiroyasu, T. Kakuta and S. Tasaka, "Study on Penetration of Diesel Fuel Spray," (in Japanese), J. JSME, Vol.44, No.385, 1978, p.3208-3219.
10. H. Rottengruber, U. Wiebicke, G. Woschni and K. Zeilinger, "Investigation of a Direct Injecting Diesel-Engine," Proc. World Hydrogen Conference XII, 1998, p.1515-1535.

2001-01-3503

Hydrogen Combustion and Exhaust Emissions Ignited with Diesel Oil in a Dual Fuel Engine

Eiji Tomita, Nobuyuki Kawahara, Zhenyu Piao and Shogo Fujita
Okayama University

Yoshisuke Hamamoto
The University of Kitakyushu

ABSTRACT

Hydrogen is expected to be one of the most prominent fuels in the near future for solving greenhouse problem, protecting environment and saving petroleum. In this study, a dual fuel engine of hydrogen and diesel oil was investigated. Hydrogen was inducted in a intake port with air and diesel oil was injected into the cylinder. The injection timing was changed over extremely wide range. When the injection timing of diesel fuel into the cylinder is advanced, the diesel oil is well mixed with hydrogen-air mixture and the initial combustion becomes mild. NOx emissions decrease because of lean premixed combustion without the region of high temperature of burned gas. When hydrogen is mixed with inlet air, emissions of HC, CO and CO_2 decrease without exhausting smoke while brake thermal efficiency is slightly smaller than that in ordinary diesel combustion. In particular, both smoke and NOx are almost zero and HC is low when the injection timing is significantly advanced although the engine operation becomes unstable.

INTRODUCTION

Regulations on exhaust emissions such as NOx, particulate matter, hydrocarbon, carbon monoxide, even carbon dioxide, from engines are being tightened. It is very important to improve thermal efficiency and to reduce exhaust emissions. The excellent thermal efficiency of diesel engines certainly has advantages for conserving energy and solving the greenhouse problem. However, a diesel engine has a problem of the trade-off, that is to say, a difficulty to reduce both smoke and NOx, simultaneously.

Hydrogen is considered as a future fuel without carbon dioxide emissions [1]. Hydrogen has some different features from other hydrocarbon fuels. For example, the burning velocity is so fast that very rapid combustion can be achieved. The limits of flammability are very wide and

an engine can be operated with a wide range of air/fuel ratio including very lean condition. Lean combustion yields to low NOx emissions because the temperature becomes low. Self-ignition temperature is so high that a certain ignition source is needed while the minimum ignition energy is considerably small. This means that it is very difficult to ignite hydrogen just by the compression process. The ignition source may be spark, glow plug or other energy. There are many studies on spark-ignition engines [2][3]. Diesel-type engines were tested by some researchers, however, it was very difficult to operate with just compression due to high self-ignition temperature [4]. Therefore, glow plug is often used and Wong tried to use a ceramic part to retain heat as ignition source [5]. Ikegami et al. investigated the hydrogen combustion with a special injector that is equipped with a leak structure and a glow plug [6] and obtained moderate operation.

Recently, dual fuel engines using a pilot diesel fuel were developed in view of the minimizing resources of petroleum, and/or utilizing a co-generation system because of cleaner exhaust gas [7-10]. In particular, compressed natural gas is considered as primary gaseous fuels. Many research efforts have been done for these concepts [11-13] and investigated the performances and emission characteristics. The smoke exhausted from an engine is reduced enormously, while the emissions of hydrocarbon increase in light load and knock occurs in heavy load. When hydrogen is burned, it is expected that no hydrocarbons are exhausted even in part loads.

Dual fuel diesel engines associated with hydrogen have been studied by some researchers. Gopal et al. investigated the performance in a conventional single cylinder four-stroke diesel engine with hydrogen as inducted under wide range of the engine operation [14]. The results showed that the thermal efficiency obtained were comparable with pure diesel operation and up to half the engine's energy requirement could be derived from hydrogen. They pointed that the main problem area was the onset of knocking well before stoichiometric region and obtained satisfactory results. Li et al. added 3-7 % of hydrogen from an intake port to a normal diesel engine and obtained less smoke [15]. Mathur et al. found that hydrogen could be substituted for diesel oil up to the point where 38% of the total fuel mixture with no resulting loss in the system efficiency and only nominal loss power output [16]. In order to prevent knock, some diluents of nitrogen, helium or water were used. Nitrogen was the most effective to engine performance. Hydrogen of 10-15% in total energy was aspirated in the intake port by Varde et al. [17]. The results shows that smoke is reduced at part load while the effect of aspiration is limited at full load. Lambe et al. optimized the design of the engine for a pilot diesel fuel engine and obtained 80% reduction of carbon dioxide and smoke and 70% of nitrogen oxides [18]. Patro developed a simplified heat release model from the pressure history and the second derivative of the pressure shows to give an acceptably accurate diagnosis of the hydrogen fueled diesel engine combustion process [19]. Liu and Karim et al. studied the effect of property of the gaseous fuel (methane, propane, ethylene or hydrogen) with air on ignition delay in dual fuel engines [20]. It was shown that the changes due to gaseous fuel admission in the temperature and pressure levels during the delay period, the extent of energy release due to pre-ignition reaction processes, variations in the parameters of external heat transfer to the surroundings and contribution of residual gases are the most important factors that determine the ignition delay characteristics of dual fuel engines. The observed values were evaluated while using detailed reaction kinetics for the oxidation of the gaseous fuel and employing an experimentally based formula for the ignition of the liquid pilot. Then it was suggested that the ignition delay could be correlated in terms of the type of gaseous fuel, its concentration and other operation conditions.

Another new trend is premixed combustion in diesel engine [21], of which concept is like a homogeneous charge compression ignition (HCCI) engine [22]. An ultra lean mixture is compressed and autoignited with longer period of ignition delay. Recently, this concept was applied to diesel engines. Very early injection timing leads to enough time when the fuel mixes with air because the temperature of the air is too low to ignite. Both nitrogen oxides and smoke are reduced enormously although the operation is limited only in light loads. This concept is very interesting in that nitrogen can be reduced very much.

In this study, hydrogen was inducted into the cylinder of a diesel engine from an intake port and ignited with diesel oil that is injected into the engine cylinder. This study is connected with our previous study on ignition delay of diesel oil injecting into the mixture of air and gaseous fuel (propane, methane or hydrogen) [23]. The injection timing was changed within a very wide range

including extremely early injection timing. The heat release rate was obtained from the pressure in the cylinder and exhaust emissions, including smoke, the nitrogen oxide (NOx), hydrocarbon (HC), CO and CO_2 were measured. The relations between the heat release rate and exhaust emissions are discussed. Das discussed hydrogen combustion in internal combustion engine and when the engine is applied to a practical use, undesirable combustion, such as pre-ignition in the combustion chamber and backfire into the intake manifold often occur [24]. This study focuses on the very low exhaust emissions when the injection timing is extremely advanced although these problems may remain.

EXPERIMENTAL APPARATUS

A four-stroke diesel engine with a single cylinder was used for this study. The bore and stroke are 92 and 96 mm, respectively. The compression ratio is 17.7, and the type of combustion chamber is deep dish. Here, the injection timing can be changed over a very wide range from 62.7 degrees before TDC to 3.2 degrees after TDC. The schematic diagram of the combustion chamber and the nozzle used here were shown in Fig.1. A pressure transducer is equipped in the cylinder head to determine the cylinder. The valve lift was detected with a gap sensor in the injector.

Gaseous hydrogen was inducted into the engine cylinder from an intake port and diesel oil was injected in the cylinder from four nozzles. For reference, the ordinary diesel condition that only air is inducted into the cylinder was also performed. The injection timing of the diesel oil, and the amounts of diesel oil injected and hydrogen inducted, were changed. The signals of pressure in the cylinder, TDC and the valve lift were stored in a digital recorder. Exhaust emissions of HC, CO and CO_2 were determined simultaneously (Horiba, MEXA 554J). The value of the hydrocarbon was translated as hexane. The NOx concentration was measured with a chemiluminescent type analyzer (Shimadzu, NOA-305). Smoke was measured with a Bosch type meter (Zexel, DSM-10). The engine speed was 1000 rpm. The total equivalence ratio was set to 0.3, 0.4 and 0.5. The ratio

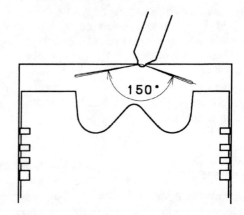

Fig.1 Configuration of combustion chamber of the test engine

Table 1 Specifications of the engine and experimental conditions

Bore	92 mm
Stroke	96 mm
Compression ratio	17.7
Combustion chamber	Deep dish
Nozzle Number of hole	4 holes (ϕ0.26 mm)
Injection angle	150 degrees
Injection timing	62.7 degrees BTDC-
	3.2 degrees ATDC
Inlet gas	Air, Hydrogen+air
Equivalence ratio	0.3, 0.4, 0.5
Engine speed	1000 rpm
Temperature of engine oil	333 K
Temperature of cooling water	333 K

of hydrogen was changed variously. The temperatures of cooling water and engine oil were kept at 333 K as presented in Table 1.

PRESSURE ANALYSIS AND HEAT RELEASE RATE

ORDINARY DIESEL ENGINE CONDITION

The heat release rate was determined by analyzing the pressure history in the cylinder using one region model with the first law of thermodynamics. Figures 2 and 3 show the pressure history and heat release rate, changing the injection timing in the case of ordinary diesel operations without hydrogen.

The overall equivalence ratio, ϕ_t, was 0.40 and the engine speed was 1000 rpm. The mass flow rate of

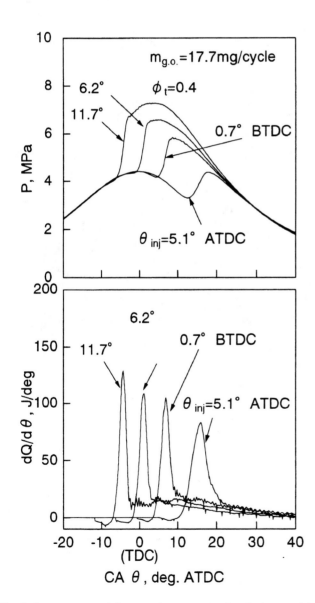

Fig.2 Pressure and heat release rate (ϕ_t=0.4, θ_{inj}=11.7 degrees BTDC~ 5.1 degrees ATDC)

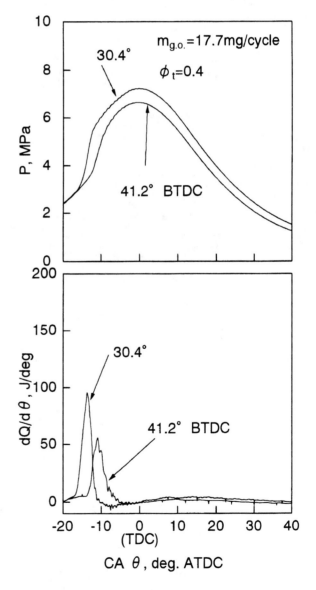

Fig.3 Pressure and heat release rate (ϕ_t=0.4, θ_{inj}=41.2 and 30.4 degrees BTDC)

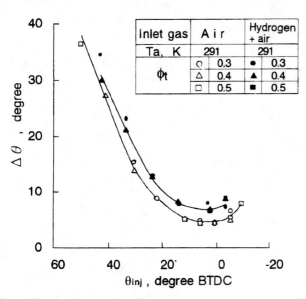

Fig.4 Comparison of ignition delay of dual fuel condition with ordinary diesel condition

diesel fuel was 17.7 mg/cycle. The injection timing of diesel oil, θ_{inj}, was changed from 41.2 degrees BTDC to 5.1 degrees ATDC. When the injection timing, θ_{inj}, is 11.7 degrees BTDC, the value of the first peak is the largest. As the injection timing is earlier than 30.4 degrees BTDC, the value of the first peak becomes smaller without the second peak and the heat release rate shows the characteristics of premixed combustion as described in ref.[21]. This is because the region of the very lean premixed combustion increases due to the longer ignition delay as shown in Fig.4. The ignition delay increases with advancing injection timing very early.

DUAL FUEL CONDITION

Figures 5 and 6 show the typical pressure history and heat release rate in changing the injection timing for hydrogen-air mixture instead of only air as intake gas from the intake port. The amount of the injection of diesel oil, m_{go}, is 4 mg per cycle and this means that the equivalence ratio of diesel oil corresponds to 0.08. The

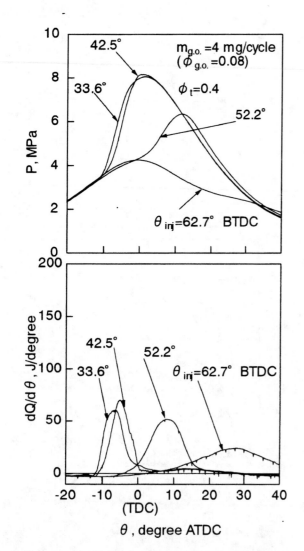

Fig.5 Pressure and heat release rate (ϕ_t=0.4, hydrogen-air mixture, θ_{inj}=62.7deg.BTDC~33.6deg. BTDC)

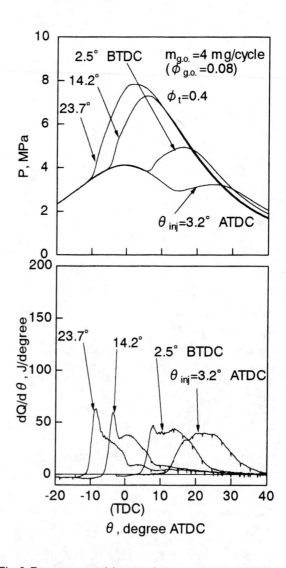

Fig.6 Pressure and heat release rate (ϕ_t=0.4, hydrogen-air mixture, θ_{inj}=23.7deg.BTDC~3.2deg. ATDC)

overall equivalence ratio including hydrogen was 0.4. The injection timing of diesel oil, θ_{inj}, was changed between 62.7 degrees BTDC and 3.2 degrees ATDC.

When hydrogen is mixed with air, the characteristics of the combustion are summarized in comparison with the ordinary diesel operation as follows: (1) Before the injection timing of 33.6 degrees BTDC, there is much time when the diesel oil is mixed with air-hydrogen mixture. That is to say, a very lean mixture is prepared in the cylinder and the diesel oil becomes the ignition sources in the wide range of the cylinder. Therefore, the combustion becomes very slow and the heat release rate shows only one mild peak. (2) The maximum value and the gradient of the heat release rate are smaller in general. (3) The heat release rate is similar to the ordinary diesel operation at the injection timing between 2.5 and 23.7 degrees BTDC, while the second peak mainly corresponds to the combustion of hydrogen that starts from the ignition points due to diesel oil. (4) In θ_{inj}=3.2 degrees ATDC, the heat release rate shows only one mild peak. As shown in Fig.4, the longer ignition delay promotes the diffusion of diesel oil and then lean premixed combustion occurs. (5) The ignition delay is larger under all the conditions as some researchers indicated before [19][20]. The mole fraction of oxygen in air induced from the intake port and the temperature of gas in the cylinder affects the ignition delay. The mole fraction of oxygen is smaller because the volumetric ratio of hydrogen is larger in this experiment. This is one of the reasons why the ignition delay is longer. The ratio of specific heats for hydrogen is almost the same as air. However, the heat loss from the gas to the wall increases due to higher thermal conductivity of hydrogen. Therefore, the gas temperature might be smaller and this may lead to increase in ignition delay. (6) The maximum pressure is larger. (7) In extremely early injection timing, cycle-to-cycle fluctuation increased and the engine operation became unstable.

EXHAUST EMISSIONS AND BRAKE THERMAL EFFICIENCY

Exhaust emissions, such as smoke, NOx, HC and CO, and brake thermal efficiency, η_e, were obtained in changing the injection timing as shown in Figs. 7 and 8. The overall equivalence ratio in total, ϕ_t, was set to be 0.3, 0.4 and 0.5, while the amount of the diesel oil in dual fuel operation was 4 mg/cycle, which corresponded to an equivalence ratio of 0.08. Under the condition without hydrogen, there are no data at injection timing between 10 and 50 degrees BTDC, and 10 and 30 degrees BTDC because of knocking at ϕ_t=0.5 and 0.4, respectively. In the case at hydrogen-air mixture, there are no data at the injection timing between 60 and 5 degrees BTDC in ϕ_t=0.5.

In straight diesel conditions, it is said that smoke increases with increase in load, that is to say, equivalence ratio. When hydrogen was mixed with air, smoke was not exhausted because hydrogen does not

include carbon atoms in its molecular structure as described in previous papers [15][16][17]. It is considered that smoke may be formed from the combustion of the diesel oil that is injected in the cylinder. Therefore, the smoke could not be observed because the amount of diesel oil was very small under dual fuel combustion.

When the injection timing was advanced from near TDC, NOx increased and had the maximum value around 20 deg. BTDC for both cases as shown in Figs. 7 and 8. When the value of NOx was large, the heat release rate was also large. When the injection timing was advanced more, the value of NOx decreased gradually. This is because the regions of high temperature gas decrease due to the well-mixing of diesel oil with air [21]. In dual fuel combustion, the value of NOx was larger in general in comparison with equivalence ratio for ordinary diesel combustion because the maximum value of pressure was large as shown in Figs. 2, 3, 5 and 6.

Hydrocarbon is exhausted because the combustion of diesel oil is incomplete. The values of the hydrocarbon showed the minimum value near θ_{inj}=20 degrees BTDC in the normal diesel combustion in changing equivalence ratio. The relation between hydrocarbon and the nitrogen oxide is a trade-off. When hydrocarbons decrease, the nitrogen oxide increases because the combustion ends more completely. Under hydrogen-air conditions, the levels of the hydrocarbons were almost the same as those in the ordinary diesel combustion. The values of CO emissions presented almost the same tendency for every equivalence ratio in ordinary diesel combustion while they presented almost zero and showed very small value of 0.05% in θ_{inj}=52 degrees BTDC and ϕ_t =0.3 in dual fuel combustion.

The brake thermal efficiency in hydrogen-air mixture is slightly smaller than that in ordinary diesel combustion under the condition of ordinary injection timing. In very advanced injection timing, the brake thermal efficiency becomes slightly small in ϕ_t=0.4 and the combustion becomes incomplete, causing increases in CO and HC and shows almost the same as those in ordinary injection timing. However, both smoke and NOx are almost zero, and HC presents low when the injection timing is 50degrees BTDC in the hydrogen-air mixture although the engine operation became unstable.

Figure 9 shows the characteristics of the exhaust emissions in changing the ratio of hydrogen to diesel oil at θ_{inj}=3.5 degrees ATDC. The ratio is expressed on the basis of equivalence ratio as follows:

$$\psi = \phi_{hydrogen}/\phi_t$$

The overall equivalence ratio was set at 0.3, 0.4 and 0.5. The injection amount of diesel oil was changed to 4, 8, 12 mg/cycle for each equivalence ratio. In ψ=0 (straight diesel operation), the data at the injection timing of 5 deg. ATDC were used. The value of smoke decreased with increasing hydrogen fraction, ψ, because the diesel oil injected into the cylinder decreased. In particular, smoke showed almost zero in the ratio of hydrogen more than 40% for $\phi_t=0.4$ and 60% for $\phi_t=0.5$. In $\phi_t=0.3$, the formation of the nitrogen oxide did not decrease very much even when the hydrogen fraction increased. On the other hand, in $\phi_t=0.4$ and 0.5, the nitrogen oxide increased and presented the maximum value near $\psi=0.6$. NOx emissions increased or decreased under various conditions.

The hydrocarbons showed the similar values in changing the ratio of hydrogen, while they decreased slightly with increasing the overall equivalence ratio. The CO_2 and CO concentrations decreased with increasing hydrogen ratio. The lowest value of CO was 0.01% in this study, which was the minimum resolution for the measurement equipment. The brake thermal efficiency, η_e, was unchanged with in changing hydrogen fraction from zero to 0.8 in $\phi_t=0.5$.

CONCLUSIONS

Hydrogen mixed with air was induced from an intake port and diesel oil was injected into the cylinder in a four-

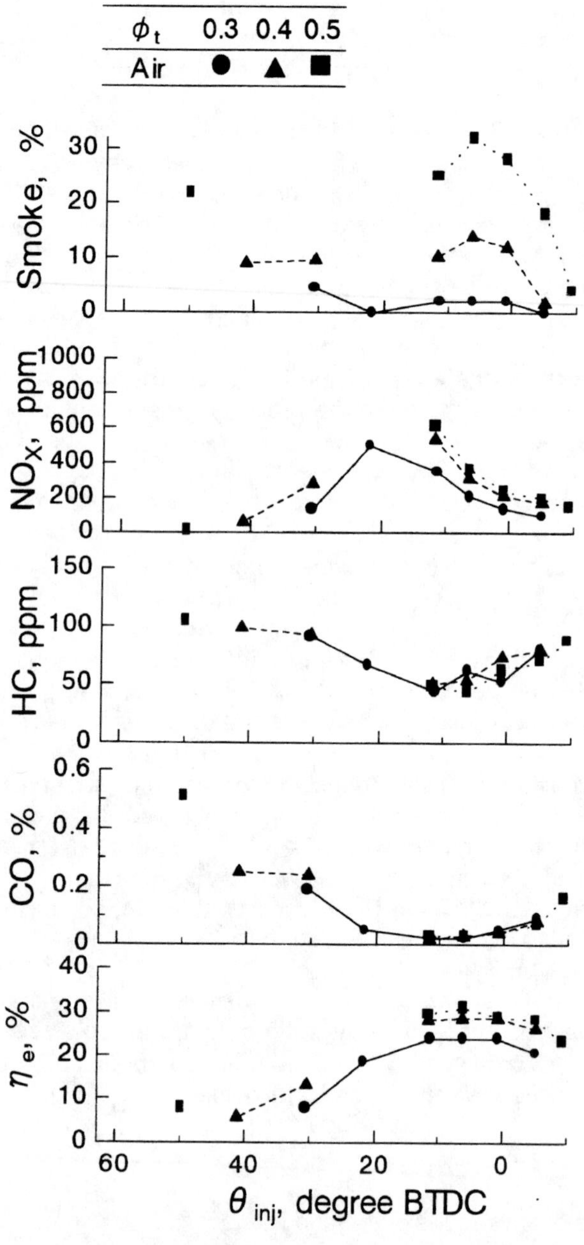

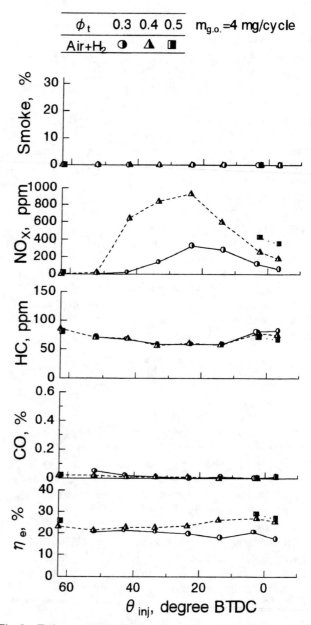

Fig.7 Exhaust emissions of smoke, NOx, HC, CO and brake mean thermal efficiency (ordinary diesel operation)

Fig.8 Exhaust emissions of smoke, NOx, HC, CO and brake mean thermal efficiency (dual fuel operation with hydrogen)

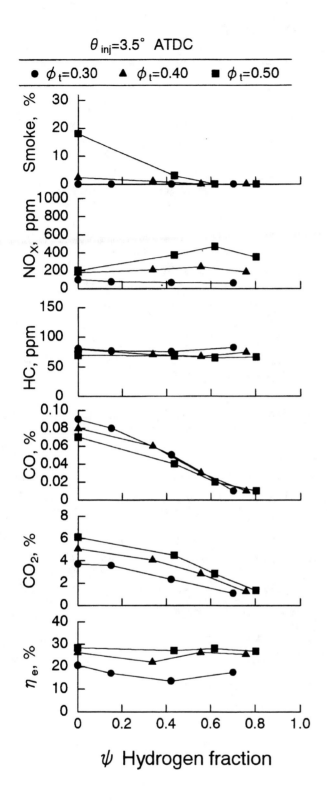

$\theta_{inj} = 3.5°$ ATDC

Fig.9 Exhaust emissions (Smoke, NOx, HC, CO and CO2) and brake thermal efficiency, η_e, changing in hydrogen fraction, ψ.

stroke single cylinder diesel engine. The ratio of the hydrogen was changed from 0 to 80 %. The pressure in the cylinder and the exhaust emissions were measured and their relation was discussed. The main results obtained in this study are as follows:

(1) In θ_{inj}=3.5 degrees ATDC, when hydrogen is mixed with inlet air, smoke enormously decreases. It shows almost zero in the ratio of hydrogen more than 40% for ϕ_t=0.4 and 60% for ϕ_t=0.5. Other emissions of HC, CO and CO2 decrease with increasing ratio of the hydrogen. Brake thermal efficiency is slightly smaller than that in ordinary diesel combustion. NOx emissions increase or decrease under various conditions.

(2) In particular, both smoke and NOx are almost zero while hydrocarbons are almost the same when the injection timing of diesel oil is extremely advanced about 40 or 50 degrees before TDC in ϕ_t =0.3 and 0.4 although the engine operation becomes unstable. Therefore, no smoke, low NOx, very low CO2 and relatively low hydrocarbons are achieved when hydrogen fuel is added to the air from an intake port.

(3) The ignition delay increases when hydrogen is inducted from the intake manifold. This is mainly because the mole fraction of hydrogen is so large that oxygen in the intake air is reduced.

(4) When the injection timing of diesel oil into the cylinder is extremely advanced, the diesel oil is well mixed with air or hydrogen-air mixture and the initial combustion becomes mild. NOx emissions decrease because of lean premixed combustion without the region of high temperature of burned gas such as seen in ordinary diesel operations.

(5) In dual fuel combustion with hydrogen, the maximum value and the gradient of the heat release rate are smaller in general than those in ordinary diesel combustion.

REFERENCES

1. Norbeck, J.M., Heffel, J.W., Durbin, T.D., Tabbara, B., Bowden, J.M. and Montano, M.C., Hydrogen Fuel for Surface Transportation, (1996), Society of Automotive Engineers, Inc.
2. Karim, G.A., and Taylor, M.E., SAE Paper, (1973), No.730089.
3. Furuhama, S. and Yamane, K., Combustion Characteristics of Hydrogen-Fueled Spark Ignition Engine, Trans. of JSAE, (1973), p.54.
4. Homan, H.S., Reynolds, R.K., De Boer, P.C.T. and McLean, W.J., Hydrogen-Fuelled Diesel Engine

without Timed Ignition, Int. J. of Hydrogen Energy, Vol.4, (1979), p.315.

5. Wong, J.K.S., Compression Ignition of Hydrogen in a Direct Injection Diesel Engine Modified to Operate as a Low-Heat-Rejection Engine, Int. J. of Hydrogen Energy, Vol.15, (1990), p.507.

6. Ikegami, M., Miwa, M. and Shioji, M., A Study on Hydrogen Fuelled Compression Ignition Engines, Int. Journal of Hydrogen Energy, 7, (1982), p.341.

7. Karim, G.A., The Dual Fuel Engine, Automotive Engine Alternatives (Ed. Evans, R.L.), (1987), pp.83-104, Plenum Press.

8. Tesarek,H., Investigation Concerning the Employment Possibilities of Diesel - Gas Process for Reducing Exhaust Emissions, Especially Soot (Particulate Matters), SAE Paper, (1983), No.831197.

9. Karim, G.A., The Dual Fuel Engine of the Compression Ignition Type - Prospects, Problems and Solutions - A Review, SAE Paper, (1983), No.831073.

10. Karim, G.A., An Examination of Some Measures for Improving the Performance of Gas Fueled Diesel Engines at Light Load, SAE Paper, (1991), No.912366.

11. Wong, W.Y., Midkiff, K.C. and Bell, S.R., Performance and Emissions of a Natural Gas Dual-Fueled, Indirect Injected Diesel Engine, SAE Paper, (1991), No.911766.

12. Cole, J.J., Kienzle, E., Wells, A.D., Meyer, R.C., Development of a CNG Engine, SAE Paper, (1991), No.910881.

13. Grosshans, G. and Litzler, M., Pielstick Experience with Dual-Fuel Engines and Configuration, Trans. ASME, J. Engng. Gas Turbine and Power, Vol.110, (1988), pp.349-355.

14. Gopal, G., Srinivasa Rao, P., Gopalakrishnan, K.V. and Murthy, B.S., Use of Hydrogen in Dual-Fuel Engines, Int. J. of Hydrogen Energy, 7-3, (1982), pp.267-272.

15. Li, J.-D., Lu, Y.-Q. and Du, T.-S., Improvement on the Combustion of a Hydrogen Fueled Engine, Int. J. Hydrogen Energy, 11-10, (1986), pp.661-668

16. Mathur, H.B., Das, L.M. and Patro, T.N., Hydrogen Fuel Utilization in CI Engine Powered End Utility System, Int. J. Hydrogen Energy, Vol.17, (1992), p.369.

17. Varde, K.S. and Frame, G.A., Hydrogen Aspiration in a Direct Injection Type Diesel Engine - Its Effects on Smoke and Other Engine Parameters, Int. J. Hydrogen Energy, 8-7, (1983), pp.549-555.

18. Lambe, S.M., and Watson, H.C., Optimizing the design of a hydrogen engine with pilot diesel fuel injection, Int. J. of Vehicle Design, Vol.14, No.4, (1993), pp.370-389.

19. Patro, T.N., Combustion Study of Hydrogen Fueled DI Diesel Engine: Simplified Heat Release Analysis, Int. J. Hydrogen Energy, (1993), pp.231-241.

20. Liu, Z. and Karim, G.A., The Ignition Delay Period in Dual Fuel Engines, SAE Paper, (1995), No.950466.

21. Takeda, Y., Nakagome, K., Niimura, K., Emission Characteristics of Premixed Lean Diesel Combustion with Extremely Early Staged Fuel Injection, SAE Paper, (1996), No.961163.

22. Thring, R.H., Homogeneous Charge Compression Ignition (HCCI) Engines, SAE Paper, (1989), No.892068.

23. Tomita, E., Hamamoto, Y., Siagian, A., Piao, Z. and Fujita S., Ignition Delay of Light Oil Spray into Gaseous Fuel and Air Mixture in a Dual Fuel Engine, Proc. of the Sixth International Symposium on Marine Engineering (ISME Tokyo 2000), (2000), pp.505-510.

24. Das, L.M., Hydrogen-Oxygen Reaction Mechanism and Its Implication to Hydrogen Engine Combustion, Int. J. Hydrogen Energy, 21-8, (1996), pp.703-715.

CONTACT

Eiji TOMITA,
Dept. of Mechanical Engineering,
Okayama University,
Tsushima-Naka 3-1-1
Okayama 700-8530
Japan
Phone +81-86-251-8049
FAX   +81-86-251-8266
email: tomita@mech.okayama-u.ac.jp

2001-01-0250

A Numerical Study to Control Combustion Duration of Hydrogen-Fueled HCCI by Using Multi-Zone Chemical Kinetics Simulation

Toru Noda
Nissan Motor Co., Ltd.

David E. Foster
University of Wisconsin-Madison

ABSTRACT

An engine cycle simulation code with detailed chemical kinetics has been developed to study Homogeneous Charge Compression Ignition (HCCI) combustion with hydrogen as the fuel. In order to attain adequate combustion duration, resulting from the self-accelerating nature of the chemical reaction, fuel and temperature inhomogeneities have been brought to the calculation by considering the combustion chamber to have various temperature and fuel distributions. Calculations have been done under various conditions including both perfectly homogeneous and inhomogeneous cases, changing the degree of inhomogeneity. The results show that intake gas temperature is more dominant on ignition timing of HCCI than equivalence ratio and that there is a possibility to control HCCI by introducing appropriate temperature inhomogeneity to in-cylinder mixture.

INTRODUCTION

HCCI can be one of the most promising candidates for next generation combustion of automotive engines. Experimental results[1-5] by many researchers show its advantage of ultra low NOx emission and high thermal efficiency at low load. Because of these potential benefits, many researchers are trying to understand the governing fundamental principles and operating ranges of HCCI through computational attempts[6-9] to predict its behavior. But controlling HCCI is a big challenge since an explicit ignition controlling method like spark plugs is not used. With the self-accelerating nature of its kinetics, HCCI combustion tends toward instantaneous heat release with rapid pressure rise, which is not acceptable in a real engine, in terms of noise and vibration.

On the other hand, many efforts have been made to promote combustion in SI engines by trying to extend the lean limit of stable combustion. It is important to realize that a shorter combustion duration than that of conventional stoichiometric SI engine is not necessary to obtain higher

thermal efficiency. As long as the timing of combustion is adequately controlled around the TDC and stable combustion is achieved in every cycle, good thermal efficiency can be achieved. Too short of a combustion duration may bring increased heat transfer through the combustion chamber walls and result in a trade-off correlation between the ideal cycle and heat transfer in real engines.

From both points of view, noise and vibration or efficiency, it is important to achieve appropriate combustion duration with HCCI combustion. But as long as perfectly homogeneous mixture is assumed in HCCI, the chemical kinetics are most dominant in controlling combustion duration. Unless we can control the chemical reaction rates, it seems that some sort of inhomogeneity is essential to achieve appropriate combustion duration.

Therefore the objective of this study is to obtain some insight or ideas about what kind of inhomogeneities are suitable for HCCI combustion to achieve proper combustion duration.

CALCULATION METHOD

SIMULATION CODE – An engine cycle simulation code with detailed chemical kinetics has been developed by linking the Chemkin[10] libraries to a piston-cylinder simulation. Using hydrogen as the fuel, the kinetics[11] consists of 23 reactions and 11 species including the extended Zeldovich mechanism[12]. The combustion chamber is considered zero dimensional with inhomogeneities of fuel or temperature treated as different zones, or components. The different components (zones) are not given spatial locations within the combustion chamber and transport of mass, temperature or chemical species between zones is not considered. In other words, zones or components are related to each other only through the total pressure[7]. Woschni's correlation[13] is used for the heat transfer model. Although this simulation is non-dimensional, a flat head cylindrical combustion chamber is assumed and heat transfer is calculated with this area and a mean gas temperature.

MODEL ENGINE - Table 1 shows model engine specifications. They are very similar to an automotive SI engine except for the high compression ratio and the fuel.

Table 1 Model Engine Specification

Engine Type	4-Stroke
Fuel	Hydrogen
Displacement	500 cm^3
Bore	86 mm
Stroke	86 mm
Compression Ratio	20 : 1
Intake Valve Close	50 deg. ABDC
Exhaust Valve Open	60 deg. BBDC
Engine Speed	1600 rpm

All of the following calculations for the various conditions were done with the cycle simulation code. For homogeneous cases, the number of components was set to one (single zone model), for inhomogeneous cases, a 10 or 30 component model was used.

RESULTS

EFFECTS OF EQUIVALENCE RATIO AND GAS TEMPERATURE ON IGNITION TIMING – First a perfectly homogenous mixture was assumed to investigate effects of equivalence ratio and intake gas temperature on ignition timing. Here ignition timing is defined as the crank angle where 10 percent of the total fuel energy has been released. Fig. 1 shows that equivalence ratio of the mixture does not have a strong effect on ignition timing for the same intake gas temperature. In contrast, as shown in Fig. 2, ignition timing is greatly influenced by intake gas temperature with the same equivalence ratio.

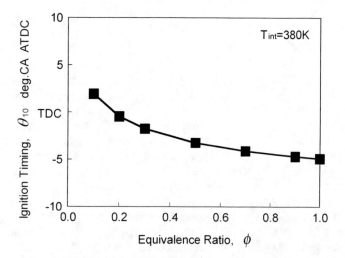

Fig. 1 Effect of Equivalence Ratio on Ignition Timing

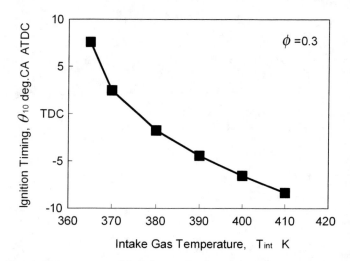

Fig. 2 Effect of Intake Gas Temperature on Ignition Timing

EFFECTS OF FUEL AND GAS TEMPERATURE INHOMOGENEITY ON COMBUSTION DURATION – Mixture fuel or temperature inhomogeneity was taken into account by dividing the combustion chamber into components. A top hat distribution of fuel or gas temperature was assumed, as shown in Fig. 3. Each component had the same amount of fuel but the amount of air varied between components, according to its gas temperature and equivalence ratio. For the results presented below, the definition of combustion duration was taken as the interval between the crank angles where 10 and 90 percent of the fuel energy has been released. As would be expected from the above results of homogeneous cases, larger fuel inhomogeneity makes combustion duration longer, but the effect is small (Fig. 4). This can be interestingly compared with the results shown in Fig. 5, where combustion duration is greatly extended with a certain degree of temperature inhomogeneity.

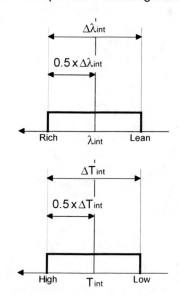

Fig. 3 Fuel/Temperature Distribution Function

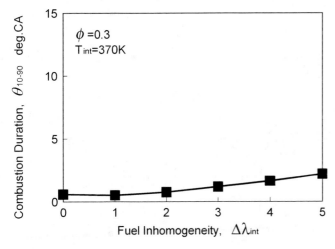

Fig. 4 Effect of Fuel Inhomogeneity on Combustion Duration

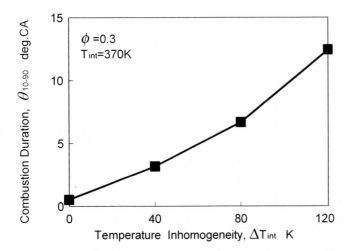

Fig. 5 Effect of Temperature Inhomogeneity on Combustion Duration

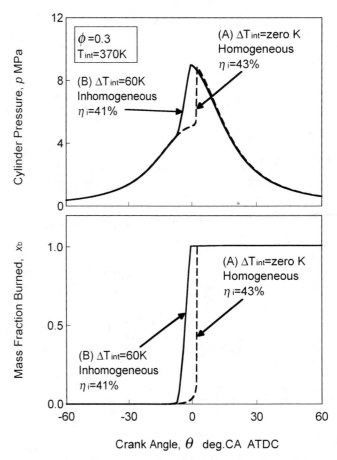

Fig. 6 Cylinder Pressure and Mass Fraction Burned of Homogeneous (A) and Inhomogeneous (B) Case

Fig. 6 shows cylinder pressure and mass fraction burned for the homogeneous and inhomogeneous gas temperature cases under the same equivalence ratio and mean gas temperature. With uniform intake gas temperature (A), combustion takes place slightly after TDC and the pressure goes up instantaneously. In the inhomogeneous case (B) with the width of temperature distribution being 60K, a longer combustion duration of approximately 7 degrees CA and milder pressure rise is predicted. However the thermal efficiency goes down, from 43 to 41 percent, due to the earlier ignition timing.

GAS TEMPERATURE INHOMOGENEITY AND MEAN TEMPERATURE - Although longer combustion duration can be achieved by introducing some temperature inhomogeneity, both ignition timing and combustion duration have to be simultaneously controlled to make combustion take place around the TDC. Mean intake gas temperature as well as the degree of temperature inhomogeneity must be controlled adequately for high thermal efficiency.

Fig. 7 shows effects of combinations of temperature inhomogeneity and mean intake gas temperature on combustion duration, NO emission and thermal efficiency. Fig. 7 is a kind of HCCI phase space, plotting inhomogeneity versus mean temperature. In the bottom - right region of the map, where temperature inhomogeneity is not so large and mean temperature is high, all components ignite one after another until complete combustion is attained. In the bottom - left region, no combustion takes place in any component due to too low of a temperature. In the remainder of the phase space, combustion begins from the hottest components, followed by next hottest etc. However, in some region of the space, some of the colder components do not ignite leaving an amount of the fuel unburned. This occurs as one moves outward from the bottom-right hand corner of the figure.

Within the complete combustion region depicted in the Fig. 7, combustion duration becomes longer with more temperature inhomogeneity. NO emissions are higher at the upper-right area where the temperature inhomogeneities result in some very high component temperatures. Thermal efficiency is highest in the bottom-left side of the complete combustion region. But the bottom-left border between complete combustion and partial/no combustion is very steep. Thermal efficiency goes down as it moves to the upper-right area, showing similar trend with NO emission.

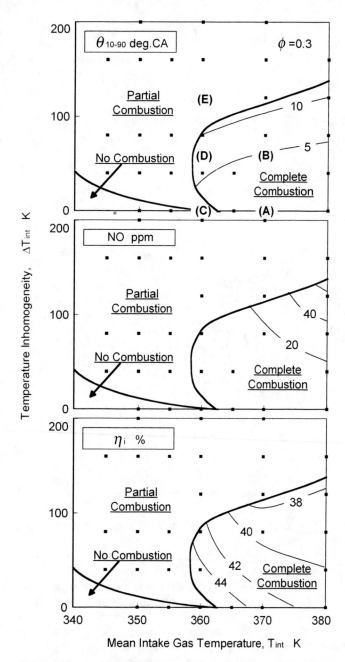

Fig. 7 Combustion Duration, NO emission and Thermal Efficiency Map for Mean Temperature and Temperature Inhomogeneity

The points (A) and (B) in Fig. 7 show the combinations of temperature inhomogeneity and mean gas temperature which have already been presented in Fig. 6. At point (A), with a perfectly homogeneous gas temperature, thermal efficiency is high and NO emission is low but the combustion duration is too short. By moving to point (B), with a temperature inhomogeneity of 60K, combustion duration becomes longer but the thermal efficiency goes down, as already mentioned above. With a 10K lower mean gas temperature than (A) and perfectly homogeneous mixture, point (C), no combustion is observed except a very small heat release from some preliminary reactions. By moving from (C) to (D), which was done by introducing the same amount of temperature inhomogeneity as between (A) and (B), complete combustion is again attained. For this case a combustion duration longer than condition (B) is

attained and an efficiency equivalent to that of (A) is predicted. If too much inhomogeneity is introduced, for example point (E), some of the colder components do not ignite and the combustion ends up being incomplete.

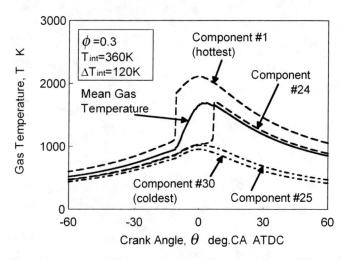

Fig. 8 Mean and Component Gas Temperature of Partial Combustion (E)

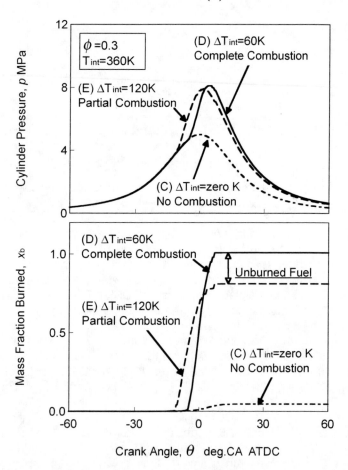

Fig. 9 Cylinder Pressure and Mass Fraction Burned of No Combustion (C), Complete Combustion (D) and Partial Combustion (E)

Fig. 8 shows the temperature history of each component for case (E). The component #1, which is the hottest, ignites first then followed by next hottest. The colder components ignite one after another due to the compression from the piston motion and energy release of the reacting gases.

These are the important temperature drivers to help colder components self-ignite. The component #24 successfully ignites by this effect but some of the cold components do not reach a high enough temperature to react and remain unburned, for example components #25 and higher.

Comparisons of cylinder pressure and mass fraction burned of case (C), (D) and (E) are shown in Fig. 9 and Fig. 10 shows temperature distribution functions of each case in Fig. 9. For case (E), with too large of a temperature inhomogeneity, combustion starts earlier than case (D) and cylinder pressure goes up mildly, but almost 20 percent of the total fuel does not complete reacting and stays unburned.

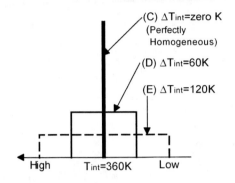

Fig. 10 Temperature Distribution Functions of case (C), (D) and (E) shown in Fig. 9

TYPE OF TEMPERATURE INHOMOGENEITY - Fig. 7 shows that both temperature inhomogeneity and mean temperature must be controlled within a very narrow operating region. To explore the possibility of making the system more robust, different forms of the temperature inhomogeneity were explored. The logic followed in developing a hypothetical temperature distribution is given below.

1. Some hot components are needed to insure ignition, but the amount should be small to avoid negative work during compression and to prevent NO formation.

2. After TDC, reacting components need to release sufficient energy so that unburned components are compressed to autoignition, making up for the decrease in the mean gas temperature by the descending piston.

An attempt to develop such a temperature distribution was made. Another set of calculations was carried out with the temperature distribution function shown in Fig. 11. The results are shown in Fig. 12 as a map of combustion duration. Complete combustion is attained in wider region as compared with Fig. 7 and longer combustion duration exists within the region. Fig. 13 shows pressure and mass fraction burned of the case (F) shown in Fig. 12.

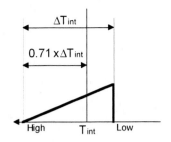

Fig. 11 Temperature Distribution Function

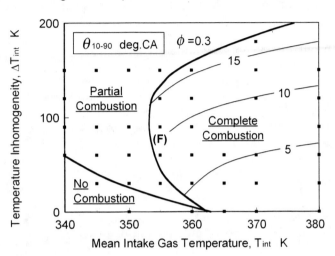

Fig. 12 Combustion Duration Map for Mean Temperature and Temperature Inhomogeneity

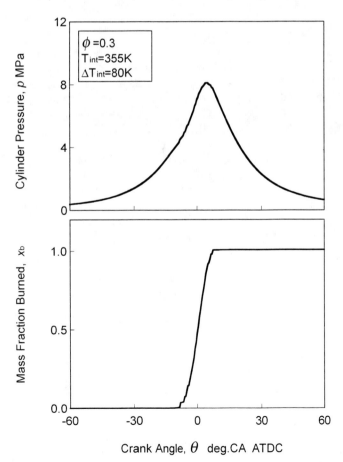

Fig. 13 Cylinder Pressure and Mass Fraction Burned of Case (F)

DISCUSSION

It is easy to concoct temperature distributions numerically and run them in a simulation. Indeed, the challenge is to investigate experimentally the proposed concepts. In this work we have attempted to keep the temperature profiles investigated within realistic bounds. That is, the lowest temperature likely to be available is atmosphere temperature (300K), and the highest temperature available is likely to be from the exhaust gas of the previous cycle, so called internal EGR. Most of the region shown in Fig. 7 and Fig. 12 is within these criteria.

The authors do not offer a means for attaining the proposed temperature inhomogeneity shown in Fig. 11. The objective of this study was to obtain insight about impact of inhomogeneities on HCCI combustion. The predictions shown in Fig. 12 indicate that a small hot core and a distribution of relatively colder components are needed to attain the largest HCCI operation regime. We interpret this to mean that strong stratification of internal EGR and fresh intake air, or some sort of localized ignition enhancement will be required.

Hydrogen was used as the fuel in this study. Hydrogen does not exhibit the cool flame and negative temperature characteristics that typical hydrocarbon fuels exhibit. This may lead to a different conclusion about the necessary inhomogeneities of mixture. The authors will investigate this in the future.

CONCLUSION

In this study, a multi-component engine cycle simulation code including chemical kinetics has been developed. A hydrogen-fueled HCCI simulation was evaluated under various conditions including fuel and temperature inhomogeneity. The results show,

1. The effect of gas temperature on ignition timing is more dominant than that of equivalence ratio. Introducing gas temperature inhomogeneity is a better strategy for controlling combustion duration than introducing air fuel inhomogeneity, although fuel inhomogeneity seems easier to accomplish using some sort of direct injection technology.

2. Both gas temperature inhomogeneity and mean gas temperature must be controlled to achieve proper combustion duration without degradation of thermal efficiency and NO emission.

3. A temperature inhomogeneity consisting from a small hot region and greater amount of relatively colder region is more suitable to make the system more robust.

ACKNOWLEDGMENTS

The authors wish to acknowledge Nissan Motor Company for the financial support of Mr. Noda during his tenure at the Engine Research Center.

REFERENCES

1. Najt, P. M., Foster, D. E., "Compression-Ignited Homogeneous Charge Combustion", SAE Paper 830264, 1983
2. Onishi, S., Jo, S. H., Shoda, K., Jo, P. D., Kato, S., "Active Thermo-Atmosphere Combustion (ATAC) – A New Combustion Process for Internal Combustion Engines", SAE Paper 790501, 1997
3. Thring, R. H., "Homogeneous-Charge Compression-Ignition (HCCI) Engines", SAE Paper 892068, 1989
4. Aoyama, T., Hattori, Y., Mizuta, J., Sato, Y., "An Experimental Study on Premixed-Charge Compression Ignition Gasoline Engine", SAE Paper 960081, 1996
5. Ryan, T. W., Callahan, T. J., "Homogeneous Charge Compression Ignition of Diesel Fuel", SAE Paper 961160, 1996
6. Kong, S.-C., Ayoub, N., Reitz, R. D., "Modeling Combustion in Compression Ignition Homogeneous Charge Engines", SAE Paper 920512, 1992
7. Aceves, S. M., Flowers, D. L., Westbrook, C. K., Smith, J. R., Pitz, W., Dibble, R., Christensen, M., Johansson, B., "A Multi-Zone Model for Prediction of HCCI Combustion and Emissions", SAE Paper 2000-01-0327, 2000
8. Fiveland, S. B., Assanis, D. N., "A Four-Stroke Homogeneous Charge Compression Ignition Engine Simulation for Combustion and Performance Studies", SAE Paper 2000-01-0332, 2000
9. Yamasaki, Y., Iida, N., "Numerical Simulation of Auto-Ignition and Combustion of n-Butane and Air Mixtures in a 4 Stroke HCCI Engine by Using Elementary Reactions", SAE Paper 2000-01-1834, 2000
10. Reaction Design, http://www.ReactionDesign.com/
11. Westbrook, C. K., Dryer, F.L., "Chemical Kinetic Modeling of Hydrocarbon Combustion", Progress in Energy and Combustion Science, Vol. 10 No.1, 1984
12. Lavoie, G. A., Heywood, J. B., Keck, J. C., "Experimental and Theoretical Investigation of Nitric Oxide Formation in Internal Combustion Engines", Combust. Sci. Technol., vol.1, pp.313-326, 1970
13. Woschni, G., "Universally Applicable Equation for the Instantaneous Heat Transfer Coefficient in the Internal Combustion Engine", SAE paper 670931, SAE Trans., vol. 76, 1967

HCCI Combustion of Hydrogen, Carbon Monoxide and Dimethyl Ether

Toshio Shudo and Yoshitaka Ono

Hydrogen Energy Research Center, Musashi Institute of Technology

ABSTRACT

Homogeneous charge compression ignition (HCCI) combustion enables higher thermal efficiency and lower NOx emission to be achieved in internal combustion engines when compared with conventional combustion systems. Control of proportion of high cetane number and low cetane number fuels is an effective technique for controlling ignition timing and load in HCCI combustion.

The aim of this paper is to analyze the characteristics of the HCCI combustion of hydrogen, carbon monoxide and dimethyl ether (DME) in a single cylinder engine. A mixture of hydrogen and carbon monoxide with a composition of 67% hydrogen and 33% carbon monoxide called methanol-reformed gas (MRG) was used as the low cetane number fuel and DME as the high cetane number fuel. Both MRG and DME can be reformed from methanol in endothermic reactions. The endothermic reactions make waste heat recovery in fuel reforming possible by using the heat from the exhaust gases. Experiments were conducted in which the proportion of the fuels, equivalence ratio and compression ratio were varied to control ignition timing.

The results show that the HCCI combustion of hydrogen, carbon monoxide and dimethyl ether has a high thermal efficiency over a wide operating range. HCCI combustion has a high power output that is comparable to that of SI combustion despite lean combustion conditions, because of the high thermal efficiency.

INTRODUCTION

It is expected that the use of homogeneous charge compression ignition (HCCI) combustion in internal combustion engines will result in a higher thermal efficiency and lower NOx emissions compared with conventional combustion systems. An adjustment of the proportion of high cetane number fuel and low cetane number fuel has been reported as a method for controlling the ignition timing and load in HCCI combustion [1]. However, this method is not commercially used because of the inconvenience of storing two kinds of fuels.

On the other hand, dimethyl ether (DME) is a clean alternative fuel for compression ignition engines because of its high cetane number and smokeless combustion properties [2-7]. DME can be reformed from methanol, a process that also produces a thermally decomposed gas with 2 mol of hydrogen

and 1 mol of carbon monoxide. Because the methanol-reformed gas (MRG) has a low cetane number, the ratio of DME and MRG can supposedly be used to control the ignition timing in the HCCI combustion of these fuels. In addition, the reactions to produce DME or MRG are endothermic. Therefore, fuel reforming using the heat from exhaust gases from internal combustion engines is a feasible technique for increasing the overall thermal efficiency. The usage of a single liquid fuel also eliminates the obvious inconvenience of having to carry two fuels.

The aim of this research is to experimentally analyze the characteristics of the HCCI combustion of hydrogen, carbon monoxide and DME in a single cylinder engine. The experiments were conducted by varying the proportion of the fuels, equivalence ratio and compression ratio to control the ignition timing.

EXPERIMENTAL APPARATUS

The engine was a four stroke cycle single cylinder spark ignition engine with a bore of 85 mm and a stroke of 88 mm. The compression ratio was set at 8.3, 9.7, 12.4 and 15.8. Hydrogen and carbon monoxide were employed as the low cetane number fuel and DME as the high cetane number fuel. A model gas of 67% hydrogen and 33% carbon monoxide was used to simulate the MRG. Both the model MRG gas and the DME were supplied from high-pressure cylinders. Fuel gases were measured with mass flow meters (Oval) and continuously supplied into the intake manifold.

The in-cylinder pressure was measured with a piezo-electric type pressure transducer (AVL GM12D) installed in the cylinder head. Data from 100 cycles were averaged for each experimental condition and used to calculate indicated mean effective pressure, indicated thermal efficiency, apparent rate of heat release, degree of constant volume and others. The thermal efficiency was evaluated using the indicated thermal efficiency and the power output with the indicated mean effective pressure. The engine speed was set at 1000 rpm and the volumetric efficiency including fuel gas for all experiments was 75%. Experiments were performed with a maximum rate of the in-cylinder pressure rise of 300 kPa/degree CA, which was the combustion limit in the richer mixture side.

The concentration of CO in the exhaust gas was measured with an NDIR analyzer and the THC concentration with an FID analyzer. The sensitivity of the FID analyzer was different to that used in ordinary hydrocarbon combustion because the exhaust gas from DME combustion contains relatively large amounts of oxygenated compounds. The correction factor for THC measurement by FID analyzer in DME direct injection diesel combustion has been reported elsewhere [8]. However, the correction factor was not used in this study, because the system did not use diesel combustion and it was fuelled with not only DME but also hydrogen and carbon monoxide. Therefore, if all of the measured THC in the exhaust gas was DME, the real concentration of THC would be larger than the results in this research by 65%.

ANALYSIS OF THERMAL EFFICIENCY

Indicated thermal efficiency η_i of an internal combustion engine is expressed by Equation (1) using theoretical thermal efficiency of Otto cycle η_{th}, degree of constant volume η_{glh}, combustion efficiency η_u, and cooling loss ratio ϕ_w [9].

$$\eta_i = \eta_{th} \cdot \eta_{glh} \cdot \eta_u \cdot (1 - \phi_w) \qquad (1)$$

This research evaluated characteristics of the cooling loss ratio and the degree of constant volume from indicator diagram.

An index Q/Q_{fuel} was introduced to analyze the cooling loss characteristics. This index is the normalized cumulative apparent heat release divided by the supplied fuel heat in a cycle. The cumulative apparent heat release Q is influenced by the cooling loss from burning gas to the combustion chamber wall and can be related to the cumulative real heat release $Q_B (= \eta_u \cdot Q_{fuel})$ and the cumulative cooling loss Q_C as follows:

$$Q = Q_B - Q_C \qquad (2)$$

Q/Q_{fuel} can be written as

$$Q/Q_{fuel} = (Q_B - Q_C) \cdot \eta_u / Q_B \qquad (3)$$

Q/Q_{fuel} can be rewritten using the cooling loss ratio ϕ_w defined as $\phi_w = Q_C / Q_B$ as

$$Q/Q_{fuel} = (1 - \phi_w) \cdot \eta_u \qquad (4)$$

Therefore, Q/Q_{fuel} increases with a decrease in the cooling loss ratio ϕ_w or with an increase in combustion efficiency η_u. If the combustion efficiency η_u is given by the exhaust gas analysis, the cooling loss ratio ϕ_w can be quantified from the heat release analysis [10-12]. This study qualitatively analyzed the cooling loss ratio characteristics by using Q/Q_{fuel}. The cumulative apparent heat release Q was obtained by integration of apparent rate of heat release $dQ/d\theta$ calculated as follows.

$$dQ/d\theta = (V \cdot dP/d\theta + \gamma \cdot P \cdot dV/d\theta)/(\gamma - 1)$$
$$- P \cdot V/(\gamma - 1)^2 \cdot d\gamma/d\theta \qquad (5)$$

where, V is in-cylinder volume, P is in-cylinder pressure, θ is crank angle and γ is specific heat ratio.

The degree of constant volume η_{glh} was calculated from apparent rate of heat release in the following equation [9]:

$$\eta_{glh} = 1/(\eta_{th} \cdot Q) \cdot \int (1 - ((V_h + V_c)/V(\theta))^{1-\gamma}) \cdot dQ/d\theta \cdot d\theta \qquad (6)$$

where, V_h is stroke volume, V_c is clearance volume.

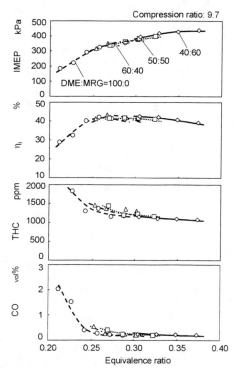

Figure 1 IMEP, indicated thermal efficiency and exhaust emissions in HCCI combustion of DME and MRG

RESULTS AND DISCUSSIONS

Influence of DME:MRG Ratio

Figure 1 shows the influence of the DME:MRG ratio on power output, thermal efficiency and exhaust emissions in HCCI combustion. The compression ratio was set at 9.7, the DME:MRG ratio was 100:0, 60:40, 50:50 and 40:60%. A larger MRG ratio enables richer mixture combustion and increases the indicated mean effective pressure. The HCCI combustion demonstrated high indicated thermal efficiency over a wide range of equivalence ratios. THC and CO exhaust emissions tended to decrease with equivalence ratio.

Figure 2 shows the indicator diagrams for the four DME:MRG ratios in Figure 1 at optimum equivalence ratios for thermal efficiency. Though the DME:MRG ratio does not influence the timing of the first stage heat release by cool flame, the second stage heat release by hot flame is largely influenced by the ratio. An increased MRG content retards the second stage heat release and can control effectively the ignition timing of HCCI combustion. These are the reasons for the extension of the operational range by the addition of MRG to the HCCI combustion of DME shown in Figure 1. The value of Q/Q_{fuel} following combustion is influenced almost singularly by the cooling loss from the burning gas to the combustion chamber wall. Therefore, a rapid decrease in Q/Q_{fuel} signifies a larger cooling loss. Figure 2 shows that the DME:MRG ratio hardly influences the characteristics of cooling loss. A high Q/Q_{fuel} during the combustion period suggests a high combustion efficiency and small cooling loss in HCCI combustion.

Influence of Compression Ratio

Figure 3 shows the influence of compression ratio on HCCI combustion with a DME:MRG ratio of 40:60%. The compression ratio ε was set at 8.3, 9.7, 12.4 and 15.8. The operable range of equivalence ratio shifts to the leaner side with increasing compression ratio. The compression ratio of 9.7 results in the highest indicated thermal efficiency of the four ratios tested. Figure 4 shows the results of the optimal equivalence ratios for each ME:MRG ratio. A compression ratio 9.7 is the optimum for indicated thermal efficiency. Compression ratios higher than 9.7 reduce the thermal

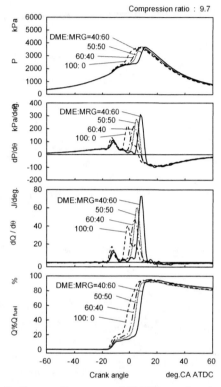

Figure 2 Indicator diagrams of HCCI combustion of DME and MRG

Figure 5 shows the indicator diagrams for each compression ratio. The DME:MRG ratio and equivalence ratio were set at optimal values for thermal efficiency. Higher compression ratios advance the timing of heat release by both cool and hot flames. Operations with a lower compression ratio exhibit a larger heat release, because the retarded heat release eliminates the rapid increase in pressure even in richer mixture conditions. This is the reason for the higher power output with lower compression ratio in Figure 4. Figure 5 indicates that the higher compression ratio causes a smaller value of Q/Q_{fuel}. This signifies an increase in the cooling loss ratio by the increased compression ratio. The increased cooling loss ratio is supposed to be one reason for the decreased thermal efficiency in the higher compression ratios. Hydrogen, the major constituent of MRG, burns with a high cooling loss from the burning gas to the combustion chamber wall because of the intense convection by high burning velocity and the shorter quenching distance for hydrogen [10]. The cooling loss characteristics of hydrogen supposedly influence the increased cooling loss ratio by the increased compression ratios.

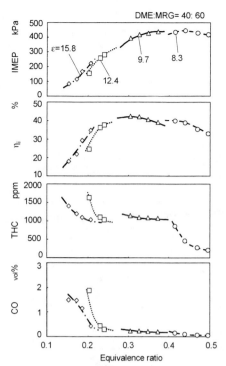

Figure 3 Influence of compression ratio on HCCI combustion of DME and MRG (DME:MRG=40:60)

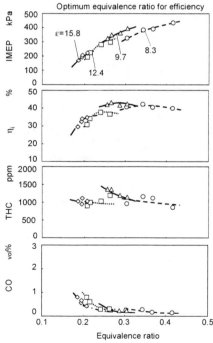

Figure 4 Influence of compression ratio on HCCI combustion of DME and MRG

Figure 6 shows the influence of compression ratio on the factors influencing the thermal efficiency such as the degree of constant volume. The DME:MRG ratio was set at 40:60% and equivalence ratio was at the optimum for indicated thermal efficiency. The theoretical thermal efficiency η_{th} was calculated with each compression ratio and a constant specific heat ratio of 1.4 in the Otto cycle. The indicated thermal efficiency η_i decreases with Q/Q_{fuel} for compression ratios higher than 9.7, and decreases by the theoretical thermal efficiency η_{th} for compression ratios lower than 9.7. Here, there is no significant change in the degree of constant volume η_{glh}, accordingly the characteristics of the indicated thermal efficiency η_i are dominated mainly by the theoretical thermal efficiency η_{th} and the cooling loss ratio ϕ_w.

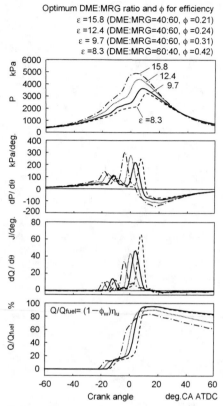

Figure 5 Indicator diagram of HCCI combustion for different compression ratios

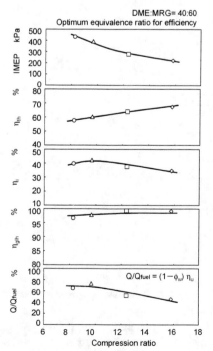

Figure 6 Influencing factors on thermal efficiency for different compression ratios

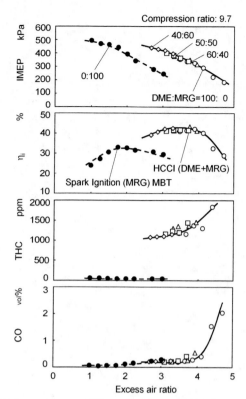

Figure 7 Comparison of SI combustion and HCCI combustion

Comparison of HCCI and SI Combustion

HCCI and homogeneous charge spark ignition combustion were compared for a compression ratio of 9.7. The spark ignition (SI) combustion was fuelled with neat MRG, because of the necessity for a high octane number. Ignition was provided by a spark discharge in the central position of the combustion chamber, and set at the optimum timing for indicated thermal efficiency. Figure 7 shows the results. The thermal efficiency of HCCI combustion is significantly higher than that of SI combustion in lean mixture conditions. Because of the high thermal efficiency, HCCI combustion has a relatively high power output that is comparable to SI combustion in spite of the lean combustion. Conditions leaner than an excess air ratio of 4.0 resulted in a relatively low CO exhaust emission from the HCCI combustion. Although the THC exhaust emission from HCCI combustion is higher than that of SI combustion, the emission decreases with excess air ratio.

Figure 8 shows a comparison of HCCI and SI combustion for four different compression ratios. The results are plotted against DME ratio. The excess air ratio was set at the optimum for indicated thermal efficiency in each compression ratio and DME:MRG ratio. SI combustion exhibits the highest indicated thermal efficiency at the highest compression ratio. Contrary to this, HCCI combustion exhibits high thermal efficiency at a relatively low compression ratio of 9.7 over a wide range of DME:MRG ratios. The thermal efficiency of HCCI combustion is significantly larger than that of the SI engine. In addition, HCCI combustion exhibits a high power output for compression ratios of 8.3 and 9.7. The power output is comparable to that of the SI combustion of MRG. Conversely, HCCI combustion tends to emit higher levels of

THC and CO. However, the level of CO emissions from HCCI combustion at compression ratios of 8.3 and 9.7 is equal to that from SI combustion.

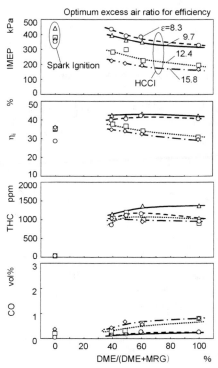

Figure 8 Comparison of SI combustion and HCCI combustion for different compression ratios

Overall Thermal Efficiency based on Methanol

DME(CH_3OCH_3) and MRG($2H_2+CO$) can be produced from methanol in the following reactions.

$$2CH_3OH \rightarrow CH_3OCH_3 + H_2O \text{ (endothermic)} \quad (7)$$

$$CH_3OH \rightarrow 2H_2 + CO \text{ (endothermic)} \quad (8)$$

The DME has a higher heating value than methanol by 4 % and MRG than methanol by 20% in ideal reforming conditions because both of the reactions noted above are endothermic. A high overall thermal efficiency can be achieved by using the exhaust gas heat during fuel reforming. Figure 9 shows the overall thermal efficiency based on LHV of liquid methanol for HCCI and SI combustion under various operational conditions. The index η_r, which is the calculated ratio of the heating value of the reformed fuels to that of liquid methanol, is introduced to show the feasible increase in the fuel heating value by ideal fuel reforming. This study estimated the overall thermal efficiency by using the product of engine efficiency η_i from experiment and the η_r calculated for ideal conditions. The η_r is 120% for neat MRG and 104% for neat DME. The production of MRG increases η_r than that of DME, therefore, conditions with a greater amount of MRG exhibit advantages in terms of exhaust heat recovery. The effect of the increase in heating value by fuel reforming in SI combustion is greater than in HCCI combustion because of the fuel proportions. However, the overall efficiency of HCCI

combustion fuelled with DME and MRG is higher than that of SI combustion of MRG, and is a high value.

Therefore, it is suggested that HCCI combustion fuelled with DME and MRG is an efficient and clean combustion system.

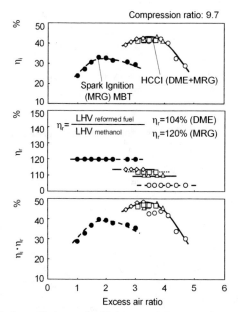

Figure 9 Overall thermal efficiency of HCCI combustion and SI combustion based on methanol

CONCLUSION

This research experimentally analyzed the characteristics of HCCI combustion fuelled with DME and MRG in a single cylinder engine. The research also investigated the feasibility of waste heat recovery in methanol reforming to produce DME and MRG and as an alternative to the carrying of two fuels. The conclusions from this study are as follows:

1. The timing of the second stage heat release by the hot flame can be controlled in an HCCI combustion engine fuelled with DME and MRG by varying the proportion of the two fuels. Controlled ignition timing enables the HCCI operation at wide range of loads.
2. An increased compression ratio advances the timing of the heat release by both cool and hot flames and increases the cooling loss from the burning gas to the combustion chamber wall. The optimum compression ratio for thermal efficiency is relatively low in the HCCI combustion of DME and MRG because of the cooling loss.
3. Use of the optimum compression ratio and proportion of DME and MRG in HCCI combustion results in a higher thermal efficiency and comparable power output as compared with SI combustion fuelled with MRG.
4. Waste heat recovery in methanol reforming to produce DME and MRG has a potential to achieve higher overall thermal efficiency.

ACKNOWLEDGMENTS

The authors would like to thank Mr. Hideki Omori, a student at the Musashi Institute of Technology, for his help with the experiments.

REFERENCES

1. Furutani, T., et al., "A New Concept of Ultra-Lean Premixed Compression Ignition Engine", Proceedings of the 12th Internal Combustion Engine Symposium, No.45, (in Japanese with English summary) (1980).

2. Murayama, T., et al., "A Study of Compression Ignition Methanol Engine with Converted Dimethyl Ether as an Ignition Improver", SAE Paper 922212, (1992).

3. Sorenson, S.C., et al., "Performance and Emissions of a 0.273 Liter Direct Injection Diesel Engine Fuelled with Neat Dimethyl Ether", SAE Paper 950064, (1995).

4. Kajitani, S., et al., "Engine Performance and Exhaust Characteristics of Direct-Injection Diesel Engine Operated with DME", SAE Paper 972973, (1997).

5. Igarashi, T., et al., "Auto Ignition and Combustion of DME and n-Butane/Air Mixtures in Homogeneous Charge Compression Ignition Engine", Transaction of JSME, Vol.64, No.618, B, (in Japanese with English summary) (1998).

6. Suzuki, T., et al., "Combustion Characteristics of Compression Ignited Engine with DME Low-Pressure Fuel Injection", Proceedings of JSAE, No.74-99, (in Japanese with English summary) (1999).

7. Chen, Z., et al., "Study on Homogeneous Premixed Charge CI Engine Fueled with LPG", Proceedings of JSAE, No.38-00, (in Japanese with English summary) (2000).

8. Konno, M., et al., "The Unburned Emissions from a CI Engine Operated with Dimethyl Ether", Transaction of JSME, Vol.67, No.659, B (in Japanese with English summary) (2000).

9. List, H. *Thermodynamik der Verbrennungskraftmaschinen*, Springer, (1939).

10. Shudo, T., et al., "Analysis of Thermal Efficiency in a Hydrogen Premixed Spark Ignition Engine", Proceedings of ASME, AES-Vol.39, (1999).

11. Shudo, T., et al., "Analysis of Degree of Constant Volume and Cooling Loss in a Hydrogen Fuelled SI Engine", SAE Paper 2001-01-3561, (2001).

12. Shudo, T., et al., "Analysis of the Degree of Constant Volume and Cooling Loss in a Spark Ignition Engine fuelled with Hydrogen", I Mech E International Journal of Engine Research, Vol.2, No.1, (2001).

13. Shudo, T., et al., "Influence of Specific Heats on Indicator Diagram Analysis in a Hydrogen-fuelled SI Engine", Elsevier JSAE Review, Vol.22, No.2, (2001).

14. Shudo, T., et al., "Combustion Analysis in a Hydrogen-Fueled Engine with Indicator Diagram", Transaction of JSME, Vol.67, No.654, B, (in Japanese with English summary) (2001).

15. Shudo, T., et al., "Combustion Characteristics of H_2-CO-CO_2 Mixture in an IC Engine", SAE Paper 2001-01-0252, (2001).

16. Shudo, T., et al., "Combustion and Cooling Loss of H_2-CO-CO_2 Mixture Reformed from Methanol", Transaction of JSAE, Vol.32, No.4, (in Japanese with English summary) (2001).

17. Shudo, T., et al., "Characteristics of an Internal Combustion Engine Fueled with Hydrogen Carbon Monoxide Mixture", Transaction of JSAE, Vol.31, No.4, (in Japanese with English summary) (2000).

18. Shudo, T., et al., "Combustion Characteristics of Pyrolysis Gas of Automobile Shredder Dust in an IC Engine", Proceedings of the 8th European Automotive Congress, SAITS 01002, (2001).

19. Shudo, T., et al., "Analysis of Direct Injection Spark Ignition Combustion in Hydrogen Lean Mixture", International Journal of Automotive Technology, Vol.2, No.3, (2001).

20. Shudo, T., et al., "Combustion and Emissions in a Methane DI Stratified Charge Engine with Hydrogen Premixing", Elsevier JSAE Review, Vol.21, No.1, (2000).

21. Shudo, T., et al., "Combustion Promotion and Cooling Loss in Hydrogen Premixing to Methane DI Stratified Charge Combustion", Transaction of JSAE, Vol.32, No.1, (in Japanese with English summary) (2001).

22. Tanford, C., et al., "Equilibrium Atom and Free Radical Concentrations in Carbon Monoxide Flames and Correlation with Burning Velocities", Journal of Chemical Physics, Vol.15, No.7, (1947).

23. Lewis, B., et al., "Combustion, Flames and Explosion of Gases", Academic Press, (1961).

NOMENCLATURE

HCCI:	homogeneous charge compression ignition
SI:	spark ignition
DME:	dimethyl ether
MRG:	methanol-reformed gas (67% H_2 + 33% CO)
IMEP:	indicated mean effective pressure
LHV:	lower heating value of fuel
THC:	total hydrocarbon
CA:	crank angle
ε:	compression ratio
ϕ:	equivalence ratio
λ:	excess air ratio
γ:	ratio of specific heats
θ:	crank angle
P:	in-cylinder pressure
V:	in-cylinder volume
V_h:	stroke volume
V_c:	clearance volume
$dP/d\theta$:	rate of pressure rise
$dQ/d\theta$:	apparent rate of heat release
Q:	cumulative apparent heat release
Q_{fuel}:	heat of fuel supplied in a cycle
Q_B:	cumulative real heat release
Q_C:	cumulative cooling loss
ϕ_w:	cooling loss ratio, Q_C/Q_B
η_u:	combustion efficiency
η_i:	indicated thermal efficiency
η_{th}:	theoretical thermal efficiency
η_{glh}:	degree of constant volume
η_r:	degree of fuel heat increase in fuel reforming

CONTACT

Toshio Shudo, Ph.D., Assistant Professor
Hydrogen Energy Research Center
Department of Energy Science and Engineering
Musashi Institute of Technology
Tamazutsumi 1, Setagaya-Ward, Tokyo, Japan
Phone: +81 3 3703 3111 ext. 3509
Facsimile: +81 3 5707 2127
E-mail: shudo@herc.musashi-tech.ac.jp
URL: http://www.herc.musashi-tech.ac.jp/shudo/home.html

HYBRID VEHICLES

2000-01-0993

Research and Development of a Hydrogen-Fueled Engine for Hybrid Electric Vehicles

Yasuo Nakajima, Kimitaka Yamane, Toshio Shudo and Masaru Hiruma
Musashi Institute of Technology

Yasuo Takagi
Nissan Motors Co., LTD.

ABSTRACT

Hybrid electric vehicle with internal combustion engine fueled with hydrogen can be a competitor to the fuel cell electric vehicle that is thought to be the ultimately clean and efficient vehicle. The objective in this research is to pursue higher thermal efficiency and lower exhaust emissions in a hydrogen-fueled engine for the series type hybrid vehicle system. Influences of compression ratio, surface / volume ratio of combustion chamber, and boost pressure on thermal efficiency and exhaust emissions were analyzed.

Results showed that reduction of the surface / volume ratio by increased cylinder bore was effective to improve indicated thermal efficiency, and it was possible to achieve 44% of indicated thermal efficiency. However, brake thermal efficiency resulted in 35.5%. It is considered that an improved mechanical efficiency by an optimized engine design could increase the brake thermal efficiency largely. In hydrogen-fueled engine, the highest thermal efficiency was obtained in lean mixture condition with excess air ratio around 2.5. NOx emission in the condition was less than 10ppm.

It was cleared that high thermal efficiency and low NOx exhaust emission can be achieved simultaneously in a hydrogen-fueled engine operated in conditions for series type hybrid electric vehicles.

INTRODUCTION

Utilization of electric vehicle is considered as a solution for problem of environmental pollutants from vehicles. Because of the disadvantage of weight of battery cells, the global trend of electric vehicle system is toward the hybrid electric vehicle system with combination of battery cell and another power plant.

In the series type hybrid system, engine is operated at steady conditions of engine speed and load. Utilization of hydrogen-fueled engine for a power plant of the hybrid

system is feasible for high thermal efficiency and almost zero NOx, CO, and HC emissions.

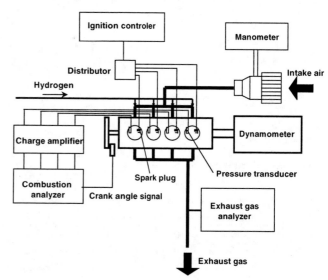

Figure 1.   Tested engine

EXPERIMENTAL APPARATUS

Tested engines in this research were a water-cooled 4-stroke cycle 4-cylinder engine (Nissan GA13, displacement of 1295cc) and a water-cooled 4-stroke cycle 3-cylinder engine (Nissan KA18 modified version of KA24 with 4 cylinders, displacement of 1791cc) modified into hydrogen-fueled engines. The GA engine had compression ratio variation of 9.5 and 12. The KA engine had compression ratio variation of 9.5, 12, 14, and 16. Hydrogen is continuously supplied into intake manifold.

Combustion pressure data were recorded with a piezoelectric type pressure transducer (AVL GM12D) and used in analysis of combustion characteristics. Ignition timing was set at the optimum MBT in all cases. NOx, CO, CO_2, THC, and O_2 concentrations in exhaust gas were analyzed with a low concentration type exhaust gas analyzer (Horiba Co. MEXA9100). Full scales of CO and CO_2 concentration of the analyzer were 10ppm and 100ppm.

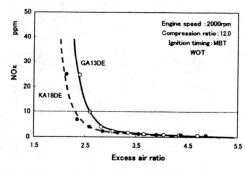

Figure 2. NOx emission characteristics

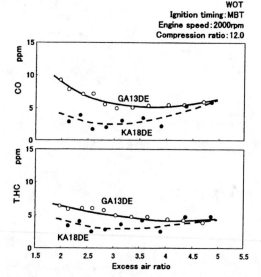

Figure 3. Emission characteristics of CO and THC

RESULTS AND DISCUSSIONS

EXHAUST EMISSIONS – Fig.2 shows the influence of excess air ratio on NOx emission. In both GA and KA engines NOx emission is lower than 10ppm in leaner condition over excess air ratio of 2.5. HC and CO emissions from hydrogen-fueled engine are extremely low as shown in Fig.3, because those emissions are originated only from lubrication oil consumption.

IMPROVEMENT OF INDICATED THERMAL EFFICIENCY – There are some measures to improve indicated thermal efficiency. Increase in compression ratio, reduction of heat loss, combustion improvement, and supercharging are supposed to be effective for the improvement of indicated thermal efficiency. For the first, influence of compression ratio on thermal efficiency was experimentally studied. Fig.4 shows the influence of compression ratio on thermal efficiency. In both engines, indicated thermal efficiency tends to increase with increase in compression ratio. In this case, the increase in compression ratio causes increase in surface / volume ratio of combustion chamber. The increase in the S/V ratio is considered to deteriorate the improved indicated thermal efficiency to some extent as shown in Fig.5. At the same compression ratio, KA engine with larger bore and smaller S/V ratio has higher thermal efficiency compared with GA engine due to the reduction of heat loss.

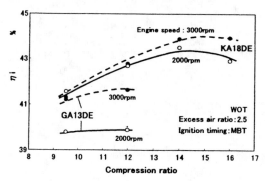

Figure 4. Influence of compression ratio on thermal efficiency

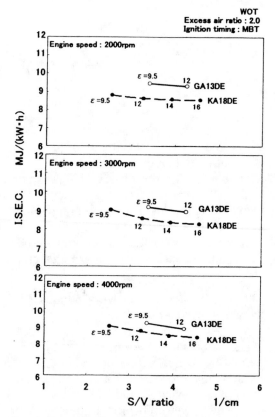

Figure 5. Influence of compression ratio and combustion chamber S/V ratio on thermal efficiency

Fig.6 shows the influence of excess air ratio on thermal efficiency at different engine speeds. The compression ratio was 12. The difference of thermal efficiency between KA and GA engine decreases with increase in engine speed. Fig.7 shows the influence of excess air ratio on indicated thermal efficiency and NOx emission in KA engine. Thermal efficiency was the highest at 2000rpm, and higher engine speeds resulted in deterioration of thermal efficiency. Though influence of the engine speed on NOx emission is small. Thermal efficiency at higher engine speed is highest in richer mixture condition with slightly higher NOx emission. Therefore, the lower engine speed has the advantage of both higher thermal efficiency and lower NOx emission in KA engine.

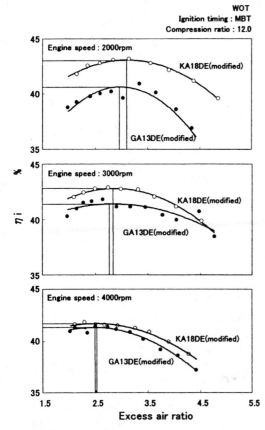

Figure 6. Influence of excess air ratio on thermal efficiency for different engine speeds

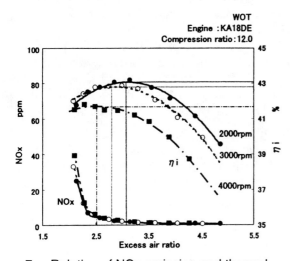

Figure 7. Relation of NOx emission and thermal efficiency

Influence of turbo charging on thermal efficiency is shown in Fig.8. The turbo charging improves thermal efficiency by 2% at excess air ratio around 2.5. The turbo charging with constant intake air temperature does not increase NOx emission. Even in the case of turbo charging, excess air ratio is dominant factor of NOx emission.

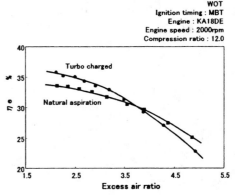

Figure 8. Influence of turbocharging on thermal efficiency

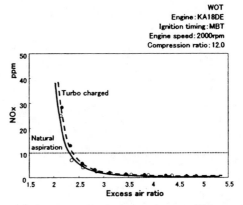

Figure 9. Influence of turbocharging on NOx emission

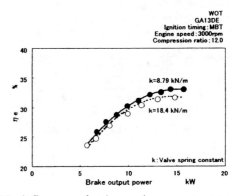

Figure 10. Influence of valve spring constant on thermal efficiency

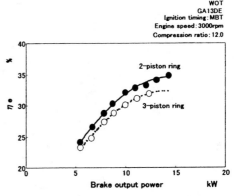

Figure 11. Influence of piston ring number on thermal efficiency

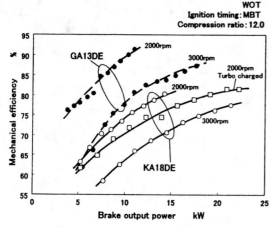

Figure 12. Characteristics of mechanical efficiency for different operating conditions

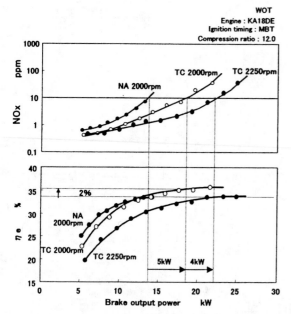

Figure 13. NOx emission and thermal efficiency vs. output power

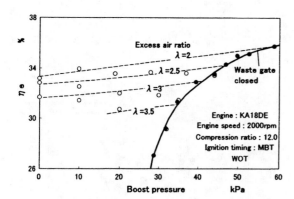

Figure 14. Characteristics of thermal efficiency for boost pressure

IMPROVEMENT OF MECHANICAL EFFICIENCY – Improvement of mechanical efficiency is effective to increase brake thermal efficiency. Engines in the series type hybrid electric vehicle system are used in stable condition of relatively low engine speeds. Because of the stable condition, friction losses can be reduced largely. A method to reduce friction is adoption of softer valve springs. Fig.10 shows the influence of valve spring stiffness on brake thermal efficiency. The engine used in this experiment was GA engine. Valve springs with the half stiffness improved brake thermal efficiency by 1 % at engine speed of 3000rpm. At higher engine speed, the effect decreases.

Decreased number of piston rings to reduce friction between piston and cylinder liner is also effective. Fig.11 shows the influence of decrease in number of piston rings on mechanical efficiency. This experiment was done in GA engine. In this experiment, one of two compression rings was removed form each piston. With this modification, ring gaps were reduced from 0.3mm to 0.1mm to avoid increase in blow-by gas. This method improved thermal efficiency by 3%. Further optimization of ring width and ring tension could improve thermal efficiency.

Fig.12 shows mechanical efficiency of operation with the softer valve spring and the reduced piston rings. Results of turbo charging in KA engine are also shown in the figure. In both engines mechanical efficiency is higher in lower engine speeds. GA engine shows higher mechanical efficiency than KA engine in all cases.

CHARACTERISTICS OF OUTPUT POWER – Fig.13 shows effect of turbo charging on power output and NOx emission. Turbo charging at engine speed of 2250rpm can achieve power output of over 20kw and NOx emission less than 10ppm simultaneously.

Fig.14 shows relation between boost pressure and brake thermal efficiency. Engine speed was set at 2000rpm. Brake thermal efficiency increases with increase in boost pressure at constant excess air ratio.

CONCLUSION

Results in this study can be summarized as follows.

1. This research showed that hydrogen-fueled engine with high thermal efficiency and low exhaust emissions has high potential as a power source of series type hybrid vehicle system.

2. Hydrogen-fueled engine has the highest indicated thermal efficiency at ultra lean condition of excess air ratio around 2.5 to 3 with extremely low NOx emission. Therefore, NOx, CO, and HC emissions can be maintained to several ppms without deterioration in thermal efficiency.

3. The highest brake thermal efficiency was 35.5% in spite of indicated thermal efficiency of 44%. Increase

in mechanical efficiency could improve the brake thermal efficiency to around 40%.

ACKNOWLEDGMENTS

The Original Industrial Technology R&D Promotion Program from the New Energy and industrial technology Development Organization (NEDO) of Japan supported this study.

Authors appreciate Dr. S. Furuhama the former president, Mr. T. Iguchi a former student, and Mr. T. Numata a student of Musashi Institute of Technology for their helps in this research.

REFERENCES

1. Nakajima, Y., et al., "Evaluation of a Lean Burn Hydrogen Fueled Engine for Hybrid Electric Vehicles", Proceedings of the 12th World Hydrogen Energy Conference, (1998).

2. Iguchi, T., et al., "Study of Hydrogen Engine for a Hybrid Vehicle", Proceedings of the SAE of Japan 1998 Fall Congress, (1998).

3. Nakajima, Y., et al., "Research and Development of a Hydrogen Fueled Engine for Hybrid Electric Vehicles", Proceedings of the SAE of Japan 1999 Spring Congress, (1999).

Hydrogen Fueled Engines in Hybrid Vehicles

Jay Keller and Andrew Lutz
Sandia National Laboratories

ABSTRACT

This paper describes the motivation for developing hydrogen-fueled engines for use in hybrid electric vehicles of the future. The ultimate motivation for using hydrogen as an energy carrier is carbon management. However, air quality concerns also provide motivation for developing hydrogen-fueled vehicles. For this reason, we discuss the position of the hydrogen-powered hybrid vehicle within the California Air Resources Board requirement for Zero Emission Vehicles.

We describe the expected performance of an electrical generation system powered by a four-stroke, spark-ignited, internal combustion engine for a hydrogen-powered hybrid vehicle. The data show that the engine-out emissions of NO_x will allow the vehicle to operate below the Super Ultra-Low Emission Vehicle standard set by the California Air Resources Board. The engine can run on either hydrogen or blends of hydrogen and natural gas. The engine can be optimized for maximum efficiency with low emissions. The target design for the engine-generator set is an indicated thermal efficiency of 47 % with emissions below 5 PPM of NO_x. We estimate the overall efficiency will be 40 %, yielding 60 miles per equivalent gallon of gasoline.

INTRODUCTION

The objective of this paper is to motivate the use of hydrogen-fueled internal combustion engines (ICE) in the near future. The paper describes the development of a spark-ignited, engine generator set—referred to as an SI-GenSet—that is optimized for maximum efficiency and minimum emissions. The engine can run on either hydrogen or hydrogen blended with natural gas, providing a cost effective and highly efficient power source for use in hybrid vehicles. In addition, since hydrogen promotes stable combustion at leaner equivalence ratios than other fuels, the engine can operate at temperatures below where thermal-NO formation occurs, thereby providing low NO_x emissions without the need for post-combustion cleanup.

Before describing the engine-generator concept, we discuss the motivations for hydrogen use with respect to concerns over air quality, greenhouse gas emissions, and energy supply and security. We provide motivation for hydrogen end-use in general, but specifically as a clean fuel for use in engines. We present a comparison of the theoretical efficiency of a hydrogen engine to a fuel cell in the context of a hybrid vehicle with a representative driving cycle. Lastly, we present data for the thermal efficiency and NO_x emissions of a hydrogen-fueled engine tested in our laboratory.

MOTIVATION FOR USING HYDROGEN

Interest in alternative fuels and technologies is motivated by concerns for air quality, greenhouse gas accumulation, and security of energy supply. Air quality is a regional problem, focused in areas where emissions are concentrated by population density and geographical containment. Consequently, local air quality boards are responsible for bringing their region into compliance with federal standards. In contrast, greenhouse gas emissions are a global concern, subject to international negotiation and agreements, such as the Kyoto Protocol. Energy supply is also a global concern, but given national imbalances in consumption and supply, the near term issue to the consuming nations is security of the supply. While hydrogen usage can impact all three of these problems, the economic and engineering factors must be examined to determine whether or not it provides the best solution compared to other alternatives.

Hydrogen use cannot be motivated solely by energy security and air quality concerns. Other alternative fuel technologies can solve these problems. For example, shifting from petroleum to natural gas as an energy source reduces the energy security problem in the United States, at least for the near term. Natural gas vehicles also reduce emissions of pollutants. Fleets of light-duty, natural-gas fueled vehicles are already in use by private and government organizations, and heavy-duty natural gas vehicles are also available.

PRIMARY MOTIVATION FOR HYDROGEN: CARBON MANAGEMENT - The primary motivation for using hydrogen as an energy carrier is that it provides a means of managing carbon dioxide emissions. Converting the end-use technologies to hydrogen allows the consumption of hydrocarbon fuels with large-scale carbon management schemes in place at the point of hydrogen production. In addition, once the supply infrastructure and end-use technologies for using hydrogen are in place, then the evolution towards hydrogen production from renewable energy resources becomes transparent to the user.

The most common and cost effective way to produce hydrogen today is steam reforming of hydrocarbon fuels, specifically natural gas. Williams [2] discusses the cost and viability of natural gas reformation with CO_2 sequestration as a cost-effective way to reduce CO_2 emission levels. He argues that if the infrastructure and technology to use hydrogen were in place, the additional cost of natural gas reformation and subsequent CO_2 sequestration would be minimal.

The magnitude of the carbon management problem is shown in Fig. 1, which presents the carbon dioxide (CO_2) emissions in the United States, along with the Kyoto Protocol level (dashed line). The impact of the Kyoto Protocol on the United States (to first approximation) is to reduce the CO_2 emission rate to 7 % below that of 1990. As shown in Fig. 1, the CO_2 emission rate continues to climb at about 1.3 % per year. The impact of this increase is that in order to meet the Kyoto Protocol today, an 18 % reduction in the current emission rate is required—not 7 %, as it would have been in 1990. At the current rate of increase, the U. S. will be 33 % above the Kyoto Protocol by 2020. Achieving the Kyoto Protocol represents a significant challenge.

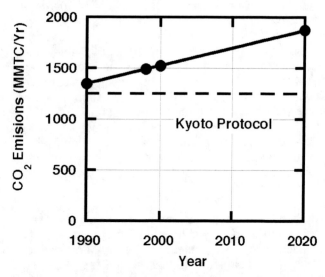

Figure 1. Carbon dioxide emissions plotted with the Kyoto protocol; source AEO2000 [1].

To understand which end-uses produce the carbon dioxide emissions, Fig. 2 presents the fraction of emissions by energy sector. By a striking coincidence, the light duty vehicle sector accounts for about 19 % of the current emissions, which means that if the entire fleet were converted to zero CO_2 emissions, then the U. S. would comply with the Kyoto Protocol. Obviously, this conversion would be impossible to accomplish immediately, but the point is that the contribution of the light-duty vehicle fleet to the CO_2 budget is significant. Consequently, controlling CO_2 emissions in this sector is

crucial to reducing the U. S. CO_2 emissions rate to the Kyoto Protocol levels.

However, converting to hydrogen-fueled vehicles does not reduce CO_2 emissions if hydrocarbon feed-stocks are used to generate the hydrogen without carbon management. In fact, using hydrogen can actually increase the emission of CO_2 over a conventional solution if the fuel cycle does not pay attention to carbon management. For example, a diesel-fueled hybrid vehicle will emit less CO_2 than a hydrogen-fueled hybrid vehicle if the hydrogen is made from a hydrocarbon fuel without the resulting CO_2 being sequestered or managed [2]. Under the Partnership for New Generation Vehicles (PNGV), the three major automobile manufacturers have developed concept cars achieving 80 miles/gallon of gasoline equivalent, which is nearly 3 times the performance of conventional vehicles. Each model is a hybrid vehicle with a diesel engine. While these vehicles can reduce CO_2 emissions greatly, the diesel engines are going to have trouble passing the increasingly stringent requirements on emissions of particulates, hydrocarbons, and NO_x.

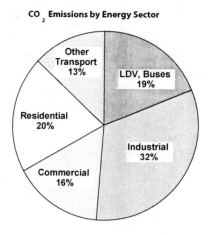

Figure 2. Carbon dioxide emissions as percentage of total by energy sector; source: AEO2000, US/DOE/OTT. LDV is light duty vehicles.

The advantages of hydrogen as a fuel are that it provides a means of carbon management, at the point of production, and it helps alleviate air quality concerns at the point of use. Therefore, a secondary, but still important, motivation for using hydrogen is that it can help reduce emissions of air pollutants.

AIR QUALITY MOTIVATION FOR HYDROGEN - While air quality alone does not motivate the use of hydrogen-fueled vehicles, it can play an important role in bringing about the transition from hydrocarbon fuels. Air quality, being a regional issue, is legislated most stringently in California by the Air Resources Board (CARB). In 1990, CARB mandated a schedule requiring manufacturers to begin producing Zero Emission Vehicles (ZEV) by 1998. Since the leading technology—the battery powered electric vehicle—was not sufficiently developed by this date, CARB delayed the schedule, but established a Memorandum of Agreement with the major manufacturers to continue development of the electric vehicles on a demonstration scale. After hearing testimony from manufacturers, environmental groups, and electric vehicle owners in September 2000, CARB voted to uphold the current mandate. The next milestone in the program is for the major manufacturers to produce ZEV's equal to 10 percent of light-duty vehicle sales in model year 2003—some 20,000 vehicles. The CARB staff report states that this will be an expensive and challenging goal for the manufacturers to meet. [3]

The 1996 modifications to the ZEV mandate provide room for technologies other than battery vehicles through an accounting mechanism called the partial ZEV allowance. This allowance offers partial credit for vehicles that meet a set of requirements, including controlling emissions to Super Ultra Low Emissions Vehicle (SULEV) standards. The mandate allows a manufacturer to fulfill 6 % of the ZEV requirement using partial ZEV's, as long as the remaining 4 % are ZEV's or full ZEV's (partial ZEV's whose credits add up to 1). The partial ZEV allowance includes a baseline credit of 0.2, so for example, a manufacturer can replace 6 % of its ZEV requirement by producing 30 % of its vehicles with the baseline PZEV allowance.

Hydrogen-fueled vehicles can play a significant role in helping manufacturers comply with the ZEV mandate. For example, a hydrogen-fueled engine in a conventional drive train, achieving SULEV standards, qualifies for the baseline credit of 0.2, plus additional credits of 0.3 for producing zero emissions of non-methane organic gases (NMOG) and 0.2 for having low fuel cycle emissions, for a total allowance of 0.7. This is the largest partial ZEV allowance for the minimum departure from a conventional gasoline-fueled vehicle.

Alternatively, a hybrid version of this vehicle could increase the partial ZEV allowance even further. The advanced components of a hybrid drive train qualify for another 0.1 credit. In addition, the zero-emission vehicle miles traveled (VMT) category, which provides the 0.3 allowance for zero NMOG, offers an alternative sliding scale credit based on providing a battery-only range between 20 and 100 miles; if the battery range were 100 miles, the vehicle would receive the maximum 0.6 allowance for this category. The significance of this is that a hybrid electric vehicle with hydrogen-fueled engine and a battery-only range of 100 miles could qualify as a full ZEV.

Despite this fact, manufacturers will most likely decide that adding the batteries necessary to provide the zero-emission VMT range does not lead to the best hybrid vehicle design. The power management systems in existing hybrid vehicles, such as the Honda Insight and Toyota Prius, do not provide the capability of operating on batteries alone. Secondly, the batteries required to supply a significant range in electric-only operation will add considerable cost and weight to the vehicle. We used the Advisor [8] model (Advanced Vehicle Simulator) provided by the National Renewable Energy Laboratory to simulate performance of a generic hybrid vehicle on the federal urban driving cycle. Using the default parameters provided by the model for a hybrid vehicle, we estimate that the batteries required for the minimum 20-mile range will weigh about 700 lbs. Since the electric vehicles already demonstrated by the auto manufacturers have a range of about 100 miles, it stands to reason that they will most likely satisfy the 4 % ZEV requirement using these vehicles.

HYDROGEN-FUELED ENGINE FOR HYBRID VEHICLE

In order for the hydrogen-fueled engine to succeed in hybrid vehicle applications, several engineering and economic factors must be favorable. The potential market share for the vehicles must be large enough to offset the initial development costs and the capital investment. The vehicle must be energy efficient to offset the cost of hydrogen production. The emissions must be near zero, or significantly below existing standards. The power density must be sufficiently high to be packaged in a vehicle. And lastly, the specific power must be sufficiently high to keep the mass of the vehicle to a minimum.

Figure 3 shows a comparison of the characteristics of different energy conversion technologies. The values presented are for the overall system needed to employ each technology. For example, the fuel cell systems include the stack, fluid-handling equipment, and necessary cooling hardware. These data exclude any fuel processing equipment.

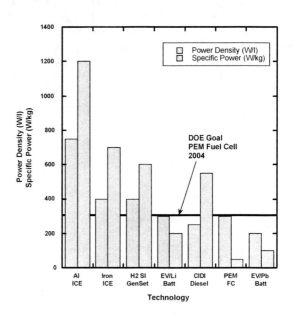

Figure 3. Power densities and specific power for selected energy conversion technologies. Technologies are: aluminum block 4-stroke; SI-ICE, iron block 4-stroke SI-ICE; hydrogen lean burn SI generator set (SI GenSet);

electric vehicle (EV) with lithium battery; compression ignition direct injected (CIDI) diesel; proton exchange membrane (PEM) fuel cell; and EV with lead-acid batteries. The horizontal line is the DOE goal for fuel cells in 2004.

The main conclusion from the data in Fig. 3 is that a hydrogen engine exceeds the power density and specific power of the goal set by the DOE Partnership for New Generation Vehicles (PNGV) for a PEM fuel cell system in the year 2004. Figure 3 also shows that a hydrogen-fueled engine is equivalent to a CIDI diesel engine in specific power and power density. However, the power density and specific power examined here are global estimates of the technology's capability; the system efficiencies need to be compared in the context of the specific application.

HYDROGEN USE IN ENGINES COMPARED TO FUEL CELLS IN HYBRID VEHICLES - Next, we examine the system efficiencies of generic hybrid vehicles powered by either a hydrogen-fueled engine or a fuel cell, compared on an appropriate driving cycle. While a thorough study is beyond the scope of this paper, we have performed two evaluations: a first order analysis assuming component efficiencies on a simplified driving cycle, and analysis using the Advisor model [8] that includes an urban driving cycle.

The hybrid vehicle drive trains containing a hydrogen-fueled engine and a fuel cell are depicted in Fig. 4. The vehicle schematics show the power routing and the relevant efficiencies of each major component. In this schematic, the upper row of components represents the base load operation. For example, the base load operation of the engine converts the fuel directly into traction through the engine and transmission. The lower row of components represents the peaking and regenerative-braking modes. In these modes, energy recovered from either the regenerative brakes or energy stored from the engine is used to drive the electric motor and provide traction. The efficiencies assumed for each component are indicated inside the boxes, and the local product efficiencies along each pathway are indicated on the arrows.

This first order analysis uses efficiency targets for the hydrogen-fueled engine and the fuel cell system. The efficiencies for the fuel cell at full (50 %) and half (55 %) load taken by linear interpolation from the DOE PNGV goals for year 2004. The engine efficiencies of 36 % and 40 % at full and half loads, respectively, are based upon conservative estimates using our laboratory data (see Fig. 6).

The analysis assumed that the result of a typical driving cycle is that 50 % of the drive power is supplied by the base load system, 30 % is supplied by the peaking system, and 20 % is recovered by the regenerative breaking. Applying these factors to the efficiencies for these paths as shown in Fig. 4 gives an overall efficiency of 48 % for the fuel cell hybrid, and 38 % for the engine

hybrid. The result is that a hybrid vehicle powered by a hydrogen-fueled engine is within 80 % of the hybrid vehicle powered by a PEM fuel cell operating at the DOE Year 2004 goal.

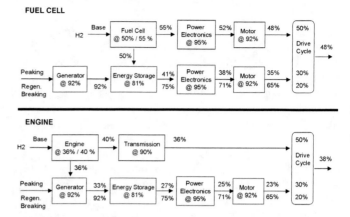

Figure 4. Schematic of hybrid vehicle system efficiency.

To explore further the expected efficiencies of hybrid vehicles powered by fuel cells versus an engine, we used the Advisor model [8] to simulate the performance using the federal urban driving cycle. The Advisor model defines a vehicle by choosing a platform for the drive train and identifying the individual components with their performance parameters. The model includes specific vehicle platforms and parameters, such as for the Toyota Prius, which are calibrated to test data. We chose this model to represent a hybrid vehicle powered by an engine—albeit gasoline fueled, but the efficiency map is similar to what we expect from a hydrogen-fueled engine. Advisor also provides a generic model for a hybrid vehicle powered by fuel cells. This model includes a more realistic fit of efficiency as a function of load than the linear interpolation of the DOE goals used above; however, the default values for the maximum efficiency and the efficiency at full-power are about the same.

Running these two vehicles through the federal urban driving cycle gives overall system efficiencies of 14.2 % for the fuel cell vehicle, compared to 12.3 % for the Prius. In this comparison, the ratio of overall system efficiencies for an engine hybrid to a fuel cell hybrid is 87 %. This larger ratio than predicted by the first order analysis is due to the fact that the Prius used energy from the batteries to support the stop-and-go driving, while the fuel cell hybrid actually stored some excess energy. The Advisor result supports the observation that a hybrid vehicle being sold today approaches the same efficiency as the fuel cell vehicle of the future. Converting the engine in today's hybrid vehicles to run on hydrogen offers the additional advantage of providing a means of carbon management and reducing pollutant emissions even further.

PERFORMANCE OF A HYDROGEN FUELED ENGINE

To demonstrate the capabilities of a hydrogen-fueled engine, we present laboratory data on the efficiency and NO_x emissions.

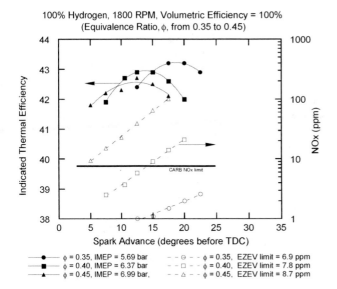

Figure 5. NO_x and thermal efficiency measured on a research engine over a range of equivalence ratio (ϕ) and Indicated Mean Effective Pressure (IMEP). EZEV represents the proposed Equivalent Zero Emissions Vehicle limit for NO_x, which is the same as the SULEV limit of 0.02 gm/mile.

The successful control of the combustion in a hydrogen-fueled engine has demonstrated extremely low NO_x emissions (<5 PPM) [4]. Figure 5 shows data taken from our research engine over a wide range of operating conditions, all with engine-out NO_x levels below the California SULEV regulation (0.02 gm/mile). Controlling the emissions by operating with a lean charge to keep the temperature below where thermal-NO formation occurs results in a system that does not need a catalytic converter. Inherently, this control strategy produces a system whose emissions performance will not degrade with time.

Figure 6 shows the indicated thermal efficiency as a function of compression ratio for the theoretical maximum, and data points from our research engine (circles). Like other researchers [4], we expect that with increasing compression ratio the performance of this system will start to fall off, causing a local maximum that deviates from the shape of the theoretical curve. While more work needs to be done to predict accurately the efficiency of this system, the peak efficiency most likely occurs at a compression ratio in the range of 18:1 to 20:1. Under the lean conditions—equivalence ratio of 0.4—this mixture autoignites at a compression ratio of 22:1, placing an upper limit on the compression ratio.

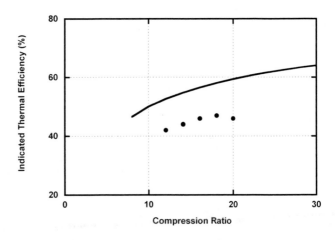

Figure 6. Measured indicated thermal efficiency (points) and the theoretical indicated thermal efficiency (curve) as a function of the compression ratio at an equivalence ratio of 0.4.

For the compression ratio of 20:1, as shown in Fig. 6, the indicated efficiency is 47%, which is 80% of the theoretical maximum. Assuming parasitic losses of 15% (conventional wisdom for engines is 10% at full load [5]), this yields a break thermal efficiency of 40%, which is our goal [6,7]. Our previous work has shown that this particular combustion chamber configuration is not optimized for efficiency, so further improvements may be achieved.

Since this engine operates lean and the fuel and air are premixed, the flame temperature is kept below the thermal-NO threshold, thereby controlling NO_x emissions below SULEV regulations. Alternatively, the same temperature control can be achieved by using exhaust gas recirculation, in which case the tailpipe NO_x emissions can be reduced further by adding a NO_x reduction catalyst to the exhaust system. Other researchers have demonstrated this approach and obtained engine-out values similar to the work reported on here, but they further reduce their tailpipe emissions to about 0.13 ppm in some cases. Note that NO_x values in some of our cities on bad days exceed this value; for example, Riverside California reaches values on the order of 0.3 ppm. Even though this approach yields extremely low NO_x emissions, recent results suggest that N_2O forms by converting NO in the catalyst to N_2O resulting in elevated N_2O levels out of the tailpipe. N_2O is a greenhouse gas that is about 300 times worse than CO_2. Further investigations into this issue are required.

CONCLUSION

This paper presented the motivation for using a hydrogen-fueled engine in a hybrid electric vehicle to address the problems of carbon management and local air quality. We performed some preliminary analysis to demonstrate that a hydrogen fueled engine can provide the necessary power with a drive-cycle efficiency approaching that of the fuel cell vehicle of the future. Lastly, we present laboratory engine data demonstrating the thermal efficiency and low NO_x emissions obtained when running on hydrogen.

Fueling the engine with pure hydrogen eliminates carbon species, including CO_2, from the exhaust, except for trace species originating from the lubricating oils. The engine tests demonstrated that control over the combustion temperature can be achieved by operating the hydrogen-fueled engine at lean equivalence ratios to maintain engine-out NO_x concentrations below the most stringent regulations. The engine-out NO_x values would qualify a hybrid vehicle to meet the super ultra low emissions vehicle (SULEV). This means a hydrogen-fueled engine in a hybrid configuration can qualify for partial ZEV allowances to fulfill a manufacturer's requirement under the California ZEV mandate. In addition, the data for indicated efficiency of the engine suggest that the estimated brake thermal efficiency will be about 40 %, which approaches the PNGV goal of 44 %.

ACKNOWLEDGMENTS

This work was performed at the Combustion Research Facility, Sandia National Laboratories and was supported by the U. S. Department of Energy through the Office of Power Technologies by the means of the Hydrogen Program Office.

REFERENCES

1. Annual Energy Outlook, AEO, 2000.
2. Williams, Robert H., "Fuel Decarbonization for Fuel Cell Applications and Sequestration of the Separated CO_2," PU/CEES Report No. 295, January 1996.
3. State of California California Environmental Protection Agency, Air Resources Board, Staff Report, "2000 Zero Emission Vehicle Program Biennial Review", August 7, 2000.
4. Caris, D. F., and Nelson, N. N., "A New Look at High Compression Engines," SAE Transactions, **67**, 112-124, 1959.
5. Heywood, John B., "Internal Combustion Engine Fundamentals," McGraw Hill, Inc. 1988, pp 712.
6. Van Blarigan, P., "Development of a Hydrogen Fueled Internal Combustion Engine Designed for Single Speed/Power Operation", SAE Paper 961690, 1966.
7. Van Blarigan, P., "Advanced Hydrogen Fueled Internal Combustion Engines," Energy & Fuels, **12**, 72-77, 1998.
8. "Advisor: Advanced Vehicle Simulator," Aaron Brooker, Terry Hendricks, Valerie Johnson, Kenneth Kelly, Tony Markel, Michael O'Keefe, Sam Sprik, Keith Wipke, Desikan Bharathan, Steve Burch, Matthew Cuddy, Dave Rausen, available from the National Renewable Energy Laboratory, August (2000).

Development of a Hydrogen Engine for a Hybrid Electric Bus

Jonathan Fiene, Troy Braithwaite, R. Boehm and Y. Baghzouz
UNLV Center for Energy Research

Terry Kell
Kell's Automotive

ABSTRACT

A report is given about the development of hydrogen-fueled spark-ignition V-8, used to drive a generator in a hybrid bus. This application is a parallel hybrid, so the engine is applied at constant speed. Various modifications include the application of electronic engine control, tuned exhaust and intake headers, and the design of the cam are noted in the paper. In a portion of the work, water injection was applied to boost performance levels. Some operational data for the engine are given.

INTRODUCTION

A hydrogen-fueled hybrid electric bus was built for the Atlanta Olympics with funding from the US Department of Energy. An all-electric bus was modified to include a Ford V-8 engine altered to utilize gaseous hydrogen fuel via a constant volume injection system. Two of four original battery packs were removed from the bus and the resulting space was used to mount two hydride beds for hydrogen storage.

In 1999, the bus was moved to be under direction of the DOE Nevada Operations Office and a contract was arranged with the UNLV Center for Energy Research to make a series of modifications to the bus. Several operational problems were found to exist with the bus system. From simulations of the vehicle dynamics, it was determined that the engine had to be able to produce at least 50 kW, but preferably up to 70 kW to sustain bus operations and charge the onboard battery packs. The engine was removed from the bus and a series of performance tests were conducted on it. These tests

concluded that this engine fell short of the required power output. This was largely due to the amount of exhaust gas recirculation employed (50%) to control backfires, pre-ignition, and NOx. Ultimately it was decided to develop a new engine for the power conversion unit that utilized newer technology. This paper discusses various aspects about this engine.

To start this project, research was conducted into prior research to find information on previous approaches that had been tried. Several reports are found in the literature about hydrogen engine development. Mathur and Das[1] reported work on the use of hydrogen on a variable compression engine that utilized port injection. Engine speeds (1000-2000 rpm) and compression ratios (6-11) were varied while measuring engine performance. This research concluded that timed manifold injection and higher compression ratios better utilized hydrogen's properties to increase performance and efficiency of hydrogen fueled engines.

Comparisons between a rotary (Mazda two-rotor engine with a displacement of 1306 cc and a compression ratio of 9.4) and a conventional internal combustion engine (Mazda 4 cylinder engine of 1998 cc and compression ratio of 9.8) were made by Morimoto et al.[2]

Shudo et al.[3, 4] reported on a constant rpm study of a 4-cylinder spark ignition gas engine modified from an automobile engine (Nissan CA20S, bore 85 mm, stroke 88 mm, compression ratio of 8.5). Hydrogen and methane fuel, separately, were used in the study. The conclusions here were that the thermal efficiency of an engine operated on hydrogen is dominated by cooling losses in the combustion chamber. It is further indicated that the cooling loss can be reduced by retarding the ignition timing and increasing the excess air ratio.

A Ricardo Variable Compression Engine (506 cc) was used to study the effects of combustion duration on various aspects of efficiency and emissions when fueled with hydrogen[5]. It was concluded that the combustion duration should be 4 to 6 ms for optimal performance in both power and economy. These researchers controlled this parameter by varying the equivalence ratio.

Naber and Sieber[6] reported use of hydrogen in a diesel engine

A description of a German bus project is given by Knorr et al.[7] The engine used (an H 2866 UH) was developed from a standard 12l, six-cylinder natural-gas engine using an 8:1 compression ratio. This engine utilized sequential injection into the manifold runners via rotary valves in hydrogen mode and conventional injectors in gasoline mode. No information was given concerning the engine's operating parameters.

A paper that reported results of both an analytical study as well as an experimental study was given by Van Blarigan and Keller.[8] An Onan single cylinder diesel engine with a 14:1 compression ratio was used for this work. Their work concluded that it is possible to achieve EZEV emissions standards by operating the under lean conditions at the sacrifice of power output, and high efficiency can be obtained by using a high compression ratio.

Reports and studies of fuel injectors or fuel injection for hydrogen engines have been given by a few workers, including Glasson and Green, [9] Das, [10] Das et al., [11, 12] Heffel et al., [13] and Guo et al.[14]

A review of hydrogen fueled engine studies has been given by Billings.[15] Included in this discussion are the effects of water injection as well as several other issues.

The conclusions reached when all of these papers are considered together were mixed. For example, several different values for the compression ratio have been applied, without a clear view of an optimal range. It is true that the effect of compression ratio on the efficiency of a basic Otto cycle is will known. What is not described in the literature to any extent is the affect of

compression ratio on all of the operational parameters of importance, including control of pre-ignition, emissions, and fuel economy. Undoubtedly this parameter is affected by flow passage and combustion chamber design, but little information is given on the basic parameter choice.

In addition, it was found that fuel injection, while used in several of these studies, has not been clearly defined in terms appropriate applications. It is also clear that commercially-available units are limited. Almost no mention of the various problems experienced with pintel-based designs was noted in the literature. Although a few researchers have touched upon it, water injection does appear to be quite beneficial from both stabilizing operation as well as decreasing emissions. However, limited quantification of this statement appears in the literature noted above. Also, with the constant speed application, tuning of the exhaust and intake is a possibility. However, none of the papers reviewed used this type of application. Finally, current advances in computer control of the various aspects of engine operation were not really dealt with in these reports of studies.

MAIN SECTION

OVERVIEW

A Chevrolet 7.5 L spark ignition engine was used as a basis for this study. To utilize hydrogen fuel, several components of the engine needed to be redesigned. Three factors were very much of concern in the design of this engine. These include, first and foremost, stable operation of the engine. Secondly, the minimization of the NOx emissions was also of great concern. Finally, engine efficiency was also important in an attempt to increase the relatively short range demonstrated earlier with the bus.

The majority of the engine modifications were made to the air/fuel induction system. Included in this was an attempt to create a tuned intake effect. Custom cylinder heads with smaller ports were used as well as a new intake manifold which incorporated tuned runners to maximize the volumetric efficiency at the specified RPM. Custom mounts for the fuel injectors were also incorporated into the intake system.

Other modifications that were made to the engine included a new camshaft that would maximize torque at a significantly lower speed than a stock camshaft and the incorporation of a water injection system in the intake manifold. The compression ratio of the engine was also increased to 14:1 to take advantage of hydrogen's properties to increase the thermal efficiency. The complete engine is shown in Figure 1.

Figure 1: Hydrogen Internal Combustion Engine

ELECTRONIC FUEL INJECTION

At the beginning of the project, a search for manufacturers of fuel injectors for hydrogen applications was carried out but was not too fruitful. Some companies had units under development. Also, researchers who had used conventional fuel injectors for hydrogen applications had found high failure rates with those units.

The first and only injectors that we discovered for operation in a pure hydrogen environment were made primarily as metering valves. These units were quite large to allow the necessary flow of hydrogen gas. The units were used with custom mountings installed on top of the intake manifold. These injectors were specifically designed to overcome some of the malfunctions present when using ordinary automotive injectors with hydrogen fuel. Custom computer controlled driver circuitry and software was utilized to precisely meter the fuel. The EFI system could be adjusted from a PC while the engine was running to alter the air/fuel ratio, change injector timing and modify other important parameters. Furthermore, the control software allows for the adjustment of individual injector opening times to compensate for any variations in unit dimensions and supply pressure between injectors.

Although the rated response times were slower than comparable conventional automotive injectors were it was determined that these valves would be sufficient for our application. Since the valves were not made primarily for automotive use, they did not fit in a typical injector housing. This required the design and manufacturing of custom mounts for the injectors to mate to the intake manifold. The two figures below depict the custom injector mounts as designed using Pro/Engineer solid modeling software. To accommodate all eight injectors, four of these housings were manufactured from aluminum using CNC machining techniques.

Figure 2: Computer model of injector housing

After initial testing of the engine using these injectors, it was determined that a higher flow rate was necessary to inject more fuel into the cylinder in a shorter period of time. To facilitate this, the injectors were disassembled and the port area enlarged using computer controlled machining to ensure an even increase among all of the injectors. A feedback regulator has been incorporated into the hydrogen supply system that allows for compensation of the fuel delivery pressure with respect to the manifold absolute pressure. This has been done to maintain a constant pressure differential across the injector. Prior to the incorporation of this regulator it was observed that the injectors would have difficulty opening when the manifold pressure was very low (which created the highest-pressure differential). After several months of testing, the injectors have proven reliable against any malfunctions associated with galling or freezing and have proven to be consistent in their operational characteristics.

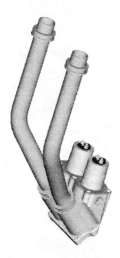

Figure 3: Computer model of injector housing showing injectors and intake tubes.

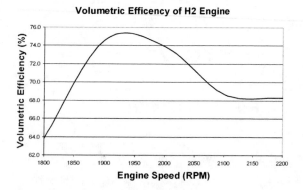

Figure 4: Volumetric efficiency of the hydrogen engine using tuned intake system

ENGINE SPEED CONTOL

While the engine is designed to operate at nearly constant speed (near the peak in Figure 4), clearly some variation in the speed will result from normal operation. We have attempted to control this under normal operation to ±100 rpm. This is set by several operational aspects. First, a solid-state governor is included which attempts to control speed to this band in all conditions except on the initial transient. In addition, the generator load for the engine has a frequency control included.

INTAKE / EXHAUST SYSTEM

The use of a gaseous fuel such as hydrogen or natural gas in a conventional engine presents some interesting challenges. One of these challenges lies in the fact that the fuel takes up considerably more volume within the combustion chamber than does a liquid fuel such as gasoline. This typically reduces the amount of air that can be drawn into the cylinder by the same amount, thus decreasing the power output of the engine. For a stoichiometric mixture of liquid gasoline to ignite, a volumetric ratio of about 5% fuel to air is required. In contrast, for a stoichiometric mixture of hydrogen to ignite, a volumetric ratio (at atmospheric pressure and temperature) of approximately 39% fuel to air is necessary. The volumetric efficiency at wide open throttle of an engine run on gasoline and then on hydrogen should therefore decease by this ratio from around 95% for gasoline to 61% for hydrogen. This corresponds to a decrease in power of around 38%. To attempt to recover this loss in power, the use of custom intake ports and runners has been employed to create a natural supercharging effect. The results of this customization are shown for a stoichiometric air/fuel ratio in Figure 4. This effect, called "tuning" in the literature, is well known. However, for variable speed engines it tends not to be too effective. In our case of constant speed operation, it offers potential for improved engine operation.

Figure 5: Air intake system

To accomplish this "natural supercharging" some basic principles of wave dynamics were employed. The port diameter was actually decreased to increase the velocity of the air traveling inside the intake tubes. The major alteration over a typical intake manifold is the use of long runners between the common plenum and the intake valve. The centerline length of this tube was established from calculations based upon the harmonic oscillation of air in the tubes. The desired effect was to have air pushed into the cylinder when the intake valve was opening, thus creating a slight "natural supercharging" effect. For the case of the exhaust, the length is designed such that the expansion wave occurs at the exhaust valve as the latter opens. This enhances the removal of exhaust gases from the cylinder. As can be observed in Figure 4, the volumetric efficiency appears to have a peak around 1900 to 1950 RPM, which is in accordance with the speed for which the tuning was calculated. It can also be observed that the volumetric efficiency of nearly 75% is significantly higher than the calculated maximum efficiency of 61% for a typical air intake system. Figure 5 is a photo of the finished system.

The cylinder heads used on this engine are custom castings based upon stock General Motors' big-block heads. They are cast from aluminum with smaller intake and exhaust port diameters that were designed to match the diameter of the intake and exhaust tubes. This is desired to increase the kinetic energy that is stored within the gasses over what would be generated by using stock heads. The combustion chamber is cast with a volume of 76 cubic centimeters. When this was combined with domed-top pistons a compression ratio of 14:1 was achieved. This higher compression ratio is obviously improves the thermal efficiency of the engine compared to a standard value. In the literature hydrogen engines operating with a higher compression ratio have been seen to have brake thermal efficiencies as high as 47%. We have also seen efficiencies in this range.

CAMSHAFT

The camshaft being used is one that has been custom ground for this application. Again, because of the constant speed operation of the engine, the cam can be optimized for this speed of operation. Also because the engine is using hydrogen as a fuel, there needed to be modifications of the profile of the cam to accommodate

differences the differences of hydrogen from other fuels. The more significant differences for hydrogen are its faster flame speed and smaller quench gap. Because of these differences, the intake valve has to be closed more quickly than in an engine that burns a conventional fuel. Valve overlap was also reduced from conventional engine designs to help minimize the possibility of hot exhaust gasses being brought into contact with the hydrogen as it is being injected into the combustion chamber.

ENGINE COMPUTER CONTROL

Through the use of a proprietary system developed by Kell's Automotive, a computer handles the data sampling of a large number of operational parameters for the engine and controls nearly every adjustable aspect of its operation. The system is responsible for controlling ignition timing, injector timing, and injector opening duration as well as controlling the throttle position to maintain stable operation. A fuel mapping system is included to adjust the injector opening based upon engine speed and manifold pressure. The computer maintains the air/fuel ratio by monitoring the oxygen present in the exhaust and subsequently adjusting the fuel delivered by the EFI system. Through the use of this system, any of the above named parameters can be adjusted from a standard PC through a fiber-optic interface while the engine is running.

PRE-IGNITION

A common problem to many researchers working with hydrogen fueled internal combustion engines is the tendency of the fuel to pre-ignite. Although this can be caused by a number of different parameters, the main factor is a heat source. Hydrogen has relatively low ignition energy. This means that a small but very hot source can pre-ignite the air/fuel mixture.

Under low load conditions there was no observed pre-ignition problems. Once the load was increased to a level equivalent to approximately 25 kW, the gases would begin to pre-ignite. Since it was desirable to run the engine at higher load levels, various methods were incorporated into the engine design to bring down any high temperature sources. The first alteration made was to use spark plugs with a colder heat range. Doing this showed some improvement. Having realized this improvement and knowing that spark plugs would not foul in a hydrogen environment spark plugs with very low heat ranges and special types of ground electrodes were obtained. This change reduced the pre-ignition problem even more, producing around 35 kW before pre-igniting. Clearly, controlling the heat inside the combustion chamber is a major issue that has to be addressed in order to operate an internal combustion engine on hydrogen reliably.

Since the desired maximum power output of the engine was around 70 kW, other techniques were necessary to decrease the cylinder temperatures. Having consulted previous research on hydrogen engines, it was determined that water injection would be a suitable method for reaching the desired output levels[15]. The water that is injected into the intake air stream serves to cool the intake air and absorbs a very large amount of the thermal energy released when the fuel is ignited. Eight custom injectors were designed and built to inject atomized water into each intake tube. This ensured even distribution of the water between the cylinders. A storage tank and electronic control valve were incorporated along with a regulator to control the water flow to the nozzles. With as little as 2 gallons of water per hour to the engine, it could operate at levels exceeding the 70 kW desired. A photograph of the arrangement is shown below in Figure 6.

Figure 6: Water injection system

Although not a part of the current design, a water recovery system could be incorporated in the exhaust stream consisting of a heat exchanger and a reservoir to condense the water out and contain it. The water available is in excess of 12 gallons of water per hour from the exhaust when the engine is operating at full load. Most of this water recovery would be from the hydrogen and oxygen combining at combustion and some would be the water that was originally injected.

As suggested in the literature[9], there is a good possibility that decreasing the compression ratio could have positive effect on decreasing the pre-ignition phenomena. This factor will be investigated subject to funding availability.

METHODS TO CONTROL NOX

There were several different techniques used to control NOx. Results from simulations and studies that were performed to help increase mixing effects were incorporated into the design of various engine components. These engine components included the cylinder heads and the use of fuel injectors. Other simpler methods that could be effected with the engine in operation, such as air/fuel ratio, ignition timing and injector timing, were also studied.

As noted earlier, fuel injection was used because it can precisely meter the flow of fuel, and it generates a mixing effect of the air/fuel mixture as it passes into the cylinder. Previous tests that were performed on the original engine exhibited large temperature variations between cylinders, which was noted to produce higher NOx than when the cylinders were at uniform temperature. Testing of the new engine has shown that there is a correlation between temperatures relative to each cylinder. By changing the fuel injector duration, the temperatures between cylinders were stabilized within the range of 100-200 °F.

Minimizing emissions, as expected, did not coincide with running conditions that optimized thermal efficiency. However, since NOx is formed in conditions of high temperature and high pressure, the methods incorporated to prohibit the pre-ignition had the benefit of significantly decreasing the emissions levels. As the water to hydrogen mass ratio is increased, a decrease in the level of NO and NO_2 is observed which is in agreement with previous research. While increasing the water/hydrogen mass ratio is effectively decreasing the equivalence ratio (leaning out the air/fuel mixture), it also has a significant effect on the production of NOx.

It is theoretically possible to eliminate completely all emissions by moving these two parameters to the point that the temperature within the cylinder never exceeds the critical temperature for NOx formation. At this point the water/hydrogen ratio is nearly 10. For our engine this is nearly 20 gallons or water per hour. Leaning out the air/fuel mixture also has drawbacks since the power decreases at nearly the same rate as the formation of NOx. To address this problem a catalytic converter was used to treat the exhaust gases.

To utilize a three-way catalytic converter effectively, the air/fuel mixture must be very near stoichiometric. By adjusting the air/fuel ratio in close proximity to stoichiometric, the optimal operating point for the catalyst was found to be at an equivalence ratio around 1.06. At this point the NOx levels were below those which could be achieved through the use of any other method. A special high-resolution analyzer has been incorporated into testing capable of measuring NO and NO_2 in the parts per billion range.

ENGINE MOUNT MODIFICATIONS

Due to the change from the original Ford engine to the new GM based engine, necessary modifications needed to be made to the mounting bracket that held the engine and generator in the back of the bus. In addition, extra room was necessary to fit the tuned intake system into the compartment. To facilitate this, the mount was lowered by six inches. At the same time, the rear opening of the bus was modified to allow a great deal more access and to facilitate installation of the engine as well as other prime movers that may be incorporated later.

PERFORMANCE

The engine was extensively tested using a water brake dynamometer connected to a standard PC. A plot of the maximum corrected power and the volumetric efficiency versus engine speed was generated and is shown in Figure 7. The table following that shows additional data.

Although these values are a significant improvement over the engine that was in the bus, further improvement of the emissions is being undertaken through the optimization of the water-to-hydrogen ratio. It is also theorized that the compression ratio may be too high for our operating conditions and this should be investigated.

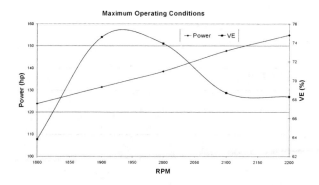

Figure 7: Maximum power and volumetric efficiency

Table 1
Some Operational Data

Operating Speed	1900±100 rpm
Equivalence Ratio	1.06
Max. Torque	>300 ft-lb
Corrected Max. Power	>120 hp
Operating Torque	273.3 ft-lb
Corrected Operating Power	105.1 hp
Brake Mean Effective Pressure	89.5 psi
Fuel Consumption	15.6 lb/hr
Volumetric Efficiency	55.3 %
Emissions of NOx	<12 ppm

CONCLUSION

A conventional spark ignition engine has been modified to use hydrogen gas as fuel for a constant-speed application on hybrid-electric bus. These modifications have included the use of a special cam, special design for the intake and the exhaust, computer control and fuel injection. Use of a catalyst brought NOx emissions to very satisfactorily low levels. At a compression ratio of 14, some problems with preignition were found to occur. Water injection was shown to be quite effective in improving this aspect of performance. The engine has proven to be quite efficient and suitable for the bus application.

ACKNOWLEDGMENTS

The support of the US Department of Energy, through the National Hydrogen Program and the Nevada Operations Office is gratefully acknowledged.

REFERENCES

1. H. B. Mathur and L. M. Das, "Performance Characteristics of a Hydrogen Fuelled S.I. Engine using Timed Manifold Injection," INTERNATIONAL JOURNAL OF HYDROGEN ENERGY, 16, pp. 115-127, 1991.

2. Kenji Morimoto, T. Teramoto, and Y. Takamori, "Combustion Characteristics in Hydrogen Fueled Rotary Engine," SAE Paper 920302, 1992.

3. Toshio Shudo, Y. Nakajima, and T. Futakuchi, "Analysis of Thermal Efficiency in a Hydrogen Premixed Spark Ignition Engine," PROCEEDINGS OF THE ASME ADVANCED ENERGY SYSTEM DIVISION, AES-Vol. 39, ASME, pp. 351-354, 1999.

4. Toshio Shudo, Y. Nakajima, and T. Futakuchi, "Thermal Efficiency Analysis in a Hydrogen Premixed Combustion Engine," JSAE REVIEW, 21, pp. 177-182, 2000.

5. J. A. A. Yamin, H. N. Gupta, B. B. Bansal, and O. N. Srivastava, "Effect of Combustion Duration on the Performance and Emission Characteristics of a Spark Ignition Engine using Hydrogen as a Fuel," INTERNATIONAL JOURNAL OF HYDROGEN ENERGY, 25, pp. 581-589, 2000.

6. J. D. Naber and D. L. Sieber, "Hydrogen Combustion under Diesel Engine Conditions," INTERNATIONAL JOURNAL OF HYDROGEN ENERGY, 23, pp. 363-371, 1998.

7. H. Knorr, W. Held, W. Prumm, and H. Rudiger, "The MAN Hydrogen Propulsion System for City Buses," INTERNATIONAL JOURNAL OF HYDROGEN ENERGY, 23, pp. 201-208, 1998.

8. P. Van Blarigan and J. O. Keller, "A Hydrogen Fuelled Internal Combustion Engine Designed for Single Speed/Power Operation," INTERNATIONAL JOURNAL OF HYDROGEN ENERGY, 23, pp. 603-609, 1998.

9. N. D. Glasson and R. K. Green, "Performance of a Spark-Ignition Engine Fuelled with Hydrogen using a High-Pressure Injector," INTERNATIONAL JOURNAL OF HYDROGEN ENERGY, 19, pp. 917-923, 1994.

10. L. M. Das, "Fuel Injection Techniques for a Hydrogen Operated Engine," INTERNATIONAL JOURNAL OF HYDROGEN ENERGY, 15, pp. 833-842, 1990.

11. L. M. Das, R. Gulati, P. K. Gupta, "Performance Evaluation of a Hydrogen-Fueled Spark Ignition Engine using Electronically Controlled Solenoid-Actuated Injection System," INTERNATIONAL JOURNAL OF HYDROGEN ENERGY, 25, pp. 569-579, 2000.

12. L. M. Das, R. Gulati, P. K. Gupta, "A Comparative Evaluation of the Performance Characteristics of a Spark Ignition Engine using Hydrogen and Compressed Natural Gas as Alternative Fuels," INTERNATIONAL JOURNAL OF HYDROGEN ENERGY, 25, pp. 783-793, 2000.

13. James W. Heffel, Michael N. McClanahan, and Joseph M. Norbeck, "Electronic Fuel Injection for Hydrogen Fueled Internal Combustion Engines," SAE Paper 981924, 1998.

14. L. S. Guo, H. B. Lu, and J. D. Li, "A Hydrogen Injection System with Solenoid Valves for a Four Cylinder Hydrogen-Fuelled Engine," INTERNATIONAL JOURNAL OF HYDROGEN ENERGY, 24, pp. 377-382, 1999.

15. Roger E. Billings, THE HYDROGEN WORLDVIEW, 2nd Edition, International Academy of Science, 2000.

CONTACT

Robert Boehm, Professor of Mechanical Engineering and Director of the Energy Research Center at the University of Nevada, Las Vegas, NV 89154-4027. boehm@me.unlv.edu

1999-01-2906

Final Report for the H2Fuel Bus

W. D. Jacobs, L. K. Heung, T. Motyka and W. A. Summers
Westinghouse Savannah River Company

J. M. Morrison
Southeastern Technology Center

would provide a daily range of at least 120 miles in a hybrid hydrogen/electric-operating mode. The project developed reduced engine tail-pipe emissions, with NOx measured at less than 0.2 ppm. In addition todemonstrating the inherent safety of a solid-state hydrogen storage system, the project also addressed permit, liability, and safety issues, including a safety risk assessment of the metal hydride storage system. State-of-the-art technology in battery system management was likewise demonstrated.

INTRODUCTION

The Savannah River Bus Project involved the development, manufacturing, and testing of the world's first hybrid hydrogen-electric transit bus, the H2Fuel Bus. The project was designed to successfully transfer technologies developed at the Department of Energy (DOE) Savannah River Site (Aiken, South Carolina) to commercial use in a public transit bus. The project was a technology verification activity to establish the technical feasibility, economic viability, and environmental benefits of hydrogen as a transportation fuel. Solution of hydrogen infrastructure issues (e.g., refueling, liability, safety, etc.); public awareness and public acceptance of hydrogen as a vehicular fuel were additional objectives.

The H2Fuel Bus was developed through a partnership of federal and local governments, universities, non-profit organizations, and industry. Based on favorable findings of a feasibility study conducted in 1994, development and manufacturing activities were carried out during 1995-1997. The resulting H2Fuel Bus is a 33-foot long, 27-passenger transit bus that was shipped to Augusta, Georgia at the end of March 1997 for demonstration and evaluation in passenger service. The bus was operated in the city of Augusta by the Augusta-Richmond County Public Transit Department (ARCPT) until the completion of the Augusta project in September 1998. Technical and management support during this period was provided by the Westinghouse Savannah River Company (WSRC) and the Southeastern Technology Center (STC).

ABSTRACT

The H2Fuel Bus is the world's first hydrogen-fueled electric hybrid transit bus (see Figure1.). It was a project developed through a public/private partnership involving several leading technological and industrial organizations, with primary funding by the Department of Energy (DOE). The primary goals of the project were to gain valuable information on the technical readiness and economic viability of hydrogen fueled buses and to enhance the public awareness and acceptance of emerging hydrogen technologies. The bus completed its field-testing and was placed into public service on September 4, 1998 by Augusta Public Transit in Augusta, Georgia. The bus employs a hybrid Internal Combustion (IC) engine/generator and battery powered electric drive system, with onboard storage of hydrogen in metal hydride beds. The initial operating results demonstrated an overall energy efficiency (miles/BTU) twice the range of a similar diesel-fueled bus, while doubling the range of an all-electric vehicle by providing in-transit recharging of the batteries. Subsequent data showed that the power controller was not optimized for maximum battery life and, therefore, some efficiency was lost. Correction of that condition

Following operations in Augusta, the Department of Energy transferred the bus in October 1998 to the National Hydrogen Test and Demonstration Center in Nevada for use as a national test platform. The future program is still under development but is expected to include improving the technology and taking the bus to the next stage of development.

MAIN SECTION

The starting platform for the H2Fuel Bus was a Type Q electric bus manufactured by Blue Bird Body Co., Fort Valley, Georgia. The Georgia Technology Research Institute (GTRI), Cobb County, GA was selected to integrate the unique hydrogen and data acquisition systems (DAS) into the bus. The rear seat and window were removed to make room for a compartment containing the engine and generator set, used for onboard charging of the batteries, and two large radiators with fans to control engine coolant temperature. Two of the four battery packs in a standard Type Q bus were omitted to make room for two containers of similar size and weight containing the metal hydride (MH) material, a lanthanum-nickel-aluminum intermetallic compound, for onboard storage of hydrogen fuel. WSRC developed and provided the MH material, while the other special components such as the engine, generator, external battery charger, battery management system, hydrogen gas for refueling, etc., were provided by industrial suppliers. The DAS was designed and built by GTRI. The bus modifications were completed in March 1997 and the bus was transferred to Augusta, GA to begin demonstration and evaluation. The total program cost, excluding in-kind contributions from several of the non-federal participants, was about $3.5 million, funded primarily by DOE.

The Augusta portion of the project began April 1, 1997 and continued through September 30, 1998. The primary project goals were accomplished during this period:

- The bus was extensively demonstrated at the three-day annual meeting of the National Hydrogen Association in Tyson's Corner, VA in March 1998, where rides were provided for over 200 people representing technical organizations, Congress, public schools, and the media.

- The bus was operated in passenger service in the City of Augusta during September 1998. A number of runs were made on two different days during which passengers were admitted and discharged along a standard route covering the downtown area.

- The above achievements as well as numerous test runs conducted throughout the City of Augusta contributed substantially to enhancing the public awareness and acceptance of hydrogen as a transportation fuel.

- The successful development and demonstration of the H2Fuel Bus represents a significant step forward in the transfer of government technology, developed for defense purposes, into the commercial sector.

Technical benefits of the program include the demonstration of an overall energy efficiency twice that of a similar diesel-fueled bus, while doubling the range of an all-electric vehicle by incorporating onboard recharging of the batteries. Very low tailpipe emissions were also achieved by use of a three-way catalytic bed and lean fuel operation of the IC engine, with NOx measured at less than 0.2 ppm. A range of 89 miles was demonstrated under actual conditions; under more optimal conditions, a range of more than 120 miles, approximately sufficient for a full transit day, should readily be achievable. Hydrogen refueling was accomplished successfully and routinely at commercial gas supply facilities. Safety issues were addressed by conducting a comprehensive assessment of the metal hydride system, which concluded that the safety and risk associated with the onboard storage of hydrogen were comparable to or better than those of a similar vehicle fueled by diesel or compressed natural gas (CNG). Liability concerns were resolved by purchasing general and excess liability insurance from commercial policy underwriters.

Overall, the operation of the hydrogen storage and engine systems proved to be satisfactory and relatively trouble-free. A number of difficulties, however, were encountered with the electrical and instrumentation systems. These included typical infant mortality failures of various components; design inadequacies; manufacturing deficiencies; reliability failures of standard commercial components; and insufficient battery life caused by non-optimal operation prior to installation of a battery management system. Collectively, these difficulties increased costs, limited the amount of time the bus was available for demonstration and service, and interfered significantly with efforts to demonstrate the desired level of performance. Nonetheless, they represented a valuable contribution in terms of lessons learned and potential improvements in future generations of similar hybrid electric buses for public transit use.

CONCLUSIONS AND RECOMMENDATIONS

The results of the H2Fuel Bus operation in Augusta clearly lead to a positive conclusion regarding the technical feasibility of the concept. The onboard hydrogen fuel storage system performed better than expected, operating nearly trouble-free while exhibiting a storage capacity in excess of the 15 kg design value. The operation of the hydrogen fueled engine/generator set (genset) likewise showed that onboard recharging of batteries to increase operating range and time is a realistic and achievable goal. Refueling at commercial hydrogen supply facilities is a practical operation that can be accomplished in a short time, e.g., two hours or less. Safety and liability concerns are manageable, as evidenced by a comprehensive safety and risk analysis and the insurability of the vehicle through standard commercial underwriting practices. Efforts to demonstrate the bus in public venues and educate passengers concerning the benefits of the H2Fuel Bus were highly successful and met with ready acceptance.

These achievements notwithstanding, it is not surprising that significant improvements are required for the H2Fuel Bus to become commercially viable. As with any first-of-a-kind prototype, the $3.5 million cost of the present bus is too high to be economical. As indicated, the costs of the hydrogen vehicle conversion and the metal hydride system must be reduced substantially, and special charges such as project management and Olympics operation (an early demonstration goal of the program that proved non-attainable) must be eliminated, to make the concept competitive. A rough assessment (Table 1) was made in early 1997 to identify target costs for the various systems that might result in an economical vehicle. This assessment indicated that a competitive design might be achieved by taking advantage of lessons learned from the first prototype, by taking an aggressive design and value engineering approach to reduce costs, and by increasing rates of production. The results obtained with the H2Fuel Bus in Augusta reinforce the conclusion that these improvements could be accomplished in perhaps two more design evolutions.

Table 1. Potential H2Fuel Bus Recurring Costs

Bus Generation	Second	Third	Fourth
Description	Lessons Learned	Value Eng'd	Lmt Vol Prodtn
Lot Size, no. of buses	1	10	100
Costs, $K			
Electric Bus	248	180	140
Engine Modifications	83	30	5
Engine/generator Set	22	22	15
Metal Hydride Storage Tanks	300	100	50
Mechanical Equipment	20	20	5
Electrical/Electronic Equipment	30	30	5
Manufacturing Assembly	100	60	0
Engineering Support & Testing	100	20	50
Recurring Project Management	150	40	0
Subtotal	1053	502	270
Manufacturing Allowances/Markup. 30%[1]	0	150	80
TOTAL RECURRING COST, $K	**1053**	**652**	**350**
Non-Recurring Cost, $K	500	800	0

(1) Includes amortization of new equipment, fabrication support, tooling, markup, etc.

Two recommendations are suggested for advancing the program further toward the development and validation of commercially viable, hydrogen fueled hybrid electric tran-

sit bus. First, the DOE decision to transfer the H2Fuel Bus to the National Hydrogen Test and Demonstration Center in Nevada for the next stage of its development is strongly endorsed. Assuming the availability of adequate time and funds for continued development and testing, the outcome of that program might well yield the technical and design basis information required for an eventual decision on commercialization of the concept. The second and related recommendation is to pursue the investigation and evaluation of improved metal hydride materials for onboard hydrogen storage. The goal of such an effort would be to develop a material capable of storing more fuel with less material and at lower overall cost than the material used in the present bus. Significant improvements in these factors are required to reduce the overall cost of metal hydride storage systems to the levels needed for commercial success.

SYSTEM DESCRIPTION

1. OVERALL H2FUEL BUS SYSTEM – Operational experience with this vehicle was centered in two major areas. The first area was determining the characteristics of operation of the metal hydride fuel storage system and the hydrogen fueled internal combustion engine (see Figure 2.). The second was the operation of the electric systems including the auxiliary power unit, batteries, battery charger/management systems, and electric drive system.

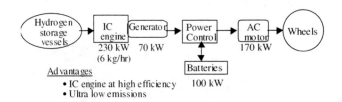

Figure 2. Schematic of the hydrogen-powered hybrid power system.

2. METAL HYDRIDE HYDROGEN STORAGE SYSTEM – The hydrogen fuel system stores hydrogen on a Lanthanum-Nickel-Aluminum (LANA) type alloy[1]. Metal hydrides are intermetallic compounds that undergo reversible chemical reactions with hydrogen. They absorb hydrogen when cooled and release it in a controlled manner when heated. The hydrogen is supplied to the engine by recycling the waste heat of the engine coolant system to release hydrogen from the metal alloy. The hydrogen released from the beds is regulated and then metered to the engine through a Constant Volume Injection™) unit. Hydrogen is stored on the bed at ambient temperatures and near ambient pressures.

To utilize this system the bus required several modifications. The first conversions began with removal of two of the four original battery packs. These packs were replaced by a LANA type alloy metal hydride storage sys-

1. $Lm_{1.06}Ni_{4.96}Al_{0.04}$ (Lm =La 55.7%, Ce 2.5%, Pr 7.7%, Nd 34.1%)

tem. The metal hydride storage system was designed to occupy the same space and have approximately the same weight as the batteries removed. This system permits storage of approximately 15.2 kg of hydrogen. The hydride has the characteristic of 1.27 weight percent loading (approximately 6000 scf) at pressures less than 500 psig. This offers intrinsic safety over high-pressure systems. The system is also designed to limit hydrogen release from a line break. Due to the thermodynamics of the hydrogen absorption on the hydride material, the bed cannot generate more than 74 scfm, which is approximately twice the maximum consumption rate. This rate ensures that the bed will not overpressurize in a fire scenario.

The patented hydrogen storage system consists of 24 tubular vessels that contain the metal hydride powder (see Figure 3.). Each vessel consists of 14 chambers that are packed with aluminum foam to enhance heat transfer, a porous steel filter, and coolant channels to enhance absorption or release of the hydrogen from the metal hydride material (Figure 4). The metal hydride material is segregated into chambers to prevent migration and damage that could result due to material swelling which results from absorption. The beds were designed to ASME standards and carry a pressure vessel rating of 500 psig.

Figure 3.   One of the two "boxes" containing the on-board hydrogen storage vessels.

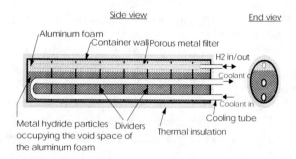

Figure 4.   Schematic of the hydride hydrogen storage vessel

The alloy choice was based on the engine operating temperature, expected ambient conditions, and bed temperature limits. The hydride bed temperature is the highest temperature at which the generation of hydrogen is expected during operation and the ambient condition affects the cooling water temperature used to cool the beds during refueling.

When operating at full power, the hydrogen combustion engine requires hydrogen feed rate of 6 kg/hr at about 100 psig. pressure. At this rate, the heat required for hydrogen desorption is calculated to be 19800 kcal/hr. The hydride storage system must be able to transfer this much heat from the coolant to the metal hydride during operation. The heat transfer rate for the system was enhanced by addition of aluminum foam pieces. The foam increases the heat transfer coefficient of the metal hydride bed by a factor of 5. The heat transfer coefficient of 426 kcal/hr/m^2/°C is more than twice the required value determined in laboratory experiments. From this result, a single box is capable of meeting the designed feed rate of 6 kg/hr.

A schematic of the hydrogen system piping is shown in Figure 5. The two aluminum boxes, each containing 24 cylindrical tubes of metal hydride, are connected in parallel so that the engine can be supplied from either or both boxes. A reducing valve is used to adjust hydride bed pressure, typically 125-150 psig, down to 100 psig for inlet to the engine. The pressure relief valves depicted inside the two boxes were replaced with new 500 psig ASME Code compliant valves in April 1998. The new relief valves are larger than the original valves and were installed on the outside of the boxes due to space constraints. Each box is equipped with an excess flow valve (essentially a check valve) to minimize loss of gas from one box from a breach in the other box and subsequent depressurization.

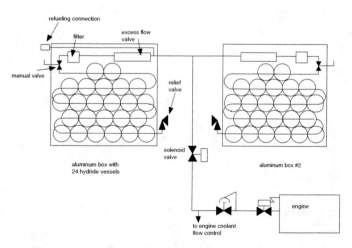

Figure 5.   Hydrogen Piping Schematic.

A manual valve on each box, operated from the external side of the bus, permits operating and refueling each box either simultaneously or individually. A single refueling connection mounted on the bus exterior is used to serve both boxes. Refueling gas is added through a special nozzle (Sherex[R]) which attaches to the refueling connector. Hydrogen bed pressure is used to adjust a three-way diversion valve to circulate either more or less engine coolant through the hydride beds, hence, increasing or decreasing hydrogen desorption from the beds, the remainder bypassing directly to the engine radiators.

3. HYDROGEN FUELED AUXILIARY POWER UNIT – The Auxiliary Power Unit (APU) is a standard industrial, multi-fuel 7.5 liter V-8 internal combustion engine (Ford Model 875) with modifications including the Constant Volume Injection (CVI™) unit, heads and pistons. The CVI™ is a mechanically timed metering device timed to the intake stroke of the engine. The standard engine was modified with high compression heads and pistons, which increased the compression ratio for higher thermal efficiency. The heads have better flow characteristics to improve volumetric efficiency as well. To reduce engine knock, the exhaust from one half of the engine was cooled and recycled into the engine intake manifold. The exhaust from the other half of the engine was directed to a water separation unit and the remaining gases were passed through a three-way catalyst bed to reduce NOx emissions. To reduce the NOx emissions, the catalyst bed temperature must be limited to 130° C; this required further modification of the engine by replacing the standard radiator with a larger, lower temperature system. In this case, a ducted fan cooling system was chosen. All of the above modifications were developed and made by Hydrogen Components Inc. (HCI) of Littleton, Colorado, except for the cooling system, which was accomplished by Georgia Tech Research Institute. The system was optimized to reduce the engine/generator set from cycling or hunting to achieve full speed (2500 rpm).

4. BATTERY AND ELECTRIC SYSTEMS – The electrical system modifications began by removing two of the four original battery packs. Then an APU was installed. The system chosen was an AC generator whose output was rectified to match the charging voltage and current range of the existing battery system. The voltage from the APU or the batteries was inverted by the Northrup-Grumann controller system and used by an AC traction motor. The system was enhanced with a regenerative braking system.

A BADICHEQ™ battery management system was added in January 1998. Advantages include knowing the state of all of the batteries in the system, warning of potential trouble conditions, and if needed, charging individual batteries to help restore them to a normal state. Preliminary testing of the battery system indicated that problems due to deep cycling and thermal overheating of the batteries caused some cells to fail prematurely. Discussions with the battery manufacturer and the bus manufacturer led to the installation of the BADICHEQ™ system. Temperature compensation of the off-board charging system was also installed. After collecting data for several runs, the vari-ous manufacturers were consulted to adjust the operating parameters.

The BADICHEQ™ system has several interlocks, which are designed to prevent overcharging the batteries, or causing damage due to operation in conditions that have not been reviewed. The first interlock requires that both battery packs be installed before the APU is permitted to operate. This prevents dumping excess current into only one pack and stressing the batteries. Similarly, the second interlock will not permit the regenerative braking feature to be used when the APU is operating, since the APU/regenerative braking combination has the potential to overcharge the batteries.

The H2Fuel Bus is equipped with an external battery charger (Enerpro) for overnight charging of the batteries. The Enerpro requires a 480-volt, 100-amp power supply. A full charging cycle takes six to eight hours to complete. A recent modification made by the Enerpro battery charger manufacturer was the development of a circuit board that will permit the BADICHEQ™ to control the battery charging profile dynamically. This allows for optimum charging and protection of the batteries. This project was under development too late to be incorporated into the H2Fuel Bus, but should be considered for future projects.

Currently, the battery charging is conducted from a static profile determined from the manufacturer of the batteries and the charger. The only dynamic feature of the system is that it is temperature compensated and will shutdown the charger if the pack temperatures exceed 50°C. This has had limitations. The charger must see both temperature probes before it will allow a charging profile to begin. If one probe fails, the temperature compensation must be bypassed to allow the charger to function. A failed probe in November of 1997 resulted in bypassing the temperature compensation. Combined with a failing battery, it resulted in over charging a battery to the point that it became a resistive heat load and heated the pack to 65°C (as monitored by the BADICHEQ™ thermistor for the pack).

The history of the battery packs was one of initial deep discharges and poor thermal management. This left the packs in less than optimum condition before the BADICHEQ™ installation. After the BADICHEQ™ monitoring began, most thermal management issues were corrected, except for a battery that fails (most failures provide sufficient warning to permit scheduled replacement without damage occurring to adjacent batteries). Further, the BADICHEQ™ system enabled better diagnostics and maintenance of the remaining batteries in the packs. It greatly aided the diagnosis and change out of failing batteries. At the current time, one battery has failed and one or more are severely degraded, indicating that battery replacement is a priority to assure successful future operation. The battery manufacturer, the BADICHEQ™ representative and Blue Bird Body Company have also recommended this.

5. DATA ACQUISITION SYSTEM – The Data Acquisition System (DAS) used in the bus project was made up of four independent subsystems: 1) Northrup-Grumman's electric bus monitor, 2) the GTRI system designed to monitor the Internal Combustion (IC) engine and the hydrogen fuel storage system parameters, 3) the Enerpro battery charger system for off-board charge control and data collection, and 4) the BADICHEQ™ battery management system which monitors the battery packs. Each system provided limited independent verification of data collected by other systems.

The Northrup-Grumman system interfaced with a laptop computer through a program called "BUSMON". The hardware interface was an optically isolated RS-232 connector located under the driver's console. The program is executed just prior to the energization of the bus systems. The program begins data collection automatically. The data displayed on the screen provided a listing of bus physical data such as speed, distance traveled, power consumption, and critical system temperatures. The screen also displayed system alarms and interlock states. Much of the data was calculated from instantaneous values, which the program logged, and was integrated to recreate information displayed during an operational run. This system could be interfaced to the GTRI system if the data protocol to the proprietary system is obtained.

The GTRI system was based on the National Instruments LabWINDOWS/CVI™ software and SCXI™ (Small Computer External Interface) signal conditioner. A laptop computer provided visualization of the data, alarms, trends, and data storage. The system was configurable for future expansion and upgrade of instrumentation. It was based in the G-programming language. It allowed the programmer to create virtual instrumentation and visual interfaces for the system.

The Enerpro system provided data for the off-board charging of the main bus batteries. The system provided both data collection and temperature protection. The charge profile used was static and was defined by the manufacturers of the batteries and the charger.

The final data system added to the bus was the BADICHEQ™ battery management system, previously described in section 4.

H2FUEL BUS HAZARD ANALYSIS

As with most hydrogen fueled vehicle demonstration projects, management and regulator safety expectations must be met. To accomplish this a systematic and comprehensive hazard analysis process was completed. The evaluation methods and criteria used in this analysis reflect the Department of Energy's graded approach for qualifying and documenting nuclear and chemical facility safety. The scope of the analysis included transit operations, maintenance, cleaning, battery recharging, and vehicle storage.

The safety evaluation demonstrated that the operation of the H2Fuel bus represents a "moderate" risk. This is the same risk level determined for operation of conventionally powered transit buses in the United States. By the same criteria, private passenger automobile travel in the United States is considered a "high" risk. The safety of passengers on the H2Fuel bus was shown dependent somewhat on the training and proficiency of the bus operator.

The evaluation also identified several design and operational modifications that resulted in improved safety, operability, and reliability. The hazard assessment methodology used in this project has widespread applicability to other innovative operations and systems, and the techniques can serve as a template for similar projects. Details of the analysis are contained in References 4 and 5.

CHRONOLOGY OF AUGUSTA OPERATIONS

Integration of the hydrogen system (APU, metal hydride storage units, and associated gas and coolant piping, fittings, and controls) and the computerized data acquisition system (DAS) was completed by GTRI in March 1997. The bus was transferred by truck to the Augusta Public Transit (APT) maintenance facilities on March 31, 1997. Work began immediately by WSRC, APT, and STC to checkout and tests the bus in preparation for entry into passenger service. A number of test runs were conducted in the period April to July 1997, which provided an opportunity for the Augusta team to become familiar with the bus systems. Several difficulties and relatively minor electrical failures occurred during this period. In July 1997, a short circuit in the rectifier and voltage regulator section of the generator caused extensive damage requiring several months to repair.

The bus was restored to operating condition in October 1997 but problems began to develop with the external battery charger as well as the main batteries. These problems were addressed with assistance from the bus and battery manufacturers, including replacement of selected batteries, rebuilding the two battery packs, and modifications to the battery charger. These efforts continued through the end of December 1997 and culminated in awarding a contract to Neocon Technologies to install a computer based battery management system, trade-named BADICHEQ™. This was installed in January 1998.and worked well; however, problems with the batteries, which by that time were known to have been overstressed in the early months of the program, continued. In March 1998 the bus was demonstrated very successfully for three days at the National Hydrogen Association annual meeting in the Washington, DC area, following which the bus was taken out of service for installation of new hydride bed pressure relief valves and vent piping.

Bus operability was restored in May 1998, but battery and other problems were again encountered over the next several months. The goal of public passenger service was achieved in September 1998, but further passenger runs were discontinued shortly thereafter in order to conserve as much of the remaining battery life as possible for the impending transfer of the bus to DOE-NV. This transfer took place on October 22, 1998.

DISCUSSION OF RESULTS

1. OVERALL EXPERIENCE – The bus was operated for a total of about 1050 miles. All but about 180 miles of this was recorded after delivery of the bus to Augusta, GA on March 31, 1997. The bus was tested under a variety of conditions, ranging from the flat inner city, to areas of steep grade, to highway conditions. It exhibited good operating characteristics in all these situations.

Some modifications had been made to the basic motor controller software, during bus construction, to compensate for the two battery packs that were removed from the original four-pack design to make room for the metal hydride storage units. The bus tended to limit the power to the traction motor if the demand exceeded the available battery power. This limitation was most pronounced on a steep grade and usually at reduced battery pack voltages. Another option for controlling this condition was to start and operate the APU prior to arriving at the grade.

When battery life became an issue, it was decided to operate the APU in a constant run mode rather than the original on-off mode of operation. This preserved battery lifetime, but reduced the efficiency and overall availability of the APU (2-4 hours of constant APU operation versus the expected 8 hours of intermittent operation).

Most of 1997 after bus delivery was devoted to test runs, characterization of bus systems, and replacement of failed and deficient electrical components. By the end of 1997, it was apparent that an aggressive effort was needed to conserve remaining battery life; therefore, the Badicheq BADICHEQ™ system was installed in January 1998. An outage was also taken in 1998 to modify the hydrogen pressure relief system. In addition to test runs, successful operation in 1998 included both the demonstration runs at the National Hydrogen Association meeting in March and the use in passenger service in September. The bus was operated for two days in passenger service and then removed from transit duty for preparation for transport to Nevada after September 30, 1998.

2. DETAILS OF PERFORMANCE – Overall, the hydrogen system operated very well. Some difficulties were occasionally experienced with the excess flow valves, which proved to be very sensitive to pressure differentials between the overpressures of the two hydride beds. These were relatively minor inconveniences, however, that could be handled procedurally by careful manipulation of the manual cutoff valves located on the exterior of the bus.

The hydrogen fuel storage system performed better than expected. The initial design estimates of hydrogen capacity were about 15.2 kg, or 1.27 weight percentage. In actual refueling and operation, after a number of loading-unloading cycles the metal hydride was found to hold more hydrogen than expected, perhaps as much as by a factor of two. This increase in storage capacity for the given conditions is under investigation in laboratory bench-scale testing at SRTC to better understand the phenomenon.

The engine/generator (APU) efficiency is limited by the Northrup-Grumman controller's ability to direct power. A review of the data available indicates that the APU is capable of generating 70 kW (nameplate rating). However, the output of the APU was typically only 9 kW avg. (20 kW peak). This is less than 30 percent as efficient as anticipated, or an overall engine efficiency of ~10 percent. This finding is consistent with a survey of various engine/generator combinations during the early feasibility study, which indicated that high generator efficiency is achieved only at near full loads, dropping to 50% or less at 25% load.

The bus consumed 2.63 kWh/mi. on average (15.70 kWh/mi. peak). This consumption is higher than the original projection of 1.53 kWh/mi. It is believed that the difference is caused by the power controller system not being optimized with the APU. There is some variability in the data; on one run, for example, the bus motor consumed 28.3 kWh avg. (90.3 kWh peak) over a 9-mile run, corresponding to an average of 3.14 kWh/mi. Power appeared to be drawn preferentially from the batteries rather than from the generator output under final (i.e., late-1998) conditions. To change this would require reprogramming the Northrup-Grumman controller. The data from initial bus operations, where the batteries were deeply discharged, showed a greater demand signal and increased generator output, hence, improved fuel economy. The initial data confirmed the achievement of the design objective of efficiency twice that of a conventional diesel bus. A task remaining at the end of the Augusta project therefore is to improve battery life while maintaining high generator efficiency.

The generator set can provide only a portion of the running power under normal operating conditions, while maintaining good battery operating practices. The bus' range could be significantly increased if this condition were optimized.

The overall average measured APU hydrogen consumption rates are 67 scf/mile and 49 scfm at 100% load. The calculated corresponding bus range is 30 miles in an all-electric mode or approximately 120 miles in the hybrid mode based on the design storage of hydrogen and optimized controller operation. This calculated range would be increased if the indicated higher-than-design storage of hydrogen late in the project life can be confirmed. The record of 89 miles actually achieved in one trip (batteries charged and full hydrogen fuel load) was not representative of expected performance in the current bus configuration. The record high mileage was achieved at the expense of battery life. The batteries were severely deep-discharged and damage resulted. The result was the eventual requirement to replace 14 batteries and a reduced lifetime of all remaining batteries. At the conclusion of the Augusta project, at least two more batteries were near failure in the passenger side pack.

3. SUGGESTED ACTIONS AND IMPROVEMENTS FOR A NEXT-GENERATION BUS

– Based on the observed performance, the following actions and improvements are recommended for either a next generation vehicle or a significant upgrade of the current H2Fuel Bus.

a. Hydrogen Storage

1. The ultimate capacity of the metal hydride bed should be investigated by developing isotherm data for the hydride bed material, along with determining its other characteristics.

2. Improving the hydrogen refueling times will require study of the cyclic nature of the hydrogen absorption. The cyclic pattern may be the result of a flow limiter restricting hydrogen flow. The loading of hydride beds plumbed in series represents an area for improvement. With the series arrangement, heat is transferred from the loading bed to the other bed in the system. Also, the coolant for the beds does not circulate simply through the beds, but mixes with the balance of the bus coolant.

3. The amount of hydrogen desorbed from the bed (hence engine run time) is limited by the pressure/temperature of the hydride beds. It may be possible to increase the temperature of the engine exhaust gas to the catalyst bed to permit higher engine operating temperatures, and therefore higher temperature of coolant flowing to the hydride beds, without affecting the NOx emissions.

b. Auxiliary Power Unit

1. The APU performs as intended, but work on increasing the fuel efficiency should continue. The consumption of hydrogen by the engine changes only 20% between full load and idle conditions.

2. The Northrup-Grumman controller should be optimized to increase the load demand signal to the generator and preferentially draw power from the generator (if available). This may change the operating scheme from continuously running the engine to shutting off the engine and operating intermittently in the all-electric mode.

Regenerative Braking

1. The regenerative braking system should be reevaluated. The system designed by Northrup-Grumman performs as intended, but it has the capability to overcharge the batteries and damage them if braking is applied too fast and the current to the battery packs is not limited. For this reason, the existing regenerative braking was temporarily disabled for many of the bus runs.

2. An energy buffer such as an ultracapacitor bank appears to be a promising method of correcting the above shortcoming and should be investigated. It offers the ability to quickly store a charge and then meter it into the system at the normal voltage without damage to the batteries. An added benefit is that it has flexible space requirements and can be scaled to the voltage requirements of an existing system.

Batteries

1. The battery system was not initially optimized and battery pack instability reduced the life of at least 15 batteries. The new BADICHEQ™ battery management systems have alerted operators to potential problems so planned work outages could be taken. This has proven to be a valuable addition and this type of system is strongly recommended as part of the base design of a next-generation vehicle.

2. The type of batteries used in a next-generation bus should be evaluated for potential upgrading. The current batteries do not have the cycle life required for a transit vehicle and their cost is high. Alternative batteries should be evaluated to determine if a different type is available that might offer greater cycle life at lower cost.

3. Future work should include incorporating the BADICHEQ™ monitoring and controls to replace the off-board charger control, creating a dynamic charge profile.

4. The same BADICHEQ™ control modification should be incorporated into the APU operation and regenerative braking systems (assuming energy storage and discharge upgrades).

5. If the present H2Fuel Bus were to be refurbished, a full complement of new batteries should be purchased and installed. The useful lifetime of the existing batteries is questionable and the effort to replace individual batteries is labor intensive. Replacement of individual batteries also tends to unbalance the pack more than having all new batteries in place.

Miscellaneous

The electric wiring of the present bus should be reviewed for compliance to National Electric Codes and Standards and corrections made to remove any deficiencies found.

1. The Data Acquisition System (DAS) should be modified to provide a "fuel" (i.e., onboard hydrogen) gauge display for the driver similar to the one supplied with the BADICHEQ™ system for the batteries. This will ensure the driver has full awareness as to the amount of energy reserves onboard.

2. The DAS should be redesigned to make it more rugged; to withstand the thermal and vibration conditions expected in normal service. All programs should be ported and run from one computer. All ports and connectors should be standardized. A distributed I/O methodology and topology is recommended.

3. All onboard sensors should be standardized to the minimum number of common signal formats, if possible. This would simplify future modifications and testing.

4. All power supplies should be standardized to 12 vdc as supplied by the bus electrical system.

5. Off-board 120-vac electric power should be provided to the bus and outlets supplied within the cabin area to support maintenance activities.

6. A 120-vac to 12-vdc rectifier and bypass should be placed onboard to allow operation of all installed instruments and equipment without requiring startup of the bus or APU.

ACKNOWLEDGEMENTS

In addition to the companies represented by the authors of this report, many public, private, and academic organizations played important roles in the conduct and success of the H2Fuel Bus project in Augusta, GA. In particular, the authors would like to acknowledge the important efforts and contributions made by the following project partners and participants:

U.S. Department of Energy – Primary project sponsor and funding support

Georgia Tech Research Institute – Hybrid bus design, integration and testing

Augusta Richmond County Public Transit – Bus owner; operation and maintenance

Blue Bird Body Co. - Electric bus manufacturer

Hydrogen Components Inc. – Hydride vessel construction; engine conversion to hydrogen fuel

Education, Research and Development Association of Georgia Universities – Subcontract administration

Air Products and Chemicals, Inc. – Hydrogen refueling in Atlanta, GA

Air Liquide America Corp. – Hydrogen refueling in Augusta, GA

Energy Research and Generation, Inc. – Aluminum foam for metal hydride beds

Power Technology Southeast, Inc – Engine and generator supplier

Northup-Grumman Corp – Bus electrical system controller

Electrosource, Inc. – Battery supplier

Neocon Technologies, Inc. – Battery management system supplier

Enerpro, Inc. – External battery charger manufacturer

In addition to the above, several companies and individuals made generous financial contributions to Southeastern Technology Center to promote public awareness and involvement activities associated with the H2Fuel Bus project. The donations made by Westinghouse Electric Corporation, Georgia Power Corporation, Dr. Earl J. Claire, and Dr. Clanton Mosely are specifically acknowledged and appreciated.

This report was prepared concerning work done under Contract No. DE-AC0996SR18500 with the U. S. Department of Energy.

REFERENCES

1. WSRC-MS-98-00470, "Performance Testing of a Hydrogen-Fueled Electric Hybrid City Transit Bus," W. D. Jacobs, L. K. Heung, T. Motyka, W. A. Summers.

2. Heung, L. K., (1988), "Heat Transfer and Kinetics of a Metal Hydride Reactor," *Metal-Hydrogen Systems*, vol. II, 1451-1461, Edited by R. Kirchheim, E. Fromm, and E Wiche, Stuttgart, Germany.

3. Heung, L. K., (1997), "On-board Hydrogen Storage System Using Metal Hydride," HYPOTHESIS II Conference held at Grimstad - Norway, August 18-22, 1997.

4. WSRC-TR-96-0202, "H2Fuel Bus Project Hazard Analysis—Rolling Operations (U)," D. A. Coutts, G. L. Hovis, J. K. Thomas, A. G. Sarrack, T. T. Wu, and J. C. Chattin, Westinghouse Savannah River Company, Aiken, SC, July 1996.

5. WSRC-TR-96-0415, "H2Fuel Bus Hazard Analysis—Operations, Maintenance and Storage (U)," D. A. Coutts, J. K. Thomas, G. L. Hovis, and T. T. Wu, Westinghouse Savannah River Company, Aiken, SC, February 1997.

DEFINITIONS, ACRONYMS, ABBREVIATIONS

APU: Auxiliary Power Unit
BADICHEQ™: Battery Diagnosis and Checking
CVI™: Constant Volume Injection
DAS: Data Acquisition System
DOE: U.S. Department of Energy
IC: Internal Combustion
LANA: Lanthanum Nickel Aluminum Alloy
NOx: Nitrogen Oxides

ATTACHMENT 1

METAL HYDRIDE BED DESIGN DATA

Metal Hydride Alloy $Lm_xNi_{(5-y)}Al_y$:

Lm is Lanthanum-rich mischmetal

$x=1.05\pm0.01$, $y=0.03$

Anneal:975 °C, 12 hours, inert atmosphere

Hydrogen desorption pressure: 13.2 atma @ 60 °C

Desorption plateau slope $\partial(\ln P)/\partial(H/M)$: 0.26 @ 60 °C

Hydrogen capacity $\partial(H/M)$: 0.86 @ 40 °C

Form: -10 mesh particles of crushed alloy

Fuel Containers:

Two

24 vessels each (48 total vessels)

Combustible gas detection in each container

Vent through top of bus

Hydride Storage Vessel:

Metal hydride: 26.6 kg

Cylindrical: 3.5" OD, 60 " long, 0.065" thick walls

Twelve chambers

316L stainless steel

10-micron sintered metal tube filters for hydrogen

U-shaped tubes for heating/cooling liquid

Maximum working pressure: 500 psig

Pressure relief through top of bus

Hydride Storage System:

Total weight: =< 5000 lbm

Hydrogen capacity: >= 12 kg

Required hydrogen purity :
   >= 99.99%

Hydrogen pressure:
   >= 2 atm-idle;
   >= 6 atm-full speed

Hydrogen pressure/flow sufficient to start engine @ 0 °C

Refueling Conditions:

Ambient temperatures (~25 °C)

H_2 pressure: ~100 psig
   (< 500 psi)

Time: =< 8 hours

ATTACHMENT 2

PERFORMANCE

	Pre-Conceptual	Actual	Optimized
Range	> 100 miles	89 miles	> 120 miles
Fuel Consumption	~ 6 kg H_2/hr (40 scf/min)	~ 7.3 kg H_2/hr (49 scf/min)	~ 6 kg H_2/hr (40 scf/min)
Speed	0 to 55 mph	0 to 55 mph	0 to 55 mph
Acceleration	0-30 mph in 20 sec	0-30 mph in 20 sec	0-30 mph in 20 sec
Gradability	Start on 12% grade	Start on 12% grade	Start on 12% grade
Emissions CO_x	Near 0	Near 0	Near 0
NO_x	=< 0.1 gm/bhp-hr	=< 0.0004 gm/bhp-hr (0.15 ppm)	=< 0.0004 gm/bhp-hr (0.15 ppm)

ATTACHMENT 3
INITIAL TEST DATA

Date	4/16	4/18	4/22	4/23	4/24	5/22	6/3	6/17	6/26	Totals	Design
Hubometer	173 (od)	268.1	309.6	352.8	439-481	481.5	534.4	609.5	679.1	679.1	
Tot Mi	45	41.8	39.6	86.5	42	39.6	38.8	63.4	38.7	435.4	120 mi/day
Mi-Batt	16	2.8	18.8	31.3	10-20	9.9	18.6	12.1	4.3	128.8	40
Mi-H2	29	39.1	20.8	55.2	22-32	29.7	20.2	51.3	34.4	306.6	80
Eng Runtime (min)	63	57	82	114		28	75	107	90	636	160
Fuel (SCF)	1099	1837	3646	3367	1692	1174	853	2341	4735	20744	**5200**
SCF/Mi	37.9	47	175.3	61	48.3	39.5	42.3	45.6	137.6	68	**65**
SCFM	17.4	32.2	44.5	28.5		41.9	11.4	21.9	52.6	33	**40**
Mi/Min	0.46	0.69	0.25	0.48		1.06	0.27	0.48	0.38	0.5	
Efficiency H2/Diesel										1.9-2.3	

Test Data gathered in 1997 upon arrival of H2Fuel Bus to Augusta, GA. Data was based on a Depth Of Discharge (DOD) of 80+% on the batteries.

Operation of a Hydrogen-Powered Hybrid Electric Bus

W. D. Jacobs, L. K. Heung, T. Motyka and W. A. Summers
Westinghouse Savannah River Company

ABSTRACT

The H2Fuel Bus, the world's first hydrogen-fueled electric hybrid transit bus (see Figure1.). It was a project developed through a public/private partnership involving several leading technological and industrial organizations, with primary funding by the Department of Energy (DOE). Using the bus, the primary goals of the project are to gain valuable information on the technical readiness and economic viability of hydrogen fueled buses and to enhance the public awareness and acceptance of emerging hydrogen technologies. The bus has been in test operation mode by the Augusta Public Transit in Augusta, Georgia, since April 1997. The bus employs a hybrid Internal Combustion (IC) engine/battery/electric drive system, with onboard storage of hydrogen in metal hydride beds. Initial operating results have demonstrated an overall energy efficiency (miles/BTU) of twice that of a similar diesel–fueled bus while nearly doubling the range of an all-electric battery powered vehicle. The project development also reduced the bus' tail-pipe emissions with NOx measured at less than 0.2 ppm. The project in addition addressed permit and safety issues as a result of the safety risk assessment undertaken regarding the metal hydride storage system. Demonstrating the inherent safety of a solid-state storage system, the project also demonstrated the state of technology in electrical system management.

Figure 1. The H2Fuel Hydrogen-Powered Bus

INTRODUCTION

The Savannah River Bus Project involves the development, manufacturing, and testing of the world's first hybrid hydrogen-electric transit bus, the H2Fuel Bus. The project seeks to successfully transfer technologies developed at the DOE Savannah River Site (Aiken, South Carolina) for defense applications to commercial use in a public transit bus. The project is a technology verification activity that seeks to establish the technical feasibility, economic viability, and environmental benefits of hydrogen as a transportation fuel. Solution of hydrogen infrastructure issues (e.g., refueling, liability, safety, etc.), public awareness and public acceptance of hydrogen as a vehicular fuel are additional objectives. The H2Fuel Bus was developed through a partnership of federal and local governments, universities, non-profit organizations, and industry. The H2Fuel Bus is a 33-foot long, 27-passenger transit bus that will be operate for regular passenger service for the city of Augusta, Georgia.

MAIN SECTION

Operational experience with this vehicle was centered in two major areas. The first area was the characteristic of operation of the metal hydride fuel storage system and the hydrogen fueled internal combustion engine (see Figure 2.). The second was the operation of the electric systems including the auxiliary power unit, batteries, battery charger/management systems, and electric drive system.

HYDROGEN STORAGE/INTERNAL COMBUSTION ENGINE- – The hydrogen fuel system stores hydrogen on a LANA alloy. Metal hydrides are intermetallic compounds that undergo reversible chemical reactions with hydrogen. They absorb hydrogen when cooled and release it in a controlled manner when heated. The hydrogen is supplied to the engine by recycling the waste heat of the engine coolant system to release hydrogen from the metal alloy. The hydrogen released from the beds is regulated and then metered to the engine through a constant volume injection unit. Hydrogen can be stored on the bed at ambient temperatures.

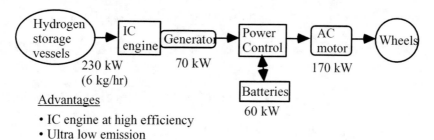

230 kW
(6 kg/hr)   70 kW

Batteries
60 kW

170 kW

Advantages
- IC engine at high efficiency
- Ultra low emission

Figure 2. Schematic of the hydrogen-powered hybrid power system

This system required several modifications. The first conversions to the bus began with removal of two of the four original battery packs. These packs were replaced by a Lanthanum-Nickel-Aluminum (LANA) type alloy metal hydride storage system. The metal hydride storage system was of the same approximate weight as the batteries removed and permit storage of 15.2 kg of hydrogen. The hydride has the characteristic of 1 weight percent loading (approximately 6000 scf) but at pressures less than 300 psig. This offers intrinsic safety over high-pressure systems. The system is also designed to limit hydrogen release in the event of a line break. Due to the thermodynamics of the hydrogen absorption on the hydride material, the bed cannot generate more than 74 scfm, which is approximately twice the maximum consumption rate.

The patented hydrogen storage system consists of 24 tubes that contain the metal hydride powder (see Figure 3.). Each vessel consists of nine chambers that are packed with aluminum foam to enhance heat transfer, a porous steel filter, and coolant channels to enhance absorption or release of the material (see Figure4.). Material is segregated in the bedchambers to prevent

migration and damage that could result due to material swelling which results from absorption. The alloy choice was based on the engine operating temperature, ambient conditions expected, and catalyst bed temperature limits.

Figure 3. "Box" of the On-Board Hydrogen Storage Vessels

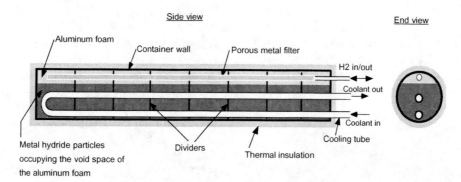

Figure 4. Schematic of the Hydride Hydrogen Storage Vessel

The Auxiliary Power Unit (APU) is a standard 7.5 liter V-8 internal combustion engine with modifications including the Constant Volume Injection (CVI) unit, heads, and pistons. The CVI is a mechanically timed metering device, timed to the intake stroke of the engine. The standard engine was modified with high compression heads and pistons, which increased the compression ratio for higher thermal efficiency. The heads had better flow characteristics to improve volumetric efficiency as well. To reduce engine knocks, the exhaust from one half of the engine was cooled and recycled into the engine intake manifold.

The exhaust from the other half of the engine was directed to a water separation unit and the remaining gases are passed through a three-way catalyst bed, reducing methane, carbon monoxide, and NOx emissions. To reduce the NOx emissions, the catalyst bed temperature must be limited to 130°F; this required further modification of the engine by replacing the standard radiator with a larger, lower temperature system. In this case, a ducted fan cooling system was chosen. All of the above modifications were developed and made by Georgia Tech Research Institute. The system was optimized to reduce the engine/generator set from cycling or hunting to achieve full power rpm's (2500 rpm).

ELECTRIC SYSTEMS- – The electrical system modifications began by removing two of the four original battery packs. Then an APU was installed. The system chosen was an AC generator whose output was rectified to match the charging voltage and current range of the existing battery system. The voltage from the APU or the batteries is then inverted by the Northrup-Grumann system and used by an AC traction motor. The system has been enhanced with a regenerative braking system.

The most recent modification has been the inclusion of a BADICHEQ battery maintenance system. Advantages include the knowing the state of all of the cells in the system, warning of potential trouble conditions, and if needed, charging individual cells to help restore them to a normal state. Preliminary testing of the battery system indicated that problems due to deep cycling and thermal overheating of the batteries caused some cells to fail prematurely. Discussions with the battery manufacturer and the bus manufacturer led to the installation of the system. Temperature compensation of the off-board charging system was also installed. After collecting data for several runs, the various manufacturers were consulted to adjust the operating parameters.

CONCLUSION

The hydrogen fuel storage system has performed satisfactorily. It still requires some modifications to improve hydrogen-loading times and simultaneous loading of parallel beds. The loading times may be addressed with a chilled water supply.

The engine runtime on hydrogen is limited by the pressure/temperature of the hydride beds. It may be possible to adjust the catalyst bed configuration to permit higher engine operating temperatures.

Future work remains to improve the hydride alloys to increase their weight percent loading.

The APU performs as intended, but work on increasing the fuel efficiency will continue. The operational range of the bus has proven to be increased to nearly double the initial all-electric bus configuration, but only under specific conditions. Future work will continue to improve the range of the bus.

The APU consumption rates measured to date are 67 scf/mile and 49 scfm at full load. The resulting range numbers are 30 miles in all electric mode, and approximately 80 miles in the hybrid mode.

The electric drive system proved to be the most challenging of the operational issues experienced, specifically the batteries. The battery system was not initially optimized and battery pack instability reduced the life of 8 batteries. The new charge management systems have alerted operators to potential battery problems so planned work outages could be taken. Future work will also include incorporating the BADICHEQ monitoring and controls to replace the off-board charger control, creating a dynamic charge profile. The same control modification should be incorporated into the APU operation and regenerative braking systems.

ACKNOWLEDGEMENTS

Georgia Tech Research Institute – Hybrid bus design, integration and testing

Augusta Richmond County Public Transit – Owner/operator

Blue Bird Bus Co.—Electric bus manufacturer

Hydrogen Components Inc. – Hydride vessel construction, engine conversions

Southeastern Technology Center; Education Research and Development Association of Georgia Universities; Air Products and Chemicals, Inc.; Energy Research and Generation, Inc.; Power Technology Southeast, Inc.; Northup-Grumman Corp.; Air Liquide America Corp.; Electrosource, Inc.; Neocon Technologies, Inc.; Enerpro, Inc.

REFERENCES

1. WSRC-MS-98-00470, Performance Testing of a Hydrogen-Fueled Electric Hybrid City Transit Bus,
2. Heung L. K. (1988) Heat Transfer and Kinetics of a Metal Hydride Reactor, *Metal-Hydrogen Systems*, vol. II, 1451-1461, Edited by R. Kirchheim, E. Fromm, and E Wiche, Stuttgart.
3. Heung, L. K. (1997) On-board Hydrogen Storage System Using Metal Hydride, HYPOTHESIS II conference held at Grimstad - Norway, August 18-22, 1997.

ADDITIONAL SOURCES

DEFINITIONS, ACRONYMS, ABBREVIATIONS

BADICHEQ: Battery Diagnosis and Checking
IC: Internal Combustion
LANA: Lanthanum Nickel Aluminum alloy
Nox: Nitrogen Oxides

The Development of Gas (CNG, LPG and H₂) Engines for Buses and Trucks and their Emission and Cycle Variability Characteristics

Stanislav Beroun
Techn. Univ. Liberec – Czech Republic

Jorge Martins
Univ. Minho – Portugal

ABSTRACT

Few diesel vehicles engines from Czech production have been converted into gas (LPG, CNG) fuelled SI engines. The higher power engines ($p_{eMAX} > 1$ MPa) used the concept $\lambda \gg 1$ in turbocharging-intercooler and with oxidation catalysts mode with total (max) efficiency of about 38%. For lower power engines ($p_{eMAX} < 1$ MPa) the more suitable concept was using $\lambda = 1$ in naturally aspirated (atmospheric) mode with 3-way catalysts. Most of the work included the development of the engines to work on Natural Gas (CNG) and LPG for buses and medium trucks. Some laboratory tests were done also with Hydrogen and Natural Gas on the laboratory single-cylinder engine.

The measurements consists of standard emissions (CO, HC, NOx, CO_2, PM) but Polycycle Aromatic Hydrocarbons (PAH and their carcinogenic derivatives PAH_{carc}) were also investigated. Combustion stability was assessed by means of Coefficient of MIP (indicated mean effective pressure) Variation ($VARp_i$) for lean and very lean gas fuelled engines. The effect of Rate of Heat Release on $VARp_i$ was also investigated.

Finally, the development of the combustion chamber design for a bus gas engine during a period of 7 years is shown and discussed, showing the improvements in various areas such as total efficiency, power and emissions.

INTRODUCTION

The conversion of diesel engines to gas fuelled SI engines has been performed with the intention of reducing harmfull emissions. To do it, it is necessary to change the basic working cycle from compression ignition (CI) to spark ignition (SI), including mixture formation, ignition system, control system and exhaust after-treatment.

EMISSIONS

The gas fuelled SI engine (with the correct control and adjustment) has, in comparison with diesel engine, the following advantages:

- much lower particular matter (PM),
- lower gaseous pollutants (NO$_x$, HC, CO),
- lower operating noise.

Gas fuelled engines are very important for urban transport, municipal good transport and other specificities such as garbage collection. Small cars and small transport vans could also benefit from the low pollution nature (and also low cost) of such technology.

At present the main interest for gaseous fuels lies with LPG and NG (in liquid or gas form – LNG or CNG), with a large number of small vehicles (cars) running in Europe, mainly in Holland [1,2], Italy and France. Throughout Europe (with the exception of Spain) the number of such cars are on the rise, with an increasing network of fuel pumps selling gaseous fuels (mainly LPG).

These gaseous fuels have different physical and chemical properties from the more usual liquid fuels: they have a simpler molecule structure which gives them a good chance for lower production of the exhaust pollutants, namely particulate and polycyclic aromatic hydrocarbon (PAH) emissions. For using as motor fuels, this type of fuels have advantages such as a very good ability for easy mixture formation, both outside and inside the engine cylinder, a wide range of air-fuel ratios (leading to lean or extremely lean mixtures), high heating value and antiknock resistance.

These properties enable this SI engine to work with relative high compression ratios and therefore a higher total efficiency (but the efficiency SI engines will not be grater than a CI diesel). The lower mass content of carbon of the gas fuels further reduces CO_2 emissions. Other advantages of the use of such gaseous fuels is the reducing emissions during vehicle refuelling and operation and lower fire risk of the vehicle.

The controlled exhaust emissions of gas fuelled engines are also very low (when compared against emission from more conventional liquid fuels) at low and very low ambient temperature (during cold start or low atmospheric temperature) and their dependence on temperature (engine and ambient) is very small.

COMBUSTION VARIABILITY

Combustion variability of Internal Combustion Engines is a result of complicated thermochemical and thermodynamic processes in engine cylinder along with mixture formation and burning processes. The major influence on variability is the velocity of the oxidising reaction at the beginning of combustion [5]. The value of variability of the diesel engine is low, while engine variability for spark-ignition is relative high over most of engine operating parameters.

The low variability operating cycle of the diesel engine and its high stability is shown by a statistical function and is a result of the great number of ignition focuses and expansion in complete burned mixture (the number of focus of ignition is in the region of 10^3 - 10^4 and ignition energy is about 200 J). On the other hand, spark-ignition engine has just one ignition focus from where the flame proceeds randomly (ignition energy is about 10-200 mJ), leading to large differences (between successive cycles) in ignition focus formation and also in velocity of flame, resulting in the high variability of such engines.

Combustion variability is an old problem of SI engines. Comparing successive cycles, there are large differences in the values of maximum cylinder pressure and position (degrees after TDC) of maximum pressure. Combustion variability of SI engines has a negative influence on specific fuel consumption and exhaust emissions. The variability can be expressed by different parameters, but its most significant indicator is the variability of the indicated mean effective pressure, where the variabilities of other parameters of the cycle are included.

During the development of the gas fuelled engines great attention must be given to the combustion stability and combustion variability, and this is more important when dealing with extremely lean mixture. The combustion variability can be expressed by the mean indicated pressure variability (VARpi). To calculate the variability it is necessary to analyse the indicated cycle. This is done by accessing values of in-cylinder pressure (in sufficiently large numbers, i.e. more than 150 cycles) with basic statistical tools. Variability criterion is the ratio of the standard deviation to the mean value of the evaluated parameter, so the variability of indicated mean effective pressure (VARpi) can be expressed as:

$$W_p = \frac{\overline{\sigma_p}}{\overline{p_i}}$$

$\overline{p_i}$ ... middle value of indicated mean effective pressure of working cycles at sample,

$\overline{\sigma_p}$ ... standard deviation of indicated mean effective pressure at sample.

At first sight the combustion variability is influenced by changes of maximum cylinder pressure: for each operating regime the $VARp_{MAX}$ can also be used as standard for combustion variability. Modern spark-ignition engines were successfully developed for all categories as a result of combustion process optimisation. This was achieved by intensive research work and led to a sufficiently high engine efficiency and low exhaust emissions. The combustion variability is an important instrument for property identification for the main stage of the working cycle and a tool for assessing objectives during engine development.

Variability of the indicated mean effective pressure proved to be linked to the emission of unburned hydrocarbons. Very lean mixtures tend to increase the variability and this parameter is significantly affected by the design of the combustion chamber [3,6], through the increase in turbulence.

Experimental Apparatus for VARpi measurement

To assess the variability it is necessary to measure very precisely the instant pressure inside the combustion chamber and the location (crankshaft position) of the measurement. The measurements need reliable sensors and dedicated software for data analysis. A schematic view of the measuring rig and the attached calculating system can be seen in Fig.1. To evaluate a value for VARpi it is necessary to measure, at the least, 150 cycles, with pressure taken at 1 degree of crankangle interval.

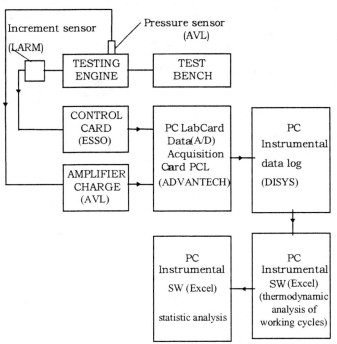

Figure 1: Schematic view of measuring and calculating system for research work on the piston engines (Research Laboratory on Technical University of Liberec)

Data from Research Laboratory shows that diesel engines achieve very low combustion variabilities: for all regimes VARpi $\leq$ 0,8% and VARp$_{MAX}$ $\approx$

0,3%. The modern SI gasoline (MPI) engines have VARpi $\approx$ 2% and VARp$_{MAX}$ $\approx$ (8-9)%.

THE GAS FUELLED SI ENGINES - CONCEPT SOLUTION

Gas fuelled SI engines with low emissions can be produced from two different concepts:

- stoichiometric mixture combustion ($\lambda=1$) with three-way-catalyst in exhaust system.

- very lean mixture combustion ($\lambda \gg 1$), usually with two-way (oxidising) catalyst in exhaust system and turbocharged, to achieve sufficient performance.

Table1 shows the selected characteristic parameters of gas fuelled SI engines for buses and medium trucks from Czech production. All of these gas fuelled engines were authority tested EEC 49 (13 fixed operating regimes), certificating their power and emissions parameters.

Table1 – Data for developed engines

Engine	1-LPG $\lambda \gg 1$	2-CNG $\lambda \gg 1$	3-CNG $\lambda \gg 1$	4-CNG $\lambda=1$	5-CNG $\lambda=1$	EURO II / III
V [dm^3]	11,95	11,95	11,95	3,59	3,59	-
ε	10,8	10,4	11,5	10,5	10,5	-
power [kW]	180	175	210	58	65	-
p_{eMAX} [MPa]	1,11	1,00	1,21	0,72	0,88	-
PM	0,03	0,02	-	-	0,01	0,15/0,1
NO$_X$	3,72	5,15	3,45	0,51	0,22	7,0/5,0
HC	0,26	0,95	0,94	0,16	0,11	1,1/0,7
CO	0,55	0,28	0,23	0,65	0,49	4,0/2,5
CO$_2$	780	640	625	800	770	-

Note: types 1-LPG, 2-CNG and 3-CNG turbocharging with intercooler
type 4-CNG natural aspirated
type 5-CNG turbocharging without intercooler
exhaust emissions in [g/kWh] (according to ECE 49)

The gas fuelled engines are produced in small numbers and are used at Most and Litvinov (North Czech region) where 85 city buses with gas fuelled engines type 1-LPG are run. At Havirov (North Moravian region) another 35 citybuses of type 2-CNG (lower power alternative without intercooler) are used.

The gas fuelled SI engines for extremely lean combustion are provided with a control system for mixture strength which is a function of load. This solution ensures very good operating properties (acceleration, fuel consumption) and very low exhaust emissions. The schematic layout of the gas fuelled SI engines types 1-LPG, 2-CNG and 3-CNG is shown in Fig.2 and the characteristic of mixture strength control is shown in Figure 3.

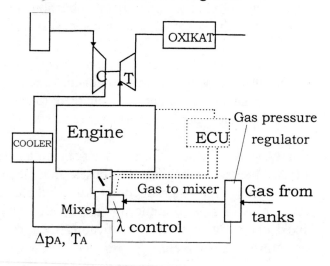

Figure 2: Schematic view of gas fuelled SI engine

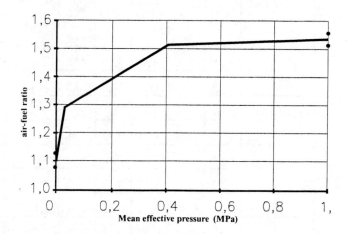

Figure 3: Mixture strength as a function of engine load working cycle.

For the optimisation of the combustion process of these engines a few procedures were applied for analysing the important parameters affecting the variability of the working cycle.

It was confirmed that the indicated mean effective pressure variability reduction is a way to increase engine efficiency, and also to reduce exhaust emissions, specially unburned hydrocarbons. It was also confirmed that the very lean mixture combustion is related to variability increase.

A significant effect of design of the combustion chamber on parameters affecting variability of working cycles was found. The influence of combustion chamber design (turbulence) on combustion process is shown in Figure 4: the considerable reduction of unburned hydrocarbons was achieved by turbulent specific intensity (squish swirl) in combustion chamber. Pressure measurements shown that VARpi was reduced.

Figure 5 shows the traces of in cylinder pressure from a LPG engine (VARpi = 2% with $p_{i\ min}$ = 0,883 MPa, $p_{i\ med}$ = 0,951 MPa, $p_{i\ max}$ =0,996 MPa, VARp$_{max}$ = 8,3% with p_{min} = 3,105 MPa, p_{med} = 4,277 MPa, p_{max} = 5,183 MPa).

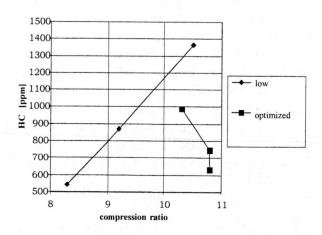

Figure 4: Optimisation of compression ratio and combustion chamber form (type 1-LPG, n = 1300 rpm, p_e = 1,1 Mpa, before oxicat)

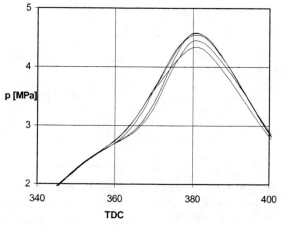

Figure 5: In cylinder pressure (5 following cycles) of a gas fuelled engine at n = 1300 rpm, p_e = 0,814 MPa, λ = 1,45

This type of detailed measurement on gas fuelled SI engines and analysis of the indicator diagram shown that the combustion chamber design and turbulent specific energy (intensity squish swirl) at the end of the compression stroke have a great influence on the variability of the working cycle. This knowledge led the working team to make design changes of the combustion chamber [6]. The piston head of a conventional diesel engine was redesigned when the engine was converted to LPG operation. Different piston head designs were tested, as shown in Figure 6. The different designs were used in city buses in the period from 1992 to 1999 (design A:1992-1994; B: 1995-1997; C: 1998-1999).

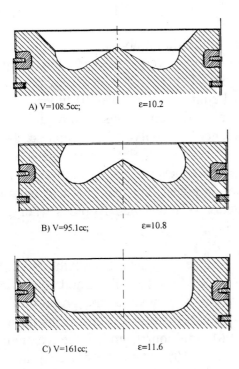

A) V=108.5cc;   ε=10.2

B) V=95.1cc;   ε=10.8

C) V=161cc;   ε=11.6

Figure 6: The combustion chamber design changes (CNG)

The first head (design A) had a combustion chamber quite open, with low turbulent energy. Subsequent designs had an increase in turbulence intensity in the combustion chamber, allowing higher compression ratios.

For comparison of the total turbulent energy at the combustion chamber a simplified calculating model was made.The results show the effect of design changes of the combustion chamber on turbulent intensity (Figure 7, 8 and 9). In these figures the engines were run at 1400 rpm, full load, giving a maximum of specific turbulent energy of 0.62 m^2/s^2

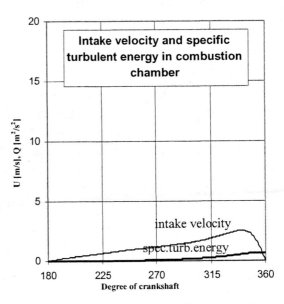

Figure 7: Turbulent properties of comb. chamber type A

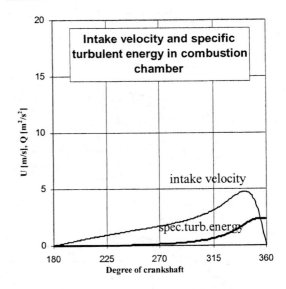

Figure 8: Turbulent properties of comb. chamber type B

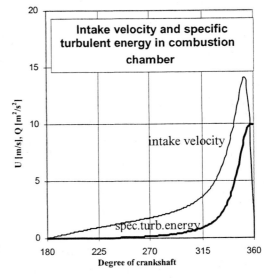

Figure 9: Turbulent properties of comb. chamber type C

239

(type A), 2.35 m²/s² (type B) and 9.96 m²/s² (type C). At 2000 rpm the values for specific turbulent energy were of 1.26 m²/s² (type A), 4.80 m²/s² (type B) and 20.2 m²/s² (type C).

The gradual development of the combustion chamber design contributed to the improvement of power, economy and emission parameters of the bus gas fuelled engines. That is evident from Figure 10, which shows the plot of total efficiency of the engines on engine load for two different speeds. The last developed engine (type C) worked at idle with the almost stoichiometric mixture ($\lambda \cong 1.02 - 1.03$) and at full load with very lean mixture:

$$n = 1400 \ rpm, \quad p_e = 1.1 \ MPa \quad , \text{ mixture } \lambda \cong 1.45$$
$$n = 2000 \ rpm, p_e = 0.88 \ MPa \quad , \text{ mixture } \lambda \cong 1.49$$

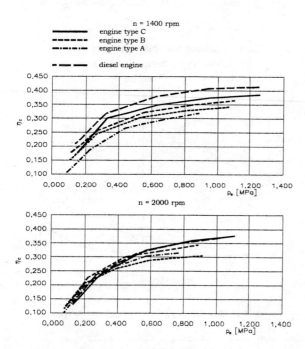

Figure 10: Total efficiency vs. engine load for different engine designs

At optimal adjustment and with oxicat KEMIRA that gas engine shown very good emissions and economy. The results of emissions test according to EEC 49 are:

NOₓ	4.87 g/kWh
HC	0.18 g/kWh (THC)
CO	0.03 g/kWh

NO_x 4.87 g/kWh
HC   0.18 g/kWh (THC)
CO   0.03 g/kWh

The specific emission CO_2 at this test was 601.5 g/kWh, giving a specific fuel consumption of = 28.7g CNG/kWh at test EEC 49.

At conditions of the test EEC 49, this engine gave a total efficiency of 33.5%. At n = 1400 rpm and p_e = 1.1 MPa the total efficiency was η = 37.1% and for nominal output (n = 2000 rpm and p_e = 0.88 MPa) the total efficiency was η = 35.4%.

The selected gas fuelled engines from Table1 (types 1-LPG, 2-CNG and 5-CNG) were tested for lawful emissions and for other noxious elements in the exhaust. Attention was paid to the identification of various organic matter groups at the exhaust gases. The same research was done on the original diesel engines with liquid fuel (1/2-DE, 5-DE). A selection of the results of this research work can be seen in Figs.11, 12 and 13 [4].

The results definitely show that the gas fuelled SI engines are a very ecological drive for the motor vehicles. The gas fuelled SI engines have, against diesel engines, more than 10 times lower particular matter (PM) emissions. More important is the difference in polycyclic aromatic hydrocarbons (PAH) and their carcinogenic derivatives (PAHcarc) between gas fuelled SI engines and diesel engines: the gas fuelled engines have a reduction of more than 10 times when compared to diesel engines [2,4].

The PAH emissions exist at the exhaust gases in both free (gaseous) form and connected to PM: the research shows that the proportion of PAH on PM depends on total PM emissions from the engine. The PM emissions from gas fuelled engines contain only about 10% of total PAH emissions; PM from diesel engines contains as much as 50% of total PAH emissions. Considering that PM emissions represent a great health risk (breathing), the conversion of diesel engines to gas fuelled SI engines represent a very important ecological action [2,4].

In order to improve the gas fuelled engines, research should be focused on the technical solutions for:
- fuels systems, method of mixture formation,
- internal aerodynamics in engine cylinder (optimisation of turbulent combustion chamber) as an instrument for combustion variability minimisation,

- ignition systems (namely for extremely lean mixture combustion),
- exhaust pollutant reduction and efficiency increasing,

exhaust catalytic systems: efficiency increasing (namely for NOx), reliability and life extension,

- control systems (power, mixture formation, idle, pressure boost, high working safety and high stability adjustment.

HYDROGEN FUELLED ENGINE

At our research laboratory an experimental single cylinder engine was modified to work on a hydrogen-air mixture (Figure 14). The hydrogen was injected directly into the cylinder head, either during compression stroke or during intake stroke. The injection system allowed also the use of CNG, so a comparison between the two fuels could be done.

Figure 15 shows a comparison of the engine performance (variability criterion) when fuelled by CNG and hydrogen [5]. The figure also includes VARpi values obtained in modern gasoline and diesel engines (year 1998). Hydrogen mixture can be easily burned in a SI engine with an extreme lean mixture (point D : λ ≈ 4; VARpi is in this case very high, at ≈25%). Figure 15 shows different characteristics between hydrogen and CNG combustion. For the same conditions (compression ratio, combustion chamber design, mixture formation method and mixture strength), the hydrogen combustion variability is lower than CNG combustion variability. For very lean mixtures, the hydrogen fuelled engine shows a lower increase in variability, whereas the CNG fuelled engine has a large increment of VARpi.

Indicated mean effective pressure variability VARpi depends on combustion velocity in engine cylinder. Points A and B shows approximately the same value for VARpi (≈2,4%). Conditions in points A and B are the same in respect to mixture formation method. Although points A and B can be distinguished from the way fuel is injected and from the mixture strength. A thermodynamic analyses shows that the burning velocity in both cases is almost the same, at similar circumstances, as can be seen from Figure 16. This figure shows the rate of heat release at points A and B, plotted against the

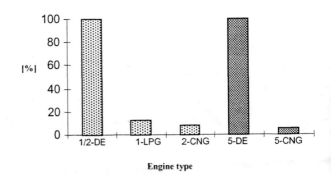

Relative emissions of PM

Figure 11: Diesel engines 1/2-DE and 5-DE fulfil the emissions regulation EURO II. Emissions [g/kWh] measured [4] at conditions specified by procedure ECE 49, engines from Table1.

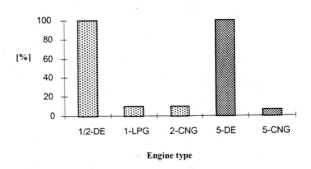

Relative emissions of PAH

Figure 12: PAH emissions contains 3 - 7 nuclear polyaromatic hydrocarbons. The samples from exhaust gases were caught by special pack, filters, adsorption substances at glass tubes and condensed matter (after sample freezing) for chemical analysis at special laboratory of organic chemistry [4]. Emissions [mg/kWh] measured at conditions specified by procedure ECE 49, engines from Table1.

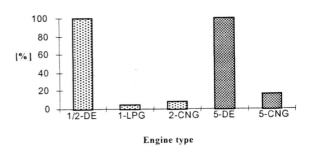

Relative emissions of PAHcarc

Figure 13: PAHcarc contains emissions of benzo(a)anthracen, benzo(k)fluoranten, benzo(a)pyren, dibenzo(a,c)anthracen and indeno(1,2,3-cd) pyren. PAHcarc emissions [μg/kWh] measured at conditions specified by procedure ECE 49 [2] engines from Table1.

stage of burning (0.5%, 5% and 50% of the total amount of heat in the supplied fuel). Both lines are almost coincident, showing similar burning velocities.

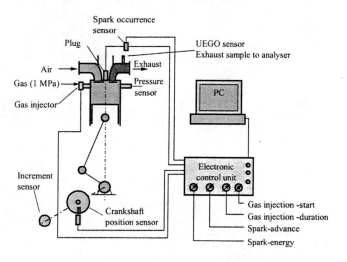

Figure 14: The research single cylinder engine with accessories and with electronic control unit (schematic drawing of test facility)

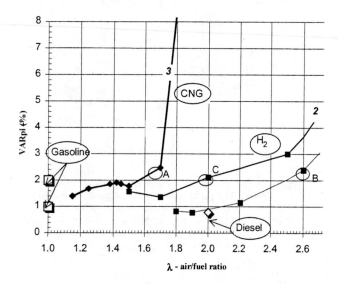

Figure 15: VAR$_{pi}$ on mixture richness for hydrogen mixtures combustion (1- injection during intake, 2- injection during compression) and for CNG combustion (3- injection during intake). Research single-cylinder engine (see Figure 14), compression ratio ε = 10, spark-advance α = 5°BTDC) [5].

Hydrogen fuelled internal combustion engines produces almost only one exhaust pollutant, NO$_x$, which is dependant on mixture strength (as for the other fuels). Figure 17 shows this dependence for hydrogen fuel (injection during compression and intake) and CNG for the single cylinder direct injection engine.

The low NO$_x$ emissions from hydrogen mixtures combustion can be attained with air excess ratio of $\lambda \approx 2,2$-$2,5$. Mixture formation by hydrogen

injection during compression stroke induces very low NO$_x$ production already with $\lambda \approx 2,0$ (curve 2). This is a consequence of variability increasing (see Figure 15), with which the indicated efficiency reduction is coupled. Low-emission internal combustion engines for extremely lean hydrogen mixtures combustion should be turbocharged, in order to obtain sufficiently high performance, and engine control can be partially achieved by boost pressure (turbocharging) control.

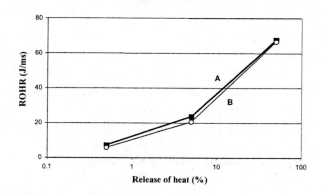

Figure 16: Rate of heat release for CNG and Hydrogen at various stages of combustion[5] - for the same operating conditions like on Figure 15.

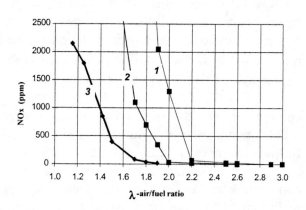

Figure 17: NO$_x$ emissions with hydrogen (1,2) and CNG (3) - for the same operating conditions like on Figure 15 [5].

CONCLUSIONS

The gas fuelled SI engine is very attractive for municipal and residential driving. Although its variability (when LPG or CNG are used) is higher than the diesel engine where it is based, the regulated and unregulated emissions from gas fuelled engines are much lower than those. Hydrogen proved to be a very good fuel for use with very lean mixtures mode.

ACKNOWLEDGEMENT

The publication was created with the connection to the project Czech Ministry of the Education No. LB00B073.

REFERENCES

1. HOLLEMANS, B., VOOGD, A., WEIDE, J.: Europan Technologies to Reach ULEV Emissions for Gaseous Fuels in Combination with Fuel Consumption Reduction. Metha-Motion-European Conference. London, 1995.

2. HOLLEMANS, B.: LPG a Clean and Efficient Motor Fuel - Regulated and Non Regulated Emissions of a Commercially Attrective LPG Vehicle. The Commomnwealth Conference and Events Centre. London, 1996.

3. CHMELA, F.,G., KAPUS, P., E., CARTELLIERI, W., P.: Alternative Fuels for Automotive Engines. CATEC Symposium, Beijing, 1997.

4. BEROUN, S., BLAŽEK, J., POSPÍŠIL, M.: Emission of solid particulates and unburned hydrocarbons by diesel and gas engines. MOTORSYMPO 99, Brno, 1999.

5. SCHOLZ, C.: Combustion variability of spark ignition hydrogen burning engines. MOTORSYMPO 99, Brno, 1999.

6. BEROUN, S.: Importance of the Combustion Chamber Form for the Quality of the Gas Engines. VIIth International Scientific Conference MOTAUTO, Sofia, 2000.

Hybrid and Conventional Hydrogen Engine Vehicles that Meet EZEV Emissions

Salvador M. Aceves and J. Ray Smith
Lawrence Livermore National Lab

ABSTRACT

Hydrogen-fueled, spark-ignited, homogeneous-charge engines offer an alternative for providing Equivalent Zero Emission Vehicle (EZEV) levels, along with a range and performance comparable to today's automobiles. Hydrogen in a spark-ignited engine can be burned at very low equivalence ratios, so that NO_x emissions can be reduced to less than 10 ppm without a catalytic converter or EGR. HC and CO emissions may result from oxidation of engine oil, but by proper design are negligible (a few ppm). Lean operation also results in increased indicated efficiency due to the thermodynamic properties of the gaseous mixture contained in the cylinder and due to reduced heat transfer. The high effective octane number of hydrogen allows the use of a high compression ratio, further increasing engine efficiency.

In this paper, a time-dependent engine model is used for predicting hydrogen engine efficiency and emissions. The model uses basic thermodynamic equations for the compression and expansion processes, along with an empirical correlation for heat transfer, to predict engine indicated efficiency. A friction correlation and a supercharger/turbocharger model are then used to calculate brake thermal efficiency. The model is validated with many experimental points obtained in a recent evaluation of a hydrogen research engine.

The validated engine model is then used to calculate fuel economy and emissions for three hydrogen-fueled vehicles: a conventional, a parallel hybrid, and a series hybrid. All vehicles use liquid hydrogen as a fuel. The hybrid vehicles use a flywheel for energy storage. Comparable ultra capacitor or battery energy storage performance would give similar results. This paper analyzes the engine and flywheel sizing requirements for obtaining a desired level of performance. The results indicate that hydrogen lean-burn spark-ignited engines can provide a high fuel economy and Equivalent Zero Emission Vehicle (EZEV) levels in the three vehicle configurations being analyzed.

INTRODUCTION

Hydrogen has very special properties, including a very high laminar flame speed, a high effective octane number, and no toxicity or ozone-forming potential [1]. Homogeneous-charge spark-ignited piston engines can be designed to take advantage of these characteristics. The high laminar flame speed allows the use of very low equivalence ratios (as low as 0.2), reducing NO_x emissions to near-zero levels without requiring a catalytic converter, that may deteriorate with time. The use of low equivalence ratios also increases the indicated efficiency, and reduces the need for throttled operation [2]. The engine can have a high compression ratio, due to the high octane number of hydrogen.

This paper analyzes the applicability of hydrogen homogeneous-charge spark-ignited piston engines to Equivalent Zero Emission Vehicles (EZEV).[1] The analysis uses an engine model that is calibrated to match the data obtained in a recent experiment [4], and a vehicle simulation code that can be applied to calculating fuel economy and emissions [5]. The engine model is used to generate optimized engine performance maps for supercharged operation, with constraints of limited NO_x emissions, limited manifold pressure and maximum mean piston speed.

The engine maps are applied for predicting fuel economy and emissions for a conventional vehicle, a parallel hybrid vehicle, and a series hybrid vehicle. All these vehicles use liquid hydrogen as a fuel, and hybrid vehicles use a flywheel for energy storage. Although liquid hydrogen is not a common fuel, its safe

---

[1] Equivalent Zero Emission Vehicles are defined here as those that generate less emissions when operating inside the Los Angeles Basin than the power plant emissions generated as a result of electric car operation. These emission levels are being considered for approval by CARB [3]. Proposed EZEV levels are: 0.0025 g/km (0.004 g/mi) NMHC; 0.11 g/km (0.17 g/mi) CO; and 0.012 g/km (0.02 g/mi) NO_x. These values are equal to one tenth of CARB ULEV standards.

use has been demonstrated by BMW for a number of years [6]. The parallel hybrid operates almost identically as a conventional car, with a flywheel for regenerative braking and for complementing the power of the engine during sudden accelerations. The series hybrid vehicle engine operates in an on-off mode to keep the flywheel charged. When the engine is off, the flywheel provides all the energy for transportation and accessories.

ENGINE MODEL

The engine model uses first principles and correlations to predict piston engine efficiency and power output. The engine model is a lumped (zero-dimensional), time-dependent model which solves the basic differential equations for the compression and power strokes. An empirical equation [7] is used for calculating heat release. The heat release equation has three free parameters which determine the shape and duration of the heat release function as a function of crank angle. These three parameters are determined for each experimental run by using an optimizer [8]. The optimizer finds the combination of the three parameters that minimizes the differences between the experimental pressure trace and the pressure trace calculated by the model. The results have been very satisfactory. The relative errors in matching the pressure traces at MBT timing have been of the order of 0.5%, with a maximum error of 1% over all engine speeds, equivalence ratios and manifold pressures.

The engine model uses Woschni's correlation [9] to estimate engine heat transfer. It was found during the analysis that the heat transfer correlation underpredicts heat transfer losses. Woschni's correlation includes two constants, C_1 and C_2, which determine the effect of fluid flow on heat transfer. To obtain a better match with the experimental data, the original values of these two constants were multiplied by 1.8. The increased heat transfer coefficient reduces the predicted engine efficiency by about 1% with respect to the engine model with the unmodified Woschni's correlation. It is noted that Woschni's correlation has been applied in the past to diesel and gasoline engines, so some adjustment to the coefficient is to be expected as a result of using a different fuel. It is encouraging that the adjustments to those coefficients, which are not directly proportional to heat transfer, have a small effect on engine efficiency.

The engine model includes a friction model and a supercharger/turbocharger model to predict brake thermal efficiency. The friction model uses a detailed correlation [10]. Supercharger and turbocharger performance are calculated by using a thermodynamic model and assuming a constant (0.7) isentropic

efficiency for both the turbine and the compressor. Selection of a supercharger to optimally increase the power as required to meet the power demands is outside the current experience of the authors. However, a detailed supercharger map could be incorporated into the model if further refinement is desired. A water-cooled intercooler is assumed with a thermal effectiveness of 0.7. Based on reported data for typical engines [2], volumetric efficiency is assumed to vary linearly from 85% at low engine speeds, to a maximum of 95% at 4000 rpm, down to 90% at 5000 rpm.

The engine model is validated by comparing the calculated results with the experimental results obtained in a recent engine evaluation [4]. The engine used in the experimental evaluation is an Onan engine which was modified by incorporating a head containing two spark plugs, along with the original two valves. The intake valve was modified with a 1.5 mm high, 18 degree shroud to add some swirl which resulted in improved efficiency. The combustion chamber is a simple right circular cylinder with no squish and a flat top piston. This geometry has been shown to be the most efficient shape for reducing heat transfer losses in lean-burn engines [11]. Engine characteristics are listed in Table 1, along with the range of conditions used in the experiment.

Table 1. Modified Onan engine characteristics and experimental conditions.

Bore, mm	82.55
Stroke, mm	92.08
Displacement, cm³	493.0
Geometric compression ratio	14.0
Experimental range for equivalence ratio	0.2-0.5
Experimental range for engine speed, rpm	1200-1800
Experimental range for volumetric efficiency, %	90-215

Figure 1 shows a comparison between experimental and calculated indicated efficiencies, as a function of equivalence ratio, for all the experimental points at MBT timing. Engine speeds are indicated with different symbols. A 0.39 equivalence ratio was selected for most supercharged runs. Figure 1 shows gross indicated efficiency, defined as including only work done on the piston during the compression and expansion strokes [2]. The experimental engine was supercharged by directly supplying the fuel and air from pressurized sources. Pumping work is accounted for as a part of the engine friction model.

Figure 1 shows that the model predicts absolute values as well as trends with good accuracy for engine indicated efficiency, over the whole range of operating conditions, with the maximum error of the order of 1%.

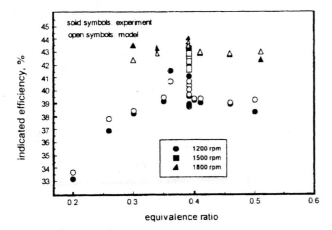

Figure 1. Indicated efficiency for the Onan engine as a function of equivalence ratio. The figure includes both the experimental results and the model predictions.

No validation is done for brake thermal efficiency, because the Onan engine used in the experiment has substantially more friction per cylinder than a current automotive engine, for which the friction correlation applies.

Figure 1 shows that indicated efficiency increases as a function of engine speed, as a consequence of reduced heat transfer losses. The variation of indicated efficiency with equivalence ratio is best observed for 1200 rpm operation, for which the greatest fuel/air range was used. Indicated efficiency reaches a maximum near a 0.40 equivalence ratio. Increasing the equivalence ratio from this point results in a decreased indicated efficiency, due to increased heat transfer and a decreased specific heat ratio ($\gamma = c_p/c_v$) for the gas products in the cylinder [2]. Decreasing the equivalence ratio from the optimum point increases the timing losses, due to slower heat release, thereby reducing the indicated efficiency. Supercharged operation results in small indicated efficiency gains due to slightly lower heat transfer losses per unit mass of fuel at the higher densities. Supercharged operation has a larger effect on brake thermal efficiency by increasing the output work relative to the frictional work.

The engine model also includes a correlation for calculation of NO_x emissions. Experimental results (Figure 2) have indicated that emissions of NO_x are mainly a function of engine equivalence ratio, being very insensitive to engine speed or intake pressure. The correlation of NO_x as a function of equivalence ratio predicts emissions accurately. The model uses a correction for supercharged and turbocharged operation, to take into account the higher intake temperature resulting from the compression process

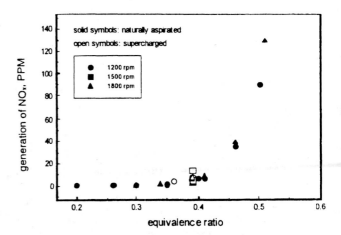

Figure 2. Emission of NO_x in parts per million for the Onan engine as a function of equivalence ratio, for all the experimental points at MBT timing obtained in the analysis. Engine speeds and supercharged operation are indicated with different symbols.

and the less than perfect effectiveness of the intercooler.

The engine model is then applied to predicting the engine performance that results if a 4-cylinder (1.97 liter) and a 3-cylinder (1.48 liter) hydrogen engines are built with the same cylinder characteristics of the Onan engine. The applicability of the 4-cylinder and 3-cylinder engines to conventional and hybrid vehicles is discussed later in this paper. The geometry and dimensions of the engine cylinders are not changed in the analysis, because small changes in geometry may result in significant changes in efficiency. It is expected, however, that larger engine cylinders will improve engine efficiency.

Using the engine model for predicting vehicle performance requires extrapolating from the engine speeds used in the experiment (1200-1800 rpm) to engine speeds that are required for vehicle operation. A maximum engine speed of 5000 rpm is assumed. Extrapolation is done by assuming that the heat release correlations developed from the experimental data apply throughout the desired range of engine speeds, along with the previously cited volumetric efficiency assumptions. It is recognized that this extrapolation may result in errors. Errors are due to turbulence variations with engine speed which influence heat release rate, thus changing both timing losses and heat transfer losses. However, Figure 1 shows that the model predicts the efficiency trends with good accuracy for the range in which experimental data exist, and it is considered that the model can do a reasonable job at predicting efficiency for high and low engine speeds. In addition to this, engines in conventional and hybrid vehicles are most often operated at low to moderate speeds, which is the range for which the model has been validated. For the conventional and parallel hybrid vehicles analyzed in a later section of this paper the mean engine speed is 1500 rpm for the urban cycle and 2700 rpm for the highway cycle (see Figure 8). Maximum engine speed for the driving cycles is 3300 rpm. The series hybrid vehicle is set to operate at a constant 2400 rpm during the driving cycles.

Engine brake thermal efficiency is required for applying the engine code to vehicle calculations, and brake thermal efficiency is calculated with a friction and a supercharger/turbocharger model that have not been validated for this particular application. While the model cannot replace experimental runs, it is considered that the brake thermal efficiency calculated with the model gives a good idea of the performance that can be obtained with such an engine. Engine emissions are very insensitive to engine operating conditions other than maximum temperature within the cylinder, and it is therefore expected that the engine NO_x model can provide accurate predictions for emissions levels throughout the operating range.

The engine model is applied to generate engine emissions and performance maps, necessary for predicting vehicle performance. A gasoline engine has only one degree of freedom for controlling the output torque at any given speed: the inlet manifold pressure. This is due to the use of three-way catalysts that require near-stoichiometric operation for high conversion efficiency. A hydrogen engine has two degrees of freedom, because equivalence ratio can also be varied. Generating an engine map therefore requires determining a control strategy that specifies how to adjust these two parameters to obtain the desired torque for any given engine speed. In this analysis, an optimizer [8] is used to determine the combination of equivalence ratio and inlet manifold pressure that satisfies the torque requirement while providing the maximum engine brake thermal efficiency. A constraint is used in the optimization: engine NO_x emissions are less than 10 ppm under all operating conditions. An engine generating 10 ppm of NO_x is well below the EZEV standards (a tenth of ULEV), provided that it is installed in a vehicle with a fuel economy of 40 mpg or higher. While emission levels for a car are set in grams per mile and not in ppm, setting a maximum allowable ppm limit guarantees that the vehicle is intrinsically an EZEV, regardless of how the vehicle is driven.

Lean operation results in low power output, and therefore turbocharged or supercharged operation is required for providing adequate power output for the chosen displacement. Both supercharged and turbocharged operation have been considered for generating the engine performance map. The performance maps for both cases are very similar, with turbocharged operation having a slight efficiency advantage over supercharged operation. Only the results for supercharged operation are shown in this paper. Supercharged operation is preferred to turbocharged operation due to the lag time that may exist in turbocharged operation for the conventional and parallel hybrid vehicles.

Figures 3 and 4 show the predictions for engine efficiency and emissions maps, for the 4-cylinder supercharged engine. Performance maps for the 3-cylinder engines can be obtained to within a good approximation by multiplying the torque scale by 3/4. Not everything scales by a factor of 3/4 when the number of cylinders is decreased (for example, the number of bearings). However, the effect of these changes is small, and the scaled maps give a very good approximation for the 3-cylinder engine maps.

Figure 3 shows lines of constant brake thermal efficiency (in percent) as a function of engine speed and engine torque. The figure also shows a dotted line corresponding to the conditions at which the engine generates 10 ppm of NO_x, and a dashed line that indicates the maximum torque that can be obtained within the upper bounds of equivalence ratio (0.5) and inlet pressure (2 bar) used in the analysis. The 10 ppm NO_x curve is the lower of the two, and therefore sets the limit on the maximum torque and power that can be obtained from the engine. The maximum power approaches 60 kW at 5000 rpm. The contour lines in this and the following figures spread beyond the 10 ppm line, showing the potential power gains obtained by relaxing this restriction. A square in the figure indicates the approximate range of experimental conditions covered.

Figure 3 shows that the engine is predicted to have a broad area of high efficiency, for intermediate speeds and high torques. The efficiency drops for lower speeds due to increased heat transfer losses, and for higher speeds due to increased friction. As expected, the efficiency drops to zero as the load is reduced. However, the drop occurs more slowly than in conventional engines, because the equivalence ratio can be reduced as the load is reduced, resulting in lower throttling losses.

The predicted engine map is similar in shape to engine maps for recent gasoline engines (see, for example, [12]). However, engine maps are most often drawn with efficiency as a function of speed and power, so the maps presented in this paper have to be transformed before a direct comparison is possible. Current gasoline engine maps show that efficiency drops with increasing speed at low torque conditions,

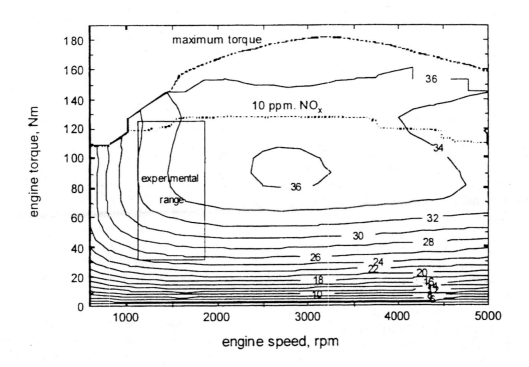

Figure 3. Contour lines of constant brake thermal efficiency (in percent) as a function of engine speed and engine torque. The dotted line corresponds to the conditions at which the engine generates 10 ppm of NO_x; the dashed line indicates the maximum engine torque that can be obtained within the constraints of maximum equivalence ratio (0.5) and inlet pressure (2 bar); and the square indicates the approximate area in which the experimental data were taken.

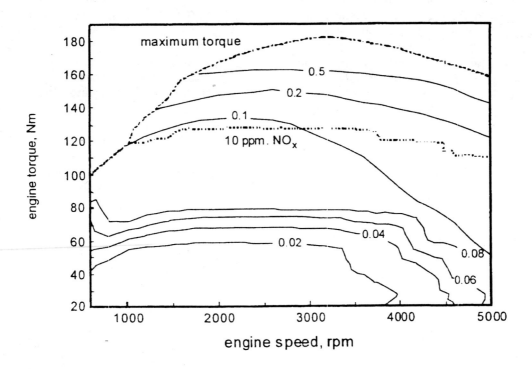

Figure 4. Contour lines of NOx emissions in g/kWh as a function of engine speed and torque. The figure includes a dotted line for the 10 ppm NO_x limit, and a dashed line for the maximum engine torque that can be obtained within the constraints of maximum equivalence ratio (0.5) and inlet pressure (2 bar).

while Figure 3 shows that efficiency is almost independent of speed at these conditions (lower right corner of the map). These differences may be due to lower heat transfer and pumping losses in the lean-burn engine compared to the gasoline engine.

Figure 4 shows contours of NO_x emissions in g/kWh as a function of engine speed and torque. Emissions are less than 0.1 g/kWh for most of the map, with higher emissions obtained at high speed and high torque. The line of 0.1 g/kWh roughly corresponds to the limit of 10 ppm at intermediate speeds. Emission levels are near zero (<0.02 g/kWh) over the low load range, which is the range at which the engine is operated most of the time during city and highway driving in conventional and parallel hybrid cars (see Figure 8). Emissions increase slowly as the torque increases, until the restriction of 10 ppm is approached. When this restriction is approached, the operating conditions in the engine are adjusted so that the 10 ppm line is pushed as high as possible by increasing the manifold pressure without further increases in equivalence ratio, at the cost of some losses in efficiency. This explains the great distance between the 0.08 and the 0.1 g/kWh curves shown in the figure. Emission levels shown in the figure are expected to be valid over the lifetime of the engine, since no catalytic converter, which may deteriorate with time, is used to control emissions.

Intake pressure and equivalence ratio are selected by the optimizer to obtain the optimum brake thermal efficiency at every operating point. At low loads, the equivalence ratio is as low as 0.25 and the intake pressure is 0.5 bar. The equivalence ratio is reduced to keep a relatively high manifold pressure, therefore reducing pumping losses. Full information about the control strategy, including maps of equivalence ratio and intake pressure as a function of engine torque and speed, can be obtained from a previous publication [13].

VEHICLE ANALYSIS

The engine efficiency and NO_x maps presented in the previous section are now used in predicting vehicle fuel economy and emissions for a conventional and parallel and series hybrid vehicles. Vehicle power train components are also selected to meet minimum performance requirements (acceleration, hill climb and range). This is accomplished by incorporating the engine maps into an existing vehicle simulation code [5].

The main characteristics of the three vehicles are listed in Table 2. All the vehicles have good aerodynamics and low rolling friction, and use a cryogenic liquid hydrogen storage. The conventional

and the parallel hybrid cars use a 5-speed transmission, and the series hybrid uses a single-speed transmission. Transmission efficiency is considered independent of driving conditions. All the vehicles are required to meet the following minimum performance specifications: acceleration from 0 to 97 km/h (0-60 mph) in no more than 10 seconds; climbing a slope of at least 6% at a constant 97 km/h (60 mph) speed with a 272 kg payload; and minimum range of 483 km (300 miles).

Both hybrid vehicles use a flywheel for energy storage. A flywheel is selected because it has the potential of providing a high efficiency and power output. A detailed model of a flywheel has been developed. This model includes a performance map of efficiency as a function of state of charge and power output, as well as a correlation of stand-by losses as a function of state of charge.

A brief description of the operating strategies for the parallel and series vehicles is as follows: The parallel hybrid vehicle operates very similarly to a conventional car, except that a flywheel and a traction motor are used for complementing the power of the engine during sudden accelerations, and for regenerative braking. The use of the flywheel for power peaking allows a reduction in the size of the engine, which may therefore operate more efficiently than the engine used in the conventional car at the low-power conditions that constitute most of the urban and highway driving cycles. The engine is sized to provide the required performance during long hill climbs, for which the flywheel cannot provide the required energy. A low-capacity, low-power flywheel is enough for this application. The control strategy used for the flywheel consists of keeping it near 50% state of charge. In this way, the flywheel is always ready to provide energy for a sudden acceleration, and to absorb energy during regenerative braking. It is noted that there are many possible parallel hybrid control strategies. The 50% state of charge for the flywheel is chosen as a simple strategy that accomplishes both engine downsizing and regenerative braking. It is likely that better parallel strategies may exist.

The series hybrid vehicle operates with an engine in an on-off mode, with no mechanical link between the engine and the wheels. An electric motor provides all tractive power. When the engine is running, it drives a generator that supplies electricity to both the electric motor and an energy storage system. A flywheel is used for energy storage. When the storage system is fully charged, the engine is turned off, and the storage system provides all the energy required for traction and accessories. Series hybrid vehicles have high fuel economy because the engine is operated at a high efficiency condition without ever idling. Spark ignition engines can be run unthrottled thus avoiding pumping

Table 2. Main parameters for hydrogen-fueled conventional, parallel hybrid and series hybrid vehicles.

Vehicle parameter	conventional	parallel	series hybrid
frontal area, m^2	2.04	2.04	2.04
aerodynamic drag coefficient	0.24	0.24	0.24
coefficient of rolling friction	0.007	0.007	0.007
transmission efficiency	0.94	0.94	0.95
transmission gears	5	5	1
accessory load, W	1000	1000	1000
engine idling speed, rpm	600	600	-
launch engine RPM, maximum effort acceleration	3600	3600	-
regenerative braking	no	yes	yes
fraction of available energy recovered by regen. braking, %	-	70	70
maximum one-way flywheel efficiency, %	-	96	96
generator type	-	permanent magnet	permanent magnet
motor type	-	AC induction	AC induction
energy storage device	-	flywheel	flywheel
hydrogen storage	liquid	liquid	liquid

Table 3. Results of the analysis for the conventional, parallel hybrid and series hybrid hydrogen-fueled vehicles.

Vehicle parameter	conventional	parallel hybrid	series hybrid
test weight, (empty weight + 136 kg)	1136	1127	1238
number of engine cylinders	4	3	3
engine displacement, liters	1.97	1.479	1.479
hydrogen tank storage capacity, kg	7.2	5.7	4.8
motor maximum short-term torque, Nm	-	48	190
motor maximum speed, rpm	-	11000	11000
flywheel energy storage, kWh	-	0.2	1
flywheel maximum power, kW	-	20	100
average engine efficiency, urban cycle, %	21.4	21.9	36
average engine efficiency, highway cycle, %	22.7	25.7	36
fuel economy[1], urban cycle, km/liter (mpg)	15.2 (35.8)	20.0 (47.0)	24.4(57.4)
fuel economy[1], highway cycle, km/liter (mpg)	22.0 (51.8)	26.5 (62.4)	29.8(70.0)
fuel economy[1], combined cycle, km/liter (mpg)	17.7 (41.6)	22.5 (52.9)	26.6(62.5)
NO_x emissions, urban, 10^{-3}g/km (10^{-3}g/mile)	2.08 (3.35)	2.24 (3.61)	12.4(20.0)
time for 0-97 km/h (0-60 mph), s	10.0	10.0	10.0
max. climbing slope at 97 km/h (60 mph), %	12.8	9.0	6.0
vehicle range, combined cycle, km (miles)	483 (300)	483 (300)	483 (300)

losses. Engines in series hybrids operate most of the time at a low power, high efficiency condition, at constant speed and load. When additional power is required during long hill climbs, the engine can be operated at a high power level, possibly at a lower efficiency. The flywheel provides the power for sudden accelerations, and the engine is sized for providing the required performance for long hill climbs. On-off engine operation results in decreased engine efficiency, due to the repeated engine cold starts. This reduction in efficiency is taken into account in this analysis by introducing an energy penalty (10 Wh) every time the engine is restarted. This is done to account for increased friction due to engine cold operation. Emissions are not expected to increase considerably due to on-off engine operation because no catalytic converter is used to control emissions.

The results of the analysis are listed in Table 3. Empty weight is set to 1000 kg for the conventional car, which is used as base case. This low weight is chosen because research programs such as PNGV (Partnership for a New Generation of Vehicles) are likely to result in weight reductions for future cars.

Weight for the other two vehicles is calculated by replacing components, and adding or subtracting the weight of the components. A 30% structural penalty is added to the difference in power train weight, to take into account the need for a heavier structure that results from a heavier power train. The parallel hybrid car has almost the same weight as the conventional car. The weight of the added components (flywheel and motor) is compensated by the reductions in weight that result by downsizing the engine, transmission and hydrogen tank. The series hybrid vehicle weighs about 100 kg more than the other two vehicles, primarily due to its need for a flywheel with greater energy storage capacity.

The conventional vehicle requires a 4-cylinder engine to provide the required acceleration performance. This engine is therefore oversized for providing the required hill climbing performance (12.8% instead of 6%). The hybrid vehicles only require a 3-cylinder engine, since the flywheel can be used for providing the power for maximum effort acceleration. Engines for the hybrid vehicles are sized for the hill climb requirement. However, the previously imposed restriction of keeping the cylinder size and geometry constant in the analysis results in an oversized engine for the parallel hybrid. It would be necessary to reduce the size of the individual cylinders to exactly obtain the specified hill climb performance. The mass of hydrogen in the tank (and therefore the mass and volume of the hydrogen tank) is adjusted to meet the desired range. The internal tank volume for the conventional car is about 100 liters (27 gal) [14]. The parallel hybrid only requires a 20 kW, 0.2 kWh flywheel for power augmentation, compared to 100 kW and 1 kWh for the series hybrid.

Fuel economy is listed in the table in gasoline-equivalent units. Fuel economy for the series hybrid is highest because the engine always operates at peak efficiency (36%). The difference in fuel economy between the parallel hybrid and the conventional car is due in part to the higher average engine efficiency that results from downsizing the engine, and in part due to the regenerative braking that can be done in the hybrid.

While all vehicles have a high fuel economy, the fuel economy of the series hybrid vehicle is lower than the 33.5 km/l (79 mpg) predicted for a hydrogen series hybrid vehicle in a previous work by the authors [5]. The results are different because the engine model used in this analysis is based on experimental data for a particular engine (Onan), while the previous work indicates improvements that are likely to be obtained in a future optimized hydrogen engine.

The most desirable feature of the vehicles analyzed in this paper is their low emissions. Emissions are measured only during the urban cycle for EPA certification. Therefore only these values are reported

in Table 3. The conventional car has the lowest emission levels. Emissions for the conventional car are projected to be a factor of 60 lower than the CARB ULEV requirements, and therefore a factor of 6 lower than EZEV. The reason for the emissions to be so low is that the engine is operated most of the time at low torque, generating much less than the 10 ppm maximum allowable NO_x (Figure 8). Emissions out of the parallel hybrid vehicle are slightly higher than the conventional car, because the downsized engine has to operate at a higher load. The series hybrid vehicle operates at peak efficiency, which occurs near the limit of 10 ppm of NO_x. Emissions out of the series hybrid engine are therefore higher than for the other two vehicles, exactly at the EZEV limits. Emissions out of the series hybrid could be reduced by changing the engine operating conditions, at the cost of a slight loss in efficiency, to allow for a margin of error during the certification process.

Two important facts should be emphasized about the emissions generated by these vehicles. First, no catalytic converter is required, and therefore emissions do not increase during the life of the car due to catalytic converter or emissions control system failure. Second, emissions are intrinsically low. No matter how aggressive a driver may drive, emissions are always under 10 ppm NO_x, eliminating high off-cycle emissions that are common in gasoline cars.

Figures 5, 6, 7 and 8 are used to illustrate some of the results. Figure 5 shows flywheel state of charge for the series hybrid and the parallel hybrid vehicles along the

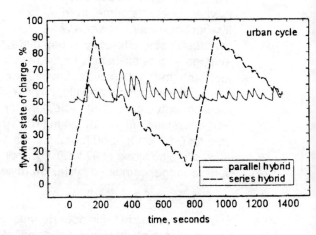

Figure 5. Flywheel state of charge for the series hybrid and the parallel hybrid vehicles along the urban driving cycle. For the series hybrid, the engine runs at constant power until it fully charges the flywheel. For the parallel hybrid, the flywheel is kept at about 50% state of charge, ready to store energy during regenerative braking, or provide energy for a sudden acceleration.

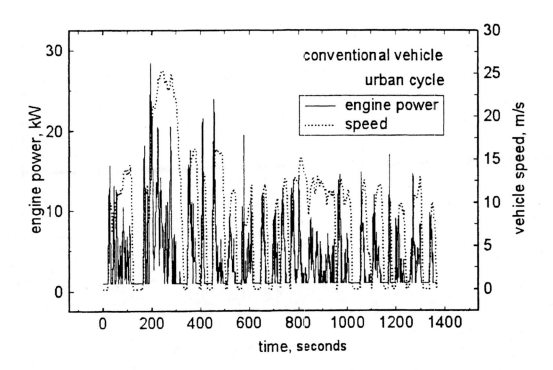

Figure 6. Engine power for the conventional vehicle during the urban cycle. The driving cycle speed is indicated by a dotted line.

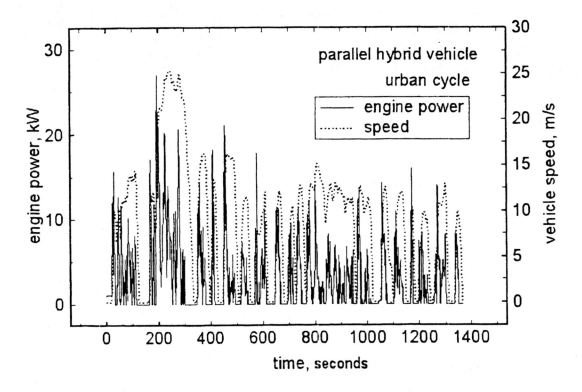

Figure 7. Engine power output for the parallel hybrid vehicle. The driving cycle speed is indicated by a dotted line.

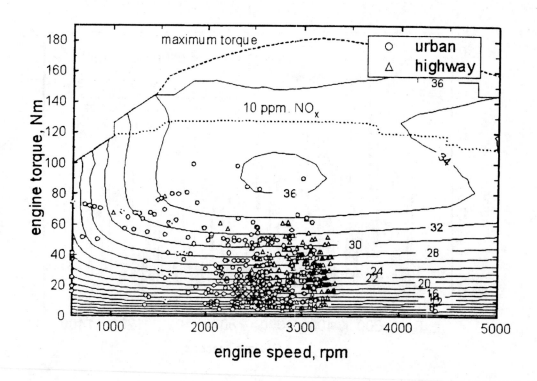

Figure 8. A representative set of engine operating points for the conventional car, for both the urban and highway cycles, superimposed on the engine efficiency map (repeated from Figure 3).

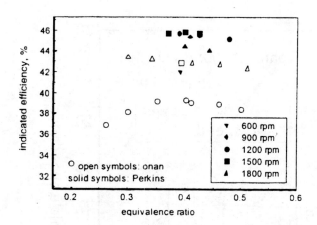

Figure 9. Indicated efficiency as a function of equivalence ratio and engine speed for the Perkins engine (solid symbols) and the Onan engine (open symbols, repeated from Figure 1 for comparison).

urban driving cycle. For the series hybrid, the engine runs at constant power until it fully charges the flywheel, which takes about 180 seconds. The flywheel then provides all the energy requirements, discharging in about 600 seconds. The engine is then turned on

again to repeat the cycle. For the parallel hybrid, the flywheel is kept at about 50% state of charge, ready to store energy during regenerative braking, or provide energy for a sudden acceleration. The flywheel state of charge increases due to regenerative braking, and drops when power is extracted from the flywheel. Flywheel stand-by losses are projected to be of the order of 80 W and also tend to reduce the state of charge.

Figure 6 shows engine power for the conventional vehicle during the urban driving cycle. The driving cycle speed is indicated by a dotted line. Minimum engine power output is 1000 W, due to the need to provide for accessories. Figure 7 shows the nearly identical engine power output for the parallel hybrid vehicle. Power out of the two engines is very similar, except that peak power out of the parallel hybrid is reduced by 1 to 3 kW due to the power provided by the flywheel. Another difference between the output powers is that minimum power out of the parallel hybrid engine is zero, since the flywheel can now provide the accessory loads.

Figure 8 shows a representative set of engine operating points for the conventional car, for both the urban and highway cycles, superimposed on the engine efficiency map (repeated from Figure 3). The points indicate that the engine operates at low torque during most of the time, therefore generating very low emissions. Efficiency could be improved by relaxing the 10 ppm NO_x emission limit, and this limit could be relaxed while still achieving EZEV levels. However, a vehicle that can emit more than 10 ppm of NO_x would not be intrinsically an EZEV, and would generate high NO_x if driven aggressively at high torque.

EXPERIMENTAL HYDROGEN ENGINE EVALUATION UPDATE

An experimental evaluation of a new hydrogen engine (Perkins) has been recently done [4]. The main characteristics of this engine are listed in Table 4. The cylinder is bigger than for the Onan engine (834 cm^3 compared to 493 cm^3), therefore reducing heat transfer losses. Preliminary results are shown in Figure 9, which gives indicated efficiency as a function of equivalence ratio and engine speed for the Perkins engine and the Onan engine (repeated from Figure 1 for comparison). The indicated efficiency for the Perkins engine is higher than for the Onan engine, and it is very insensitive to equivalence ratio and engine speed. Future runs will attempt to increase the engine efficiency by optimizing the turbulence within the cylinder to minimize heat transfer and timing losses.

Table 4. Perkins engine characteristics and experimental conditions.

Bore, mm	91.44
Stroke, mm	127.0
Displacement, cm^3	834.0
Geometric compression ratio	14.04:1
Experimental range for equivalence ratio	0.36-0.47
Experimental range for engine speed, rpm	600-1800
Experimental range for volumetric efficiency, %	90-110

CONCLUSIONS

This paper presents the development and validation of a time-dependent engine model, and its application to conventional and hybrid vehicles. The engine model is applied to a hydrogen engine which has been experimentally tested. The model predicts accurately engine efficiency and NO_x emissions over the full range of experimental operating conditions. The validated model is then used to generate engine performance and emission maps for supercharged engine operation. The performance maps are then incorporated into a vehicle evaluation code to obtain performance and emissions for hydrogen-fueled conventional, parallel and series hybrid vehicles. All

vehicles are specified to meet minimum acceleration, hill climb and range requirements that make them comparable to current automobiles. Analysis of these vehicles yields the following results:

1. Emissions out of the conventional car are projected to be a factor of 60 lower than the CARB ULEV requirements, and therefore a factor of 6 lower than EZEV. The engine control strategy presented in this paper guarantees that the conventional vehicle achieves EZEV emissions levels regardless of how the car is driven. The conventional vehicle also has a high fuel economy.

2. The parallel hybrid vehicle uses a downsized engine and a small flywheel for power augmentation and regenerative braking. Emissions are also well within EZEV range. Fuel economy falls between the conventional and the series hybrid cars.

3. The series hybrid vehicle has the highest fuel economy. Emissions are higher than for conventional cars, at the EZEV limits. Emissions in a series hybrid vehicle are intrinsically independent of driver's input.

These results indicate that lean-burn hydrogen spark-ignited engines are an alternative to providing EZEV emissions, while at the same time providing a range, acceleration and hill climbing performance comparable to conventional cars. These vehicles are intrinsically EZEV, since emissions are not controlled with a catalytic converter that may degrade, and are also independent of the driving habits of the driver.

REFERENCES

1. Smith, J.R., 1994, "Optimized Hydrogen Piston Engines," Proceedings of the 1994 International Congress on Transportation Electronics, Convergence 1994, SAE, pp. 161-166.

2. Heywood, J.B., 1988, Internal Combustion Engine Fundamentals, McGraw-Hill, New York.

3. California Air Resources Board, Mobile Sources Division, 1995, "Proposed Amendments to the Low-Emission Vehicle Regulations to Add an Equivalent Zero-Emission Vehicle (EZEV) Standard and Allow Zero-Emission Vehicle Credit for Hybrid Electric Vehicles," Preliminary Draft Staff Report, CARB, El Monte, CA, July 14.

4. Van Blarigan, P., 1996, "Development of a Hydrogen-Fueled Internal Combustion Engine Designed for Single Speed/Power Operation," Proceedings of the 1996 SAE Future

Transportation technology Conference and Exposition, Vancouver, BC.

5. Aceves, S.M., and Smith, J.R., 1995, "A Hybrid Vehicle Evaluation Code and Its Application to Vehicle Design," SAE paper 950491.

6. Pehr, K., 1996b, "Aspects of Safety and Acceptance of LH2 Tank Systems in Passenger Cars," International Journal of Hydrogen Energy, Vol. 21, pp. 387-395.

7. Ferguson, C.R., 1986, "Internal Combustion Engines, Applied Termosciences," John Wiley and Sons, New York.

8. Haney, S.W., Barr, W.L., Crotinger, J.A., Perkins, L.J., Solomon, C.J., Chaniotakis, E.A., Freidberg, J.P., Wei, J., Galambos, J.D., and Mandrekas, J., 1992, "A SUPERCODE for Systems Analysis of Tokamak Experiments and Reactors," Fusion Technology, Vol. 21, p. 1749.

9. Woschni, G., 1967, "Universally Applicable Equation for the Instantaneous Heat Transfer Coefficient in the Internal Combustion Engine," SAE Paper 670931.

10. Patton, K.J., Nitschke, R.G., and Heywood, J.B., 1989, "Development and Evaluation of a Friction Model for Spark-Ignition Engines," SAE Paper 890836.

11. Olsson K., and Johansson, B., 1995, "Combustion Chambers for National Gas SI Engines Part 2: Combustion and Emissions," SAE paper 950517.

12. Thomson, M.W., Frelund, A.R., Pallas, M., and Miller, K.D., 1987, "General Motors 2.3L Quad 4 Engine," SAE Paper 870353.

13. Aceves, S.M., and Smith, J.R., 1996, "Lean-Burn Hydrogen Spark-Ignited Engines: The Mechanical Equivalent to the Fuel Cell," Proceedings of the 18th Annual Fall Technical Conference of the ASME Internal Combustion Engine Division, ICE-Vol. 27-3, Alternative Fuels, Edited by J.A. Caton, pp.23-32.

14. Peschka, W., 1998, "Liquid Hydrogen, Fuel of the Future," Springer-Verlag, Vienna, Austria.

FUEL CELLS

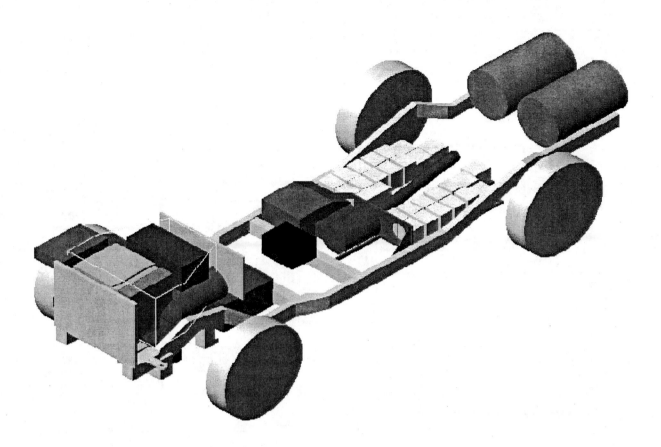

Figure 4. Component Packaging Diagram

existing rear drive shaft, while the other would connect to the front drive shaft from the transfer case. Because of the motor size and a lack of ground clearance for the front motor, this design was considered unusable. So the motor that would have plugged into the front axle was replaced with a transaxle motor. The rear motor was then integrated into the battery box.

The motor controllers or inverters need to be close to the motors that they drive to reduce radiated electrical noise. To reduce heavy wiring for the system, the inverters need to be located close to the batteries and the contactor box. With the previously mentioned restraints and useable space under the vehicle, one inverter was located under the vehicle between the transaxle and contactor box. The other inverter was integrated into the rear power module, next to the rear drive motor. This layout allows the remaining fuel cell systems to be located in the engine bay of the vehicle, and the hydrogen storage tanks are located where the spare tire would normally exist. The spare tire is relocated to the interior of the vehicle.

FUEL CELL SYSTEMS

Virginia Tech has chosen to use a proton exchange membrane (PEM) fuel cell system. A PEM fuel cell requires oxygen from air, hydrogen and cooling water delivered at the appropriate pressure, temperature, flow-rate and humidity. The components required to deliver these fluids plus the fuel cells are what make up the fuel cell system, as shown in Fig. 5.

HYDROGEN STORAGE AND DELIVERY OPTIONS

Several hydrogen storage methods were considered to provide our vehicle with adequate storage to maximize the mass of hydrogen that can be stored on board the vehicle. The four types of storage that were explored were metal hydrides, cryogenic liquid, reformed hydrocarbon technologies, and compressed gas. Metal hydrides were not used because the tanks have relatively low energy density, or the ability to store many kg of hydrogen per kg of tank weight. Metal hydrides also require cooling to fill, and heating to extract the hydrogen, which complicate vehicle integration. Liquid hydrogen was also not chosen because it must be stored at 20 K, which has its own safety concerns. Maintaining the hydrogen at this temperature would induce a large parasitic loss due to energy required for refrigeration. Additionally, unavoidable boil-off could cause standby losses of up to 5% per day. A review of hydrogen storage costs can be found in James et al. (1994).

Hydrocarbon reformers have the advantage of using existing fueling infrastructure, and significantly increase vehicle range. Hydrocarbon reformers were not used because of the lack of development of a practical vehicle reformer system. Current reformer technology is not

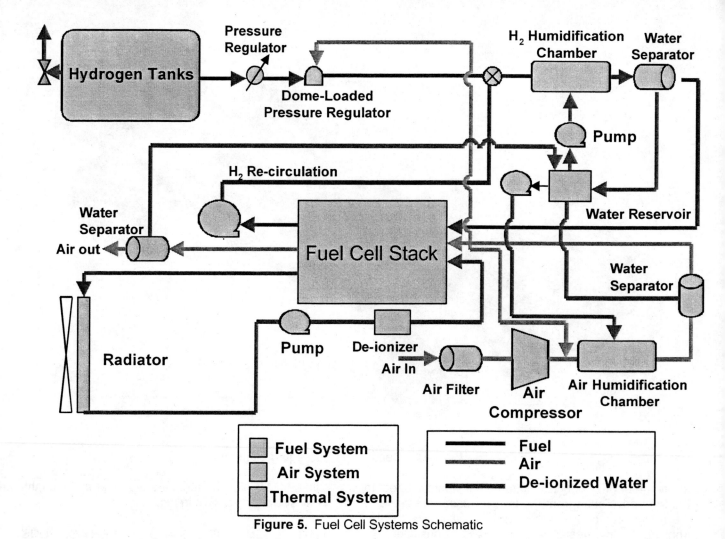

Figure 5. Fuel Cell Systems Schematic

■ Fuel System	━━ Fuel
■ Air System	━━ Air
■ Thermal System	━━ De-ionized Water

small or lightweight enough for integration into our vehicle. We decided to use compressed gas tanks to store our hydrogen, mainly because this method has the highest energy density (kg hydrogen/ kg tank) currently available. In addition, there is no refrigeration or energy required to transport the hydrogen from the stored state to the fuel cell.

The tanks selected are provided to the FutureTruck Competition by Quantum Technologies through the U.S. Dept. of Energy. The tanks are rated at 35 MPa (5000 psi) and achieve a storage density of 7.5 % kg H2/kg. Due to safety considerations and limitations on available sizes, we were limited to two tanks, each of 61 l (16 gal) capacity, to fit under the rear of the vehicle between the frame rails. This tank size and mounting location limits the total mass of hydrogen that can be stored to 3 kg. Range with this limited storage is about 120 km (75 mi). The stock gasoline tank is approx. 130 l (34 gal) 105 kg (230 lb), which provides the stock vehicle with over 500 miles of range.

Our design for a hydrogen tank mounting system is illustrated in Figure 6.

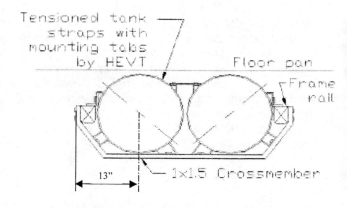

Figure 6. Hydrogen Tank Mounting

1999-01-0529

The Hydrogen Fuel Option for Fuel Cell Vehicle Fleets

Venki Raman
Air Products and Chemicals Inc.

ABSTRACT

Hydrogen is the ideal fuel for the proton exchange membrane fuel cells that are being developed for transportation applications. However, since no hydrogen fuel supply infrastructure currently exists, much of the effort in fuel supply for the fuel cell vehicles is directed at developing on-board fuel processors to generate a hydrogen-rich gas from liquid fuels such as methanol and gasoline. The use of pure hydrogen as the fuel significantly simplifies the vehicle design, and provides greater fuel economy and environmental benefits. While on-board processing of liquid fuels may be a good transitional strategy to commercially introduce large numbers of fuel cell personal automobiles in the near term, the advantages of hydrogen and the relative ease with which it can be supplied for centrally fueled vehicles should favor their adoption for fleet applications.

This paper describes the feasibility of building a hydrogen fueling infrastructure based on delivered hydrogen and small scale on-site production of hydrogen at the fuel station. A first of a kind hydrogen fueling station to support regular transit operations of three fuel cell city buses in Chicago was built based on an adaptation of industrial liquid hydrogen pumping technology.

INTRODUCTION

With the rapid rate of technical advances in fuel cells, and with major resources of the automotive industry being directed at commercialization of fuel cell propulsion systems for transportation it looks more promising than ever, that fuel cells will soon become a viable alternative to internal combustion engine technology.

The proton exchange membrane (PEM) fuel cells currently being developed for near term use in vehicles need hydrogen as a fuel. However no hydrogen fuel infrastructure exists today. This has given rise to the alternative of generating the hydrogen onboard the vehicle from other fuels such as methanol, or gasoline. Thus the fueling choices facing the fuel cell vehicle (FCV) developers today are to develop a hydrogen fuel supply infrastructure, or the on-board fuel processors that enable use of the existing liquid hydrocarbon fuel infrastructure.

The direct hydrogen fuel option entails building an entirely new fuel infrastructure at additional cost, but it vastly simplifies the design of the FCV, and leads to a truly zero emission vehicle (ZEV). Also it enables the implementation of long term strategies such as carbon dioxide capture, and the use of renewable energy in transportation with hydrogen as the carrier. However, hydrogen storage onboard the vehicle is more difficult than the simple liquid fuel systems we are currently accustomed to. To provide a vehicle with sufficient range, hydrogen needs to be stored as a liquid at -253 °C (-423 °F), or as compressed gas at 25 MPa (3,600 psi) or higher, or reversibly adsorbed on lightweight carbon materials. At the current state of development, compressed hydrogen is the most readily implemented option due to its similarity to the growing compressed natural gas fuel infrastructure. Liquid hydrogen fueling needs to be engineered to make it a fully automated process for use in the mass market. Another drawback with liquid hydrogen use for personal cars is the inevitable boil-off losses associated with long periods of non use of the vehicle. Some technical breakthroughs are required to make carbon adsorbents a viable alternative for hydrogen storage.

The on-board fuel processing option is attractive because a well developed gasoline, and diesel infrastructure exists in the developed countries, and can be used with little or no additional investment for fuel cell vehicles. However, the vehicle design becomes much more complex and extremely challenging. Highly compact fuel processors to generate a hydrogen-rich gas need to be developed. These devices need to be capable of quick start-up, have rapid response to load changes, and of course, have very high reliability. Also the fuel processors need to be significantly less expensive than conventional reformers since they will be idle much of the time and thus represent a very poor utilization of capital. In addition, significant capital investments will be required in refinery modifications to produce very low sulfur "fuel cell friendly" gasolines.

With the large infrastructure investments already in place to distribute large quantities of gasoline, it is not surprising that most of the auto manufacturers and oil companies are devoting a great deal of energy to developing onboard fuel processing technology. If successful this represents a near term option to the widespread introduction of fuel cell cars.

Technical and economic studies [1,2] suggest that fuel processors will add more weight to the vehicle, and reduce fuel cell peak power output, which then requires larger fuel cells and larger motors to obtain the same performance in terms of drive train power to vehicle weight ratio. The resulting extra weight in turn requires larger drive train components, further compounding the weight problem. Thus for the same level of performance, hydrogen FCVs will be simpler in design, lighter weight, more energy efficient, and lower cost than those with onboard fuel processors.

Given the advantages of direct hydrogen fuel cell vehicles, the option of hydrogen fueling merits serious consideration. In particular, these advantages of direct hydrogen, and the relative ease of implementing appropriate supply systems for centrally fueled vehicle fleets, will strongly favor the adoption of hydrogen in this market segment. This paper will show how hydrogen fuel infrastructure can be implemented, and provide an example of how existing know-how in the industrial hydrogen markets was adapted to set up a fuel station for fuel cell buses.

CURRENT INDUSTRIAL HYDROGEN INFRASTRUCTURE

The production, and distribution of hydrogen are well established industrial technologies, with about 100,000 metric tons per day being commercially produced and consumed worldwide. Most of the hydrogen produced is consumed "captively" within the producing facility to produce ammonia and methanol, and to refine oil. In North America, and Europe, a very small portion of the total hydrogen production (less than 10% in the U.S.) is produced in central plants and distributed to customers. The current industrial hydrogen distribution infrastructure consists of several hundred miles of high pressure gas pipelines, and fleets of hundreds of trucks delivering compressed hydrogen gas in tube trailers, or liquid hydrogen in cryogenic tankers. This so called "merchant" hydrogen is used in numerous industrial applications as a chemical reactant to manufacture specialty chemicals, as a reducing agent in metals processing, semiconductor and glass manufacture, and as a coolant for large electric power generators. Although the energy content of the hydrogen produced per day worldwide is about 15 million GJ or about 1 percent of the world's energy demand, an insignificant amount is actually used as a fuel. The only readily identifiable fuel use of hydrogen is its use as a rocket propellant which represents about 0.1 percent of the hydrogen produced.

The current U.S. industrial gas hydrogen production and distribution base of about 2,000 metric tons per day is equivalent to that needed to support 40,000 fuel cell buses or 3 million medium sized cars. Thus if hydrogen is to become a major transportation fuel, then a much expanded production and distribution system would be needed. However, there are no major technical hurdles to extending the current technologies to provide hydrogen as a transportation fuel.

LIQUID HYDROGEN-BASED FUEL STATION FOR TRANSIT BUSES

The Chicago Transit Authority (CTA) began running a demonstration fleet of three PEM fuel cell buses in September 1997. The buses operate on hydrogen gas stored at 25 MPa (3,600 psi) in roof-mounted lightweight composite tanks. A hydrogen fuel station system was implemented by adapting commercial hydrogen supply technologies to the specific needs of this emerging application. The station has facilities for turn-key operations consisting of receiving, storing, processing and transferring hydrogen to the buses. The fuel transfer to the bus is via a fully automated dispenser. The fueling is performed by CTA personnel, with adequate training in the safe handling of hydrogen. The station has a fast-fill capability i.e. fueling of three buses is accomplished within a 2 hour period. This required a maximum delivery capacity of 2,100 Nm^3 per hour (80,000 SCFH) to achieve a final settled storage pressure of 25 MPa at ambient conditions ranging from -18 to 38 °C (0 to 100 °F) and 10 to 100% relative humidity. Each bus consumes about 400-525 Nm^3 (15,000-20,000 SCF) of hydrogen per day.

For this location, it proved to be most economical to deliver liquid hydrogen, and to use cryogenic liquid pumps to achieve the high storage pressure needed for the hydrogen on the bus. The cryogenic pump efficiently pumps liquid hydrogen from the storage tank pressure of about 1 MPa (150 psi), through a vaporizer to refuel the buses with gaseous hydrogen at 25 MPa. This compact liquid hydrogen pump requires less energy than required to compress gas to the same pressure. With the addition of a minimal amount of high pressure gas buffer storage it was possible to initiate bus fueling immediately while the cryogenic compressor was still cooling down and before it could begin to pump hydrogen to the bus. The hydrogen fuel station has been in service since September 1997, and is now routinely fueling the three fuel cell buses on a daily basis in Chicago.

Figure 1 is a schematic drawing of the fuel station system. The fuel station is comprised of two major systems: a fuel preparation system to receive, store, compress, and vaporize the hydrogen; and a fuel transfer system, to accomplish the transfer of the high pressure hydrogen to the tanks onboard the buses. The station design is based on a patented Cryogenic Hydrogen Compressor (CHC) [3] which has gained wide applications for high pressure hydrogen supply in the chemical industry. A modified design was developed for the fuel station application and designated CHC-6000.

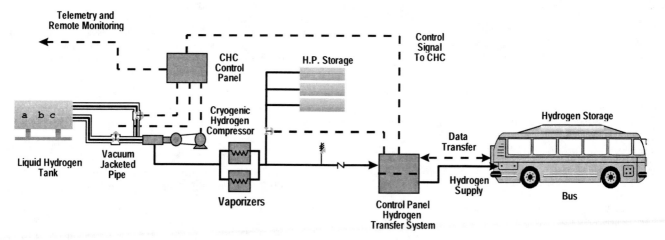

Figure 1. Schematic of Fuel Station

FUEL PREPARATION SYSTEM – The purpose of this equipment is to receive and store liquid hydrogen at a pressure of about 1 MPa, and to supply gaseous hydrogen at pressures of up to approximately 31 MPa (4,500 psi) to the fuel transfer system. Liquid hydrogen is transported to Chicago from a plant located about 500 km (300 miles) away in specially designed 57,000 liter (15,000 gal.) cryogenic liquid semitrailers. The hydrogen is transferred to, and stored on-site in, a double walled vacuum insulated tank with a liquid capacity of 34,000 liters (9,000 gal.).

The filling station is equipped with a single CHC-6000 hydrogen compressor capable of compressing up to 2,100 Nm^3 per hour of hydrogen to a pressure of 41 MPa (6,000 psig). The CHC can draw liquid, or cold hydrogen vapor, or any combination of the two from the tank and simultaneously compress liquid and gas to the high discharge pressure. The compressor package consists of three major parts, the cold (compression) end, the warm (drive) end, and a 30 kW AC motor. The cold end is a single-stage positive displacement unit that actually compresses the liquid and gaseous hydrogen. The warm end is simply a crankcase that converts rotary motion of the motor to the reciprocating motion required by the compressor cold end. A microprocessor-based programmable logic controller (PLC) is used to continuously monitor system conditions, pressures, and temperatures, and automatically control the CHC-6000 operation. The system control panel is designed with a number of safety controls, alarms, and shutdowns. Required safety shutdown alarms were identified in extensive process hazards reviews. It is also equipped with several safety alarms to shut down the system and isolate the liquid hydrogen tank in emergency situations. The system control panel also has several external interface connections. It works closely with the control panel on the fuel transfer system to coordinate the filling of the buses. It is also equipped with a telemetry system, which automatically monitors critical process conditions in the CHC as well as the liquid hydrogen tank inventory. This system is used for automatic delivery of liquid hydrogen, remote data logging, remote troubleshooting, and dispatch of maintenance personnel.

The CHC-6000 pumps liquid hydrogen at -253 °C (-423 °F). This product needs to be vaporized and warmed to ambient temperature before it can be delivered to the bus or to the high pressure storage vessels. This is accomplished using two vaporizer banks which consist of stainless tubing with aluminum finned surfaces that extract heat from the ambient air. The gas outlet temperature is usually within about 5 °C (10 °F) of ambient temperature. To facilitate faster filling of the bus during hot weather, the vaporized gas is further cooled to -7 °C (20 °F) by injecting a small fraction of the liquid hydrogen flow, via an automatic bypass valve, directly from the pump into the exit of the vaporizers. This low temperature of the mixed product then compensates for the natural heat of compression caused by filling the bus quickly.

A bank of three high pressure gas storage vessels (each tube is 16 in. dia. x 23 ft. long) with a storage capacity of about 420 Nm^3 (16,000 SCF) at 31 MPa (4,500 psig) is used to begin the fill process of the bus while the CHC-6000 is still in the cooldown phase. The tube bank is refilled at the end of each bus refueling cycle so that it is ready for the next fill cycle.

FUEL TRANSFER SYSTEM – The hydrogen fuel transfer system is the interface between the fuel preparation system and the buses. It activates the CHC, when an operator demands hydrogen fuel, and directs the flow of hydrogen to the bus tanks or to the buffer storage tanks.

A PLC-based control panel limits access, to only those individuals who are authorized to use and maintain the equipment, through the use of personal identification numbers (PINs). A data and grounding cable, that is attached to the bus before a fueling sequence is initiated, carries data from the bus to the control system. The PLC checks to be sure that the data and grounding cable is attached throughout the refueling process, and immediately stops the refueling process if the cable is disconnected. The PLC processes data from the vehicle, and the fuel preparation system, issues control signals to access the hydrogen source, and directs the flow of hydrogen to the vehicle tanks. An air operated fuel valve starts and stops the flow of fuel automatically in response to air pressure signals issued from the PLC control system. The PLC monitors the amount of fuel transferred into the vehicle tanks and stops the flow of hydrogen when the tanks are filled to their rated capacity. A second "excess pressure" valve in series with the fuel valve, operates independently of the PLC, and closes automatically if the tank pressure exceeds the maximum permissible fuel pressure.

The control panel also displays messages to guide the operator, step-by-step, through the refueling process. If there is a problem (wrong PIN, broken ground wire, etc.), the display indicates the nature of the problem, and the corrective action required. An emergency stop button, located on the front panel of the control cabinet, disconnects power from the hydrogen fuel transfer system and the CHC, when activated.

A flexible fuel hose carries hydrogen from the fueling system to the vehicle. It is attached by a breakaway fitting that releases if the vehicle moves while the hose is connected. A vent hose, bonded to the fuel hose, carries a small flow of hydrogen from the refueling nozzle to a vent stack before the fuel hose is disconnected from the vehicle, to insure that no air enters the flexible fuel hose. The vent hose is connected to the vent stack through a plastic coupling that breaks off if the vehicle moves before it is disconnected. A 1/4 turn emergency manual shutoff valve is located at the fueling interface on the bus, and another is located next to the control panel on the hydrogen supply pipe.

SAFETY CONSIDERATIONS – Safety considerations were given the highest priority in both the design and operation of the fuel station. The three primary hazards which were addressed in the design were: fire; over pressurization of the storage tanks on the bus; and movement of the bus while refueling.

Refueling operators were chosen from the group of CTA's diesel refueling operators. The selected individuals were given both classroom and hands-on training about the properties and hazards associated with hydrogen by technical specialists. The training included a review of a hydrogen fuel station operating manual. Foremen were also trained as backup. Foremen possess the key to reset the PLC if the fuel station trips. Periodic retraining is critical to the total hydrogen fueling safety process. Having the trained refueler enter his/her authorized PIN in order to activate the fuel transfer system reduces the possibility of hydrogen refueling by an untrained person.

Special procedures exist to defuel a bus if a bus is over-filled (over pressure). Other procedures exist at the station when a bus is to be taken indoors. Fire training procedures are critical to the safety of the refueling operation.

SYSTEM OPERATION – The CHC-6000 refueling system is normally in a "standby" mode where it is ready to begin filling buses immediately. While in this mode, the storage tube and tank pressures are monitored continuously by the PLC within the CHC system control panel. Once a bus fueling sequence is initiated, a signal is received from the fuel transfer system control panel. This signal initiates a timed process to cool the CHC-6000 compressor to operating temperatures. While this is happening, the bus is being filled from the bank of high pressure gaseous storage tubes. Once pressures have equalized between the storage tubes and the bus tanks, an automatic valve is shut to isolate the storage tubes. After the cooldown period, the CHC-6000 starts and begins to fill the bus directly. Typically the bus can be completely filled within 15-20 minutes depending upon the starting pressure of the bus. Once the bus is filled, the isolation valve is reopened and the storage tubes refilled in 5-10 minutes so that it is available for the next bus.

The in-use operating experience of the hydrogen fuel station has been excellent to date. The station is routinely used to fill 3 buses per day during the week. The single CHC pump system is capable of filling 2-3 buses per hour on a continuous basis if needed. Typically, all three buses can be filled in about 55 minutes.

ON-SITE PRODUCTION OPTION FOR HYDROGEN INFRASTRUCTURE

While the foregoing example illustrates the adaptation of existing liquid hydrogen delivery technology to serve the needs of hydrogen fueling, its application will be largely limited to North America, and some parts of Europe, where liquid hydrogen is readily available. In most of the rest of the world, liquid hydrogen is not commercially available. In these areas it is anticipated that the hydrogen infrastructure will develop via small scale hydrogen generation units installed at the fuel station to produce hydrogen from other commonly available fuels such as natural gas, methanol, propane etc.

Currently, commercial steam-methane reformers (SMRs) for on-site hydrogen production are economic at a scale above about 25,000 Nm^3 per day (1 MMSCFD or 2.5 TPD). However, recent developments in hydrogen generators, driven by the needs of stationary fuel cell power plants, are offering the real possibility that distributed hydrogen generation could be competitive with centrally produced hydrogen in sizes as small as 100 Nm^3 per hour (~3,500 SCFH). Although these generators are based on the same technologies as the large central plants, e.g. SMR or partial oxidation of hydrocarbons, they may drive costs lower due to inherent design innovations that permit the use of less expensive materials of construction, and mass production of hundreds of units. This could lead to economic hydrogen production at individual fuel stations, from natural gas and liquid hydrocarbons, to refuel as few as 4-5 buses, or 30-40 cars per day. These products are currently under intense development, and are expected to be commercially introduced in the very near future. By installing a larger hydrogen generator at the fuel station instead of thousands of fuel processors on individual vehicles, a greater utilization factor is achieved on the capital, and many of the demands such as rapid startup and rapid response to transients etc. are eliminated. Also, natural gas-based fuel generators provide a ready-made way to implement hydrogen infrastructure by taking advantage of the widespread network of natural gas pipelines that exist in many countries,. Where natural gas is not available, operationally simple methanol reformers can be implemented since methanol is readily available around the world. Studies by others have shown that the investment costs per vehicle for these small stationary hydrogen fueling appliances would most likely be less than the current annual cost per vehicle required to maintain or add new gasoline infrastructure, or onboard liquid fuel processors (4).

To implement a hydrogen fuel station based on these on-site fuel processors, additional gas purification, compression, storage and dispensing equipment will be required, as shown in Figure 2. These small generators afford a pathway to a gradual build-up of the hydrogen supply infrastructure in lock-step with the growth of fuel cell vehicle sales. Thus the industry can closely match the supply and demand for hydrogen in the early years of the commercialization of fuel cell vehicles without having to risk the large amounts of capital needed for large scale hydrogen production facilities. As hydrogen-fueled vehicle technology becomes well established and accepted by the public, and hydrogen demand grows sufficiently,

additional investments in large scale production plants can be economically justified in the future.

CONCLUSIONS

Considering the advantages of direct hydrogen fueled fuel cells for vehicles, it is well worth pursuing the hydrogen fuel option.

The current technologies for the industrial supply of hydrogen, with modifications, are suitable for serving the fuel cell vehicle application. The Chicago Transit Authority project has demonstrated that sufficient experience exists within industrial gas industry to design and construct central hydrogen fueling stations for fleet applications. This system is based on liquid hydrogen and industrially proven high pressure cryogenic pump technology suitably modified for fueling requirements. This robust system design maybe replicated to provide the fueling capabilities needed to implement other such projects in regions where a liquid hydrogen supply infrastructure exists.

Where commercial liquid hydrogen distribution is non-existent, small scale on-site hydrogen production technologies, that are coming into the market, are a way to start building a hydrogen infrastructure, and to make hydrogen available widely. Furthermore these technologies provide the flexibility to match supply and demand for hydrogen in the early years of fuel cell vehicle commercialization when large investments in large scale hydrogen plants can not be financially supported.

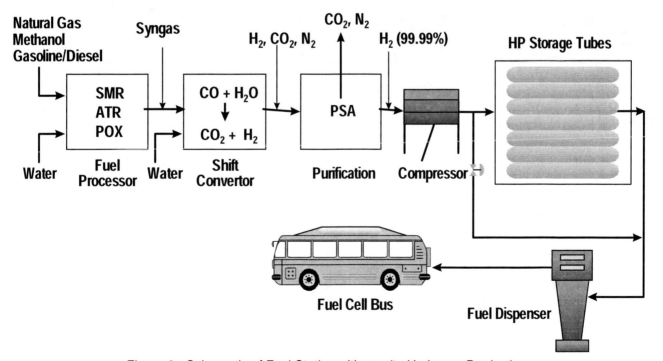

Figure 2. Schematic of Fuel Station with on-site Hydrogen Production

REFERENCES

1. Thomas, C. E., B. D. James, and F. D. Lomax Jr., "Market Penetration Scenarios for Fuel Cell Vehicles", 8th Annual U.S. Hydrogen Meeting Proceedings, National Hydrogen Association, Washington, D. C., p. 137-169, March (1997)

2. Ogden, J., Steinbugler, M., and Kreutz, T., "Hydrogen as a Fuel for Fuel Cell Vehicles - a Technical and Economic Comparison" 8th Annual U.S. Hydrogen Meeting Proceedings, National Hydrogen Association, Washington, D. C., p. 469-506, March (1997)

3. Schuck, T.W. and VanOmmeren, J. US Patent 5,243,821, Sep 14,1993 "Method and Apparatus for Delivering a Continuous Quantity of Gas over a Wide range of Flow Rates".

4. Thomas, C. E., I. F. Kuhn, B. D. James, F. D. Lomax Jr., and G. N. Baum, "Affordable Hydrogen Supply Pathway for Fuel Cell Vehicles", World car Conference, Paper 97WCC061, Riverside California, January 21, 1997.

CONTACT

Venki Raman (Phone: 610-481-8336; e-mail: ramansv@apci.com) is based in Allentown, PA, and leads worldwide efforts in new production methods and applications development for hydrogen. He has been involved in numerous hydrogen transportation projects involving internal combustion engines, and fuel cell engines. He serves as the Chairman of the National Hydrogen Association, Washington, D.C. He earned his Ph.D. in Chemical Engineering from the University of Washington, Seattle, WA.

1999-01-0530

Maximizing Direct-Hydrogen PEM Fuel Cell Vehicle Efficiency – Is Hybridization Necessary?

David J. Friedman
University of California

ABSTRACT

The question of whether or not direct-hydrogen fuel cell systems in automotive applications should be used in load following or load leveled (battery hybrid) configurations is addressed. Both qualitative and quantitative analyses are performed to determine the potential strengths and weaknesses of each option. It is determined that the amount of energy that can be recovered through regenerative braking has a strong impact on the relative fuel economy of load following versus load leveled operation. Further, it is demonstrated that driving cycles with lower power requirements will show an improvement in vehicle fuel economy from hybridization while those with higher power requirements will not. Finally it is acknowledged that the practical considerations of cost and volume must also weigh heavily into the decision between the two configurations.

INTRODUCTION

One issue that has been driving the development of fuel cells for automotive applications is their potential to offer clean and efficient energy without sacrificing performance or driving range. In the case of a direct-hydrogen proton exchange membrane (PEM) fuel cell vehicle, realizing this potential means ensuring that the complete fuel cell system operates as efficiently as possible over the range of driving conditions that may be encountered. This paper investigates the question of whether battery hybridization is necessary to maximize fuel economy, as has been employed for internal combustion engines.

The issue of hybridization is investigated based on a fuel cell vehicle employing a direct-hydrogen fuel cell system operating strategy which is optimized for efficiency.[1,2] The characteristics of this fuel cell system do not represent the details of any specific current technology, but rather show the key features that would be seen for any automotive PEM fuel cell system. This fuel cell system is incorporated as a load following power source (as with a conventional internal combustion engine vehicle) and as a load leveled power source (as with an internal combustion engine hybridized with a battery in a series configu-

ration). The two are compared with the goal of determining which one has the highest fuel economy.

Emissions maximization is not included in this paper since direct-hydrogen fuel cell systems produce only water and electricity. All of the emissions associated with a direct-hydrogen PEM system are produced in creating the hydrogen, and therefore cannot be minimized through vehicle operation, other than to minimize fuel consumption.

This paper begins with a brief description of the UC Davis simulation program that has been developed to analyze automotive fuel cell vehicles. Next, the characteristics of the mid-sized vehicle that is used to show the impacts of load following versus load leveled configurations are presented.

After the vehicle characteristics have been presented, the basic theory of hybridization is presented as a potential method to improve fuel economy when using internal combustion engines. This theory is then applied graphically to show the potential impacts of hybridization for an automotive fuel cell system used over the Federal Urban Driving Schedule (FUDS) and the proposed US06 high power driving cycle. The theory is then evaluated quantitatively through second-by-second simulations of the vehicle performance over the driving cycles.

Finally, various issues not accounted for in the simulations are addressed along with their implications relative to the choice between load following and load leveled configurations for fuel cell vehicles:

- Cost and volume tradeoffs among the various components.
- Transient response issues associated with the air supply compressor.
- Impacts of including a reformer system to supply the hydrogen from a liquid fuel.

SIMULATION PROGRAM

A realistic representation of a complete fuel cell system is needed when evaluating the fuel economy of a fuel cell vehicle. Through the support of the UC Davis Fuel Cell

Vehicle Modeling Program, a direct-hydrogen PEM fuel cell system simulation has been developed for automotive applications.

The system simulation contains component models for each of the critical fuel cell system components. The PEM fuel cell stack model is based on analysis done by the Electronic Materials and Device Research Group at Los Alamos National Laboratory.[3,4,5] The other component models are based on UC Davis research or manufacturer's data.

This results of this system simulation, in the form of an optimal operating curve for the fuel cell system, are embedded in a fuel cell vehicle (FCV) simulation. The fuel cell vehicle simulation is based on the Advanced Vehicle Technology Evaluator (AVTE) developed by UC Davis as an extension of the capabilities of the National Renewable Energy Laboratory's ADVISOR vehicle simulation tool.

The overall simulation is divided into two stages. The first stage models and then applies an optimization procedure for the key components in a direct-hydrogen fuel cell system. The second stage incorporates the information from the first (in the form of an operating curve for the fuel cell system[1,2]) and simulates the performance of a complete fuel cell vehicle (FCV) over a second-by-second driving cycle.

The simulation program has been developed within the Matlab programming environment, using both the Matlab programming language and Matlab's Simulink visual programming language. The use of Simulink allows for a simple modular design for the program and provides access to a vast library of mathematical functions and numerical calculation processes.

VEHICLE CHARACTERISTICS

The analysis performed herein is based on an advanced mid-sized vehicle similar to a Ford Taurus. The majority of the vehicle characteristics are based on the first UC Davis FutureCar, a modified 1996 Ford Taurus developed in 1995 through 1997[6]. Table 1 presents the baseline vehicle characteristics assuming a 1500 kg vehicle test weight for the load following configuration.

Table 1. Baseline Mid-Sized Vehicle Characteristics.

Test Weight excluding the Fuel Cell and/or Battery Systems	1275 kg
Test Weight in Load Following Configuration	1500 kg
Aerodynamic Drag Coefficient	0.27
Frontal Area	2.18 m^2
Rolling Resistance Coefficient	0.007
Accessory Load	500 W

Based on these characteristics, the vehicle power system has been sized to ensure that it maintains the perfor-

mance of conventional vehicles in the mid-sized class. The peak power requirement of the electric motor is 75 kW which is sufficient to accelerate the vehicle from 0 to 100 kph (0-60 mph) in twelve seconds. This implies that the fuel cell system and the combination of the fuel cell system and battery must provide at least that amount.

To ensure performance on a grade, it was determined that 35 kW would be required to for the vehicle to maintain 100 kph (60 mph) on a 6 percent grade. For the load leveled case, this was used to guide the sizing of the fuel cell system.

One of the issues associated with the tradeoffs between load leveled and load following configurations is the weight trade off between the batteries and the fuel cell system. The data for the batteries employed in these simulations are based on UC Davis testing of 88 Ah Ovonic Nickel-Metal Hydride batteries with a peak gravimetric power density on the order of 200 W/kg.

The gravimetric power density of the fuel cell system is based on the direct-hydrogen fuel cell system established by Ballard as of 1997. For this system, including PEM stacks, a high pressure/flow air compressor, a thermal and water management system, and a hydrogen storage system, the gravimetric power density is about 3 kg/kW.[7]

The weight of the hydrogen storage system is expected to be the same for both the load leveled and load following systems unless there are drastic differences in the respective fuel economies. Assuming that the fuel economy difference will not be very large, the mass of the hydrogen storage has been assumed to be accounted for in the vehicle glider weight. Since the comparisons in this paper are between the two configurations and are not intended to be compared to actual vehicles, it is the relative weight of the two vehicles that is most important rather than the absolute value.

Further details regarding the characteristics of each configuration are presented in the Analysis section.

HYBRIDIZATION FOR IMPROVED FUEL ECONOMY

As evidenced by the Partnership for a New Generation of Vehicles, hybridization is often put forth as a method for improving the efficiency and emissions associated with using internal combustion engines in vehicles. Specifically in reference to the series hybrid configuration, there are three primary advantages to hybridizing an internal combustion engine vehicle: 1) Enabling the engine to operate at a higher fraction of its peak load and therefore at higher efficiencies, 2) De-coupling the engine from the instantaneous torque and speed requirements at the wheels, and 3) The ability to capture some of the energy normally lost in braking through regenerative braking. It is because of the specific characteristics of the internal combustion engine that hybridization if effective in items one and two. This is discussed further in the next section.

HYBRIDIZING INTERNAL COMBUSTION ENGINE VEHICLES – The typical characteristics of an internal combustion engine are that the efficiencies tend to be higher at medium to high torque and at medium speeds. Thus, overall, higher engine efficiencies are obtained in medium to higher power operation, and a severe efficiency penalty can be paid if there is frequent operation at low power.

The reason hybridization provides a benefit relative to the efficiency characteristics of the engine is because typical driving is often at low power. For example, the majority of the energy use when driving on the Federal Urban Driving Schedule (FUDS, the driving cycle used by the US EPA to determine the city vehicle fuel economy), takes place in the low power regions. Figure 1 shows a power frequency diagram where the amount of energy spent at various power levels for the baseline,1500 kg vehicle, is plotted versus the power level category as a fraction of peak power. For this vehicle, a 75 kW engine is employed and over 70% of the energy spent over the driving cycle occurs at 20% of the peak power or less. Thus, over 70% of the energy use will be in the low efficiency region of the engine.

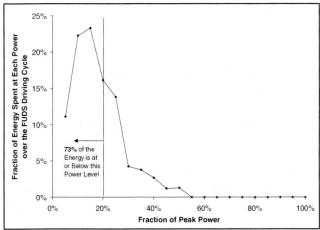

Figure 1. Power Spectrum for the Federal Urban Driving Schedule.

It is generally accepted the FUDS is not representative of the way people drive today. People tend to drive at higher speeds and accelerate faster than the FUDS would imply. The proposed high acceleration, high speed US06 driving cycle is used to investigate the impact of this behavior. The power frequency during this driving cycle is shown in Figure 2.

Over the US06 driving cycle, the power demands are significantly increased, but there is still a significant amount of energy being spent at lower powers. For example, nearly 50% of the energy is spent at or below 35% of the peak engine power. At or below 50% of the peak power, nearly 75% of the energy is accounted for. Therefore, while there is significantly more high power operation, the engine still expends a large amount of its energy in the lower power regions.

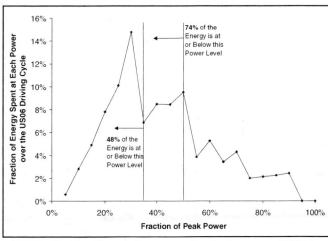

Figure 2. The Power Spectrum for the Proposed US06 Driving Schedule.

For internal combustion engine vehicles, hybridization can reduce the amount of energy spent at low power by allowing a smaller engine to be used while taking up the higher power requirements with an onboard energy storage device such as a battery. There are many ways of doing this, but one typical method is to operate the engine only under two conditions, first: when the battery state of charge is too low, and second, when the required power is above and beyond what the battery can provide. In this configuration, the engine is sized to provide power on a 6 percent grade at 100 kph and the battery is sized to provide the additional power that would be required for a 0 to 100 kph acceleration in 12 seconds. When the engine is on, is generates electricity through a generator and follows a power demand that roughly stays with the average power requirement of the vehicle. This allows the battery to be recharged and the engine to be operated at higher powers relative to its peak power – since the size of the engine will have been significantly reduced and there is no idling.

An additional benefit of hybridization in this series configuration is that the engine is de-coupled from the instantaneous torque and speed demands from the wheels. This allows the engine to be operated at each power using the torque and speed where it is most efficient, rather than that required by the transmission. In effect, the addition of the generator acts as a continuously variable transmission, enhancing the ability of the engine to operate at higher efficiencies.

The final benefit of hybridization is that the use of batteries and an electric motor allow for recovery of some of the energy lost during braking. The process of regenerative braking uses the motor to provide a negative torque to the wheels, slowing the car down and allowing the motor to act as a generator and recharge the batteries. Since about one-third of the energy required to drive a vehicle over the FUDS driving cycle is thrown away as heat when braking, regenerative braking can provide significant energy savings.

HYBRIDIZING FUEL CELL VEHICLES – The question for fuel cell vehicles is whether the same benefits that can be achieved in hybridizing an internal combustion engine exist for fuel cell systems, or if a load following configuration makes more sense due to the characteristic properties of fuel cell systems.

The characteristics of a fuel cell system for automotive use are quite different from that of an internal combustion engine. One key difference is the relationship between efficiency and load. For an optimized direct-hydrogen fuel cell system, the peak efficiency tends to be somewhere between 10 and 30 percent of its peak power output. Figure 3 demonstrates that this low power peak efficiency actually matches quite well with the energy use over the FUDS. In the case shown, about 90 percent of the energy used over the driving cycle occurs at 30 percent of the peak power or less. Therefore, except for the ten percent of the time when the vehicle requires very high power and the ten percent of the time when it requires very low power, the fuel cell system is operating at near its peak efficiency.

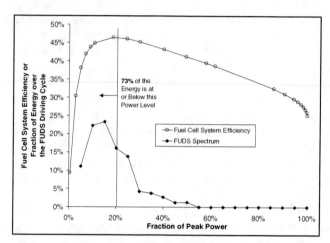

Figure 3. Comparison of the FUDS Power Spectrum and the Fuel Cell System Efficiency.

Comparing these characteristics to the first benefit of hybridizing an internal combustion engine, it becomes obvious that shrinking the fuel cell system size would only force it to operate at higher fractions of its peak power and therefore at lower, rather than higher, efficiencies. A potential mitigating factor could be that, through hybridization, the fuel cell system would never be operated in the very low power regions, below 5 percent of peak power, where the efficiency drops off rapidly.

Investigation of Figure 4, a comparison between the fuel cell system characteristics and the US06 driving cycle power/energy requirements, shows that there will likely be an efficiency penalty on the US06 cycle as well. Since this driving cycle already requires a large amount of high power operation, reducing the size of the fuel cell system would, when it is on, force it to operate at higher fractions of its peak power and therefore lower efficiencies. Thus, is seems that hybridizing a direct-hydrogen fuel cell system would tend to decrease its in-use efficiency rather

than improve it. Even in cases where the fuel cell system efficiency is not significantly reduced, there is also an additional efficiency loss associated with recharging the batteries to maintain them within the desired state-of-charge range (as is typically done for charge sustaining series hybrids) and then discharging them when the fuel cell cannot provide enough power or when operating on battery power alone.

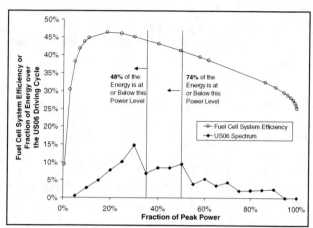

Figure 4. Comparison of the US06 Power Spectrum and the Fuel Cell System Efficiency.

The second benefit to hybridization that was mentioned, the use of the generator and associated electronics to de-couple the engine from the demands at the wheels, is also not realized in the case of a fuel cell system. Since the fuel cell is already an electrical device and require electronics to interface with the motor and controller, the only direct connection to the wheels is the power requirement, rather than a specific torque and speed (or in the case of the fuel cell system, a particular pressure and mass flow rate). Thus, the fuel cell system can be operated at the optimal efficiency for each power requirement as represented in the comparisons in Figure 3 and Figure 4.

The final benefit previously discussed is the ability of a load leveled vehicle to use regenerative braking to capture the energy normally lost to friction brakes. A hybridized fuel cell system will capture this energy up to the ability of the battery to absorb the incoming power.

The general results of hybridization of a direct-hydrogen fuel cell system with a battery can be summarized as follows:

- The potential for a decrease in in-use fuel cell system efficiency.
- The ability to capture regenerative braking energy.
- An increase in vehicle weight and complexity to incorporate the batteries.

It seems, then, that the potential for hybridization to benefit the overall efficiency of a direct-hydrogen fuel cell vehicle depends on the ability of regenerative braking to overcome the potential fuel cell system efficiency loss and the added weight and complexity. To evaluate these issues in detail, second-by-second simulations must be performed to account for the detailed characteristics of the driving cycle, the components, and the control strategy employed for the hybrid vehicle.

ANALYSIS OF FUEL CELL VEHICLE CONFIGURATIONS

Two configurations were originally chosen to compare load following to load leveled fuel cell vehicles. The load following configuration is the baseline vehicle described in the vehicle characteristics section. This configuration incorporates a direct-hydrogen fuel cell system with a 75 kW net peak power and has a test weight of 1500 kg.

The load leveled fuel cell vehicle incorporates a 35 kW net peak power fuel cell to ensure gradeability. A 40 kW nickel-metal-hydride battery pack with 25 30 Ah modules provides power when the fuel cell stack is not operating, or when additional peak power is required. The test weight of this vehicle is 1530 kg. The control strategy is set such that the battery state-of-charge (SOC) is allowed to float between 60% and 40%. Power is drawn from the fuel cell under two conditions: 1) When the battery SOC drops below 40% until it is recharged to 60%, and 2) When the motor requires more than 30 kW to meet the driving cycle. The latter condition enables the vehicle to use a small battery pack while maintaining acceleration performance through the use of the fuel cell during peak power requirements.

In general, when no power is demanded of the fuel cell system it is not turned off. Instead it is set to idle operation where a small amount of fuel is consumed to maintain minimum operating conditions and to provide any energy necessary for thermal and water management. When power is required of the fuel cell, but is below 5% of its peak power, the output is set to that minimum 5% level. This ensures that the stack is not operated in the region where its efficiency drops off significantly other than when it is idling.

The load leveled fuel cell power output is set based on the average power required from the motor over the previous 30 seconds of driving. This allows the battery to be recharged and also limits excursions into the very low and very high power regions, i.e. the averaging smoothes out the power requirements. The power averaging scheme is overridden during hard accelerations when the power required from the fuel cell system is set to the actual power necessary for the acceleration, up to the fuel cell system's peak power.

During the simulation runs, it was discovered that this acceleration power override feature caused a significant drop in fuel economy on high power driving cycles. This was due to the fact that the power demanded was often exceeding the limitations of the fuel cell system, thereby forcing it to operate at its peak power when providing 20 to 30 percent of the energy required over the driving cycle. This imposed a severe efficiency penalty. In addition, the small size of the battery pack meant that a large

portion of the regenerative braking energy could not be used since it was at power levels which were too high for the batteries to accept.

To investigate the potential mitigation of these issues, a third configuration was incorporated which used a larger battery pack. The same 35 kW fuel cell system was used, but a 75 kW nickel-metal-hydride battery pack with 25 60 Ah modules was used, creating a vehicle with a test weight of 1685 kg. In this case, the acceleration override was not employed until the motor power requirements rose above 50 kW. This significantly reduced the amount of peak power operation for the fuel system with a corresponding increase in system efficiency. Further it allowed for an increased amount of regenerative braking energy to be captured. These benefits, of course, are achieved with the penalty of added weight, volume, and cost.

SIMULATION RESULTS – Once all the cases were run, two primary issues became apparent in the comparison between load following and load leveled fuel cell systems. First, the difference in fuel economy between the two configurations was a major function of the amount of regenerative braking energy that could be captured. Second, the potential benefit or loss due to hybridization varied significantly depending on the driving conditions simulated.

Over the FUDS driving cycle, it was found that both the small and large battery pack load leveled configurations provided gains in fuel economy compared to the load following configuration. The small battery pack configuration provides an 8% increase in fuel economy while the large battery pack configuration provides an increase of 17%, as shown in Figure 5.

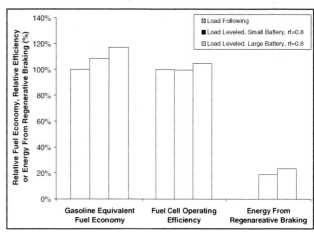

Figure 5. Simulation Results for the Three Configurations Over the FUDS Driving Cycle.

Figure 5 also shows that the efficiencies of the fuel cell systems are quite similar for the three configurations, therefore likely not influencing the difference in fuel economy to a significant degree. The three are similar because the average power required over the FUDS driving cycle is only about 5 kW. This represents 7% of the load following and 14% of the load leveled fuel cell system peak powers, regions where the fuel cell system has relatively high efficiencies. Further, there are very few severe accelerations in the FUDS cycle, leading to few excursions into high power (and therefore low efficiency) operation to maintain the performance of the vehicle.

The primary determinant of the fuel economy improvement must then be the energy captured through regenerative braking. Figure 5 shows that about 23% of the energy required to drive the small battery configuration was recaptured and stored in the vehicle battery, while 27% was recaptured for the large battery configuration. This value is highly dependent on the assumed amount of braking energy that could be made available for regeneration. For the above results, it was assumed that a maximum of 80% of the braking energy would be available, although the motor and/or battery may not be able to accept all the power available.

This initial assumption of 80% likely represents a "best technology" value for the vehicle regen fraction (rf) setting. Practical issues such as the sensitivity of the brakes, the inability of the motor to capture significant regen energy at low speed and low torque, as well as the peak torque and power limits of the motor and battery will likely require a smaller value to be used to represent a realistic vehicle. Figure 6 shows the sensitivity of the fuel economy improvement for the hybrid vehicle configurations to changes in the regen fraction. Over the FUDS cycle, it is not until the regen fraction reaches 0.2 for the large battery configuration and 0.33 for the small battery configuration, that the fuel economies of the load leveled configurations drops below that of the load following configuration. These values of the regen fraction are below what might be reasonably expected from a practical vehicle. Thus, over the FUDS cycle the hybrid configuration seems to provide some advantage in fuel economy.

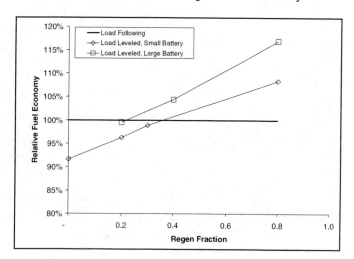

Figure 6. Crossover Regen Fraction for the Load Leveled Configurations Relative to the Load Following Configuration over the FUDS Driving Cycle.

Another feature that can be seen in Figure 6 is that the fuel economy of the large battery configuration is consistently higher than that of the small battery configuration.

This is a result of the larger battery pack's ability to absorb more of the regenerative braking power available. The smaller battery pack could only accept up to 30 kW, any regen power provided above that level would simply be dissipated in the form of heat requiring the friction brakes to provide the additional stopping power. This disparity will be larger over driving cycles with more severe braking requirements.

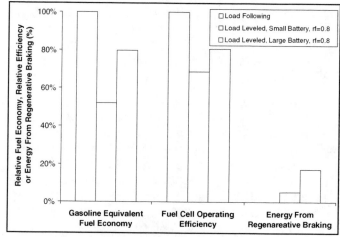

Figure 7. Simulation Results for the Three Configurations over the US06 Driving Cycle.

The simulation results for the high power US06 driving cycle depict a different conclusion relative to load following versus load leveled configurations. Figure 7 shows that the fuel economy of the large battery configuration is 20% lower than that of the load leveled configuration. The small battery configuration results are even worse, with the fuel economy being barely one-half that of the load leveled configuration.

The two primary differences between the FUDS and the US06 driving cycles are that the US06 has higher power requirements and less regenerative braking energy available as a fraction of the driving energy required. Further, the braking energy that is available is at higher powers. The latter two issues significantly reduce the amount of energy captured with regenerative braking, adding to the reduction in the fuel economy.

The result of the higher power operation is that the smaller fuel cell systems in the load following configurations are operated more often in their lower efficiency regions. Further, due to the severe accelerations, both load leveled vehicles are forced into acceleration mode where the fuel cell system is operated at or near peak power no matter what the battery state-of-charge. This drastically decreases the system efficiency over the drive cycle, especially in the case of the small battery pack which requires frequent help from the fuel cell system to stay on the driving schedule.

The effect of the high power requirements is somewhat mitigated in the large battery configuration because acceleration mode is not encountered as often. In addition, the larger battery pack enables the capture of three times more regenerative braking energy over the drive cycle. It is important to note, though, that these simulations were performed with a regen fraction of 0.8. Were a more realistic value of 0.5 to 0.6 used, the load leveled configurations would look even worse in comparison to the load following configuration.

The previous results indicate that the benefits of load following versus load leveled configurations are highly cycle dependent. The question then becomes which driving cycle is more representative of typical driving patterns. The answer to this question is beyond the scope of this paper, but some further investigation can at least provide another data point for comparison.

Another cycle that is somewhat in between the previous two and is sometimes used as a more representative driving cycle is simply the FUDS driving cycle with all the velocity requirements multiplied by 1.25. The effect is to produce a driving cycle with higher power operation like the US06, but also with a greater degree of stop and go like the FUDS (i.e. more opportunity to capture regenerative braking energy). The results for the three vehicles over this driving cycle are, as would be expected, somewhere in between the previous two.

Figure 8 indicates that with a regen fraction of 0.8, the small battery load leveled configuration is inferior to the load following configuration. On the other hand, the large battery configuration is slightly superior due to the efficient capture of regen energy and less fuel cell system operation at peak power. The lower fuel economy of the small battery configuration is due partly to more operation of the fuel cell system in the acceleration mode. It is also a result of the inability of the smaller battery to capture as much regenerative braking energy as that for the large battery configuration.

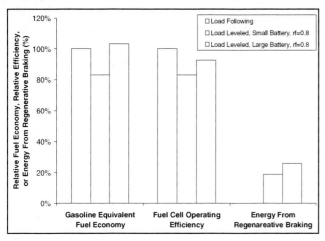

Figure 8. Simulation Results for the Three Configurations over the 1.25 X FUDS Driving Cycle.

Figure 9 investigates the sensitivity of the larger battery configuration to the value of the regen fraction. Unlike the normal FUDS cycle, the large battery pack configuration is only superior to the load following configuration when the regen fraction is 0.7 or greater. Since this is value is not likely to be achieved in practice, it can be

concluded that the load following configuration will likely be superior to the load leveled configurations on the 1.25 X FUDS cycle.

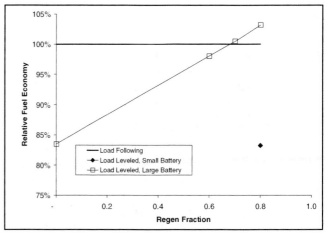

Figure 9. Crossover Regen Fraction for the Load Leveled Configurations Relative to the Load Following over the 1.25 X FUDS Driving Cycle.

ADDITIONAL CONSIDERATIONS

Besides efficiency, there are numerous other considerations that must go into the choice of whether to use a load leveled or load following configuration with fuel cell systems.

As has been alluded to earlier in this paper, the relative cost of the fuel cell system compared to the energy storage device can be a key deciding factor. If batteries or ultracapacitors which could provide the required peaking power were prohibitively expensive, hybridization would be out of the question since the gains would not likely be significant enough to compensate for the added cost. However, if the fuel cell system was significantly more expensive than the batteries, an efficiency penalty would likely be accepted to achieve lower vehicle costs by using the batteries.

Another key deciding factor is the volumetric density of the energy storage device relative to the fuel cell system. A practical fuel cell vehicle must maintain the cabin and trunk space afforded in conventional vehicles. If either the fuel cell system or energy storage device were too large, the configuration which utilized the least space would likely be chosen, even if it incurred some fuel economy penalties.

One issue that might cause a call for hybridization is the response time of the air system. Since the power demands change rapidly over the driving cycles for a load following configuration, so too would the demand for air. However, compressors will have limitations to the speed at which they can go from providing minimum air flow to providing maximum air flow for peak power. These limitations might either cause the fuel cell system power output to be limited from second to second, or might create a drop in the net efficiency of the fuel cell system. One approach to mitigate these transient response limitations could be to hybridize the vehicle. As discussed previously, this would provide two benefits:

- It would smooth out the power demand from the fuel cell system thereby giving the compressor more time to respond to changes in the power demand
- It would also be able to use battery power to provide extra power when the compressor could not ramp up the air mass flow enough to meet the power demand at optimal efficiency.

However, rather than going to the expense and loss of volume of including batteries, the air system could be hybridized rather than the entire vehicle. A hybridized air system would consist of a compressor and an accumulator which would be maintained to provide the necessary air that the compressor cannot provide in the short term. This could further allow the compressor to be optimized by operating in a narrow range of pressure and mass flow where it is most efficient. This would be possible because it would no longer be required to follow the second-by-second demands of the drive cycle.

An alternate solution would be to provide a small battery that is used to boost power to the compressor without drawing energy from the fuel cell. This could increase compressor response time while not compromising performance. It would result in an efficiency loss, but only a very small change in the weight and volume of the components.

In a similar vein, if a reformer is used to supply the hydrogen, rather than onboard hydrogen storage, additional transient issues appear. These issues are also present during start-up as the reformer will have to heat up to operating temperature before it can function properly. Again, a battery could be used to power the vehicle while the reformer is warming up and when the reformer cannot respond to the power requirements of the drive cycle. However, an alternative solution is to hybridize the reformer with the use of a small reformate storage tank.

For both the hybridized air system and the hybridized reformer system, the tradeoff between the added volume and complexity of the batteries relative to the accumulator and reformate buffer must be evaluated along with the potential efficiency impacts.

SUMMARY AND CONCLUSIONS

Overall it has been shown that hybridization is beneficial under some driving conditions, such as during the FUDS cycle. However when more realistic driving conditions are employed, the load leveled configuration either provides minimal benefits or actually creates a fuel economy penalty. These facts, along with the complexity and added cost associated with hybridization imply that, for mid-sized vehicles, load following direct hydrogen fuel cells may be a better choice.

Of course these conclusions are highly dependent on the fuel system and energy storage characteristics. The use of batteries with significantly higher peak power capability, ultracapacitors, or advanced regenerative braking systems might increase the amount of regenerative braking energy that can be captured, thereby increasing the potential benefits of hybridization. Further, if fuel cell systems can be developed with flatter efficiency curves, the efficiency losses associated with operation at higher fractions of the system peak power could be reduced.

Additional issues associated with the air and fuel supply sub-systems might also weigh in the direction of hybridization. However, it will be important to investigate the possibilities of hybridizing the sub-systems using accumulators or buffer tanks as well as complete vehicle hybridization.

ACKNOWLEDGMENTS

This work is supported by the Fuel Cell Vehicle Modeling Project at the University of California. Development of the vehicle simulation model, excluding the current fuel cell system, was supported by the National Renewable Energy Laboratory. Development of the fuel cell stack model has been done with the help of the Los Alamos National Laboratory.

REFERENCES

1. D.J. Friedman and R.M. Moore, PEM Fuel Cell System Optimization, *Proceedings of the Second International Symposium on Proton Conducting Membrane Fuel Cells*, The Electrochemical Society, Inc., Pennington, NJ (1998).

2. R.M. Moore, D.J. Friedman, A.F. Burke, in Proceedings of Commercializing FCVs '97, Frankfurt, Germany, October 20-22, 1997.

3. T. A. Zawodzinski, S. Gottesfeld, J. Electrochem. Soc., 138, 2334 (1991).

4. T. E. Springer, M. S. Wilson, S. Gottesfeld, J. Electrochem. Soc., 140, 3513 (1993).

5. T. E. Springer, T.A. Zawodzinski, M. S. Wilson, S. Gottesfeld, J. Electrochem. Soc., 143, 587 (1996).

6. B. Johnston, D. Friedman, K. Burch, C. Frazer, D. Kilmer, T. Scheiblich, D. Funston, E. Chattot, T. McGoldrick, R. Carlson, "The Design and Development of the University of California, Davis FutureCar", 1996 Future Car Challenge, Society of Automotive Engineers, PA, SAE SP-1234, 1997.

7. Ford Motor Company, *Direct-Hydrogen-Fueled Proton-Exchange-Membrane Fuel Cell System for Transportation Applications*, July 1997, DOE/CE/50389-503.

1999-01-0532

Supporting Fuel Cells with PowerBeam Surge Power Units

Chris W. H. Ellis
XKF Developments Ltd

ABSTRACT

Various attempts have been made to develop 'fuel cell only' vehicles, and these have highlighted some key limitations of fuel cell technology, namely low levels of efficiency at peak power, no support for regenerative braking and slow response to changing power demands. This paper shows how these problems can be overcome by supporting the fuel cell stack with a Surge Power Unit (SPU), which can reduce hydrogen consumption by up to 45% in urban applications, enhance driveability and lower initial capital costs.

Details are given of the European Zero Emission Hybrid (ZEH) project, which features a large urban bus and a long-haul truck drivetrain, and which will demonstrate the major benefits of supporting fuel cells with PowerBeam kinetic energy SPUs.

INTRODUCTION

The combination of low-cost off-peak electricity and high fossil fuel taxes in Europe has already created the opportunity to justify the purchase of fuel cell vehicles on the basis of lower operating costs, rather than just zero emissions, for which there is no mandatory requirement in Europe for any type of vehicle. The paper shows how the savings in operating costs will more than cover the inevitably higher capital costs of 'New Generation' commercial vehicles.

A conventional engine is necessarily a compromise between the most efficient conversion of fuel into energy and the responsive provision of a wide range of power. The fundamental hybrid electric proposition is that two specialised, optimised, power units can meet these conflicting requirements more effectively than a single unit. In the hybrid concept, the Energy Supply Unit (ESU) provides the average level of energy required as efficiently as possible, a role for which fuel cells are ideal. The complementary Surge Power Unit (SPU) responds efficiently to short-term variations in power, including acceleration and regenerative braking.

An introduction to kinetic energy storage devices in general and PowerBeam in particular is attached as an appendix.

The paper refers to the ZEH (Zero Emission Hybrid) project, which has been submitted to the European Commission for funding under the Fifth Framework programme. The ZEH project is intended to validate the claim that a ZEH bus or truck operating in Europe will have substantially lower fuel costs than equivalent conventional vehicles.

MAIN SECTION

In the ZEH project, off-peak electricity is used to produce compressed hydrogen and oxygen by the electrolysis of water. Approximately five kilowatt hours of electrical energy are required to produce one cubic metre of hydrogen at atmospheric pressure, plus the related amount of oxygen, after making full allowance for the electrolyser overheads and compression to 200 atmospheres. A typical fuel cell system is capable of generating a kilowatt hour from approximately 0.8 cubic metres of hydrogen when running at peak efficiency, which is usually at well below full power.

Until recently, most legislative measures to reduce emissions from vehicles have concentrated on passenger cars and light vans, with less severe restrictions on diesel engines and heavy commercial vehicles. Ideally, all vehicles should produce no emissions at all, especially those operating in city centres. For many years, this has been a feature of electric trams and trolley buses. However, they are relatively inflexible in operation, require large and unsightly investments in infrastructure and offer no solution for the many other urban applications. Curiously, emission standards for urban buses in Europe are only voluntary, despite the close proximity of buses to large numbers of passengers, cyclists and pedestrians. However, there is now increasing concern that some of the most serious forms of air pollution are caused mainly by older buses, trucks and taxis, most of which are diesel-powered. Fortunately, a new generation of vehicles will soon be able to deliver much lower levels of fuel consumption, particularly in the city, by combining mechanical and electrical power, in a hybrid configuration. Exhaust emissions will be radically reduced not just as a direct consequence of the sharp fall in fuel consumption but also because these hybrid designs will help a variety of engine types to meet proposed new ultra-low emission requirements, especially the demanding 'Euro IV' NOx limit of 0.25g/km, proposed for 2005. Attempts have been made over the years to solve the general problem by using battery-powered vehicles, but with no real success. Some of the latest efforts are using fuel cells, with considerable promise.

The European ZEH project will examine what is believed by many to be the ideal, 'end game', solution, applicable eventually to all sizes of cars, vans, buses and trucks. The key elements are a fuel cell stack which supplies electricity to a surge power unit and an electric motor, with hydrogen as the onboard fuel. This choice of configuration will be examined and justified in detail for a large urban bus. The central question to be addressed by the project is whether such a drivetrain has the potential to deliver savings in full-life costs which are large enough to convince most commercial vehicle operators that the inevitably higher prices of 'New Generation' commercial vehicles are easy to justify. The technical focus is on the PowerBeam surge power unit, particularly, its reliability, safety, efficiency and full-life costs.

The ZEH project has three phases. The first phase focuses on proving the feasibility of the key elements of the design, particularly the PowerBeams, and simulating the performance of the proposed drivetrain and vehicles. The second phase involves building and testing a complete drivetrain, off vehicle. The third phase integrates the drivetrain into a variant of the existing Volvo Environmental Concept Bus (ECB), and tests actual performance, with particular emphasis on hydrogen consumption. If resources permit, a turbogenerator will be added to the drivetrain simulations and off-vehicle tests, to investigate the applicability of the enhanced drivetrain to heavy duty trucks and long distance coaches.

EUROPEAN FUEL PRICES – The price of off-peak electricity delivered by Electricité de France (EDF) to industrial users is approximately 1.2 US cents per kilowatt hour. This low price is possible because of the high percentage of electricity generated by France's nuclear power stations; consequently the marginal cost of any extra base load is low, and emissions are zero, as is CO_2 output. Making an allowance to cover the capital cost of electrolysers, profit margin, etc., a 'pump price' of ten cents per cubic metre would seem easily achievable. It is unlikely that the French government would wish to place additional taxes on hydrogen produced in this manner, given the clear environmental and strategic benefits. The price of diesel fuel in France is currently almost three dollars per gallon, excluding 'value added tax' (VAT), and this is expected to rise to well over four dollars a gallon by 2005.

Consider a large conventional truck operating in France, averaging 80,000 miles per year, and burning diesel at an average rate of 8 miles per US gallon. Annual fuel costs alone will probably exceed $40,000 by 2005. The 'New Generation' replacement truck will require approximately three kilowatt hours per mile at the motor. Taking a conservative view of efficiencies in the drivetrain, this will equate to 200,000 cubic metres of hydrogen, costing $24,000, yielding an annual saving of more than $16,000.

While the offpeak price of electricity in France is exceptionally low, French diesel prices are also lower than those in several other European countries. Taking the UK as an example, the offpeak electricity price to industrial users is already less than 3.5 US cents per kilowatt hour. Further competition is likely to lower the transfer price to the hydrogen supply subsidiary of a major electricity company to less than 2.5 cents by 2005. The filling station price of diesel fuel in the UK is now over four dollars per gallon (just over three dollars per US gallon to commercial users, allowing for bulk purchase and refundable VAT), and this is expected to rise to at least six dollars per gallon by 2005. By then, the same large conventional truck will cost over $45,000 per year in fuel alone to operate in the UK. This contrasts with its fuel cell hybrid replacement, which should have annual fuel costs of less than $30,000, giving a net saving of over $15,000 per year.

If a large battery could be used, less than 5 kWh per mile might be required at the truck depot, allowing for charging and discharging losses. However, a battery-based solution is entirely impractical in these large vehicles. Using fuel cells and PowerBeams may look relatively inefficient by comparison, but it is the only solution which is actually capable of providing zero emission vehicles with adequate range while still offering users substantial overall savings.

If CO_2 output is to become the principal determinant of levels of fuel taxation, diesel fuel should carry more tax than gasoline, because a gallon of diesel produces approximately 12% more CO_2 than a gallon of gasoline. Arguably, a diesel hybrid car would need to achieve 90 mpg to meet the spirit, rather than just the letter, of the US government's 80 mpg PNGV target. This discrepancy is not a function of engine design, but results from the fundamental fact that a gallon of diesel contains considerably more carbon than a gallon of diesel. Several vehicle manufacturers are already reflecting this in the fine print of their sales brochures by showing the CO_2 output per mile or kilometre, not just the fuel consumption.

Given the higher CO_2 output of diesel fuel, it would be perfectly reasonable for the European Commission not just to recommend a move towards tax parity but to propose a diesel surcharge. This could be phased in over the next five to ten years by raising the tax on diesel faster than petrol. Diesel fuel prices might then have to rise by more than 100% in France, Germany, etc. No doubt there will be protests that this would have a knock-on effect on other prices, but, in general, the impact will be small, less than one percent. The oil companies have an interest in keeping the consumption of diesel up as the use of oil for home and industrial heating falls away, and will no doubt point out that refining gasoline results in higher levels of CO_2 production than refining diesel from the types of crude oil normally supplied to European refineries. However, this effect is usually small, and a tax premium of 10% for diesel over gasoline might be an accurate reflection of overall CO_2 production.

Today, gasoline prices in France, Italy, the Netherlands, Sweden and the UK are similar, and will soon exceed one euro per litre. Germany and Spain lag behind, but are already under strong pressure from the other countries, the European Commission and internal factions to catch up, perhaps even overtake. The Green Party in Germany has in the past proposed a tax-inflated price of 5 marks per litre (around $12 per gallon) for unleaded. With Greens now members of the newly elected German government, taxes on diesel and gasoline are expected to move sharply higher. One possible scenario is unleaded in Germany moving from its current price of 82 eurocents per litre to 1.20 by 2005, with diesel moving up from 60 eurocents to 1.30 over the same period, i.e. more than doubling in price. Note that this will almost certainly still leave gasoline and diesel prices lower in Germany than they will have become in the UK, and possibly also in France and Italy. The result will be that commercial vehicle operators across all of Europe will become even more obsessed with fuel costs than UK operators are today.

As it now seems highly probable that conventional engines and small gas turbines can be developed to meet very low emission levels when used in hybrids, then the environmental argument favouring battery-only and fuel cell vehicles becomes relatively weak and operating costs will dominate most decisions. Fuel cell technology is still the great hope, alongside fusion power, for the 'end game' when oil fuels become costly due to shortage of supply rather than taxation. However, in Europe, the opportunity is much more immediate, because cost justification is already easy, given that European end users are paying prices for oil fuels that Americans may not see for 20 years or more, while electricity prices are roughly similar in both America and Europe.

When oil eventually begins to run out, the main form of energy delivery will gradually become electrical, whether the energy source is electricity generated from nuclear fusion or hydro, solar or wind power, or still, for a few more years, from natural gas. By accident or by design, most European governments have put in place a tax regime for gasoline, diesel and natural gas which effectively simulates now the inevitable future, when fossil fuel production costs will rise to the point where electrically produced hydrogen is less expensive to produce than oil, and hydrogen will take over as the most credible long-term fuel for most vehicles.

MARKET ENTRY – By 2005, in France, the UK and probably most other European countries, there will be a clear cost justification for the transition from conventional diesel engines for long haul trucks to the proposed Power-Beam, fuel cell, turbogenerator solution. The case for similar technology in long distance coaches may look equally attractive. Local distribution vehicles cover less

miles per year, but will require smaller fuel cell stacks and may not need turbogenerators at all in many markets and applications. Consequently, these may also soon prove fully cost effective. City buses with this drivetrain should be particularly attractive, especially if governments assist early market penetration by helping with capital costs, in pursuit of zero emissions in city centres.

Meanwhile the European car market will probably have to wait until cost reduction through design and commitment to volume brings the cost premium sharply down from the level that already makes sense for high mileage commercial vehicles. However, the premium sector of the passenger car market, dominated by marques such as Audi, BMW, Cadillac, Jaguar, Lexus and Mercedes, does not obey the normal rules of cost justification, and has consequently been the traditional entry point for most innovations in car design.

In summary, the priority applications in Europe are depot-based local and regional distribution trucks and city buses, followed by long haul trucks and buses. One or more of the premium car makers may attempt to 'steal a march' on the others by demonstrating superior technology. The result will be a most impressive motor car.

HYDROGEN FROM NATURAL GAS – Natural gas in Europe generally carries a significant level of tax, even for industrial use, and probably cannot compete as a source of hydrogen against electricity at French prices. However, across most of Europe the majority of industrial hydrogen is still produced by steam reformation of natural gas, because electricity generally is not as cheap as it is in France. Given the recent commitment by several major oil companies to produce hydrogen in volume at attractive prices for the airlines, in most countries natural gas will remain competitive with electrolysis as a source of hydrogen unless carbon taxes move prices sharply upwards, and until supply shortages begin to drive gas prices higher. In most European counties the source of hydrogen for vehicles is still likely to remain natural gas for at least the next decade. Although this will not result in zero CO_2 output, it should result in a reduction of at least 50% in overall CO_2 production from heavy road transport when compared with diesel, and will still deliver zero emissions in city centres. Of course, ZEH vehicles already on the road will become progressively cleaner as electricity replaces natural gas, with no further changes required to the vehicles themselves.

Across most of the United States, natural gas will probably continue (in the short and medium term) as the most cost effective source of hydrogen. In the US, electricity prices are roughly comparable with those in most of Europe, while natural gas is much cheaper. Because US natural gas prices are so low, fuel costs should be halved or better in moving to true 'New Generation' vehicles. However, note that diesel prices in the US are around one third of the European average and approximately 25% of the UK price, so the annual costs savings will not be as dramatic, despite the higher speeds and annual mileage of American trucks. In the US, hybrid drivetrains combining turbogenerators running on natural gas with PowerBeam SPUs and electric traction may eventually play a major role.

Some US states are proposing additional financial incentives to accelerate the takeup of clean vehicles. These carrots may yet motivate US consumers to keep pace with the severely beaten Europeans!

THE ZEH DRIVETRAIN CONCEPT – In Europe, there will probably be no 'New Generation' of vehicles unless they can deliver a significant reduction in full-life costs. There is no equivalent of the Californian 'Zero Emission' mandate, nor any likelihood of one, although certain cities will probably introduce zero emission zones.

It has been well understood for decades that overnight electricity offers a radically cheaper fuel in Europe than heavily-taxed diesel, but the range and cost limitations of battery technology have frustrated the many attempts to take full advantage of this. The one fuel that governments would find it almost impossible to differentially tax for transport is electricity, and most governments would not want to try, for good environmental and strategic reasons. Road usage pricing is now almost certain to take over from fuel taxes as the preferred major source of government revenue from transport. Electricity prices will effectively set the ceiling for hydrogen prices, because hydrogen can be readily generated by electrolysis from water. This then sets the cost targets for all other forms of hydrogen generation, and should limit any attempts at taxing hydrogen supply for transport. Although it is more energy efficient to generate hydrogen directly from natural gas than to generate electricity from natural gas and then produce hydrogen by electrolysis, the end user is only interested in the final cost of the hydrogen in the vehicle, which must include distribution costs and taxes, if applicable.

The project's electricity utility partner will produce hydrogen and oxygen by electrolysis in the vehicle depot, with detailed measurement of electricity consumption and hydrogen production. Hydrogen and oxygen will be stored in compressed form in conventional high-pressure cylinders both on and off the vehicle. The project will not attempt to exploit more advanced forms of onboard hydrogen and oxygen storage, although these may be necessary for production vehicles. The demonstration vehicle will carry sufficient hydrogen and oxygen to support an adequate range for testing and demonstration, but is not expected or required to achieve the range objectives of viable products, just prove the potential efficiency.

Note the decision to exploit the almost free production of oxygen inherent in electrolysis. The alternative to storing oxygen on the vehicle is to extract it from ambient air as required, on the vehicle. There is a clear tradeoff between extraction and storage costs. If the oxygen is produced off the vehicle, the vehicle itself is marginally more efficient. On large commercial vehicles where the size and weight of the oxygen storage is not critical but annual mileages are high, onboard storage of oxygen may prove more cost effective.

The normal means of refuelling the demonstration vehicles will be by pressure hose, with the depot's high pressure storage system being connected by trained operators to the storage systems in the vehicles. If resources and time permit, onboard electrolysers may also be fitted. This will allow the vehicles to be 'recharged' electrically by their drivers wherever recharging facilities for electric vehicles have been installed. The GM Hughes inductance system is likely to become the global standard now that Toyota has declared its support, and it will be used in this project.

It is expected that the demonstrator will be based on Volvo's Environmental Concept Bus (ECB). The performance of the demonstrator will at least match that of the conventional vehicle in every respect, except range. The surge power capability will be provided by two large PowerBeams, together weighing some 500 kg. These will be mounted in the roof of the ECB, replacing the existing 900 kg of batteries. PowerBeams will form part of the vehicle structure in production buses, acting as compression struts and torsion beams in the ECB configuration. However, in the demonstrator, the PowerBeams will be structurally passive, carried as a load like the batteries they will replace. Each PowerBeam will contain two 60 kW rotors, giving aggregate surge power and regenerative braking of up to 240 kW. The available capacity of each rotor is likely to exceed one kilowatt hour in the demonstrator. Subject to confirmation in testing, this may prove sufficient at the product level, although energy capacity could be at least doubled with no increase in rotor diameter or overall PowerBeam dimensions.

An alkaline fuel cell stack will be used. Because the PowerBeams provide powerful, highly efficient surge power, some of the perceived advantages of PEM fuel cells over alkaline systems become irrelevant, and the advantages of alkaline systems can be exploited to the full. These include higher peak efficiency, the potential to operate without rare and expensive catalysts, the inherent electrolyser function (i.e., alkaline fuel cells are potentially fully reversible) and efficient sub-zero operation. For the bus, it is expected that a fuel cell stack with a peak rating of some 80 kW capable of providing 50 kW continuously at 65% efficiency will be fitted. A large truck would need more power, probably a 200/120 kW stack. Alkaline fuel cell development has reached the stage where the power density of the stack is approximately 200 kW per cubic metre. A fuel cell production cost target of $150 per kilowatt is likely to be achieved within three years.

LONG-HAUL TRUCK – If time and resources permit, the project will also explore the feasibility of extending the application of the drivetrain to a variety of heavy duty applications, in particular, long haul trucks. This will be limited to enhanced simulations and the addition of a turbogenerator to the off-vehicle drivetrain tests, and is not, at this stage, expected to involve an additional demonstrator vehicle.

While the power requirements of urban buses or delivery trucks can be met by a combination of fuel cells and PowerBeams alone, the additional power requirements of long-haul trucks can best be met by adding a turbogenerator, rather than using a larger fuel cell stack. In a typical truck application, the 320 kW motor will be chassis mounted, driving a conventional propeller shaft and rear axle through a simple gearbox. The gearbox is likely to need no more than five gears. The gears will be changed by electric actuators under computer control, with manual override. No conventional large friction clutch or synchromesh will be required.

In the long-haul application and also for local distribution in hilly regions, the final key component will be a turbogenerator capable of producing some 120 kW. While the fuel cells combine hydrogen and oxygen at relatively low temperatures to generate electricity directly, emitting only water, the gas turbine burns hydrogen and air to produce shaft power, which then drives a high-speed alternator. Because of the high combustion temperatures involved, some level of nitrous oxides could be emitted by the turbine. The key to the efficient operation of a gas turbine is a high combustion temperature, which aggravates the NOx problem. However, in this application the efficiency of the turbogenerator will not normally be critical, because the turbogenerator is likely to be in use for less than 5% of the time, as a 'power booster', and even then it will be producing only a third of the total power. Some manufacturers may eventually prefer to fit turbines running on hydrogen and pure oxygen (with water injection) to production versions of long haul trucks. This would result in high efficiency at relatively low cost, in an application where the extra weight and space of oxygen storage should not be critical. No expensive emission controls would be required, yet the vehicles would still be true ZEVs (and zero CO_2 vehicles), even with their turbines running.

A highly efficient turbogenerator, running on hydrogen and oxygen with water injection, will require approximately one cubic metre of hydrogen to produce one kilowatt hour of electricity, which is slightly more efficient than most fuel cell stacks at full power. The main advantages are that the turbogenerator will be smaller, lighter and less expensive than the extra fuel cells. An important additional advantage is that the turbogenerator provides the vehicle with a 'fast start' capability. While this will usually be unnecessary in a bus, there are applications, such as fire trucks, where it may be essential. However, the fuel cell stack will remain the primary ESU, typically generating over 90% of the energy at a peak efficiency of 60% or more, and will only rarely need to operate at full power and consequently lower efficiency. In most applications, the amount of water produced by the fuel cells will exceed the average amount required for injecting into the turbine. Consequently, there should normally be no need

for frequent refills, even with a relatively small reservoir. It may also be possible to exploit exhaust heat from the turbine to warm the fuel cell stack, accelerating start up and slightly improving efficiency.

Where a low-cost, low-weight solution is preferred, the turbogenerator may still be cost-effective burning hydrogen in air at an efficiency as low as 25%. On the other hand, at least 35% is achievable if the extra weight, space and cost of an efficient recuperator and appropriate NOx emission controls is acceptable. Note that 35% efficiency is getting close to the efficiency of the fuel cell stack at full power. The turbogenerator saves the marginal cost, weight and space of extra fuel cells. It should prove to be significantly cheaper, and much lighter and smaller, even in its efficient form.

CITY BUS – Because of its critical importance to urban transportation in the 21st century, arguably the most important 'New Generation' vehicle is the city bus. The only relatively rapid way that urban public transport can be made more competitive with cars is by improving bus services, because of the infrastructure constraints, delays and costs of the alternatives. While light rail systems are, in some cases, effective long-term additional options, they cannot offer the flexibility and convenience of well-run bus services. If manufacturers can be encouraged to make buses cheaper to operate, less polluting, quieter and more comfortable, then travellers will be attracted onto the buses, not just forced out of their cars by taxation and regulation.

Consider now a series hybrid bus, with a gross weight of some 15 tonnes. This needs a traction motor powerful enough to move the bus smoothly back into city traffic without too much delay to other users. Because bus stops may be only some 400 yards apart, regenerative braking can yield significant savings, so the traction motor needs to be equally powerful as a generator. This indicates a power requirement, both in and out, of some 250 kW for the traction motor(s) and the SPU. However the average power required, given an SPU with at least 90% in/out efficiency, is only some 50 kW. This indicates that an ESU operating near continuously at less than 60kW but at optimum efficiency would be ideal. The SPU need only provide some 3 to 5 kWh of available, usable, energy capacity, because there is no requirement for sustained periods of high power.

This is clearly more efficient than driving the bus with a conventional diesel engine, or directly with a 250 kW PEM fuel cell without an SPU. Not only can the fuel cell stack be much smaller, lighter and less expensive if it is supported by an SPU, but it will use considerably less fuel, because energy from regenerative braking can be re-cycled and the stack can run steadily at peak efficiency. If batteries are used, most types cannot accept charge at better than 50% of their nominal power rating and are hard pressed to deliver back more than 80% of the energy fed in, both of which factors significantly reduce overall efficiency, requiring a larger, more expensive battery as well as more costly recharging.

Because of the high efficiency and high power of the PowerBeam SPU, full advantage can be taken of this application's potential for energy saving from regenerative braking, which, given the low speeds and stop-go driving, can substantially reduce the net energy required per mile.

IMPACT ON FUEL CONSUMPTION – Some developers are working on fuel cell powered vehicles without SPUs. This approach may be because the fuel cells typically used give 60% efficiency at low loads, falling to 40% at high power levels. Consequently, if an overly small fuel cell stack with this characteristic is fitted alongside an SPU and expected to run continuously at or near peak power, much of the efficiency gain from the SPU will be lost. However, optimal efficiency can usually be achieved by combining a slightly more powerful stack with an SPU.

For example, in the hybrid city bus application an ideal combination might be an 80 kW (peak) fuel cell which would normally need to run at 60 kW or less and hence deliver 60% efficiency most of the time. The surge power will come from 240 kW of PowerBeams with 90% in/out efficiency, driving a 250 kW traction motor. On the other hand, a 250 kW fuel cell without an SPU might require as much as 45% more fuel, made up of some 30% from the absence of regenerative braking and a further 15% because, in the city bus application, most of the energy must be generated at peak power during acceleration and therefore at low efficiency. This would almost completely erode the cost justification for using fuel cells, particularly in markets with relatively low prices for diesel. Saving the capital cost of 170 kW of fuel cells and ancillaries will more than pay for the PowerBeams, and the fuel and fuel storage savings (both on and off the vehicle) will be the payoff. This argument also applies to most other applications. Simulations indicate a reduction of some 30% in hydrogen consumption for passenger cars in urban driving with a PowerBeam/fuel cell combination.

CONCLUSION

There is clearly a strong financial case for supporting fuel cells with appropriate surge power units. This paper has shown how the precise requirements for SPUs can be met more effectively by PowerBeams than by any known battery or ultra-capacitor technology. Soon, demonstration vehicles will be ready for testing. If the tests confirm the promise, production zero emission heavy duty vehicles will follow.

CONTACT

chrisellis@bcs.org.uk

APPENDIX

THE EVOLUTION OF KINETIC ENERGY STORAGE SYSTEMS

For more than 20 years, advanced forms of kinetic energy storage (KES) systems, based on high-speed flywheels, have been used to provide satellites with a superior alternative to batteries. Outer space has proved to be a particularly benign environment for such systems. The natural vacuum avoids the earthbound requirement for an enclosure to prevent air friction from dissipating the stored energy. In unmanned satellites, there is no need to provide a heavy containment vessel to protect against rotor failure. Gyroscopic forces are low, simplifying bearing requirements. During transport into orbit, KES systems are uncharged and therefore particularly benign - no acids, etc.

Most KES systems used in satellites have rotors wound from an anisotropic material such as carbon fibre, because of its very high tensile strength. The main mass of the rotor is concentrated in the rotor rim to maximise stored energy. Stored energy is then essentially a function of the mass of the rim, multiplied by the rim speed squared. A relatively large diameter is used, to keep the rim thin and rotational speeds reasonably low, which helps to simplify the design of the bearings and motor-generators. Theoretically, the rim should be as thin as possible, to get all of the mass to provide maximum energy. The maximum rim speed is proportional to the tensile strength of the rim material in the circumferential direction, divided by the density of the material. Consequently, relatively low density, high strength materials such as carbon fibre are used for rotor rims, in preference to metals.

However, the design tradeoffs are more complicated when developing a ground-based, or 'static' system. A vacuum vessel is necessary, and also some means of containing the rotor in the event of bearing failure. These requirements are usually met by designing a single massive casing. For static storage systems, used in load levelling and emergency backup power applications, the emphasis has been on rotor design, in pursuit of maximum capacity. Even though containment can be lavish for these systems, size and weight are still a consideration. Compared with the theoretical optimum, the diameter is usually reduced. Consequently, the thickness of the rim has to be increased and the height of the rim needs to be raised to be more than the diameter, if the energy capacity is to be kept the same and critical dynamic conditions avoided. In many implementations the rotor could be described more accurately as a 'flycylinder', rather than a flywheel. Thickening the rotor wall has the effect of reducing the rotor's specific energy (i.e.

watt hours per kilogram), compared with a unit with a theoretically optimum rotor, which would require a much larger and heavier casing.

The obvious way to install a KES system in a road vehicle is to mount one or more units in gimbals. Gimbals are required to accommodate the gyroscopic forces that would be generated if the unit were fixed in the vehicle. Without gimballing and with the rotor fixed with it's main axis vertical, turning the vehicle would have no gyroscopic effect. However, any rolling or pitching would generate powerful gyroscopic moments; these are directly proportional to the product of the rotor inertia, the rotational speed and the vehicle's rate of pitch or roll. The gyroscopic moments would result in loads in the rotor bearings which would be inversely proportional to the distance between the bearings. These loads could be unacceptably high, particularly with the narrow bearing placement of most wheel rotor implementations. However, by adding gimbals, the complete unit can be made free to pitch and roll, the axis remaining vertical as the vehicle moves. Effectively, the vehicle pitches and rolls but the rotor does not.

Designers of static systems can afford to be relatively lavish in the provision of containment. Static designs tend to focus on the specific energy of the rotor, whereas dynamic designs must be optimised for the overall performance of the vehicle, including reliability and safety. Most implementations of gimballed systems use wheel, rather than cylinder, rotors, to minimise gimballing space. However, adopting wheel rotors has the effect of increasing bearing loads, if gimballing limits are exceeded.

While the use of gimbals can work well under normal conditions, despite cost, weight and space penalties, it is potentially unsafe under severe conditions. It is difficult to design gimballing with more than 30 degrees of movement; once the limit is exceeded, the effect on the bearings is exactly the same as if the unit were fixed. Off-road vehicles frequently exceed 30 degrees, as do ordinary vehicles on some parking ramps.

The most severe problem is in an accident, where the vehicle may roll over, or pitch more than the gimbal limit. Then, either the unit could break out of the gimballing or the rotor could burst its bearings. Similarly, in a severe impact the gimballing will have to be massively strong to prevent the unit from breaking free, even though there may be no significant pitching or rolling. The conclusion must be that, while gimballing is quite satisfactory in a benign environment, it will prove extremely difficult to meet international safety requirements with this approach.

Fortunately, there is a better solution, the PowerBeam. In this case, the rotor returns to a relatively thin-walled, efficient form, because the 'height' is increased dramatically. The 'height', or now the length, of the rotor can be several times the diameter. If the energy capacities of a flywheel and a PowerBeam rotor are the same, then the bearing loads due to gyroscopic forces in the PowerBeam are

only some 30% of the peak levels in the flywheel. Put another way, the PowerBeam could have several times the energy capacity before incurring the same peak bearing loads.

A PowerBeam rotor and a conventional rotor with the same kinetic energy will develop approximately the same gyroscopic moment if disturbed at the same angular rate. However, because the bearings are much wider apart in the PowerBeam, the bearing loads are proportionally lower. Where the conventional system must be gimballed to prevent the bearings from being rapidly destroyed, the PowerBeam needs no such gimballing, with all the inherent space, weight, cost and safety advantages.

Gimballing insulates the rest of the vehicle from the effects of gyroscopic forces, at least until the gimballing limits are exceeded. If the casing becomes a fixed part of the vehicle structure and only one rotor is used, the gyroscopic forces resulting from any change in pitch or roll (vertical rotor axis) would have a direct and destabilising effect on the vehicle as a whole. PowerBeam overcomes this problem by using twin contra-rotating rotors. The gyroscopic forces in the rotors translate, through the bearings, to bending moments in the PowerBeam casing, and these moments cancel each other out within the casing. This has the superior result of insulating the vehicle from gyroscopic forces without any limits on the degree of pitch, roll or yaw.

The long rotors are placed inside a long, thick-walled containment vessel which now has the shape, size and strength for it to do double duty as one of the main structural members of the vehicle. Thus the containment vessel, which, mounted in gimbals, would be a deadweight load on the vehicle structure, now displaces some of the structural mass, and hence increases the effective specific energy of the total energy storage system. Note also that the additional space requirement for a PowerBeam may be much lower than for a gimballed unit. The PowerBeam might, for example, be placed inside what was previously the transmission tunnel, with little or no increase in its size. The gimballed unit, by contrast, is going to take up a substantial amount of extra, exclusive, space. PowerBeams are not fixed in size, capacity or power, and can scale to serve a wide range of applications. However, as one example, consider an implementation suitable for a family car. This might use a 25 cm external diameter tube with a length of some 200 cm, which connects together the front and rear suspension sub-frames to form the 'backbone' of the vehicle. This 'backbone' might contain two co-axial rotors, each 20 cm in external diameter and 60 cm long, in separate, nose-to-tail, compartments. The rotors would be mounted using hybrid ceramic ball bearings, and would counter-rotate at similar speeds. By placing the two bearings of each rotor wide apart, exploiting the beam characteristic of the rotor and containment design, the bearing loads from gyroscopic moments are minimised.

The success of PowerBeam turns, quite literally, on the latest hybrid ceramic bearing technology, which now makes possible load and speed combinations which would have been inconceivable ten years ago. Supporting long rotors with hybrid ceramic bearings removes the need for gimballing. Advantage can be taken of the necessary duplication of the rotors to use a single, massively strong, containment tube which can also make a full-time contribution to the structural strength of the vehicle. Beyond the PowerBeam casing, pitching and turning place no extra stresses on the vehicle. To meet the demanding containment requirements, the beam stiffness of the casing is much more than required just to deal with the gyroscopic moments.

Compared with a gimballed system, the PowerBeam solution is particularly advantageous in a severe accident. There are no gimbal bearings or structure to fail.

Rather than design containment to a minimum level of strength and weight, containment can be generous, because any excess will be contributing to the structural stiffness of the vehicle, providing a superior platform for the suspension and enhancing the passive safety characteristics.

By incorporating the PowerBeam casing into the structural framework or chassis of the vehicle, the theoretical promise of kinetic energy storage becomes practical, on the basis of superior safety, specific energy, volumetric efficiency, power and economy.

Progress In Development and Application of Fuel Cell Power Plants For Automobiles and Buses

Alfred P. Meyer, Joseph M. King and Daniel Kelly
International Fuel Cells, LLC

ABSTRACT

Fuel cell development at United Technologies Corporation has pioneered the successful application of fuel cell power plants for space craft electrical power and for stationary electric generation in building applications. A major effort is now underway to utilize proprietary Proton Exchange Membrane (PEM) fuel cell stack technology and fuel processing technology in power plants designed for application to vehicles. This activity is carried out at International Fuel Cells, LLC which was formed by United Technologies specifically to pursue transportation applications of its fuel cell technology. This paper reports on the status of development and demonstration activities for automobile and bus applications. A 100 kW, methanol-fueled power plant is providing power for operation of a 18 meter transit bus. This 1727 Kg power plant has achieved impressive efficiency, response and emission characteristics in testing to date. The ambient pressure power plant can be configured for operation on other fuels such as compressed natural gas and naphtha. A 50 kW hydrogen-air power plant operating at ambient pressure has achieved specific weight of 2.7 Kg per kW and efficiency ranging from nearly 50 percent at rated power to 60 percent at the part load conditions associated with most automobile operation. IFC is currently developing a 50 kW gasoline fueled power plant for application in both automobiles and buses. The status of these activities will be presented.

INTRODUCTION

Among the key issues that must be addressed in bringing fuel cell power plants into commercial practice for vehicle propulsion is establishing a hydrogen fuel source for the power plant. Hydrogen must either be generated aboard the vehicle from a stored fuel or it must be stored aboard the vehicle which is periodically resupplied from an off-board hydrogen generator. International Fuel Cells has and continues to develop power plants tailored for operating with either hydrogen source. This paper reports only the results of the work to develop power plants with on-board hydrogen generation capability.

When operated on hydrogen the fuel cell's efficiency as an energy converter is unsurpassed, its emissions are zero and its transient response is like that of a battery. When a fuel processor is introduced into the system, for converting a more conventional fuel to hydrogen, these three characteristics -- the cornerstone of the fuel cell's attractiveness and the characteristics which differentiate it from the ICE, may be seriously compromised. Therefore, fundamental among the measures of successful on-board hydrogen generation are: high overall power plant efficiency, throttle responsiveness, i.e., rapid transient response, and low exhaust emissions. These parameters and the factors contributing to achieving them for power plants IFC has under development are discussed.

TRANSIT BUS POWER PLANT DEVELOPMENT

Fleet vehicles, e.g. buses and delivery trucks, typically operate from a fixed base. Therefore, fuels may be chosen for these operations which may be commercially available but not in the consumer or retail infrastructure. These vehicles are also not as constrained by engine compactness or rapid start up time (seconds) as in a personal vehicle. Therefore, they serve as a potential avenue for earliest commercialization of fuel cell power plants in vehicles. IFC's entry into transportation power plant development and demonstration began in this venue.

The team of Georgetown University, Lockheed-Martin, NOVABus America, Booz Allen and Hamilton and IFC developed the 18-meter fuel cell powered transit bus shown in Figure 1. This program is sponsored by the US Department of Transportation's Federal Transit Administration. The power plant is fueled with methanol. It may also be configured to operate on CNG or naphtha. IFC's power plant, including all the equipment required for automatic, push-button startup and shutdown, water recovery and water treatment, weighs less than 1727Kg and occupies less than $5.6M^3$.

Figure 1. 40-Foot Multi-Fuel, Fuel Cell Powered Transit Bus

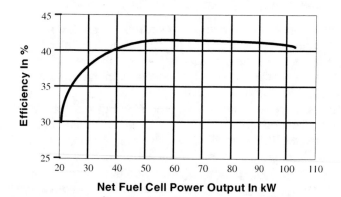

Figure 2. Measured Efficiency of IFC's 100-kW Methanol-air Transportation Power Plant

The power plant's fuel processor is based on a catalytic steam reformer (CSR). Both the fuel processor and the power plant operate at ambient pressure. These contribute to achieving high power plant efficiency. The CSR provides the highest hydrogen content product stream producible with a hydrocarbon fuel conversion process. High fuel stream hydrogen content in turn provides high fuel cell performance and efficiency. Operation at ambient pressure avoids the power losses associated with process air compression for the stack and fuel processor, again contributing to high overall power plant efficiency.

Figure 2 presents the measured overall power plant efficiency, based on the lower heating value of methanol, as a function of net power delivered to the vehicle power controller. The power plant is capable of providing 102 kW at full power. Because it requires no compressor, it draws less than 7 kW to power all power plant components and accessories. This results in an overall efficiency greater than 41% at full power. This efficiency level is comparable to the highest demonstration by any fuel cell power plant that includes a fuel processor which produces hydrogen from hydrocarbon fuels.

Careful design of the power plant's burners, used in the fuel processor to provide the energy fuel processing requires, have kept emission levels very low. Table I pre-

sents the measured emission levels of ICE's operating with both conventional and alternative fuels, an early fuel cell bus power plant operating on methanol, IFC's 100-kW fuel cell bus power plant operating on methanol and the 1998 standard established by the U.S. EPA for heavy vehicle engines.

IFC's fuel cell is at least two orders of magnitude lower in emissions than the "Standard". In the case of NO_x and particulates the measured output is undetectable. Comparison of the fuel cell data with the best diesel results, regardless of the type of fuel used or whether exhaust treatment was employed, shows a one to two order of magnitude improvement over the ICE.

Table 1. Comparison of Internal Combustion Engine and Fuel Cell Power Plant Emissions with US EPA 1998 Standards

Line	Fuel	Power Plant	HC	CO	No$_x$	PM
1	Diesel	DD Series 50*	0.10	0.90	4.70	0.04
2	CNG	DD Series 50	0.80	2.60	1.90	0.03
3	Diesel	Cummins C8.3	0.20	0.50	4.90	0.06
4	CNG	Cummins C8.3	0.10	1.00	2.60	0.01
5	Methanol	94 Fuji Fuel Cell	0.09	2.87	0.03	0.01
6	Methanol	98 IFC Fuel Cell**	<0.01	<0.02	~***	~***
7	1998 Standards****		1.30	15.50	4.00	0.05

*	With catalytic converter
**	IFC test results
***	Undetectable
****	Standard established by the U.S. EPA for Heavy Vehicle Engines

The data in lines 1 to 4 was measured in a testing program at West Virginia University. The data in line 5 was measured in the course of testing the US DOE/US DOT 13.6-meter bus by Georgetown University. IFC's power plant characteristics were measured by IFC. Comparable results for an IFC CSR power plant were obtained in measurements made for the South Coast Air Quality Management District on a power plant located on their site in Diamond Bar California.

Figure 3 illustrates the ability of IFC's 100-kW power plant to provide power for rapid acceleration. The power plant's transient response in this test was set by its control system to a rate of 20% of fuel power per second. This test was conducted at IFC to a simulated load profile. Subsequent road testing of the power plant in the bus when driven by a professional driver of the Connecticut Transit Company (**CT** Transit) prompted the assessment "This bus accelerates and handles better than our latest diesels."

IFC's commercial stationary power plants which operate using the same controller and an identical process flow design have the capability to make 50% of full power transients within one second anywhere within its load range. With an alternate controller and without changes to the system or its hardware the power plant is capable of full 100% power transients in one second. These results should dispel the misconception that a fuel cell power plant with fuel processing cannot make rapid transients. The issue is one of proper design not of capability or lack thereof in the concept.

This first IFC transportation power plant clearly shows that high efficiency, low emissions, and rapid transient response can be achieved in a fuel cell system with a fuel processing unit.

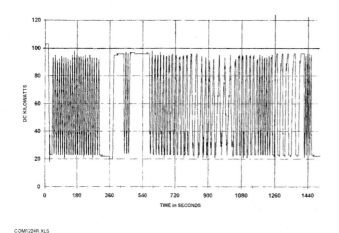

COM1224R.XLS

Figure 3. Capability of IFC's 100-kW Methanol Air Transportation Power Plant to Provide Rapid Acceleration Power

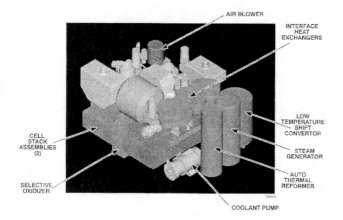

Figure 4. IFC's First Generation Multi-Fuel Automotive Power Plant

AUTOMOBILE POWER PLANT DEVELOPMENT

IFC has begun the aggressive development of a fuel cell power plant for automobile application The requirements this power plant must satisfy are more demanding in terms of weight, volume, cost, fuel type and start-up time than those of a larger commercial vehicle. The power plant concept under development is shown in Figure 4. This is the first in a series of power plants which progress successively to completely meeting automobile requirements. The first power plant of the series, Series 200, will be delivered to OEM's for testing and evaluation in the latter part of 1999.

The power plant as shown is complete with all the equipment and accessories necessary to convert gasoline to dc power. It is based on IFC's proprietary ambient-pressure hydrogen-air PEM stack and proprietary autothermal reformer. The power plant rating in its initial configuration is 50 kW.

The advantages of IFC's ambient-pressure PEM stack was previously demonstrated in the 50 kW power plant developed with US DOE sponsorship and operated by the Ford Motor Company. The results of testing at IFC witnessed by the US DOE and Ford and subsequent testing by Ford showed the power plant to have superior efficiency and specific weight and volume when compared with pressurized power plants. The results were reported at the Fuel Cell Seminar on November 17 - 19, 1998 at Palm Springs, California.[1] The PEM stack for this power plant is now undergoing testing. Figure 5 shows one of the 25 kW reformate-air stacks.

IFC selected an ATR for this power plant because it better meets the more demanding requirements of automotive application than does a CSR. Because an ATR is internally fired it is more compact, lighter, will startup faster and is capable of processing a broader range of fuels, including gasoline. However, it is not as efficient as a CSR and it does not produce a fuel stream as rich in hydrogen as a CSR. The reduction in fuel stream hydrogen content reduces fuel cell efficiency.

IFC has practiced autothermal reforming for more than 15 years. This technology had its roots in work sponsored by EPRI and the US DOE in the late 1970s to develop methods for converting diesel and other heavy fuels to hydrogen for fuel cells. Among the processes considered in the EPRI and DOE work, in addition to the ATR, were partial oxidation (POX), high temperature steam reforming (HTSR) and cyclic catalytic reforming (CCR). Of the group the ATR process was found to be the best suited for use in fuel cell systems in terms of weight, volume, cost and efficiency.

Figure 5. 25 kW Reformate-Air PEM Fuel Cell Stack for IFC's Series 200 Power Plant Under Test

IFC's autothermal reformers have processed methanol, methanol contaminated with diesel oil, gasoline, No. 2 oil, diesel fuel, military aircraft fuel and coal liquids successfully, for extended periods of time, to a hydrogen stream suitable for a fuel cell power plant. The ATR now undergoing test for the power plant is shown in Figure 6. It is one of two versions under development. As shown, the unit incorporates fuel processing reactors in addition to the ATR in a single integrated unit. As shown in Figure 4 the fuel processor is comprised of discrete reactor units. The choice of configuration is determined by the specific installation and integration requirements of an application.

Figure 6.   IFC's 50 kW Integrated ATR Fuel Processing Unit

The overall power plant efficiency target that we have set for the fully developed power plant significantly exceeds 40 percent. We believe this is achievable even with the lower potential efficiency of the ATR, relative to the CSR, because of the high performance of IFC's ambient pressure PEM stack and the low parasitic power requirements of our ambient pressure system design.

Our analytical model's predict emissions characteristics for this power plant that are the same as that of the bus power plant. The exhaust gases burned in each power plant are similar dilute gas streams and the heat recovery burner configurations in the automobile power plant will be the same as that in the bus power plant.

Our analytical models also predict transient response characteristics for the power plant that are superior to the bus power plant.   This results from the fact that the energy needed to drive the reforming reaction is provided in the ATR by burning fuel in the combustion zone of the reactor before hydrogen fuel is produced. In the case of the CSR the energy supplied to drive the reforming reaction is supplied by burning the hydrogen exhaust stream of the stack. Therefore, there is the potential for a short period of energy deficit in the reformer. This can be avoided by proper design. In the ATR it inherently does not exist.

SUMMARY AND CONCLUSIONS

The great attractiveness of the fuel cell as a potential automotive prime mover stems from its high non-Carnot limited efficiency, its zero emission characteristics and its rapid transient response. These are the inherent characteristics that differentiate it from the ICE. However, these are dependent on a pure hydrogen source stored aboard the vehicle. If the fuel cell is adapted to the current fuel infrastructure by addition of a hydrocarbon fuel processor, for it to remain attractive and to continue to have characteristics which differentiate it from the ICE, efficiency must be maintained at a high level, emissions must be kept at a low level and transient response must be kept like that of a battery.

IFC's automotive power plants achieve a high level of efficiency as a result of operation at ambient pressure, which keeps parasitic losses to a minimum, and by using fuel processors which have the highest efficiencies attainable for a given fuel type. Emissions are kept low through careful design of their energy reclamation burners and transient response is maintained through careful design of the energy transfer configuration in the fuel processor.

REFERENCES

1. Meyer, Alfred P. et. al., "Fuel Cell Systems Development for Automobiles and Commercial Vehicles", International Fuel Cells, LLC, November 17, 1998. Presented at Fuel Cell Seminar, Palm Springs, California, November 18, 1998.

Fuels for Fuel Cell-Powered Vehicles

Sean Casten and Peter Teagan
Arthur D. Little, Inc.

Richard Stobart
Cambridge Consultants, Ltd.

ABSTRACT

While it is generally agreed that the PEM fuel cell technology is best for road vehicles, the need for a source of relatively pure hydrogen poses significant challenges. There are two distinct options that are currently being considered:

- On-board processing of gasoline or methanol
- Fueling with hydrogen gas made in an off-board facility

Each option has different implications for the fueling infrastructure and for the technologies required both on- and off-board the vehicle.

In addition, various fueling strategies shift the balance of risk between fuel providers and vehicle manufacturers. Generally speaking, alternative fueling options can be seen to trade off technical risk (e.g., will it work?) for commercial risk (e.g., will anyone buy it?). In seeking a satisfactory business solution, a key issue is the balance between these two risks on the part of the vehicle manufacturer and the fuel provider. Only when this balance is struck will the industry be able to move forward and the number of vehicles grow to a reasonable proportion of the total fleet.

This paper will address the options and the implications associated with them. It will illustrate that decisions cannot be made independently between the different players in the industry.

INTRODUCTION

Fuel cell powered vehicles have been the focus of increasing public attention as multiple manufacturers have publicly asserted that they will have vehicles with fuel cell power trains on the market by 2003/2004. However, while these commercialization plans (and associated R&D investments) have led to dramatic improvements in the technology, it is not yet clear what the fuel of choice for these vehicles will be.

The hesitancy to select a given fuel has been driven in large part by the industry's focus on PEM fuel cell technology. PEM technology has been preferred for automotive applications by virtue of its high power density (which leads to compact size), low temperature (which leads to rapid start-up time) and potential for low-cost manufacture. On all three of these points, PEM fuel cells outperform all other fuel cell technologies.

However, the relative intolerance of PEM fuel cells to impurities in the fuel (particularly CO) places significant constraints on fuel selection, and forces automobile manufacturers and fuel providers to choose between two distinct system architectures, each with markedly different performance levels and technical complexity:

- An on-board reformer supplying a reformate mixture with negligible carbon monoxide, using one of gasoline, methanol or another widely available liquid fuel.
- An off-board reformer and located in a central re-fueling area using natural gas as the fuel, and utilizing a storage volume to meet the needs of a fleet or a large population of publicly owned vehicles.

There are many variations within these two extremes, but it should be readily apparent that these two designs imply dramatically different technologies within the vehicle and the fueling station, technical uncertainties and stakeholders. Each of these alternatives has its own unique advantages and disadvantages, but it is critical to realize that quick decisions are needed on all fronts if the vehicles are to realize the vehicle manufacturers' asserted schedules for commercialization.

MAKING HYDROGEN FOR FUEL CELLS

The choice of hydrogen manufacture is from a number of possible technical solutions and fuel sources:

- *On-board Reforming.* Hydrogen-rich gas is produced from liquid fuels on board the vehicle with partial oxidation, steam reforming or auto-thermal techniques.

- *Central Merchant Hydrogen.* A plant using large scale steam reforming of natural gas produces hydrogen which is transported over the road to local hydrogen dispensing stations
- *Local Reformer/Fueling Station.* A hydrogen re-fueling station with on-site hydrogen storage, compression and dispensing equipment in which the hydrogen is manufactured in a natural gas reformer (which may be based on steam reforming, partial oxidation or autothermal reforming technology)
- *Local Electrolyzer/Fueling Station.* A hydrogen refueling station with on-site hydrogen storage, compression and dispensing equipment in which the hydrogen is manufactured via electrolysis

Of these four alternatives, the Central Merchant Hydrogen and Local Electrolyzer/Fueling Station options can be quickly dismissed as being impractical for a widespread hydrogen fueling infrastructure.

While the use of merchant hydrogen is an appropriate strategy for small-scale fleet demonstrations (such as the Ballard bus in Chicago), the logistics of a large-scale fuel chain based on this strategy are daunting. There are numerous safety and codes and standards issues associated with hydrogen storage, transport and use, all of which favor that the hydrogen be generated as close to the end user as possible. It is of course possible that at some point in the future, consumers and governments will broadly accept the distribution of hydrogen as a fuel. However, it must be kept in mind that the profitability of the automotive industry is driven by economies of scale measured in production runs of 100,000 vehicles per year or greater. The successful introduction of fuel cell vehicle technology by 2004 will have to be based upon a fueling infrastructure that can be rapidly implemented – by this metric alone, central merchant hydrogen plants are unlikely to represent a viable near term alternative.

Local electrolyzers are unlikely to realize broad success based solely on the cost of the hydrogen thus produced. While access to low-cost (e.g., wholesale, base-load electricity) can be used to generate competitively priced hydrogen, the electricity prices common to most of the U.S. are considerably higher. At electricity rates common to light commercial establishments (such as would be encountered in local fueling stations), the hydrogen thus produced will be too expensive to be used as a transportation fuel. Nonetheless, electrolytic hydrogen may be a competitive in those locations where:

- Wholesale power prices are exceptionally low, such as in nuclear or renewable power plants
- Broad differences exist between on- and off-peak energy prices, and low nighttime rates can be used to generate hydrogen to be used during the day.

These markets may prove to be important once markets for hydrogen develop, but they are not sufficiently widespread at present to justify the rollout of a new vehicle based on electrolysis-derived hydrogen.

Having ruled out these two options for large-scale hydrogen production, we are therefore left with two practical options for fuel cell vehicles: on-board reforming or local natural gas reformers.

ON-BOARD REFORMING – There are several competing technologies for on-board reforming, each with distinctive advantages and disadvantages. For the purposes of this paper, it is not necessary to highlight the details of the alternatives for on-board reforming, except to note that:

- Modest differences in efficiency and cost exist between each reformer technology, but these are not so large as to overwhelmingly favor any particular technology.
- All reformer technologies include complex sub-systems for shift-reactions, water recovery and thermal management that add cost, complexity and weight to the vehicle.
- The use of dilute hydrogen as a fuel for the fuel cell requires a redesign of the flow fields (the stack cannot be "dead-ended"), which necessarily implies reduced hydrogen utilization as compared to direct-hydrogen fuel cells – this leads to reduced fuel cell efficiency and increased fuel cell size and cost.
- Fueling with carbon-containing fuels mandate that CO-control functionality be added to the fuel processor to protect the fuel cell stack.

OFF-BOARD REFORMING – From a purely technical perspective, the generation of hydrogen off-board the vehicle has several compelling advantages. Most noticeably, substantial complexity is removed from the vehicle (which has the most stringent constraints on volume, weight and cost of any operation within the fuel chain). Figure 1 shows a system configuration for a vehicle with on-board reforming, and illustrates the room for simplification afforded by off-board hydrogen generation.

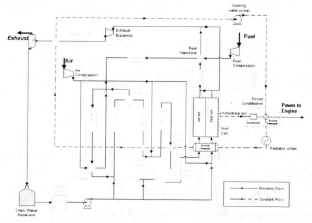

Items in **black** are required by systems with on-board reforming or direct hydrogen fueling
Items in gray are required only by systems with on-board reforming.

Figure 1. System Complexity of Direct Hydrogen FCVs vs. On-board Reforming

In addition to these on-vehicle advantages, it should be noted that the generation of pure hydrogen in large-scale, stationary chemical plants is a well-established process, while on-board hydrogen generators are currently a developmental technology. Moving the generation function off-board the vehicle therefore brings about substantial reduction in the technology risk throughout the fuel chain.

However, the complications associated with storage of hydrogen on-board the vehicle should not be overlooked. Generally speaking, there are three commercially available technologies for on-board hydrogen storage, each of which is discussed in detail below.

Compressed Hydrogen – Compressed hydrogen is a natural extension of the CNG industry, insofar as the technologies to deliver and store high pressure gases have been developed for CNG, and may be applied to hydrogen fueled vehicles. Storing hydrogen at pressures of 5,000 psi allows one to realize energy storage densities of approximately 4 MJ/liter. While this is well below the energy density of gasoline (approximately 30 MJ/liter), it can be acceptable once higher fuel economies for fuel cell vehicles are taken into account.

However, compressed hydrogen is not without its challenges in automotive environments. Most notably:

- Hydrogen is an extremely diffuse gas – even with a 3-fold increase in vehicle efficiency that one would expect from a direct hydrogen fuel cell vehicle, hydrogen storage pressures of 5,000 psi or greater are required to minimize the volume occupied by the hydrogen storage tank. This is considerably higher than the 3,600 psi pressures common in CNG vehicles.

- Hydrogen is a notoriously "leaky" gas by virtue of its small molecular diameter. This implies that all seals and valves must be redesigned to safely store compressed hydrogen gas.

- Compressor power requirements are significant, and can amount to a substantial fraction of the energy content of the hydrogen pressurized, as shown in Figure 2.

Two critical observations can be made from this figure:

- Electricity consumption for compression can noticeably reduce the overall efficiency of a fuel chain that includes hydrogen-fueled fuel cell vehicles. Note that while this plot depicts only the electricity used, its effects are magnified if one includes the primary fuel used to generate the electricity (U.S. average generation efficiency is approximately 33%).

- As on-vehicle storage pressure is reduced, the power requirements fall exponentially. This has dramatic implications for the fueling station design and cost – in other words, the choice of an on-board hydrogen storage system can have dramatic upstream impacts.

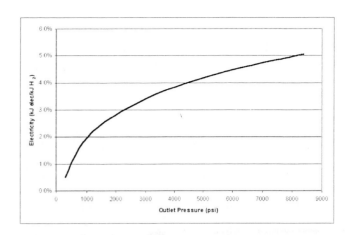

Figure 2. Power Requirements for Hydrogen Compression -- 4 stage, intercooled compressor with 85% stage efficiencies[1]

Finally, it should be noted that there remain significant safety concerns associated with the storage of high-pressure hydrogen on-board a vehicle. It is critical to realize that these safety concerns include both real *and* perceived safety issues – both of which must be addressed prior to successful commercialization.

Nonetheless, compressed hydrogen has been successfully demonstrated in multiple vehicles to date, including Ballard's fuel cell bus in Chicago and the Ford P2000 prototype.

Liquefied Hydrogen – Liquefied hydrogen storage systems press compressed hydrogen systems to their thermodynamic limits by physically liquefying the hydrogen to achieve volumetric energy densities of approximately 8.5 MJ/liter. Liquefied hydrogen systems have been demonstrated in multiple demonstration vehicles, most recently in DamlerChrysler's NECAR 4.

However, liquefied hydrogen systems face the same real and perceived safety challenges that are associated with compressed hydrogen storage. In some cases, these risks may be exacerbated by the phenomenon of hydrogen "boil-off", whereby a finite amount of hydrogen vaporizes and escapes from the tank in regular service. This poses a particular problem for vehicles that may be stored in closed garages, in which hydrogen could accumulate and create an explosive mixture in the air.

In addition, the parasitic power requirements for hydrogen liquefaction can be substantial. It has been estimated (see reference 2) that the energy requirements for liquefaction are approximately 9 kWh/kg hydrogen – or 23% of the energy content of the hydrogen thus liquefied on an HHV basis.

---

1. Note that the compressor will have to require some pressure greater than the pressure on-board the vehicle in order to rapidly refuel the vehicles in a hydrogen fueling station.

Metal Hydrides – Metal hydrides offer substantial safety and volume benefits over compressed and liquefied hydrogen storage. By storing hydrogen in a hydrogen-metal complex, they are able to realize very high volumetric densities (9 – 12 MJ/liter) that exceed those of liquid hydrogen. However, they face several key challenges in automotive applications:

- Their weight density remains quite low, on the order of 1 – 3 MJ/kg, which can significantly increase the overall weight of the vehicle

- Many of the metals used (La, Ti, Mn, Ni, Zr, among others) are quite expensive, which can increase the cost of the metal hydride to unacceptable levels *independent* of the costs of packaging, manufacture, marketing, etc.

- While some low-cost materials offer higher gravimetric densities (as much as 9 MJ/kg) and lower costs, these tend to use magnesium-based compounds which have high heats of formation and require high temperatures (>200°C) to release their hydrogen. These thermodynamic constraints make them impractical in PEM fuel cell vehicles. (See Figure 3)

- They require complex packaging for fluid flow of both hydrogen gases and liquid heat transfer materials, heat transfer subsystems, and allowances for thermal expansion and contraction during charge/discharge cycles. A schematic of a generic metal hydride system is illustrated in Figure 4.

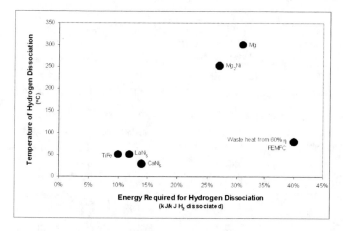

Figure 3.   Thermodynamic Requirements of Selected Metal Hydrides[2]

---

2. Note that the PEMFC data point is based on the heat available rather than the heat needed for dissociation; a metal hydride system that is viable for on-vehicle applications must be capable of operation using this waste heat – the higher temperature systems would impose a substantial efficiency penalty on the vehicle as they would have to use fuel cell-derived electricity to provide their heat of operation.

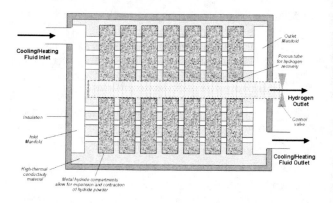

Figure 4.   Simplified Metal Hydride System Schematic

These limitations make it difficult for metal hydrides to find use in automotive applications, although they might be quite useful in other applications that require hydrogen storage.

Other Hydrogen Storage Technologies – There are a number of other hydrogen storage technologies currently in development, but none of them have yet been sold in commercial markets. One of the more intriguing materials currently in development is carbon-based storage systems, either in nanotube or nanofiber form. In some laboratories, these materials have been observed to achieve extraordinarily high hydrogen storage densities – in some cases, the reported densities have been greater than that of *solid* hydrogen. In addition, based on the material costs of carbon, these materials have the potential for low costs that are not likely to be matched by metal hydride systems. However, there are numerous outstanding questions about this technology at present, as there is almost no understanding of the theory which would allow hydrogen to realize these densities or of how to manufacture the ideal carbon "shape". Nonetheless, the observance of these densities in multiple laboratories suggests that the technology deserves closer attention.

COSTS – Multiple studies have shown that produced hydrogen is more expensive than gasoline on a $/gallon gasoline equivalent basis. (Note that this is essentially a $/unit energy cost.) Comparably, many critics of methanol have noted that the cost to retrofit existing stations to dispense methanol (coupled with the low energy density of methanol relative to gasoline) will make it more expensive than gasoline on a $/unit energy basis.

However, it is critical to realize that alternative fuels need not be price competitive on energy terms with gasoline in order to be economically competitive. Instead, they need only be competitive on a $/mile driven basis, which is the consumer's metric of value. We have developed a cost model for a hydrogen fueling station that includes an on-site, POX-based natural gas reformer and capacity for 300 fuel cell vehicles per day. Details of this model are shown in Table 1. Costs in this table are based upon production volumes of 100 stations/year.

Table 1: Costs of Hydrogen Production in a 300 car/day Fueling Station

Expense Item	Cost ($/GJ H₂)	Comments
Electricity Cost	$1.09	Assumes 6 ¢/kWh
Gas Cost	$6.47	Assumes 4 $/MMBTU
Reformer O&M	$0.43	Assumes 10% of capital cost
PSA O&M	$0.38	Assumes 10% of capital cost
Compressor system O&M	$0.03	From literature estimates of Ogden, et al.
Labor	$3.40	Assumes 3 employees at $50,000/year total labor cost
Annual capital charge	$3.88	Annual cost = 15% of total capital cost
Markup (profit, marketing, etc.)	$0.97	25% of capital cost
Total $/GJ	$17.14	Does not include tax
Total $/mile before tax	$0.02	Assumes H₂-FCEV attaining 91 mpg gasoline equivalent
Compare $/GJ for gasoline	$9.23	Assumes $1.20/gallon, 130 MJ/gallon INCLUDES TAX
Compare $/mile for gasoline ICEV	$0.04	Assumes 30 miles/gallon
Gasoline ICEV $/mile before tax	$0.03	U.S. Tax is approximately 25% of total fuel cost

Note that the costs are quite competitive based on U.S. energy rates, and are facilitated by the high vehicle efficiency of hydrogen-fueled fuel cell vehicles. However, it must be noted that the construction of such a fueling station would be an extremely capital-intensive endeavor – on the order of $1 million per station. As with any capital-intensive project, it will therefore be critical to achieve a high load-factor if the station is to be economically viable. The sensitivity analysis shown in Figure 5 illustrates that of all the assumptions in this analysis, the one that most strongly affects the economics of small-scale hydrogen production is the load factor of the fueling station.

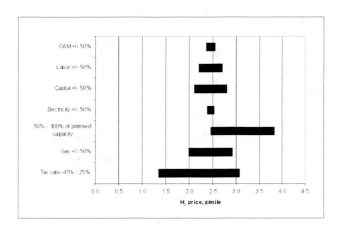

Figure 5.   Sensitivity of Hydrogen Price to Cost Model Assumptions

In this figure, the strong impact that load factor has on the overall economics of the fueling station should not be overlooked. In essence, this implies that one needs to have a careful understanding of the load profile of the fueling station before beginning construction, as much of the capital is sized for the peak load. If one is not able to realize high capacity factors for the fueling station, the cost of hydrogen very quickly rises to unacceptable levels.

Interestingly, even a relatively low load factor (e.g., <70%) would still be considerably higher than that which could be achieved by an on-board fuel processor, where use is limited to a few hours per day (and is typically at only a fraction of peak power). By increasing the duty cycle of the fuel processor (which is often <1% when housed on the vehicle), the annual cost of the entire fuel chain is thus reduced.

RISKS – There are three distinct forms of risk that must be considered in the implementation of any of these options, and each risk must be assessed from the perspective both of automobile manufacturers and of fuel providers.

1. *Technical Risk:* What is the chance that the technology will not be able to meet consumer performance requirements?
2. *Cost Risk:* What is the chance that the technology will not be able to meet consumer cost requirements?
3. *Commercial Risk:* What is the risk that markets will not exist for the technology?

Of these three types of risk, note that the first two are almost entirely under the control of a single industry. If automobile manufacturers succeed in producing low-cost, high-performance fuel cell vehicles, then they will face no technical or cost risk. Conversely, if fuel suppliers succeed in providing alternative fuels at low costs, with acceptable performance (in terms of fuel availability, refill time, etc.), then they will face no technical or cost risk.

However, commercial risk cannot be eliminated solely by the efforts of a single company or industry. If the automobile industry produces low-cost, high-performance hydrogen-fueled fuel cell vehicles, there is no guarantee that markets will materialize unless hydrogen is broadly available. Simultaneously, if hydrogen is made widely available, markets will not materialize until hydrogen-utilizing technologies are available. Indeed, this is the perennial "chicken-and-egg" problem of the hydrogen industry.

VEHICLE COSTS – In spite of significant progress in other areas, it is compressed hydrogen that offers the lowest overall vehicle cost in the near-term. Table 2 provides a comparison of the costs for multiple fueling options.

Table 2: Comparison of the Potential Costs of Alternative Fuel Cell Vehicle Technologies[3]

	Powertrain cost as a function of range		Comments
	250 km	500 km	
On-board reforming			
Fuel tank	$33	$57	Increased stack size to handle reformate may add 10-20% to the cost of the fuel cell stack
Reformer	$900 - $1,500	$900 - $1,500	
Fuel cell	$1,750 - $5,000	$1,750 - $5,000	
Total*	**$2,700 - $6,500**	**$2,700 - $6,600**	
Pressurized H₂			
Tank	$900	$1,500	Some literature sources have quoted lower values for compressed hydrogen tank costs
Fuel cell	$1,750 - $5,000	$1,750 - $5,000	
Total*	**$2,650 - $5,900**	**$3,250 - $6,500**	
Metal Hydrides			
Hydride system	$2,300 - $3,600	$4,600 - $7,200	Hydride system costs based on Toyota's TiCrV material
Fuel cell	$1,750 - $5,000	$1,750 - $5,000	
Total*	**$4,000 - $8,600**	**$6,400 - $12,200**	

* For primary drivetrain subsystems - does not include balance of plant components (air handling, transmission, etc.) or any additional batteries that may be required.

FUEL CHAIN EFFICIENCIES – IS HYDROGEN REALLY COMPETITIVE?

When hydrogen is used directly as a fuel for fuel cell vehicles, there is no doubt that the vehicle efficiency increases, and tailpipe emissions are completely eliminated. However, some of these emissions are simply moved upstream in the fuel chain if the hydrogen is derived from natural gas, and legitimate questions are frequently asked about the overall public benefit of hydrogen derived from natural gas.

Detailed fuel chain analyses carried out at Arthur D. Little have shown that hydrogen fuel cell vehicles can indeed reduce CO_2 emissions over other options when derived from natural gas. This results in large part from the higher on-vehicle efficiency of hydrogen-fueled fuel cell vehicles, which leads to less energy consumption in all upstream portions of the fuel chain. However, emissions of NOx are comparable for all fuel chains that use fuel cell vehicles, suggesting that the improvement in urban air quality will not show substantial variation amongst competing fuel options.[4] (Note that the use of *any* fuel cell vehicle reduces NOx emissions by approximately 75% relative to a gasoline IC engine meeting ULEV standards.) A summary of the calculated fuel chain emissions is shown in Figure 6.

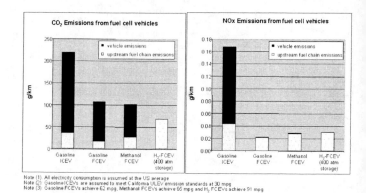

Note (1) All electricity consumption is assumed at the US average
Note (2) Gasoline ICEVs are assumed to meet California ULEV emission standards at 30 mpg
Note (3) Gasoline FCEVs achieve 62 mpg, Methanol FCEVs achieve 65 mpg and H₂ FCEVs achieve 91 mpg

Figure 6. Fuel Chain Emissions for Alternative Fuel Cell Vehicle Architectures

CONCLUSION

Having compared several alternative fueling options for fuel cell vehicles, several observations can be made:

- Alternative fuels for fuel cell vehicles are not likely to show substantially different costs. While the elimination of the reformer can lower the cost of hydrogen-fueled FCVs, the addition of a hydrogen storage system will offset large portion of these costs.

- The cost of hydrogen produced from natural gas in local fueling stations will have a comparable cost to gasoline on a $/mile basis.

- The improvements in air quality associated with fuel cell vehicles are relatively comparable for multiple fueling options. CO_2 emissions will be slightly lower for hydrogen-fueled fuel cell vehicles, but emissions of criteria pollutants will be comparable on a full-fuel cycle basis.

Given that costs and public benefits do not show substantial variation between different fueling options, there is no overriding reason to favor one vehicle architecture over another *solely on the basis of their potential public benefits*. Instead, the decision to chose a particular fuel will be based on business and strategic issues related to the relative risk, liability and commercialization timeframe of each option. This will in turn be based on the relative risk, liability and commercialization schedules desired by each of the major stakeholders – auto companies, fuel companies, governments and consumers.

It is readily apparent that with the exception of gasoline-fueled FCVs, the successful introduction of any fuel cell vehicle will require a partnership between the automobile and fuel industries to minimize the risk to each. Indeed, some of these partnerships are already being seen, in the form of the California fuel cell partnership and the Iceland hydrogen program. However, these small first steps will need to be rapidly accelerated if they are to lead to hydrogen- or methanol-fueled FCVs by 2004.

---

3. Costs for on-board reforming are based on long-term PNGV targets and prior ADL analyses. All costs are based upon high-volume production (>100,000 units/year).

4. The reduction in CO emissions closely mirrors that of NOx emissions, since most of the emissions in conventional fuel chains occurs on board the vehicle and all fuel cell architectures eliminate this pollutant from the tailpipe. However, NMOG emissions would be expected to remain higher for liquid-fueled FCVs by virtue of evaporative emissions that occur during refueling.

REFERENCES

1. Arthur D. Little, *Alternative Fuel Infrastructure Economics an Issues*, Completed for U.S. DOE under subcontract DE-AC36-83-CH10093, March 1997.

2. Berry, Gene. "Hydrogen as a Transportation Fuel: Costs and Benefits", Lawrence Livermore National Laboratory. Published by U.S. Department of Commerce NTIS, Springfield, VA. March 1996.

3. Berry, Gene D., J. Ray Smith and Robert N. Schock. "A Smooth Transition to Hydrogen Transportation Fuel". Submitted to DOE Hydrogen Program Review, Coral Gables, FL. April 19-21, 1995.

4. DeLuchi, Mark, *Emissions of Greenhouse Gases from the Use of Transportation Fuels and Electricity*, Volumes I and II, Center for Transportation Research, Argonne National Laboratory, November 1993.

5. Doss, E.D., R. Ahluwalia and R. Kumar. "Hydrogen-Fueled Polymer Electrolyte Fuel Cell Systems for Transportation". Argonne National Laboratory report ANL-98/16. August 1998.

6. Ogden, Joan M. "Infrastructure for Hydrogen Fuel Cell Vehicles: A Southern California Case Study". Presented at the '97 World Car Conference, Riverside CA. January 19-22, 1997.

2000-01-0592

Integration of Fuel Cell Technology into a Hybrid Electric Vehicle

Michael Ogburn, Alex Boligitz, William Luttrell, Brian King, Scott Postle, Robert Fahrenkrog and Douglas J. Nelson
Virginia Polytechnic Institute and State Univ.

ABSTRACT

The Virginia Tech Hybrid Electric Vehicle Team (HEVT) has integrated a proton exchange membrane (PEM) fuel cell as the auxiliary power unit (APU) of a series hybrid design to produce a highly efficient zero-emission vehicle (ZEV). This design is implemented in a 1997 Chevrolet Lumina sedan, renamed ANIMUL H_2, using an efficient AC induction drivetrain, regenerative braking, compressed hydrogen fuel storage, and an advance lead-acid battery pack for peak power load leveling. The fuel cell is sized to supply the average power demand and to sustain the battery pack state-of-charge (SOC) within a 40-80% window. To optimize system efficiency, the fuel cell is driven with a load-following control strategy. The vehicle is predicted to achieve a combined city/highway fuel economy of 4.3 L/100 km or 51 mpgge (miles per gallon gasoline equivalent).

INTRODUCTION

On May 14, 1996, Daimler-Benz introduced the fuel cell powered passenger vehicle NECAR II (New Electric CAR) (H&FCL, May 1996). This event marked the beginning of recent international activity among carmakers to develop the first fuel cell powered passenger car suitable for everyday operation. As part of the effort to develop a consumer acceptable fuel cell vehicle, the U.S. Department of Energy (DOE) selected the Virginia Tech Hybrid Electric Vehicle Team as one of the recipients of a PEM fuel cell stack provided by Energy Partners. This fuel cell has been integrated into a 1997 Chevrolet Lumina for the 1998-1999 FutureCar Challenge.

The focus of the Virginia Tech HEVT is to produce a series hybrid vehicle that meets the goals of the FutureCar Challenge and the Partnership for a New

Generation of Vehicles (PNGV). As a guideline of the required capabilities of the converted vehicle, the 1999 FutureCar Challenge specifies the following criteria:

- Standing 200 m (1/8 mi) acceleration < 15 s
- 525 km (326 mi) range
- Curb weight < 1950 kg (4300 lb.)
- Seating for five
- 250 L (8.8 ft^3) luggage capacity
- Superior energy efficiency
- Minimal emissions of NMHC, NO$_x$, CO

A vehicle powered by a fuel cell stack has the potential of addressing all of these criteria. A fuel cell is a device that harnesses the energy produced during the electrochemical reaction between hydrogen fuel and oxygen. The products of this process are electricity, heat, and water. Unlike other auxiliary power units (APU), by-products such as nitrogen oxides are not present due to the lack of combustion. The absence of carbon from the reaction excludes hydrocarbons, carbon dioxide, and carbon monoxide as possible emissions. Fuel cells can theoretically convert energy at higher efficiencies than internal combustion engines since they are not limited by inefficiencies inherent to the Carnot heat engine cycle. Given their greater efficiency and lower operating cost, fuel cell vehicles are likely to have lower life-cycle costs than gasoline or battery-powered vehicles (Sperling, 1995). Hybrid fuel cell vehicles offer improved acceleration and higher drivetrain efficiency of battery-powered vehicles, while extending the range. Daimler-Benz has shown that the fundamental technical problems of implementing fuel cells into automobiles can be resolved. General Motors has indicated that fuel cell systems could be built for about the same price as conventional systems of an internal combustion engine powered vehicle, once in mass production.

Designing a fuel cell vehicle that can meet the FutureCar Challenge criteria requires that many important design parameters must be addressed. Component selection and packaging directly affect the luggage capacity of the vehicle. Packaging also affects passenger capacity and curb weight.

Although the fuel cell provides superior efficiency and zero emissions, the efficiency of the entire drivetrain and any possible emissions from the vehicle need to be considered. A study (Cuddy and Wipke, 1996) shows that a 1% increase in APU efficiency results in a 1% improvement in fuel economy. A 1% decrease in mass results in a 0.6% increase in fuel efficiency, and a 1% decrease in aerodynamic drag results in only a 0.3% improvement in fuel economy. These results seem to indicate that improvements in APU efficiency have the greatest effect on fuel economy. Simulations verify these results for the Virginia Tech FutureCar.

A range of 525 km (326 mi) is difficult to achieve with a compressed hydrogen gas fuel system while maintaining trunk space. Other possible fuel storage systems must be researched and considered. In order to achieve the acceleration requirements, peak power of about 80 kW (107 hp) is required. The average power requirements of the vehicle are less than 18 kW (24 hp).

Based on the design constraints, HEVT identified the most critical design considerations as fuel cell integration, drivetrain selection and implementation, battery selection, fuel storage system, control strategy, component weight reduction, and packaging. Final packaging of all the major components within the Virginia Tech FutureCar is illustrated in Figure 1. This method of packaging the components provides room for five passengers and more than 250 L (8.8 ft^3) of trunk space. The overall system schematic is shown in Figure 2. This schematic effectively depicts the organization of the vehicle's propulsion system. A summary of the specifications for each of the major systems on the vehicle is included at the end of this paper.

DESIGN CONSIDERATIONS

Maintaining safety, utility, performance, and consumer acceptability while improving fuel economy and eliminating emissions called for the following approach:

- The existing Lumina body structure, interior, and overall integrity should remain intact

- The new drive system should be transparent and easy to use
- Performance and handling should be as close as possible to stock
- Hydrogen fuel used for the fuel cell eliminates local production of hydrocarbons, carbon monoxide, and carbon dioxide

The fuel cell stack provided by Energy Partners is capable of providing 20 kW (26.8 hp) peak power at 138 kPa gauge (20 psig). Due to limitations of the air delivery system, however, operation of the stack is limited to 18 kW at 69 kPa (10 psig). This limited the possible types of power train configurations that could be selected. A pure fuel cell-powered car like the NECAR II and NECAR III from Daimler-Benz (H&FCL, Jun 1996 & Oct 1997) was eliminated as a possibility. Both the NECAR II and III have fuel cell systems that are capable of producing 50 kW (67 hp), sufficient for maintaining speed while climbing grades and providing adequate acceleration. To provide temporary power for acceleration, a vehicle can also use an energy storage device, such as a battery pack, for peak power requirements.

Like Toyota's FCEV, unveiled at the 1997 Frankfurt Auto Show (H&FCL, Oct 1997), the Virginia Tech FutureCar takes on the design layout of a series hybrid electric vehicle with the PEM fuel cell acting as the APU. A series hybrid provides all driving power to the wheels through an electric motor. The electric motor draws power from a battery pack, and the APU replenishes the batteries while driving. Since the fuel cell system's net power output exceeds the average energy requirements of the vehicle, the system will be charge-sustaining. For such a design, the battery pack size, fuel storage capacity, and the weight and space of the vehicle must be well balanced. These decisions are dependent on drivetrain selection, battery type used, and the fuel storage systems that are available. The following sections describe the major decision factors and selection processes.

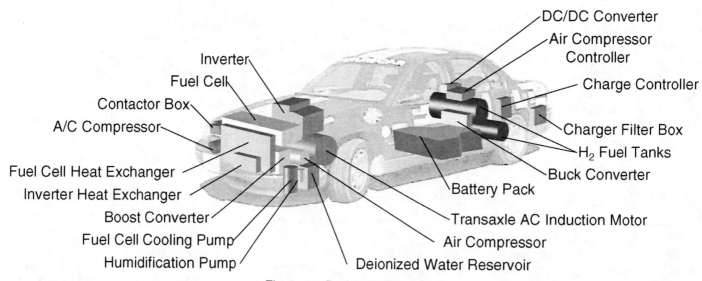

Figure 1. Packaging Diagram

DRIVETRAIN

Important design decisions such as the sizing and voltage of the battery pack in a series hybrid depend on the selection of the electric motor and controller. The desired characteristics of the electric drivetrain system are high efficiency, compact packaging, low weight, high maximum speed, and high power. Four electric drivetrains were investigated and compared: the Solectria 140 kW Dual Motor, the Hughes Dolphin 50, the AC Propulsion AC-100, and the GE MEV-75 3-phase AC induction motor used in the Ford Ecostar. Table 1 provides specifications for these drivetrains. All four electric drivetrains considered use AC induction motors.

The Hughes system does not provide the peak power needed to achieve the necessary acceleration. Although the Solectria dual motor system has the potential for a unique packaging design, the weight of the entire package is unacceptable. The AC-100 and the GE MEV-75 motors have very similar characteristics and meet the system requirements quite well.

A GE system was chosen for its transaxle packaging, continuous power output, higher maximum speed, and higher efficiency. Virginia Tech has experience using this motor with the EV2000 controller and local technical support through VPT, Inc. in Blacksburg, Virginia. This motor has a continuous power rating of 60 kW (80 hp) at 3000 rpm and a peak power rating of 85 kW (114 hp) at 4000 rpm. The maximum rotor speed is electronically governed to 13,500 rpm. Torque at the rotor is constant at 195 N-m (143 ft-lb) from 0 to 4000 rpm, and then decreases with speed to 40 N-m (30 ft-lb) at 13,500 rpm as the motor operates in near constant power mode. A planetary gear set with a gear ratio of 12.18:1 and differential are housed inside the lightweight aluminum-magnesium alloy transaxle motor case. All of these features provide a drivetrain system with ample power in a light compact package.

Table 1. Electric Motor Options

	Solectria Dual Motor	Hughes Dolphin 50	AC Propulsion AC-100	GE MEV-75
Weight kg (lb)	150 (331)	90 (198)	77 (170)	106 (234)
Nominal Power kW (hp)	36 (48)	38 (51)	41 (55)	60 (80)
Peak Power kW (hp)	134 (180)	50 (67)	100 (134)	85 (114)
Peak Torque N-m (ft-lb)	140 (103)	160 (118)	150 (111)	195 (144)
Max. Speed (rpm)	13000	9000	12000	13500
Eff. (%)	92	93	91	93

INVERTER – The General Electric EV2000 inverter that controls the AC induction motor is designed for maximum efficiency by using a two-mode, six-step technique to control current in each of the motor windings. At lower speeds and torques, the switching is pulse-width modulated over a range of frequencies, while at higher speeds and torques, the devices are hard-switched in a typical six-step scheme. Switching is performed by 3 Toshiba IGBT module pairs, rated at 400 A and 600 V. The motor control feedback loop consists of a two channel optical hall-effect tachometer sensor for basic variable slip control of the drive, as well as quadrature flux sensors which maximize drive performance by allowing the inverter to operate the motor near magnetic saturation.

The basic design provides 145 ft-lb of stall torque when supplied from a 300 A inverter. The motor/inverter efficiency is over 90% from maximum torque to torque values down to 5 ft-lb (3.5% of maximum). The inverter controller adjusts the motor flux level and basic modulation frequency to reduce the motor and inverter losses when operating at low torque. A midsize [1451 kg (3200 lb) test weight] Ecostar vehicle on the Federal Urban Driving Schedule operates with an overall energy consumption including accessory load of 235 watt-hours/mile (147 watt-hrs/ton-mile). The efficiency and performance are well matched to the requirements of a light-duty electric drivetrain at the operating areas of interest, which include part-load and half speed.

The EV2000 inverter features several safeguards to ensure maximum performance under limiting conditions. A temperature sensor embedded in the stator windings allows the inverter to reduce drive power to protect the motor under high operating temperatures. Temperature sensors within each IGBT pair allow the inverter to provide maximum performance while simultaneously guarding against device failure from overheating. Traction battery voltage is also carefully monitored under load. The inverter begins current limiting as soon as the traction battery drops to 60% of its nominal voltage. This gives the inverter a "limp-home" mode that allows the vehicle to be driven at low power with nearly depleted batteries until they can be recharged.

The EV2000 inverter is designed to emulate a conventional automobile through such features as a creep mode, which simulates the creeping action of an automatic transmission, and regenerative braking to replace friction braking. An easy-to-use software interface allows the user to customize several control options.

VEHICLE INTERFACE MODULE – The Vehicle Interface Module (VIM) is the primary interface to the General Electric drive motor controller. All signals controlling the electric vehicle drivetrain go through the VIM. These signals include accelerator position, brake pedal position, and drive selector position. The VIM also provides status outputs for fault conditions, such as low voltage, high temperature, and ground fault.

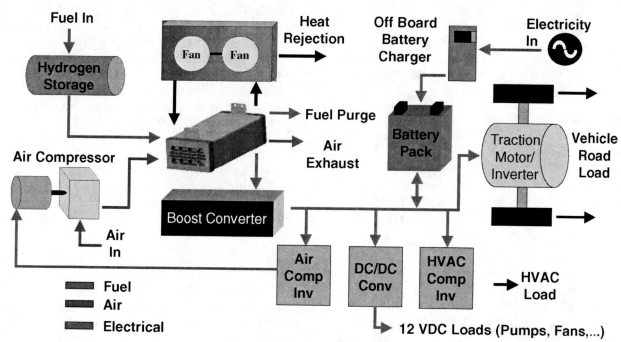

Figure 2. System Schematic

POWER REQUIREMENTS – Once the drive motor for ANIMUL H_2 was selected, several driving scenarios were simulated using ADVISOR v2.1 software, in order to determine power and charge-sustaining requirements for the vehicle. ANIMUL H_2 was modeled using a vehicle weight of 1900 kg (4190 lb), a motor rated for 60 kW continuous (85kW peak) power (94% efficient), a series load-following powertrain controller powered by twenty-eight 16 Ah batteries, and a 12 kW continuous (18 kW peak) power fuel cell (49% efficient) as the APU. A 3.4 kW load to power the air compressor and a 1 kW load to account for efficiencies and power requirements of fuel cell subsystems were applied to the above model as constant electrical loads.

ADVISOR output indicates that the vehicle can be charge-sustaining using an 18 kW (nominal) APU with a maximum accessory load of 4.4 kW. With this load, the vehicle is able to maintain a state of charge of approximately 42.5% on the FUDS driving cycle and 40.0% on the highway cycle given initial charges of 41% and 40%, respectively. The APU provides enough power to keep the vehicle charge-sustaining under nominal driving conditions. Driving up steep grades at high speeds, however, will deplete the batteries unless the APU power is increased and the average state of charge can be raised.

Acceleration performance of the preliminary design was also simulated using the above input data. Assuming power limited acceleration, the 200 m (1/8 mi) standing-start result is 11.4 s at 97 kph (60 mph). This result matches the performance of the stock vehicle at the same mass. Gradability of the modified vehicle is greater than 40% from 0 to 32 kph (20 mph), and is still greater than 7% at 112 kph (70 mph) (Senger, Merkle and Nelson, 1997).

OVERALL EFFICIENCY

An average efficiency for ANIMUL H_2 as a hybrid electric vehicle for each drive cycle can be estimated by dividing the total amount of energy required at the wheels by the total energy required as input to the fuel cell.

$$\eta = \frac{WheelEnergy}{FuelCellInputEnergy}$$

Efficiency for the drivetrain can also calculated in this manner using the energy at the battery terminals in place of the fuel cell input energy. The wheel, fuel cell, and battery energies are taken from an ADVISOR simulation that assumed a 4.4-kW accessory load and included regenerative braking. Table 3 shows the results of these calculations.

Table 3. Overall Efficiencies

Cycle	HEV	Drivetrain
City (FUDS)	17.6%	75.1%
Highway(HWFET)	22.6%	84.4%

Fuel economy values listed in Table 4 below are also taken from ADVISOR simulations. These values represent the EPA unadjusted miles per gallon gasoline equivalent of the vehicle.

Table 4. Fuel Economy

Cycle	Fuel Economy(mpgge)
City(FUDS)	49
Highway(HWFET)	53

CONTROL STRATEGY

The 1999 Virginia Tech FutureCar employs a combination of thermostatic and load-following control. With thermostatic control, the APU is used as a supplementary charger and the battery pack provides the rest of the power. A Cruising Equipment E-Meter, which monitors the battery pack SOC, and the shifter are attached to a single-board computer. When the SOC drops below 40% and the shifter is in hybrid mode, the single board computer instructs the microcontroller to start the fuel cell. After the fuel cell has operated long enough for transients to dissipate, the microcontroller turns on the boost converter, thus placing a load on the fuel cell. The boost converter uses a load-following strategy. Under light load, the boost converter operates the fuel cell at the minimum power required to sustain battery SOC. The air compressor also operates with a load-following strategy. If the stack is under a light load, the air compressor provides minimum flow. At high load, the air compressor supplies maximum flow. This allows the fuel cell to produce power at a more efficient rate and minimizes losses due to battery charge/discharge efficiencies.

The use of a fuel cell as the APU requires a controller that is capable of managing every aspect of the fuel cell's operation as well as interfacing with the user and the drivetrain. A two-controller system is used to meet this need. A Z-180-based microcontroller is used to startup, monitor, and shutdown the fuel cell. The microcontroller features direct driving of relays and solenoids from its digital ports, 18-Mhz clock speed, 11 analog-to-digital ports, and the ability to expand the number of ports and A-to-D channels with expansion boards. A single board computer reads information from the E-Meter, the Kilowatt+2, the vehicle mode selector (shifter), and the drivetrain, instructs the microcontroller to turn on the fuel cell, and provides the driver with information.

Controlling the fuel cell operation correctly requires a control algorithm that safely starts, monitors, and shuts down the fuel cell under all operating conditions. The flow chart algorithm includes monitoring the pressures and temperatures of the supply-side hydrogen and air flows, the stack cooling loop, and stack voltage in groups of four cells.

Cells are monitored in groups of four (for a total of 28 groups) to detect individual cell failures early and prevent damage to the stack. Expansion boards with analog-to-digital ports monitor cell-group voltages, and analog isolation amplifiers are used to isolate the controller from the fuel cell stack high voltage output. A cell group voltage less than the minimum for the fuel cell current indicates a possible problem with an individual cell or the fuel cell system as a whole.

Starting and stopping the fuel cell requires the controller to open and close relays for the different pumps and fans, turn on and off the buck converter, operate the air compressor at speeds that are proportional to the fuel cell load, open and close the various solenoids, and turn on and off the boost converter.

OPERATIONAL MODES – A simple driver control interface is a must for today's consumers. The HEVT driver interface continues this trend by using a transmission-style shift lever for control of pure electric mode, hybrid electric mode, fuel cell on mode, and idle charge mode as shown in Figure 4. In Pure EV mode, the fuel cell is locked out and the car will run in electric mode only for range limited by battery capacity. In Hybrid mode, the battery SOC dictates the operation of the fuel cell. When battery SOC reaches about 40%, the controller enables the fuel cell APU power until the batteries are charged to 80%.

The Fuel Cell mode is for situations when continuous operation of the fuel cell is required, such as long grade climbs. Selection of this mode locks-in a request for fuel cell operation to the APU controller. A variation of this mode is the idle charge selection, which disables the electric drivetrain and allows battery charging at a higher than normal voltage to quickly replenish the battery pack.

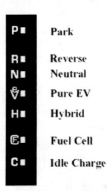

Figure 4. Mode Selector Layout

BATTERY SELECTION AND IMPLEMENTATION

The GE drivetrain requires a battery pack voltage between 180 V and 400 V. Considering the majority of the batteries on the market have a nominal voltage rating of 12 V, the battery pack should be sized to have a nominal voltage of 336 V using 28 batteries in series. Selecting the best battery technology to create a lightweight, high capacity, high power battery pack of this size is the critical design decision. Table 5 compares the general characteristics of different available battery technologies.

Table 5. Rechargeable Battery Options

Technology	Wh/kg	Wh/l	W/kg	W/l
Advanced Lead-Acid	35	71	412	955
Nickel Metal Hydride	80	200	220	600
Lithium Polymer	155	220	315	445
Sodium Nickel Chloride	90	150	100	200
Nickel Cadmium	50	150	-	-

There are advantages and disadvantages to each battery technology. Cadmium is a hazardous material. Lithium polymer technology is still very experimental and difficult to obtain. Nickel-metal hydride and sodium nickel chloride module sizes are too large and heavy to create a battery pack sufficient to power the vehicle. The only remaining alternative is advanced lead-acid batteries.

The Genesis battery from Hawker Energy Products is installed in ANIMUL H_2. These batteries have excellent power and energy density, very low internal resistance, and the ability to retain a substantial portion of their capacity under high current draw applications, as would be found with an HEV. Table 6 gives observed capacity of the 16 amp-hour Hawker batteries used in ANIMUL H_2 when discharged to 1.67 V per cell.

Table 6. Genesis 16 Ah performance

Performance Specification	Manufacturer Claimed Capacity	Observed Capacity
20 hour rate	17 Ah	20 Ah
10 hour rate	16 Ah	16 Ah
5 hour rate	15 Ah	15 Ah
1 hour rate	13 Ah	14 Ah
30 min. rate	10 Ah	11 Ah

The test data show that the Genesis battery performs well under discharge. The next test was 250 A impulse discharge performance. This test was used to verify the battery's internal resistance as a function of discharge. Internal resistance values start in the 7 to 8 mW range for a fully charged battery and increase to about 15 mW at 50% SOC. Near full discharge the battery resistance climbed to near 30 mW.

Another key feature of the Genesis batteries is their ability to retain a large portion of capacity under high current draw. It is common practice to rate a battery in terms of amp-hour capacity at a 20-hour discharge rate. However, in a high current EV application, it is more important to observe the capacity at a discharge rate between 30 minutes and one hour. As shown in Figure 5, a Genesis battery offers nearly 100% more capacity at these high discharge rates, compared to a similar sealed lead acid battery from another manufacturer.

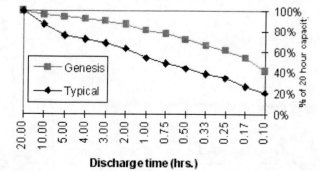

Figure 5.   Discharge Performance

Load bank test results show that the HEVT battery pack delivers approximately 3.8 kWh of energy at the one-hour discharge rate. This translates to an all-electric range of 12 miles at 30 mph and results in a APU operating window that falls in the range of 40% to 80% SOC. Finally, since these batteries are non-spillable, there is minimal risk of battery acid hazard should the batteries become damaged in an accident.

BATTERY BOX – The battery box contains twenty-eight Hawker 16 Ah batteries producing 336 V and a single 26 Ah battery for the 12 V car systems. It also contains two fuses that protect the entire high-voltage system from current above 400 amps. The battery box fits under the rear seat in the location originally occupied by the gasoline tank. Placement of the box in this location facilitates attachment of the box directly to structural members without reducing structural integrity of the vehicle or hindering access if repairs are necessary. The batteries are in a protected location outside the passenger compartment. This location also keeps the batteries in a low, central part of the vehicle to enhance stability. Finally, this location provides easy service access to the batteries. Removing the lower portion of the rear passenger seat exposes a removable lid. A drawing of the box is shown in Figure 6.

The battery box frame is welded angle aluminum with rectangular aluminum tubing. The walls are nonconductive Haysite reinforced plastic sheeting, so additional electrical insulation inside the enclosure is not required. The box is caulked with RTV silicone to prevent foreign materials from entering through the seams and to prevent accidental contact with the aluminum frame. The enclosure is designed to contain the batteries in the event of a severe collision and is mounted to the vehicle frame with steel brackets. The box frame restores the strength and rigidity of the vehicle where the rear seat pans of the car were removed. These modifications do not infringe on rear passenger comfort or safety.

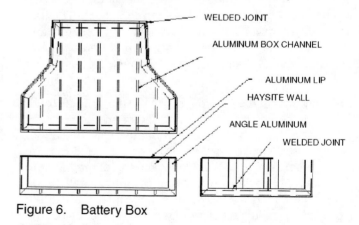

Figure 6.   Battery Box

A finite element analysis was performed to ensure that the design under consideration was sound. Using a 20g crash load on a frame using 3.8 cm x 3.8 cm x 6.4 mm (1½" x 1½" x ¼") aluminum angle and 1.9 cm x 3.2 mm

(¾" x 1/8") square tubing for the frame, the finite element model indicated that the strength of the frame was adequate to prevent failure by yielding and buckling with a minimum factor of safety of four. This analysis provided additional assurance that the frame was strong enough to stay intact in an accident.

Two six-inch diameter centrifugal fans, which move over 3,400 L of air per minute (120 cfm) each, and a three-inch 12 VDC brushless fan, which moves roughly 280 L per minute (10 cfm), provide active cooling. Vents in the front of the battery box provide ram air cooling through the box. The large fans draw this air through the battery box when the car is running and when the batteries become hot due to rapid discharge. A smaller muffin fan runs constantly to remove any hydrogen that may discharge from the batteries as a result of overcharging. All three fans are mounted on the highest portion of the battery box to remove any hydrogen as it rises to the top.

The batteries are restrained vertically using four fiberglass U-channels attached at the left and right sides of the battery box as shown in Figure 7. If this vehicle flips over in an accident, these restraints will prevent the batteries from working loose and injuring vehicle occupants.

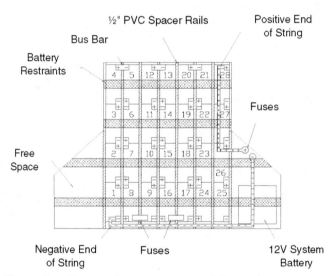

Figure 7.   Battery Arrangement and Restraint

OFF-BOARD CHARGER – The off-board charger represents a key element in consumer use of an electric vehicle. Consumers require a charger that can deliver the energy necessary to replenish the battery pack as quickly as possible. This requires that the method of power transfer involve minimum distortion of voltage and current to protect other sensitive equipment connected to the power grid. For this reason, HEVT uses an inductive charger manufactured by Hughes Power Systems.

This charger incorporates input voltage controlled switching to minimize distortion of the input AC power from the wall outlet to the charger. The flow of energy from the charger to the car uses alternating high frequency magnetic fields to transfer the power as

illustrated in Figure 8. This power coupling is done with the use of a charge paddle that is actually the primary side of a transformer coil. The vehicle charge port is the secondary coil, with a slot sized exactly for the entry of the charge paddle.

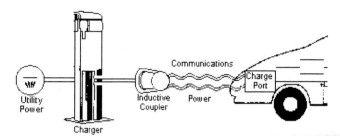

Figure 8.   Magne-Charge Power Connection

The charge port receives the alternating magnetic fields, creates a voltage waveform, rectifies it, and finally passes it on to the battery. This magnetic coupling of the hybrid vehicle to the electric power grid puts the user at zero risk of shock while charging.

The onboard charge control computer calculates battery SOC and requests the appropriate amount of power from the charger via RF communication through the charge paddle. The charger then transfers 6.6 kW of energy to the charge port. Temperature compensation is used to maximize battery longevity.

FUEL CELL STACK

The fuel cell stack is a next generation design built by Energy Partners using proton exchange membranes (PEM) from W. L. Gore, Inc. The fuel cell incorporates a small active area and advanced flow distribution. The smaller active area allows for better control of operating conditions within the stack, which results in significant improvements in performance. Advances in membrane, electrode, and collector plate materials have all contributed to the stack performance. Current densities of over 1000 mA/cm^2 (155 mA/in^2) at 0.6 V are not uncommon. At 138 kPa gauge (20 psig), the 110-cell stack shown in Figure 9 achieves 20 kW (26.8 hp) at 60 V with a power to volume ratio of 0.6 W/cm^3 and a power to weight ratio of 314 W/kg. Figure 10 summarizes the performance of the fuel cell at 69, 138, and 207 kPa guage (10, 20, and 30 psig).

Figure 9.   The Energy Partners 20 kW (27 hp) Fuel Cell

The fuel cell is operated in ANIMUL H$_2$ at approximately 69 kPa gauge (10 psig) producing 60 V DC (nominal). As Figure 10 indicates, the stack can produce anywhere from 290 to 370 A at this voltage level, depending on the operating pressure.

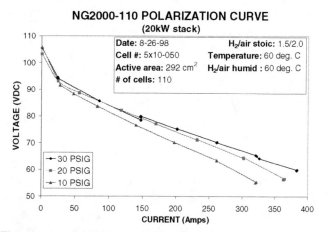

NG2000-110 POLARIZATION CURVE
(20kW stack)

Date: 8-26-98   H$_2$/air stoic: 1.5/2.0
Cell #: 5x10-050   Temperature: 60 deg. C
Active area: 292 cm^2 H$_2$/air humid : 60 deg. C
# of cells: 110

Figure 10. 20 kW (27 hp) Fuel Cell Polarization Curve

SYSTEM INTEGRATION – Successful fuel cell integration depends on the implementation of several important subsystems. This includes all fuel cell mounting and packaging, air handling components, a custom air humidification system, a custom hydrogen humidification system, a deionized water tank, a fuel cell cooling loop, and a stainless steel air-water separator. Figure 2 shows how the fuel cell subsystem is tied into other systems on the vehicle.

The design and construction of these systems is based on the required operating conditions of the fuel cell, which are: 69 kPa gauge (10 psig) air at 70°C (140°F) with 60% relative humidity, hydrogen at 69 kPa gauge (10 psig) with 60% relative humidity. The fuel cell will become contaminated if any of the components in the air, water, or hydrogen systems are made of anything other than an inert substance such as stainless steel, titanium, certain grades of rubber, or Teflon. This dictates that all of the systems have to be custom-built or ordered, since most commonly available automotive components are mild steel or low-grade rubber.

SYSTEMS

Boost Converter – The boost converter serves two critical purposes in the fuel cell system. The boost converter's primary function is to act as the conversion interface between the 60-110 V output from the fuel cell and the 336 V vehicle bus. The boost converter also applies the electrical load to the fuel cell. Using a load-following strategy, the boost converter varies the load that is applied to the fuel cell to improve efficiency of the system.

Ideally, the fuel cell stack and the battery pack should operate in the same voltage range. Unfortunately, this would require a stack with 550 cells, which is not available using current technology.

Air Handling and Humidity System – The air supply system provides the fuel cell with clean air with 60% relative humidity at a maximum temperature of 70 °C (140°F). A system schematic is shown in Figure 11.

Monitoring the pressure of the hydrogen flow into the fuel cell and adjusting air pressure accordingly regulates air pressure. Altering the speed of the compressor's motor, which runs from zero to 7,000 rpm, changes the air pressure. This is accomplished automatically by providing pressure and temperature information from sensors to a computer in the vehicle. The compressor imposes a 3.4 kW load on the vehicle electrical system when the output pressure is 69 kPa (10 psig), which is the operating point of the fuel cell in the Virginia Tech FutureCar. The airflow rate into the fuel cell is approximately 24 standard liters/sec (52 scfm).

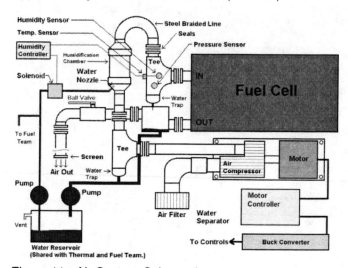

Figure 11. Air System Schematic

Humidifcation is accomplished through the use of a water injection system. Water flow to the humidity chamber is controlled by a negative feedback loop that controls the water injectors. Liquid water accumulates inside the humidity chamber as the humidified air passes through it and cools. A small reservoir collects this condensate and supplies it to the injection system and the cooling loop for the fuel cell.

Hydrogen Humidity System – The humidity chamber for the hydrogen is a small inline chamber that includes a water injection fogging nozzle. The chamber itself is a 38 mm (1.5 in) diameter CPVC pipe approximately 150 mm (6 in) long. Hydrogen enters the chamber on one end and flows straight through to the exit on the opposite end. Water is injected into the chamber through the wall of the pipe with a water fogging nozzle. The chamber also includes a drain to allow excess water to flow out to a water storage tank located below the humidification chamber. This tank will release water to the atmosphere via a solenoid when the fuel cell is not operating.

Testing has confirmed that humidification of the fuel supply to the fuel cell is not a difficult task. The warm-up period for the fuel cell is a difficult period to achieve the correct humidity. Even during this period, humidity levels of 75% are possible. This is well above the 60% reccomended for stack operation. When the stack is at its operating temperature (about 60 °C), the hydrogen vaporizes the water as soon as it is ejected from the nozzle and provides the fuel cell with the needed moisture. Therefore, the humidification chamber only drains water to the water storage tank during the warm-up period of the fuel cell.

Fuel Cell Cooling Loop – The thermal system removes excess heat from the fuel cell. A deionized water loop is used with a crossflow heat exchanger. The thermal loop consists of a 12V DC pump, deionizer, water reservoir, sensor tanks, and the heat exchanger. Figure 12 shows a schematic of the thermal system.

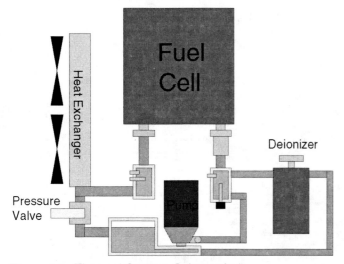

Figure 12. Thermal System Schematic

Water is delivered to the fuel cell at 76 L/min (20 gpm) and 103.4 kPa (15 psig) at the inlet. The pump is a Pacific Scientific 250 W (1/3 hp) motor with a March Pump Series 5.5 magnetic drive head. The deionization loop is in parallel with the main supply line. Deionized water is separated just before the fuel cell. About 75% of the flow continues into the fuel cell, while the remainder is sent to the deionizer. The deionizer is similar in shape, size and function to household water purification systems. This allows the user to easily service the unit. The main reservoir uses gravity to help prime the pump. Water required by the thermal loop and all humidification systems is contained in the reservoir. All thermal sensor tanks and the reservoir are made of 316 stainless steel. Sensor tanks force turbulent flow in the water for sensor readings and provide a central location for sensor mounting. Sensor tanks are located at the fuel cell inlet and outlet to obtain the pressure drop across the fuel cell stack. The heat exchanger is constructed of 316 stainless steel piping with copper fins. It removes approximately 25 kW of heat energy from the water. Air is forced across the fins using PerrmaCool low-profile DC fans.

The control system for the thermal loop monitors water temperature, pressure, flow, conductivity, and reservoir level. Table 7 lists the sensor types, parameters, and actions.

Table 7. Sensor Functions

Sensor	Type	Normal Operation	Excitation State	Controlled Action
T_{in}	Conductivity Probe	55 to 65 °C	T=57 °C T=60 °C T≥65 °C	Start Fan 1 *Start Fan 2* Shut Cell Down
P_{in}	Pressure Transducer	69 to 103 kPa (10 to 20 psi)	P<10 psi *P>20 psi*	Shut Cell Down *Shut Cell Down*
Conductivity	Conductivity Probe	0 to 4 µS	C>4	Shut Cell Down
Flow	Flow Meter	57 to 110 l/m (15 to 29 gpm)	Flow<15	Shut Cell Down
T_{out}	RTD	60 to 70 °C	ΔT>5 *Δ>7*	Warn Driver *Shut Cell Down*
P_{out}	Pressure Transducer	69 to 103 kPa (10 to 20 psi)	Δ>7 *ΔP>10*	Warn Driver *Shut Cell Down*

Water-Air Separator – The exhaust stream from the fuel cell contains air, water droplets, and steam. To maintain the water level in the deionized water tank, the water in the exhaust stream must be captured. This is accomplished by using an air/water separator in the exhaust stream. The water is pumped back to the tank using a stainless steel pump.

FUEL STORAGE AND DELIVERY SYSTEMS – The fuel delivery system shown in Figure 13 is designed to store hydrogen gas in two tanks at 24.8 Mpa (3600 psi) and deliver it at a maximum of 370 slpm (13 scfm) at 138 kPa (20 psi).

An inline shutoff solenoid downstream of the storage tanks allow the tanks to be isolated from the rest of the system when the fuel cell is not is use. Upstream of the solenoid, a tee junction leads to the fill port, which is located under the stock fuel filler door. The fill port is equipped with a check valve to prevent the escape of hydrogen to the environment.

After the fill port, a manual control regulator reduces the hydrogen flow from storage pressure as high as 25 MPa (3600 psi) down to 1.0 MPa (150 psi). The second pressure regulator, which reduces the pressure down to 69 kPa (10 psi), is a differential tracking regulator. This regulator is dome loaded with a positive bias spring on the dome to allow for pressure following with the air supply line to the fuel cell. The biased output of the regulator will compensate for the pressure losses in the hydrogen supply line between the low-pressure regulator and the fuel cell. A pressure tap from the fuel cell air supply line serves as the reference pressure for the regulator dome.

An atomizing nozzle in a humidity chamber injects water into the hydrogen flow. Liquid water entering the fuel cell would severely damage the component, therefore, a water separator in the system removes liquid water without affecting the humidity of the hydrogen. A solenoid valve located downstream of the fuel cell purges the fuel system of hydrogen during shut down.

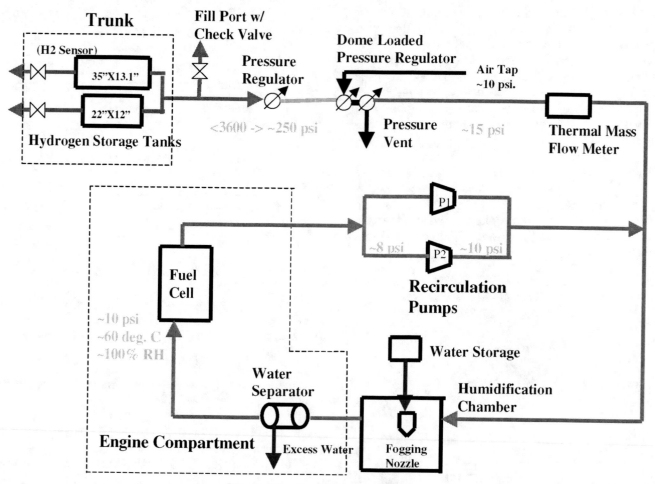

Figure 13. Fuel Delivery System

Recirculation is implemented into the fuel system by incorporating two pumps after the fuel cell to form a closed loop hydrogen system. These pumps make it possible to recirculate unused humidified hydrogen gas from the fuel cell to maintain system pressure. Hydrogen recirculation is imperative in order to maximize fuel system efficiency.

FUEL STORAGE – Fuel storage is a limiting factor in any fuel cell vehicle design. Toyota's FCEV and the NECAR III from Daimler-Benz achieve the range requirements of a modern automobile using steam-reformers to make hydrogen gas. The French-Italian-Swedish concept car FEVER has excellent range using a cryogenic system to store liquid hydrogen. Technologies such as metal hydride and carbon nanotubes are under development both here in the U.S. and abroad as possible methods to overcome hydrogen storage obstacles. Unfortunately, the HEVT was unable to obtain any of these hydrogen fuel storage systems. Thiokol conformable composite tanks that are capable of storing compressed hydrogen at 35 MPa (5000 psi) will not be available until next year. This left few design alternatives beyond compressed natural gas storage technology.

Compressed hydrogen at 25 MPa (3600 psi) contains only about 5% of the energy in the same volume of gasoline while providing twice as much energy per unit mass. The total volume required to achieve a range of 560 km (350 mi) would be more than 350 L (12.4 ft^3). Because of this volume, carrying enough hydrogen on board the car to meet the range requirement becomes very difficult while maintaining trunk space. One alternative is to store the hydrogen as liquid at low pressure. Liquid hydrogen has almost ten times the energy density as compressed hydrogen gas.

By default, the fuel storage system used consists of two tanks of compressed hydrogen. Common materials for storage canisters include mild steel, aluminum, and various composites. In an effort to keep the weight down, carbon-fiber composite tanks were acquired to store the hydrogen gas. These tanks measure 89 cm x 33 cm (35" x 13.1") for use in the trunk and 56 cm x 25 cm (22" x 10") to be placed under the trunk. These tanks provide a total volume of 71 L (2.51 ft^3) at 25 MPa (3600 psi). This capacity provides for an estimated fuel range of 110 km (70 mi). The smaller 26 L (1590 in^3) tank is placed beneath the trunk in the space previously occupied by the spare tire well. The larger 45 L (1.56 ft^3) tank is placed inside the trunk between the rear shock towers and as far back in the trunk as possible, maximizing the available trunk space.

REFUELING – Refueling gasoline-powered vehicles has become a very casual process. However, this relaxed method cannot be applied to hydrogen refueling. Special attention must be paid to the unique hazards of

302

hydrogen. One concern for refueling is static electricity igniting any leaking hydrogen. Therefore, the vehicle will need to be electrically grounded before filling the tanks. A second concern of refueling is to prevent the vehicle from driving away while the refueling nozzle is still attached to the fill port. This concern has been addressed using electric interlocks. A switch is connected to the door of the fill port. Once the door is opened, a solenoid allows hydrogen to flow into the storage tanks. More importantly, the switch prevents the drivetrain from being activated, and therefore the vehicle cannot be driven away while the refueling nozzle is still connected to the fill port.

HIGH AND LOW VOLTAGE SYSTEMS – The 336 VDC system is designed for maximum safety by using insulated enclosures for all high voltage connections. Cables and connectors are all rated for at least 150% of maximum current draw to provide an extra margin of safety during fault conditions. All high voltage cables running from the back to the front of the vehicle are encased in flexible aluminum conduit that is securely clamped to the vehicle chassis, outside of the passenger compartment. All high voltage wiring is clearly identified by a continuous band of orange marking tape. In addition, high voltage warning stickers are placed at all termination points. The contactor box is the only high voltage distribution center in the car. Housed inside this nonmetallic box are the terminations of the main battery pack to the positive and negative bus. The bus is connected to the main power contactor array and the auxiliary fuse block. Each 336 V device is connected to a magnetic blowout contactor on both the positive and negative bus designed to interrupt the line. This allows complete removal of the device from the high voltage bus.

Activation and operation of the series hybrid vehicle drive and control requires an initial startup power source. An on-board auxiliary 12 V battery provides this power for initial 12 V loads during startup. A 336V/12V DC-DC converter located in the trunk maintains the 12 V system. Some of the 12 V loads include headlights, defroster, blower fan, electric drivetrain control systems, stereo, airbags, power brakes, and power steering. A 12 V electric Delco vacuum pump has been added to replace the engine-driven vacuum source of the stock antilock braking system. The steering is electro-hydraulic power assisted, with the aid of a 12 V computer controlled pump that cuts power consumption by 80% compared to the stock power steering.

HIGH INTENSITY DISCHARGE HEADLIGHTS – In an effort to improve 12 V system efficiency, the stock halogen bulbs have been replaced with a High Intensity Discharge (HID) light system. The new HID headlights produce twice the light, consume 30% less power, and have four times the life span of the halogen bulbs. Minor modifications have been made to the stock headlight assemblies to accommodate the HID lamps. Intermediate adapters made of ABS plastic were fabricated using fused-deposition modeling rapid

prototyping. ABS is not a thermoplastic and is solid until approximately 80°C (176°F). Testing of the assemblies revealed that temperatures of this magnitude are not experienced.

HEATING, VENTILATION, AND AIR CONDITIONING

With the removal of the Lumina's gasoline engine, the vehicle is not able to use the heat dissipated from engine coolant to heat the car's cabin. ANIMUL H_2 uses a heat pump to cool and heat the passenger cabin. Adding components to the existing air conditioning loop created the system currently in the vehicle. A Sanden semi-hermetic scroll compressor, driven by an integral DC brushless motor, is the primary component of the heat pump. An inverter connected to the 336 V bus delivers power to the compressor. This system provides high efficiency and capacity. Drawing power from the vehicle's battery pack allows heating and air conditioning without requiring the fuel cell to be operating. Therefore, the HVAC system is fully functional while the vehicle is in any operating mode. All stock HVAC controls have been maintained and modified to operate the heat pump. The transition from heating to cooling mode is invisible to the operator.

HANDLING AND SUSPENSION SYSTEM

Traditionally, automotive suspension designs have been a compromise between the two conflicting criteria of road holding and passenger comfort. A vehicle with a suspension system designed for directional control and handling will inevitably suffer from a harsh ride. Conversely, a vehicle with a suspension system designed for a comfortable ride will suffer from poor directional control and handling characteristics. This compromise stems from the fact that a vehicle suspension is in essence a two-degree-of-freedom spring-mass-damper system with fixed spring and damper constants, and thus a fixed dynamic response.

HEVT, along with the Advanced Vehicle Dynamics Lab at Virginia Tech, Lord Corporation, and Koni Incorporated, has designed a semi-active suspension system using a magneto-rheological (MR) fluid to vary the damping constant of the system in real time. MR fluid changes its viscosity in the presence of a magnetic field. A coil of wire located in the damper piston is used to generate this field. Adjusting the current through the coil changes the strength of the field, thereby changing the viscosity of the MR fluid. Since the fluid is forced through an orifice in the damper piston near the coil, the net result of a change in fluid viscosity is a change in the damping coefficient of the shock absorber.

This controllable damping is best utilized in a system that can use vehicle sensory information to make adjustments in real time. Virginia Tech's system uses accelerometers mounted on the body and axle at each wheel with a feedback controller to automatically alter the damping in a true real-time sense. The controller is capable of providing different levels of damping to each wheel to improve cornering at higher speeds and reduce squat and dive. In addition, a driver-adjusted knob can be used to control the suspension damping characteristics directly. With this system, the driver can tune the ride/handling characteristics of the vehicle to his liking, and adjust them for different road surfaces or types of driving.

To further augment vehicle handling, Goodyear 225/60R16 Eagle Aquasteel EMT run-flat tires are installed on the vehicle. These tires maintain the same outside tire diameter while decreasing the sidewall height. The use of 16" wheels and lower profile tires improves handling without sacrificing ride height or ramp angle. These run-flat tires save both weight and space on the vehicle by eliminating the need for a spare tire and permit the driver to run at least 80 km (50 mi) at 88 km/hr (55 mph).

SAFETY

The dangers of hydrogen use have been well documented by scientists and engineers for many years. Numerous examples can be given of catastrophic disasters occurring, and the public concerns and fears which accompany hydrogen use. Despite these sad events, hydrogen is still used safely on a daily basis to advance the technology of the world. From the space program to chemical processing to power plants, hydrogen is used properly and securely to prevent fatal incidents. HEVT has made every effort to ensure safe development, operation, and storage of Animul H_2.

All components of the hydrogen storage and delivery system have been chosen with safety as the first concern. Flashback arrestors were added following the storage tanks and each pressure relief valve. Flashback arrestors prevent air from entering the system where hydrogen is discharged and, in case of an accident, prevent flames from reaching the hydrogen in the storage tanks.

Hydrogen detectors have been implemented in case hydrogen leaks from any of the components of the system. Detectors are located in the trunk, passenger compartment, and the fuel cell compartment. The controller monitors each detector. If hydrogen of a level of 10% the lower explosive limit is detected, an 85-decibel alarm will sound and the controller will activate a number of safety interlocks. These interlocks include opening the trunk and windows to disperse the hydrogen, sounding the horn to alert surrounding people, and shutting down the hydrogen supply.

Purging of hydrogen is required when the vehicle is shut down. When the vehicle is turned off, the bleed / purge solenoid is opened, releasing the hydrogen. All relief valves and the purge solenoid are plumbed into one pipe which is discharged through a flashback arrestor at the back of the vehicle.

Table 8. Manufacturing Cost Estimate

Part Name	Source	Prototype	Production
Production Chevrolet Lumina	General Motors	$14,000	$14,000
Transaxle Motor	Ford / General Electric	$8,000	$1,200
EV2000 Inverter	General Electric	$12,000	$1,800
Fuel Cell Stack	Energy Partners	$250,000	$2,500
Boost Converter	Soleq	$50,000	$2,000
Air Compressor	Vairex	$12,000	$750
Heat Exchangers	Super Radiator Coils	$540	$110
Cooling Fans	PermaCool	$160	$24
Cooling Pumps	March Pump	$700	$100
Humidification Pumps	Micropump & Fluid Metering	$1,000	$40
Water Separator	Andersen Separators	$1,200	$50
Deionized Water Reservoir	Fabricated	$100	$10
Micro-controller	Z-World	$4,000	$200
Fuel Cell Sensors	Omega	$1,820	$45
Hydrogen Tanks	Lincoln Composite	$3,000	$400
Fuel Lines, Regulators, & Solenoids	Parker & Tescom	$2,000	$250
Battery Pack	Hawker	$3,120	$940
HID Headlights	Bosch	$1,400	$500
HVAC Compressor	Sanden	$3,000	$750
MR Suspension	Koni / Lord	$35,000	$1,000
Run-Flat Tires	Goodyear	$600	$400
Off-Board Charging System	Hughes / Delphi	$3,000	$700
TOTAL		$406,640	$27,769

An independent safety system approach was designed into all the crucial operating systems of the vehicle. The stock Lumina airbag deployment system has been retained to protect the passengers in the event of a collision. A First Inertia crash sensor is located under the hood to disable the electric drivetrain in the event of a 6-12 g collision. A second First Inertia crash sensor shuts off the hydrogen flow at the tanks. A remote, cable operated Halon fire suppression system will cover the motor compartment, battery box, and fuel tanks in the event of an on-board fire.

Emergency disconnect switches provide a method to quickly disconnect the main high voltage system. These switches, located beside the driver's left leg and in the fuel filler door, are connected to stainless steel cables that activates a physical shutdown sequence. First the 12 V power to the contactors is disconnected, allowing any enabled contactors time to interrupt high voltage current flow. This is then followed by physically disconnecting the main 336 V system connector.

MANUFACTURABILITY AND TARGET MARKET

The 1998 Virginia Tech FutureCar was constructed and packaged with ease of manufacturability in mind. Logical subassemblies of major system components were formed in an attempt to permit the use of robotic assembly lines. The fuel cell heat exchanger, air conditioning condenser, fans, shroud and support structures are designed to be installed as one component. The fuel cell, boost converter, and support structure can be installed or removed from the motor compartment quite easily. Components that are the source of vibration such as the transaxle motor, air compressor and inverter are all integrated into the engine cradle that is isolated from the vehicle by rubber mounts. A manufacturing cost estimate is shown in Table 8 with estimates for this prototype and mass production of 100,000 fuel cell equipped Luminas.

Although the Lumina presented here has limited range, it could still be marketed in locations where local law mandates the use of zero emission vehicles. Improved range and trunk space could be achieved through the use of an alternate hydrogen fuel storage system. Many of these hydrogen fuel storage systems are still in the development phase.

VEHICLE STRUCTURE, MATERIAL SELECTION, AND WEIGHT

By design, there are no large modifications to the base vehicle structure. Major components, such as the fuel cell, boost converter, heat exchangers, inverter, electric drive motor, battery box, and fuel tanks are all integrated using the structural rails that form the frame of the vehicle. In an effort to minimize the overall weight increase, light materials were used whenever it was deemed appropriate. Light carbon fiber pressure tanks were used to store the hydrogen fuel. The battery box structure was optimized and fabricated from aluminum and lightweight plastic. Aluminum angle is used as the support structure for the heat exchangers, the fuel cell, and the boost converter. The rear stock drum brakes were replaced with disc brakes, resulting in a slight weight saving and improved braking. The brake rotors have also been cross-drilled and fitted with high-performance pads to save weight without degrading braking performance. A fiberglass hood that is an exact replica of the original metal hood now covers the engine compartment of the vehicle. All of these decisions were based on the need to keep the added weight down.

To prevent contamination of the fuel cell, components in the hydrogen, air, humidification, water-recovery, and fuel cell cooling systems must be made from inert materials such as stainless steel, titanium, Teflon, and certain grades of rubber. Most of the tubing of these systems is steel braided with Teflon coating on the inside. The heat exchanger used to dissipate the heat created by the fuel cell is made entirely of stainless steel. This protects the deionized water that is used in the cooling. Although the heat exchanger is very large and heavy due to the relatively poor thermal conduction properties of stainless steel, there are few other alternatives. The pumps used in the humidification, cooling, and water-recovery systems are made from stainless steel. The water reservoir and water separator are also made from stainless steel. The nozzles used in the humidification systems are also fabricated from inert materials. Selection of these components for their material properties surpasses weight considerations.

Due to the number of components added to the vehicle, approximately 115 kg (255 lb) of mass was added to the front axle and 245 kg (540 lb) added to the rear. The resulting vehicle weight distribution was 1021 kg (2235 lb) front and 845 kg (1860 lb) rear. New rear springs were added to keep the vehicle at GM's specified ride height due to the increased weight.

CONCLUSION

A 20 kW fuel cell can be used in a series hybrid configuration to obtain good performance from a mid-size sedan, such as a Chevrolet Lumina. This system provides performance characteristics, such as acceleration, handling, and consumer acceptability, which meet or exceed those of the stock vehicle. Although range is limited due to the simple compressed hydrogen fuel storage system, other fuel storage technologies and reformers are rapidly improving. Soon these hydrogen fuel storage and delivery systems will be small enough and light enough to provide adequate range in a mid-size sedan. The conversion of an existing conventional vehicle, and limitations of the FutureCar Challenge rules, prevent large reductions in mass needed to improve the energy efficiency of the vehicle. The estimated fuel economy is about 51 miles per gallon of gasoline equivalent energy or nearly twice (2X) the stock vehicle. To achieve the PNGV goal of 3X will require a new vehicle designed as a lightweight fuel cell vehicle. The present hydrogen powered design is a zero emission vehicle at the exhaust pipe, but the source of hydrogen and refueling infrastructure must be investigated to determine the overall effect on the environment.

REFERENCES

1. "Hydrogen & Fuel Cell Letter", Editor and Publisher: Peter Hoffman, Grinnell Street P.O. Box 14 Rhinecliff NY 12574

2. Senger, R.D, Merkle, M. A., and Nelson, D.J., 1998, "Design and Performance of the 1997 Virginia Tech Hybrid Electric FutureCar", Society of Automotive Engineers Publication SP-1359, pp. 99-113

3. Senger, R.D, Merkle, M. A., and Nelson, D.J., 1998, "Validation of ADVISOR as a Simulation Tool for a Series Hybrid Electric Vehicle", Society of Automotive Engineers Publication SP-1331, Paper 9811333, pp. 95-115

4. Cuddy, M., and Wipke, K., 1996, "ADVISOR, Advanced vehicle simulator, New opportunities for screening advanced components," Proceedings of the Annual Automotive Technology Development Customer's Coordination Meeting, Oct. 28 - Nov. 1, 1996, Dearborn, Mi.

5. Sperling, D., 1995, "Future Drive: Electric Vehicles and Sustainable Transportation," Island Press, Washington, D.C.

6. Marr, W. W., 1995, "User's Guide to Eagles Version 1.1," Published by Argonne National Laboratory, 9700 S. Cass Ave., Argonne, IL 60439.

7. Senger, R. D., Merkle, M. A., and Nelson, D. J., 1997, "Design of The 1996 Virginia Tech Hybrid Electric FutureCar," SAE Publication SP-1234, The 1996 FutureCar Challenge, pp. 101-112.

8. Gromatsky, J., Ogburn, M., Pogany, A., Pare, C., and Nelson, D., 1999, "Integration of Fuel Cell Technology into the Virginia Tech 1998 Hybrid Electric FutureCar," SAE Publication SP-1452.

CONTACT

Dr. Douglas J. Nelson
Virginia Polytechnic Institute and State University
Mechanical Engineering Department
Blacksburg, Virginia 24061-0238
(540) 231-4324
Doug.Nelson@vt.edu

HEVT - Preliminary Specifications
1999 Virginia Tech Fuel Cell Hybrid Electric FutureCar

GENERAL

Vehicle Type:
1997 Chevy Lumina 4 door sedan, PS, PB, A/C, air bags

Weight:
2000 kg (4400 lb) 57% f / 43% r

Tires:
Goodyear P225/60R16 Aquasteel EMT (run flat)

Overall Dimensions:
5.10 L x 1.84 W x 1.40 H (m)
200.9 L x 72.5 W x 55.2 H (in)

Wheelbase and Track Width:
2.73 m (107.5"), 1.50 m (59")

HEV Strategy:
Series configuration, single drive motor, powered from APU and/or battery, state of charge control with load following

Braking System:
Electronic Regenerative with vacuum assist hydraulic four wheel disc ABS

AUXILIARY POWER UNIT

Fuel Cell Stack Manufacturer:
Energy Partners, West Palm Beach, Florida

Performance: (@ 10 psig inlet)
17.5 kW @ 60 V DC full load
105 V DC no load
better than 50 % stack efficiency

Fuel Type:
Hydrogen, compressed gas storage
25 MPa (3600 psi) max pressure
70 l (18 gal) tank capacity,   1.3 gal gasoline equivalent energy

Fuel Delivery:
Pressure regulator to 10 psig, recirculation, water separation

Air Delivery:
Vairex electric drive compressor, humidification, water separation, back pressure restriction

Cooling:
Deionized water, pump, forced air radiator with thermostat on fans

Exhaust:
Warm, moist, oxygen-depleted air, pure water

Weight:
64 kg (140 lb) (stack only)

DRIVELINE

Electric Motor
Manufacturer / Model:
General Electric Drive Systems Salem, VA, Ecostar prototype

Type:
3-phase AC induction with DC/AC inverter controller

Speed range:
0-13,500 rpm @ rotor

Torque:
195 N-m (143 ft-lb) @ 0-4000 rpm

Voltage:
180-400 VDC input to inverter

Power Rating:
Continuous: 60 kW (80.5 hp)
Peak: 85 kW (114 hp)

Weight:
with inverter: 105 kg (230 lb)

Cooling and Lubrication:
Pressurized synthetic transmission oil with forced air cooling radiator,
inverter: water cooled with forced and ram air cooled radiator

Differential/Gear Reduction
Manufacturer / Model:
Integral transaxle with motor

Specifications:
2 stage planetary gearset, 12.18:1 overall reduction

BATTERY

Manufacturer / Model:
Hawker Energy Products: Genesis G12V16Ah10EP

Construction:
Advanced sealed valve regulated lead acid, Noryl flame-retardant case, classified as nonspillable

Bus Voltage:
27 modules = 324 VDC nominal

Energy Capacity:
3.6 kWhr @ C/1 rate

Weight:
174 kg (380 lb)

Cycle Endurance:
500 cycles @ 80% DOD

POWER CONDITIONING

Function:
Interface fuel cell stack output at 60V to vehicle DC bus at 324V, provide isolation and over-current protection

Model:
Custom designed boost converter using DC/DC switching technology with imbedded controls

CONTROL COMPUTER

Model:
Z-World microcontroller with isolated analog inputs/outputs

Items Controlled:
APU on/off strategy, Fuel Cell operations, Fuel Cell stack voltage monitoring, Vehicle-Driver interface

PERFORMANCE

0-50 km/h (0-31 MPH)
3.2 sec.

0-100 km/h (0-62 MPH)
11.2 sec.

Top Speed
121 km/h (75 mph)

Battery-Only Range:
20 km (12 mi) @ 30 mph

Hybrid Range (HEV):
115 km (70 mi) city/highway

Fuel Economy:
4.4 L/100 km (50 mpg) combined city/highway (gasoline equivalent) est.

Emissions:
None

For more information, contact:
Hybrid Electric Vehicle Team
Virginia Tech, Mechanical Engineering
Blacksburg, VA 24061-0238

Web site:
http://www.vt.edu:10021/org/hybridcar

Faculty Advisor:
Douglas J. Nelson
Phone: 540-231-4324
E-mail: Doug.Nelson@vt.edu

2002-01-0095

Design and Integration Challenges for a Fuel Cell Hybrid Electric Sport Utility Vehicle

Stephen Gurski and Douglas J. Nelson
Virginia Polytechnic Institute and State Univ.
Mechanical Engineering Department

ABSTRACT

Large sport utility vehicles have relatively low fuel economy, and thus a large potential for improvement. One way to improve the vehicle efficiency is by converting the drivetrain to hydrogen fuel cell power. Virginia Tech has designed a fuel cell hybrid electric vehicle based on converting a Chevrolet Suburban into an environmentally friendly truck. The truck has two AC induction drive motors, regenerative braking to capture kinetic energy, a compressed hydrogen fuel storage system, and a lead acid battery pack for storing energy. The fuel cell hybrid electric vehicle emits only water from the vehicle. The fuel cell stacks have been sized to make the 24 mpg (gasoline equivalent) vehicle charge sustaining, while maintaining the performance of the stock vehicle. The design and integration challenges of implementing these systems in the vehicle are described.

INTRODUCTION

In recent years, many initiatives have been enacted to clean the air, one of which was the proposed Kyoto treaty in Kyoto, Japan, in December 1997. This treaty was negotiated by more than 160 nations, and was designed to reduce certain greenhouse gases, mainly carbon dioxide (EIA, 1998). The agreement requires the reduction of greenhouse gas (GHG) emissions to 7% below the 1990 level (NCBA, 1999). The timeframe for this reduction to occur is 2008-2012 (TECO,1998). The main focus of this reduction will be in the energy use sector because this is the main source of greenhouse gas emissions (Hakes,1999). Under current energy production methods, that would mean a 41% reduction in energy consumption (TECO, 1998). The transportation sector contributes one-third of US GHG emissions, and is increasing faster than the industrial and building sectors.

Jay Hakes, administrator of the Department of Energy's Energy Information Administration (EIA), presented to the Energy and Natural Resource Committee of the United States Senate several ways to reduce GHG emissions and maintain the same level of energy consumption. His recommendations include a shift to nuclear and renewable sources of energy rather than fossil fuels, using fuels with less carbon intensity and use more energy-efficient technologies. With these ideas in mind, the Hybrid Electric Vehicle Team (HEVT) of Virginia Tech set out to convert an existing 2000 model Chevrolet Suburban to compete in the 2001 FutureTruck Challenge.

In an effort to address some of Mr. Hakes' suggestions as well as issues raised by the Kyoto treaty, the HEVT of Virginia Tech decided to employ a series hybrid design in their truck, named ZE*burban* (Zero Emission Suburban). This design uses a proton exchange membrane (PEM) fuel cell as its auxiliary power unit. The PEM fuel cell directly converts pure hydrogen fuel to electricity and water, which completely eliminates carbon and the possibility of forming carbon dioxide onboard the vehicle. However, the production, transportation and distribution of hydrogen generates carbon dioxide (and other greenhouse gasses) because the production process uses methane (CH_4), which is a low carbon-intensity fossil fuel (Wang and Huang, 2000). In the future, hydrogen can be produced from renewable sources, possibly eliminating greenhouse gas emissions from hydrogen production. Fuel cells have also demonstrated higher efficiencies than internal combustion engines. As part of the Kyoto treaty, the use of diesel fuel is to be reduced and eventually eliminated. Consequently, PEM fuel cell technology was chosen as an auxiliary power unit rather than a small diesel engine.

There are several advantages to using hydrogen as a fuel onboard the vehicle. Hydrogen has three times the energy content per mass as either gasoline or diesel fuel. The down side to compressed hydrogen gas is low energy content per unit volume, making it difficult to store enough fuel onboard to match the range of a gasoline or diesel vehicle.

Industrial hydrogen fuel production is relatively clean as far as emissions are concerned. Hydrogen can be produced through electrolysis, or through the use of a steam reformer. Reformers use natural gas, or methane, along with steam to produce hydrogen and carbon monoxide. The carbon monoxide is then reacted with steam (water-gas shift reaction) to produce carbon dioxide and hydrogen. For every molecule of methane used in the process, four molecules of hydrogen are produced, and only one molecule of carbon dioxide is released. This process results in a significant reduction in GHG emissions when compared with combustion of gasoline or diesel fuel. Hydrogen has 71% less GHG emissions per unit energy over the whole fuel cycle relative to gasoline (Wang, 1999).

DESIGN OBJECTIVES

The 2001 FutureTruck Challenge specifies the following performance and emissions criteria for the converted vehicle:

- A goal of two-thirds reduction of total cycle GHG emissions as compared with emissions of the stock vehicle and gasoline fuel
- Achievement of California ULEV II exhaust emissions with comparably low evaporative and running loss emissions
- Maintaining of a fully-functioning vehicle with all working accessories and options of the stock vehicle
- Towing a trailer of up to 3175 kg (7000 lb) capacity.
- 0-60 mph acceleration time of less than 12 seconds
- Seating for 8 adults
- 0.65 cubic meters (650 L, 23 ft^3) of cargo capacity

The goal of our design is to ensure that the overall vehicle meets the FutureTruck criteria. Component sizing and selection is a major concern to meet the goals of FutureTruck for two reasons. The first reason is that the components must be sized appropriately. For example a traction motor must be able to supply enough mechanical power to accelerate the vehicle from 0-60 in less that 12 seconds. The second reason is that appropriately sized components must all fit into the vehicle without compromising the current vehicle capabilities and capacities.

Further design considerations are the balance required among selecting the most efficient drivetrain, implementing the fuel cell system, selecting the batteries, providing power to vehicle components, achieving vehicle performance and safety, as well as achieving consumer acceptability and ease of mass production.

A hybrid PEM fuel cell vehicle requires three main systems: electric drivetrain, the fuel cell, and energy storage (batteries). A more in-depth analysis of component selection considers the interactions and trade-offs between the main systems and vehicle energy efficiency, performance, safety, and mass production feasibility.

VEHICLE MODELING

Vehicle modeling is used to predict the size and trade-off between the drivetrain, battery storage, and fuel cell power output. Modeling is the key to properly sizing components and maintaining vehicle performance acceptable to consumers. The series hybrid electric vehicle (HEV) model was developed using ADVISOR, an advanced vehicle simulator of the National Renewable Energy Lab (Wipke, et al.,1999). Initially, a stock Suburban was modeled and validated in ADVISOR, using available test data. The model yields the power requirements necessary to move the stock vehicle. Replacing the stock internal combustion engine with a PEM fuel cell system changed the vehicle model and mass. Keeping in mind the performance requirements of FutureTruck, the PEM fuel cell system was reflected in the model.

Figure 1 shows the energy flow and component configuration for the fuel cell hybrid electric vehicle model. This vehicle model uses a electric drivetrain model validated at Virginia Tech (Merkle, et al., 1997; Senger, et al., 1998). The fuel cell system model was scaled from a validated component model (Luttrell, et al., 1999; Fuchs et al., 2000; Ogburn, et al., 2000). The model has also been validated using test data from the 2000 FutureTruck Challenge results for *ZEburban* operating as a battery electric vehicle (Patton et al., 2001).

DRIVETRAIN SIZING

The drivetrain is sized to complete both of the EPA drive cycles, complete the towing events, and have comparable acceleration capabilities to that of the stock vehicle. Drivetrain sizing began by determining the power and energy requirements of the proposed hybrid vehicle from the stock vehicle model. The three driving cycles considered are the Federal Urban Driving Schedule (FUDS), the Highway Fuel Economy Test (HWFET), and a constant speed cycle at 105 km / hr (65 mph). These cycles are used to determine average and peak vehicle power requirements. The average vehicle energy requirements are used to select an appropriate drivetrain with average operation in a high efficiency region. For the competition, the vehicle should be capable of executing the EPA dynamometer drive cycles with a single axle. Table 1 contains the average cycle power and peak cycle power at the wheels of the vehicle for a 6800 lb. (3090 kg) vehicle weight.

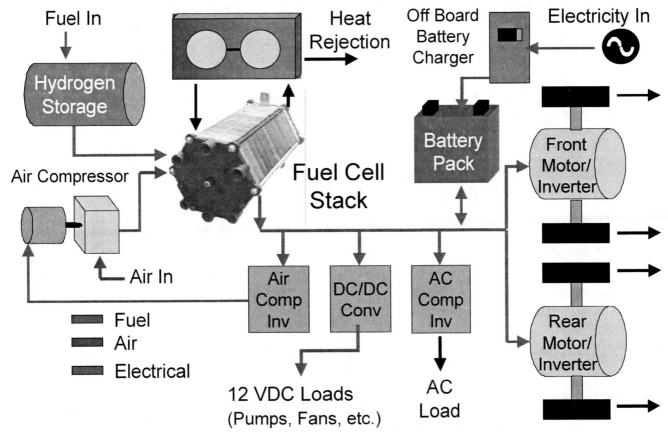

Figure 1. Energy Flow in the Fuel Cell Hybrid Electric FutureTruck

Table 1. Vehicle Road Load Power Requirements

	Average (kW)	Peak (kW)
FUDS	12.7	85.5
HWFET	20.3	68.5
Constant 65 mph	28.8	28.8

Design power requirements for the drivetrain have been determined from simulating a trailer being towed at 88 km/hr (55 mph) up a 5% grade. The combined vehicle and trailer weight was 5900 kg (13000 lbs.). The road load power required to maintain this towing on grade condition is 112 kW. The maximum power requirement of 150 kW results from the vehicle acceleration requirement of 0-60 mph in 12 s.

BATTERY SELECTION

Before we begin sizing and modeling batteries, we must specify the battery technology. There are several different types of batteries being considered for use in electric and hybrid vehicle applications. These batteries include lead acid, nickel metal hydride, lithium polymer, sodium nickel chloride, and nickel cadmium. Each of the categories in Table 2 represents specific characteristics to compare battery technology. The largest number in each category represents the best performance for that characteristic. The team first eliminated batteries that would not meet weight or size requirements.

Table 2. Rechargeable Battery Options

Technology	Wh/kg	Wh/l	W/kg	W/l
Advanced Lead-Acid	35	71	412	955
Nickel Metal Hydride	80	200	220	600
Lithium Polymer	155	220	315	445
Sodium Nickel Chloride	90	150	100	200
Nickel Cadmium	50	150	-	-

Nickel cadmium batteries were eliminated as a choice because cadmium is a hazardous material. Sodium nickel chloride batteries are too big and heavy to be considered for our vehicle platform. Lithium polymer has the best capacity comparison value, but have also been eliminated because of the lack of availability for use in HEV applications.

For use in ZE*burban*, power density (W/kg) is very important because a series HEV, like ours, needs to be able to source high power on a regular basis. Nickel metal hydride batteries have a good energy density (Wh/kg) and power density. However, nickel metal hydride batteries are very expensive and have been eliminated as a candidate. Advanced lead acid batteries have the highest power density (W/kg) of any batteries considered. These batteries are readily available, inexpensive relative to other batteries, and are used in many electric vehicle applications. Lead acid batteries also have relatively high charge/discharge efficiency for low losses in a hybrid application. After considering these options, advanced lead acid batteries were selected for use in ZE*burban*.

CAPACITY SIZING FOR BATTERY STORAGE

There are two attributes for battery sizing, energy capacity and power. These attributes are dependant upon packaging, weight, degree of hybridization and vehicle energy use. Vehicle energy use is dependant upon the vehicle characteristics, such as weight, rolling resistance, aerodynamic drag and the vehicle drive cycle. To maintain usability of the vehicle, regardless of the operational status of the fuel cell systems, the vehicle must be able to run the FUDS and HWFET drive cycles using power only from battery storage. According to the vehicle and battery models, 26 Ah Hawker Genesis batteries will provide the nominal 4 kWhr necessary to complete each driving cycle. Battery capacity sizing also depends upon the size of the fuel cell systems or the degree of hybridization (Atwood, et al., 2001). Atwood shows that the interaction between the battery capacity and fuel cell system effects fuel economy. Using these results, the 26 Ah Hawker Genesis batteries are well suited for our fuel cell vehicle application.

FUEL CELL SIZING

The Department of Energy has made available an 80 kW Honeywell PEM fuel cell stack (Fig. 2) for the FutureTruck competition. However, the 80 kW gross peak rating depends on inlet reactant supply pressures of 3 atm. The overall system efficiency and net power available at the 3 atm operating condition are lower than desirable, and can be increased by reducing the system pressure. Operating at a lower system pressure significantly reduces the amount of parasitic air compressor power needed to run the fuel cell systems, but also reduces the maximum gross power available from the fuel cell stacks. We have decided to run our

Figure 2. Honeywell PEM Fuel Cell 6-Stack Assembly

systems at 2 atm, which gives us 60 kW maximum gross power available from the stacks, or 49 kW of net power.

The fuel cell in a hybrid vehicle can be been sized to provide the average vehicle energy requirements. For cycles such as the FUDS, HWFET, or at a constant speed of 65 mph, 40 to 50 kW of net power would be ample. However this size gives a short range or reduced speed for higher power cycles such as towing or long grades. A fuel cell stack size of 100 kW gross output is needed to provide a net system power of greater than 80 kW for good towing and grade performance.

VEHICLE COMPONENT SIZING SUMMARY

Our fuel cell hybrid electric vehicle will use 60 kW gross of fuel cell stack power, a 4 kWh, 336 V nominal lead acid battery pack and 170 kW of electric drive power.

MODELING RESULTS

VEHICLE PERFORMANCE

Models of the stock vehicle 2500 kg (5400 lbs.) and the converted fuel cell hybrid vehicle 3090 kg (6800 lbs.) were run in several full power tests. These tests include accelerations both with and without a 3175 kg (7000 lbs.) trailer. The results are shown in Table 4. Based on the models used, the converted vehicle meets the performance of the stock vehicle in the areas that are considered. The converted vehicle is geared for a top speed of 130 kph (80 mph).

Table 4. Performance estimation results

	Stock	Converted
0 to 30 mph (sec)	3.9	2.9
0 to 60 mph (sec)	9.9	10.3
1/4 mile time (sec)	17.6	17.7
with 7000 lb trailer		
0 to 30 mph (sec)	8.1	6.3
0 to 60 mph (sec)	22.4	21.7
1/4 mile time (sec)	24	22.3

ENERGY EFFICIENCY AND FUEL ECONOMY

Using the modeling results from ADVISOR software, we were able to determine the vehicle energy efficiency and fuel economy. Using component sizes from above, we evaluated ZE*burban* on two different driving cycles: Fedral Urban Driving Cycle (FUDS), and Highway Fuel Economy Test (HWFET). The FUDS test includes an approximation for fuel cell cold start efficiency. Overall fuel economy (unadjusted) results are compared for four cases in Table 5. The values given for the fuel cell vehicle are for no net change in battery state of charge and a weight of 3090 kg (6800 lb). The first two columns present available measured data for the stock vehicle. The third column shows good agreement between an ADVISOR model of the stock vehicle and this data. The ADVISOR model for the ZEburban design shows a combined city/highway fuel economy of 24.2 mpgge, or a factor of 1.4 increase over the stock vehicle. This increase is due to the combination of an efficient electric drivetrain, regenerative braking, and an efficient fuel cell system running at relatively light average load.

Table 5. Fuel Economy Comparison

Driving Cycle (mpgge)	Stock Vehicle (GMTG)	Stock Vehicle (EPA)	Stock Vehicle (ADVISOR)	Fuel Cell Vehicle (ADVISOR)
City	13.5	14.9	15.2	23.3
Highway	20.0	20.6	20.3	25.2

GREENHOUSE GAS IMPACT AND EMISSIONS

Greenhouse gas emissions (GHGE) directly relate to fuel economy; an increase in fuel economy decreases greenhouse emissions. A zero emissions vehicle, such as ours, is truly not a zero emissions vehicle because of upstream GHGE from hydrogen fuel production. Fuel production and upstream GHG generated emissions are considered in our vehicle's GHG impact on the environment. Hydrogen is efficiently generated using natural gas. Natural gas is a low carbon intensity fuel that contributes less carbon dioxide during processing than other hydrocarbon fuels.

The greenhouse gas impact for our vehicle is based upon all emissions from hydrogen production and distribution from well to tank for fuel used by the vehicle. Fuel use is determined from our modeled vehicle fuel economy. The predicted fuel economy on the FUDS driving cycle is 23 miles per gallon gas equivalent (mpgge) and 25 mpgge during the HWFET cycle. Using the GREET model (Wang, et al., 1999) for hydrogen production, our vehicle design yields a reduction of 85 % in greenhouse gas emissions, exceeding the FutureTruck goal of 67% reduction. Table 6 shows the weighted comparison of greenhouse gas emissions for the FUDS and HWFET driving cycles.

Table 6. Comparison of weighted greenhouse gas emissions on EPA drive cycles

GHG Index	City	Highway
Stock Suburban	1043.7	1063.6
ZEburban	148.6	117.1

These cycles are not usually representative of consumers actual driving habits, which increase the in-

use GHG impact and fuel use of vehicles. These increased vehicle emissions are called off-cycle emissions. The efficiency of our fuel cell systems have only a 10% efficiency penalty when operating at high power conditions. Vehicle emissions do not increase during off-cycle conditions with a fuel cell system, only fuel consumption increases. Therefore our vehicle design, even during off-cycle conditions, will impact GHG emissions less than conventional vehicle technology. The ZEburban fuel cell vehicle is also locally zero emission with no evaporative emissions.

A summary of the overall specifications for the vehicle design and component selection are given below. The following sections give detailed descriptions of each of the major systems on the vehicle.

HYBRID VEHICLE CONSTRUCTION

ELECTRIC DRIVETRAIN

The desired characteristics for the electric drivetrain system are high efficiency, compact packaging, low weight, high maximum speed, and high power. The ZEburban is a series fuel cell hybrid, so the electric drivetrain is responsible for all of the motive force of the vehicle. To maintain four-wheel-drive operation and near stock performance using available size components, two separate drivetrains are used.

A dual GE system was chosen for its packaging, continuous power output, higher maximum speed, and higher efficiency. The front axle is driven by a GE MEV-75 motor mounted in the original front differential location and a GE EV2000 motor was placed in the rear of the truck behind the rear torsion bar support. The GE EV2000 uses a custom, slip-joint, driveshaft and propels the rear axle through a non-locking GM differential with a 3.73:1 reduction. This packaging configuration has the advantage of allowing the area occupied by the original transmission and transfer case to be used for other systems.

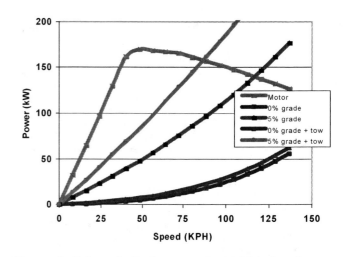

Figure 3. Drivetrain Performance for Vehicle Loads

These motors have a continuous power rating of 60 kW (80 hp) at 3000 rpm and a peak power rating of 85 kW (114 hp) at 4000 rpm. The maximum rotor speed is electronically governed to 13,500 rpm. The torque at the rotor is constant at 195 N-m (143 ft-lb) from 0 to 4000 rpm, and then decreases with speed to 40 N-m (30 ft-lb) at 13,500 rpm as the motor operates in near constant power mode. The MEV-75 boasts a planetary gear set with a gear ratio of 12.18:1, and differentials are housed inside the lightweight aluminum-magnesium alloy transaxle motor case. The EV2000 motor uses a heavier aluminum case and has a planetary gear reduction of 4.29:1 to a simple U-joint coupling. Since the final drive of the rear drivetrain is a steeper 16:1 compared with the front's 12.18:1, the limiting speed had to be determined. With the 265-75-R16 tires that the HEVT is using, the top speed of the rear motor at 13,000 RPM is 80 mph. The speed limit of the front motor is also set to a similar top road speed. This arrangement is to insure that the front motor does not try to pull the vehicle while the rear reaches its maximum rotational speed. The combination of these features provides the Virginia Tech team with an ample-powered drivetrain system (see Figure 3) in a convenient package for our vehicle.

INVERTER

Both motors utilize a General Electric EV2000 inverter. These two inverters control each AC induction motor, which is designed for maximum part load efficiency by using a two-mode, six-step technique to control current in each of the motor windings. At lower speeds and torques, the switching is pulse-width modulated over a range of frequencies, while at higher speeds and torques, the devices are hard-switched in a typical six-step scheme. Switching is performed by three Toshiba IGBT module pairs, which are rated at 400 A and 600 V. The motor control feedback loop consists of a two channel optical hall-effect tachometer sensor for basic variable slip control of the drive, as well as quadrature flux sensors, which maximize drive performance by allowing the inverter to operate the motor near magnetic saturation.

The basic design provides 145 ft-lb of stall torque when supplied from a 300-amp inverter. The motor/inverter efficiency is more than 90% from maximum torque down to torque values of 5 ft-lb (3.5% of maximum). The inverter controller adjusts the motor flux level and basic modulation frequency to reduce both the motor and inverter losses when operating at low torque. A midsize 1450 kg (3200 lb) test weight Ecostar vehicle on the Federal Urban Driving Schedule operates with an overall energy consumption, (including accessory load) of 235 watt-hours/mile (147 watt-hrs/ton-mile). The efficiency and performance are well matched to the requirements of a light-duty electric drivetrain at the operating areas of interest, which include part-load and half speed.

The EV2000 inverter features several safeguards to ensure maximum performance under limiting conditions. A temperature sensor embedded in the stator windings allows the inverter to reduce drive power to protect the motor under high operating temperatures. Temperature sensors within each IGBT pair allow the inverter to provide maximum performance while simultaneously guarding against device failure from overheating. Traction battery voltage is also carefully monitored under load. The inverter begins current limiting as soon as the traction battery drops to 60% of its nominal voltage. This design gives the inverter a "limp-home" mode that allows the vehicle to be driven at low power with nearly depleted batteries until they can be recharged.

The EV2000 inverter is designed to emulate a conventional automobile through such features as a creep mode, which simulates the creeping action of an automatic transmission, and regenerative braking to replace friction braking. An easy-to-use software interface allows the user to customize several control options.

PACKAGING DESIGN

The packaging design of the FutureTruck has an impact on the durability, serviceability and feasibility of the overall success of the vehicle. Packaging of the vehicle also has an effect on the safety and crashworthiness related issues considered with the vehicle.

OVERALL PACKAGING DESIGN

Packaging of the larger and heavier components is given the highest importance. Another guideline is to reserve space in the engine bay for components related to the fuel cell systems. Components with rotating parts (pumps, motors and fans) are mounted to the frame rather than the body to reduce noise and vibration. The overall packaging design is shown in Figure 4.

The rear power module is located in the center of the vehicle, between the rear axle and transmission mount. This places the large mass of the batteries low to the ground for decreased center of gravity and ease of serviceability. The location of the batteries also allows simple removal with a vehicle lift. Another reason that this location is chosen is for its close proximity to all of the other high voltage components such as traction motors and their subsystems.

The design of the our truck requires two traction motors. The motors can connect mechanically to the vehicle either in a single ended U joint coupling, or a transaxle halfshaft configuration. Two design configurations were considered using two single ended motors or a single ended motor with a transaxle motor. The first configuration used two single ended motors with the U joint coupling. One of the motors would connect to the

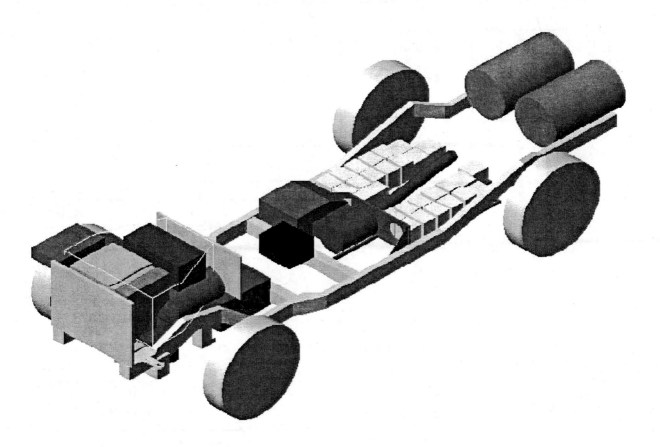

Figure 4. Component Packaging Diagram

existing rear drive shaft, while the other would connect to the front drive shaft from the transfer case. Because of the motor size and a lack of ground clearance for the front motor, this design was considered unusable. So the motor that would have plugged into the front axle was replaced with a transaxle motor. The rear motor was then integrated into the battery box.

The motor controllers or inverters need to be close to the motors that they drive to reduce radiated electrical noise. To reduce heavy wiring for the system, the inverters need to be located close to the batteries and the contactor box. With the previously mentioned restraints and useable space under the vehicle, one inverter was located under the vehicle between the transaxle and contactor box. The other inverter was integrated into the rear power module, next to the rear drive motor. This layout allows the remaining fuel cell systems to be located in the engine bay of the vehicle, and the hydrogen storage tanks are located where the spare tire would normally exist. The spare tire is relocated to the interior of the vehicle.

FUEL CELL SYSTEMS

Virginia Tech has chosen to use a proton exchange membrane (PEM) fuel cell system. A PEM fuel cell requires oxygen from air, hydrogen and cooling water delivered at the appropriate pressure, temperature, flow-rate and humidity. The components required to deliver

these fluids plus the fuel cells are what make up the fuel cell system, as shown in Fig. 5.

HYDROGEN STORAGE AND DELIVERY OPTIONS

Several hydrogen storage methods were considered to provide our vehicle with adequate storage to maximize the mass of hydrogen that can be stored on board the vehicle. The four types of storage that were explored were metal hydrides, cryogenic liquid, reformed hydrocarbon technologies, and compressed gas. Metal hydrides were not used because the tanks have relatively low energy density, or the ability to store many kg of hydrogen per kg of tank weight. Metal hydrides also require cooling to fill, and heating to extract the hydrogen, which complicate vehicle integration. Liquid hydrogen was also not chosen because it must be stored at 20 K, which has its own safety concerns. Maintaining the hydrogen at this temperature would induce a large parasitic loss due to energy required for refrigeration. Additionally, unavoidable boil-off could cause standby losses of up to 5% per day. A review of hydrogen storage costs can be found in James et al. (1994).

Hydrocarbon reformers have the advantage of using existing fueling infrastructure, and significantly increase vehicle range. Hydrocarbon reformers were not used because of the lack of development of a practical vehicle reformer system. Current reformer technology is not

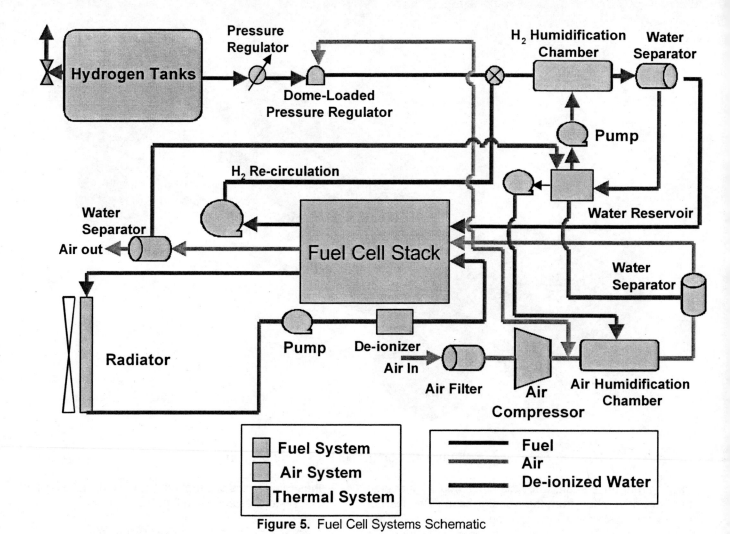

Figure 5. Fuel Cell Systems Schematic

small or lightweight enough for integration into our vehicle. We decided to use compressed gas tanks to store our hydrogen, mainly because this method has the highest energy density (kg hydrogen/ kg tank) currently available. In addition, there is no refrigeration or energy required to transport the hydrogen from the stored state to the fuel cell.

The tanks selected are provided to the FutureTruck Competition by Quantum Technologies through the U.S. Dept. of Energy. The tanks are rated at 35 MPa (5000 psi) and achieve a storage density of 7.5 % kg H2/kg. Due to safety considerations and limitations on available sizes, we were limited to two tanks, each of 61 l (16 gal) capacity, to fit under the rear of the vehicle between the frame rails. This tank size and mounting location limits the total mass of hydrogen that can be stored to 3 kg. Range with this limited storage is about 120 km (75 mi). The stock gasoline tank is approx. 130 l (34 gal) 105 kg (230 lb), which provides the stock vehicle with over 500 miles of range.

Our design for a hydrogen tank mounting system is illustrated in Figure 6.

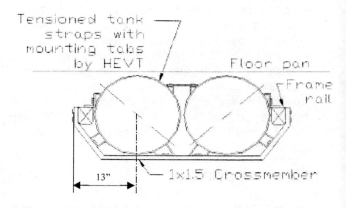

Figure 6. Hydrogen Tank Mounting

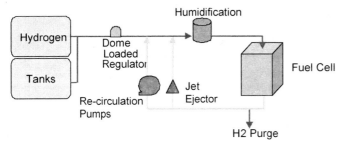

Figure 7. Fuel System Components

FUEL SYSTEM COMPONENTS

Power is extracted by the fuel cell from the electrochemical reaction of hydrogen and compressed air. Hydrogen must be delivered to the fuel cell at the correct temperature, pressure, mass flow and humidity. To achieve these requirements the team has implemented several systems: humidification, pressure regulation, re-circulation, and sensors that monitor and control hydrogen flow. The fuel system components are shown in Figure 7. The fuel system has been designed with safe hydrogen handling in mind (Kraft and Khier, 2001). Therefore, we have implemented safety systems into the design of the fuel storage and delivery system.

The inlet conditions of the fuel cell require that the pressure of the air and hydrogen be matched to prevent damage to the membrane. A dome-loaded regulator will regulate the pressure of the hydrogen in reference to a pressure of the air supply to the fuel cell. The inlet operating pressure range for the fuel cell will be maintained between 3 and 15 psi.

Another inlet condition requires that the hydrogen be humidified to at least 80% relative humidity. This will prevent the membrane from drying, which can damage it. The humidification system consists of fogging nozzles that spray atomized water into the flow of the hydrogen. Hydrogen re-circulation also contributes to humidification of the inlet hydrogen.

To increase the efficiency and range of the vehicle, hydrogen is re-circulated back into the inlet of the fuel cells. The purpose of the re-circulation is to reuse unconsumed hydrogen exiting the fuel cell stack, and re-circulate it back into the inlet of the fuel cells. There are two major components of the re-circulation system: four diaphragm pumps and one hydrogen jet ejector. The pumps and ejector are set up in parallel and are placed between the exit of the fuel cell stack and the inlet of the humidifier. The ejector re-circulates hydrogen for high loads. For the flow rate of hydrogen at high loads, a motive pressure of about 45 psig is needed for the ejector to supply hydrogen at the proper conditions to the fuel cell. At medium loads, the ejector needs a motive pressure of about 30 psig. Since the ejector will not work properly if the motive flow rate of hydrogen is too small, it will not provide sufficient re-circulation under low loads. The diaphragm pumps will begin to re-circulate during these instances.

Due to our re-circulation system, any non-reactive impurities in the hydrogen will build up in the fuel cell and hydrogen system, creating the need for a purging system. A solenoid valve is used to occasionally purge exhaust gasses through a flash arrestor. This purge valve will also vent hydrogen pressure when the system is not generating power. The purge valve is a normally open type, which opens the fuel cell stacks to the atmosphere. Opening the stacks to the atmosphere will insure that any reactants consumed in the fuel cell will not draw a vacuum and damage the membranes.

AIR AND HUMIDIFICATION SYSTEM

The air supply system provides the fuel cell with pressurized air at 80 °C and 50 – 80 % relative humidity. The pressure and flow rate vary with the fuel cell load. An overview of the air supply system is shown in Fig. 8.

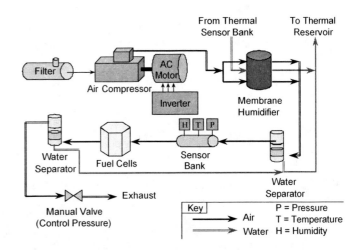

Figure 8. Air Supply and Humidification Components

The air intake filter is used to separate any particulate dust and road grime. The air is then be pressurized and supplied by an Opcon 1050A Twin Rotor compressor, driven by a low cost 10 kW, AC induction motor custom designed to be efficient for this application.

The output of the compressor is humidified by a custom Porvair membrane humidifier. Stainless steel membranes inside the Porvair unit act as a heat and mass transfer device. Heat energy from the thermal rejection system is used to drive the energy transfer required to humidify the inlet air stream. The humidifier has the capability of humidifying the air stream to 95+% relative humidity. Humidity is controlled between 50% and 80% using a solenoid valve that bypasses the humidity chamber. Water for the humidifier is taken from the thermal system by tapping into the coolant outlet of the fuel cells.

Temperature, pressure and humidity of the air is monitored at inlet of the fuel cell for control feedback to the system. After the fuel cell stack, the air stream runs through a water separator to recover any condensed water and return it to the thermal reservoir. A fixed restriction is used to maintain a backpressure that varies with flow rate during nominal operation of the system.

THERMAL SYSTEM

The objective of the thermal system is to dissipate approximately 60 kW of heat generated by the fuel cell during operation. This task is completed using a flow of deionized water into the fuel cell at approximately 75 °C, which absorbs thermal energy and exits at 85 °C. A diagram of the thermal system components is sown in Figure 9.

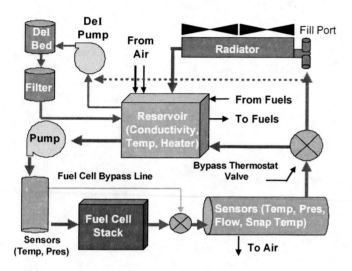

Figure 9. Thermal System Components

Our system is powered by a variable-speed centrifugal pump which has the capability of flowing up to 150 l/min (40 gpm) and overcoming 200 kPa (30 psi) of pressure loss. We have selected a low cost 0.75 kW (1hp) AC induction, industrial drive for this purpose based on the thermal properties of water and the 60 kW heat load. The heated coolant flows from the fuel cell stack through the stock aluminum radiator where fans supply a sufficient air flow to drop the temperature back to the desired inlet conditions for the fuel cell. Several secondary components are designed into the system to allow for variation in the temperature of the water.

Due to the temperature requirements of the fuel cell, a thermostat valve has been placed before the inlet to the radiator, so that when the water flow is too cold (under the expected 75 °C), the water bypasses the radiator so that it is not further cooled. This allows the system to continue heating itself through the fuel cell until it reaches the operating temperature, which increases the fuel cell efficiency.

The combination of the fuel, air and thermal systems, accounting for the parasitic power used by these systems, results in a net system power of 49 kW shown in Fig 10. The system efficiency is about 50% at the

nominal operating point of 35 kW net, and improves at lower loads. The minimum fuel cell load the hybrid system will operate at is 10 kW and corresponds to the maximum efficiency. Due to minimum speed requirements for the air compressor and coolant pump, the system efficiency drops off at lower stack power.

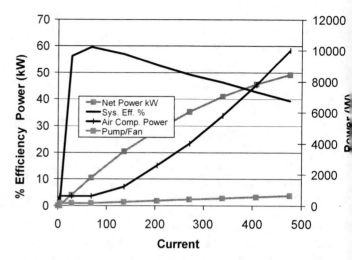

Figure 10. Fuel Cell System Part Load Characteristics

CONTROL SYSTEM

The control system is designed to start up, monitor and shutdown of the fuel cell and its subsystems. The control system drives a LCD touch screen user interface that provides communication between the driver and the control system. The control system reads all the sensors from the fuel cell subsystems and logs the acquired data. Logic in the control code determines what actions need to take place depending on the state of the individual systems.

At the heart of the control system shown in Figure 11 is a single board computer (SBC). The SBC runs the control code. Two data acquisition boards are stacked on the single board computer, enable the computer to sample signal from sensors, as well as send out control signals to the actuators in the system. All these elements are invisible to the user. A touch screen serves as interface between the driver and the vehicle. Vital information, messages and status of the system are transmitted to the user by the way of the GUI. An internal eithernet network in the ZE*burban* enable the SBC to feed to touch screen with the information the driver desires. This network makes it possible to plug additional laptops to the system, to download logged data or perform a check up of the SBC.

The operating system that the SBC uses is QNX, a real time operating system to handle data acquisition and provides a easy to use graphical environment.

For a fuel cell to perform efficiently, the air and the hydrogen fed to it need to be at very specific conditions (i.e. temperature, pressure, humidity,...). To ensure the optimal operating conditions for the system, the control system monitors 34 different sensors. Data is logged for temperature, pressure, and humidity for the reactant and coolant flows. All the data allows performing a complete energy balance on the vehicle and enabling us to further identify the behavior of fuel cells in a mobile application.

This primary use for the data is to allow the control system to make decisions. The controls system operates the 17 actuators, including pump, solenoids and actuators in the vehicle. The control code specifies the interactions between sensors and actuators for example: if current on fuel cell increases, the air compressor speed increases.

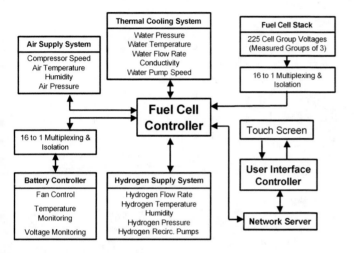

Figure 11. Control System Diagram

The six fuel cell stacks, from Honeywell, are composed of 105 cells per stack. Each cell produces a voltage between .5 V an 1.0 V. The voltage is a function of the fuel cell load and reactant conditions. The cell voltages are monitored to verify that the stack is operating properly. Cell voltages are monitored in groups of three cells, so there are a total of 210 voltages. If a group of three cells does not produce the expected voltage the control system shut down the fuel cell system. The 210 voltages are monitored through 15 multiplexing and isolation boards designed and implemented by the HEVT controls team.

Five major operating modes are identified and are controlled in the fuel cell system:

- Start-up
- Normal operating
- Shut down
- Emergency shut down
- Refill procedure

CONTROL STRATEGY

The electrical power from the fuel cell recharges the battery pack in the ZE*burban*. The control system starts

the fuel cell system if the state of charge (SOC) of the battery pack is less then 40%. When the SOC of the battery pack reaches 80%, the fuel cell system is shut down and ZE*burban* powers the drivetrain off the battery pack. While the system is on, the fuel cell load varies with bus voltage and vehicle load to minimize the amount of power processed through the battery pack. This strategy has been tested through modeling and proves to be very efficient.

BATTERY IMPLEMENTATION

The design of the battery box structure and mounting, thermal considerations, and battery safety measures in case of a vehicle accident are presented below.

MECHANICAL DESIGN OF BATTERY BOX AND INTEGRATION INTO VEHICLE

The battery box is divided into two compartments and contains twenty-eight 12 volt batteries producing 336 V, an electric motor and an inverter. It also contains three fuses that protect the entire high-voltage system from current above 800 amps.

The battery box is installed under the vehicle in the location originally occupied by the fuel tank and exhaust pipe. The batteries sit on top of a chromoly steel box-beam frame lined with haysite material. Haysite is a strong fiberglass resin material with excellent electrically insulating properties. Steel brackets welded to the vehicle frame support the battery box. There are three brackets on each side of the battery box with three 5/16" bolts in each bracket. The end of the box is supported by uprights that attach to a steel brackets on the frame cross member. The batteries are kept in place by a tie down system that will consist of four nylon straps with ratchet buckles. Each strap is looped over each battery and through tie down points on the frame on the battery box. A lightweight polypropylene lid is placed over the platform to protect the batteries from the elements.

Two small fans are used to remove hydrogen that possibly could be released from the batteries during charging. Each fan has a capacity of 17 cfm. To cool the batteries during and after operation, larger impeller-type fans pull air through the plenum system. A diagram of the battery box cooling design is shown in Figure 12. An analysis was performed to determine how much air flow was needed, and where the fans would operate. Assuming the heat lost due to conduction and radiation to be negligible, the estimated required airflow was determined to be about 50 cfm for a steady 100 A load. The operating point between the system and the fans was determined to be about 95 cfm. This is enough flow to cool the batteries at a 100-amp steady load, plus momentary increases in current during a standard driving cycle.

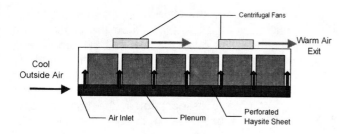

Figure 12. Battery box cooling design

The battery box consists of twenty-eight 26 Amp-hr batteries wired in series. The interconnects are made of 3/0 copper welding cable with soldered copper connectors. The welding cable is rated for 600 amps steady current. The cables are covered with insulation rated at 600 volts and 105°C. The battery and rear motor assembly, known as the rear power module is shown in Figure 13.

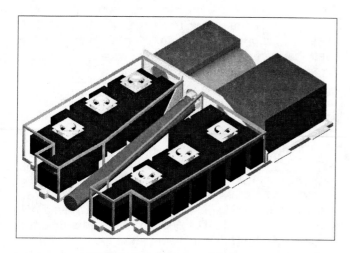

Figure 13. "Rear Power Module" Integrated battery and rear motor structure

ELECTRICAL SYSTEMS

The ZE*burban* series hybrid design requires several electrical systems. The systems are broken down into two sections: high voltage and low voltage. The high voltage system loads include the motors and their controllers, the DC/DC converter, the heating system, and the air conditioning system. The low voltage system loads include all the stock vehicle systems as well as some special 5, 12, and 15 volt control system loads. It is also important to monitor energy flow in and out of the battery pack. All of these issues are discussed in more detail below.

HIGH VOLTAGE

ZE*burban*'s high voltage electrical system varies from approximately 250 to 400 volts depending on load and state of charge of the vehicle's battery pack. The main distribution point of the high voltage is the vehicle contactor box. The box houses nine contactors that

connect high voltage to the devices. The box also contains two pre-charge relays, one for each motor inverter, and a ground fault detection system. The high voltage system can be charged with either the on board fuel cell system or the vehicle's off board charging system. This charging system uses inductive charging technology to deliver power directly to the vehicle's battery pack. To allow maintenance, the high voltage system includes a lockable manual interrupt system (MIS). This system allows the high voltage bus to be severed at the battery pack and locked so that high voltage cannot flow to the rest of the vehicle. A schematic of the high voltage system is shown in Figure 14.

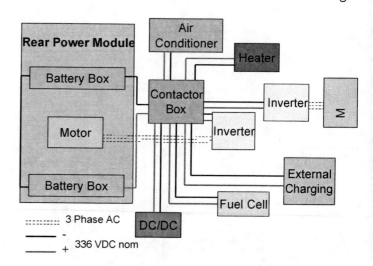

Figure 14. High Voltage Distribution and Control

The low voltage system in the truck consists solely of vehicle accessory loads. The stock accessory systems such as power locks, windows, and new electro-hydraulic steering are operated by a 3 kW DC-DC converter. This converter transforms the high voltage of the battery pack into 13.8 V DC, functionally replacing the stock vehicle's alternator. To start the vehicle, a standard 12 V battery is used to close high voltage contactors allowing the DC-DC converter to power up. A second smaller DC-DC converter supplies regulated 5, 12, and 15 V to various control circuitry throughout the vehicle.

POWER MONITORING

The power monitoring system consists of two E-meters and two current monitoring devices. The two E-meters monitor voltage and current flow of the battery pack and fuel cell system respectively to track power usage and overall battery pack amp-hour capacity. Two current monitoring devices are located in the vehicle contactor box and relay the current draw of each motor drive to that motor's respective controller.

FUEL CELL-HYBRID ELECTRIC VEHICLE SAFETY

Fuel cell hybrid electric vehicles have unique safety requirements because of the technology employed on the vehicle. Areas of concern include hydrogen storage, and high voltage. All of the stock safety features such as the air bag, and antilock braking system have been retained. Other safety features include the OnStar system, and the AutoPC, a voice activated radio.

HYDROGEN STORAGE SAFETY

A significant amount of design and development has been done to ensure that hydrogen does not escape the storage and delivery systems. However, hydrogen sensors are used to alert passengers in the event of a leak. The detectors can sense 1% volume of hydrogen in air, or 25% of the lower explosion limit. At 1%, the hydrogen detector will flash a yellow LED and sound an 80 dB alarm. When it detects 2% volume of hydrogen in air, the sensors will flash a red LED, and sound an 80 dB alarm. The alarm will trigger the opening of the rear and

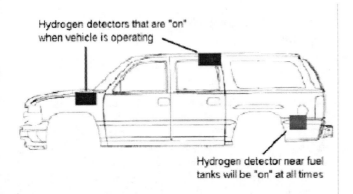

Figure 15. Hydrogen Detector Locations

side door windows. In addition, the sensors will terminate hydrogen flow at the tanks. Three hydrogen sensors are located in the engine bay, the dome of the passenger compartment, and above the fuel tanks. The locations of the hydrogen sensors are shown in Figure 15. The vehicle is equipped with service manual ball valves that prevent the flow of hydrogen from the tanks. The manual ball valves are located after the electric tank solenoids, in the rear of the vehicle.

HIGH VOLTAGE SYSTEM SAFETY

Incorporated on the vehicle are safety systems for high voltage electricity. There are two sources of high voltage electricity on the vehicle, fuel cells, and batteries. The battery pack of the vehicle uses high voltage and current contactors designed to interrupt the high voltage connection from the battery pack. These contactors are controlled four ways, as shown in Figure 16. The first is normal vehicle key operation. When the vehicle key is

turned on, the contactors are enabled, and conversely if the key is turned off the contactors are turned off. An emergency smash switch is located on the rear bumper, which turns off the contactors when pushed. The third way is a removable, watercraft kill switch where a pin is removed to disable the contactors. The contactors cannot be enabled until the pin is replaced. The last way the contactors are disabled is an inertial kill switch. In the event of a collision, the inertial switch is activated which also disconnects the contactors.

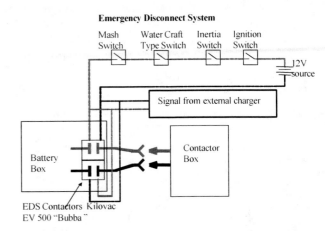

Figure 16. Emergency Disconnect System Diagram

INTEGRATION CHALLENGES

Fuel cell vehicles have barriers and challenges with specific technologies that need to be addressed in order to take these vehicles from the laboratory to a consumer's driveway. Consumer expectations of the vehicle capabilities drive many of the challenges or problems that need to be addressed and solved.

PACKAGING ISSUES

Packaging a conversion fuel cell hybrid vehicle is the most difficult task to successfully complete. The difficulty stems from the fact that a conventional technology vehicle has not been designed to accommodate fuel cell systems. A considerable amount of flexibility could be gained if a fuel cell vehicle is designed from the ground up to specifically

Figure 17. Under-hood Fuel Cell Packaging

accommodate fuel cell systems. Figure 17 shows the fuel cell systems packaged under the hood of the vehicle, with no room left to spare.

HYDROGEN STORAGE

Locating space for onboard hydrogen fuel storage is another challenge that is ongoing. Utilizing current technology compressed hydrogen storage still yielded 1/5 of the current conventional vehicle range. Additional range could be achieved by increasing the vehicle efficiency (weight reduction, aerodynamic improvements and overall fuel cell system efficiency) and increase the allocated space for hydrogen storage by reducing the size of other vehicle components.

FUEL CELL AIR SUPPLY SYSTEM

The air supply system, as previously described, uses a Opcon 1050A twin screw compressor, and a 10 kW, 10,000 rpm AC induction motor. A 10 kW industrial motor drive provides the control for the AC induction motor. The setup yielded a 60 kg system that measures approximately 0.75 m in length, and 0.25 m^2 in cross section. Figure 18. below illustrates the system.

Figure 18. Air Delivery assembly

This assembly is large, and heavy for the application that it serves. This solution is one of the few products currently available that meet the design specifications.

Another aspect of the air supply fuel cell system that needs development is the inlet air temperature and humidity control. Currently, we are working with Porvair, a porous metal material supplier, to develop a device that regulates the temperature and humidity of inlet air stream to the fuel cell system. There is a considerable amount of energy (20 kW thermal) that needs to be added to the inlet air stream to achieve the desired characteristics (~ 70 deg C at 85 % RH). The challenge is that the heat and mass transfer must occur simultaneously. The current design, seen in figure 19,

yields a device that is 0.3 m tall and 0.2 m in diameter with a weight of 10 kg.

FIGURE 19. Air Humidification chamber

The current design has temperature and humidity control issues and the weight/size needs to be reduced.

CONCLUSION

A series hybrid PEM fuel cell sport utility vehicle has been modeled, designed, and built by the Hybrid Electric Vehicle Team of Virginia Tech for the FutureTruck Challenge. The ZEburban vehicle meets most of the goals and requirements for the competition. An efficient electric drivetrain combined with vehicle accessories, low rolling resistance tires, and properly sized battery and fuel cell subsystems results in combined city/highway fuel economy of 24 mpg (gasoline equivalent) or 1.4X the fuel economy of the stock vehicle. The ZEburban meets or exceeds the performance of the stock vehicle in most areas, including towing. Due to limitations of compressed gas hydrogen storage, the range of the vehicle is limited to 120 km (75 mi). The overall single speed gearing of the electric motors limits the top speed to 130 kph (80 mph). The consumer options and safety features have been maintained, including stock vehicle interior and exterior.

The lower greenhouse gas (GHG) emissions per unit energy of hydrogen fuel combined with lower energy consumption of the vehicle results in total GHG emissions of 135 gm/mile. This represents an 80% reduction in GHG emissions, exceeding the FutureTruck goal of 67% reduction. The ZEburban fuel cell vehicle is also locally zero emission with no evaporative emissions. A ground up redesign of this SUV could provide significant weight savings and allow packaging of fuel tanks to meet consumer requirements for range.

REFERENCES

P. Atwood, S. Gurski, D. J. Nelson, and K. B. Wipke (2001), " Degree of Hybridization Modeling of a Fuel Cell Hybrid Electric Sport Utility Vehicle," Proceedings of the 2001 SAE Int'l Congress, Detroit, MI, March 5-8, 2000, SAE Paper 2001-01-0236, in Fuel Cell Power for Transportation 2001, SP-1589, pp. 23-30.

Energy Information Administration (1998), "What Does the Kyoto Protocol mean to U. S. Energy Markets and the U. S. Economy? – A Briefing Paper on the Energy Information Administration's Analysis and Report," http://www.eia.doe.gov/oiaf/kyoto/kyotobtxt.html

M. Fuchs, F. Barbir, A. Husar and J. Neutzler, D. J. Nelson, M. J. Ogburn, and P. Bryan (2000), " Performance of an Automotive Fuel Cell Stack," Proceedings of the 2000 Future Car Congress, April 2-6, Arlington, VA, SAE paper 2000-01-1529, 5 pgs.

Stephen Gurski, D. Evans, D. Knox, M. Conover, A. Harris, H. Lohse-Busch, S. Kraft, Douglas J. Nelson, (2002), "Design and Development of the 2001 Virginia Tech FutureTruck: A Fuel Cell Hybrid Electric Vehicle," Proceedings of the 2001 FutureTruck Challenge, June 4-13, 2001, Milford, MI, SAE SP-1701, 18 pages.

Jay Hakes, (1999), Energy Information Administration, Department of Energy before the Energy and Natural Resources Committee of the United States Senate," March 25, 1999, www.eia,doe.gov/neic/speches/sentest325/testv7.html

B. D. James, G. N. Baum, and I.F. Kuhn, Jr., (1994), "Technology Development Goals for Automotive Fuel Cell Power Systems, ANL-94/44, Argonne, Ill., Aug.

S. Kraft and M. Kheir (2001), "Hydrogen Safety For ZEburban And The Ware Lab," Report to FutureTruck 2001 Organizers, 16 pages plus appendices.

G. Kulp and D.J. Nelson (2001), "A Comparison of Two Fuel Cell Air Compression Systems at Low Load", SAE paper 2001-01-2547 in Fuel Cells and Alternative Fuels/Energy Systems, SP-1635, Proceedings of the 2001 SAE Future Transportation Technology Conference, Costa Mesa, Ca., pp. 81-90

W. Luttrell, B. King, S. Postle, R. Fahrenkrog, M. Ogburn, D. J. Nelson (1999), "Fuel Cell System Efficiency in the Virginia Tech 1999 Hybrid Electric FutureCar," Proceedings of the 1999 Environmental and Advanced Technology Vehicles Conference & Exposition, June 13-16, 1999, Ypsilanti, MI, published by SAE/ESD on CD-Rom, 13 pages.

M. A. Merkle, R. D. Senger, and D. J. Nelson (1997), "Measured Electric Vehicle Drivetrain Efficiencies for the Virginia Tech FutureCar," 1997 SAE Intl. Congress, Detroit, MI, Feb. 24-27, SAE Paper 97804, pp. 1-9.

National Cattlemen's Beef Association (1999) "Global Climate Treaty Fact Sheet," http://hill.beef.org/files/FSPP/gct.htm

M. J. Ogburn, D. J. Nelson, K. Wipke, and T. Markel (2000), "Modeling and Validation of a Fuel Cell Hybrid Vehicle," Proceedings of the 2000 Future Car Congress, April 2-6, Arlington, VA, SAE paper 2000-01-1566, 13 pgs.

M. J. Ogburn and D. J. Nelson (2000), "Systems Integration and Performance Issues in a Fuel Cell Hybrid Electric Vehicle," Proceedings of the 2000 SAE Int'l Congress, Detroit, MI, March 2-6, 2000, SAE Paper 2000-01-0376, in Fuel Cell Power for Transportation 2000, SP-1505, pp. 125 - 137. Also published in Fuel Cell Technology for Vehicles, SAE PT-84 2001, pp. 203-216.

D. Patton, J. Latore, M. Ogburn, S. Gurski, P. Bryan, and D.J. Nelson, , (2001), "Design and Development of the 2000 Virginia Tech Fuel Cell Hybrid Electric FutureTruck," Proceedings of the 2001 SAE Int'l Congress, Detroit, MI, March 5-8, 2000, SAE SP-1617, pp. 185-202.

R. D. Senger, M. A. Merkle, and D. J. Nelson (1998), "Validation of ADVISOR as a Simulation Tool for a Series Hybrid Electric Vehicle," Technology for Electric and Hybrid Vehicles, Proceedings of the 1998 SAE Intl. Congress, Detroit, MI, Feb. 23-26, SAE Paper 981133, SP-1331, pp 95-115.

TECO Energy (1998), "Kyoto Treaty Update,", http://www.tecom.net/energy/ENNwsVwsSmmr98c.html

Wang, M. Q., (1999), GREET 1.5 - Transportation Fuel Cycles Model: Methodology, Development, Use, and Results, Center for Transportation Research, Argonne National Laboratory, ANL/ESD-39, Vol. 1, Argonne, Ill., Aug.

M.Q. Wang and H.-S. Huang, (2000), A Full-Fuel-Cycle Analysis of Energy and Emissions Impacts of Transportation Fuels Produced from Natural Gas, ANL/ESD-44, Argonne, Ill., Jan.

Wipke, K., Cuddy, M., Burch, S., "ADVISOR 2.1: A User-Friendly Advanced Powertrain Simulation Using a Combined Backward/Forward Approach," IEEE Transactions on Vehicular Technology: Special Issue on Hybrid and Electric Vehicles, Nov. 1999.

Wipke, K., T. Markel, and D. Nelson (2001), "Optimizing Energy Management Strategy and Degree of Hybridization for a Hydrogen Fuel Cell SUV," accepted for publication at 18th Electric Vehicle Symposium (EVS-18), Berlin, Germany, October 20-24, 2001.

CONTACT

Dr. Douglas J. Nelson
Virginia Polytechnic Institute and State University
Mechanical Engineering Department
Blacksburg, Virginia 24061-0238
(540) 231-4324 Doug.Nelson@vt.edu

2000-01-0003

Fuel Choices For Fuel Cell Powered Vehicles

Paul J. Berlowitz
Exxon Research & Engineering

Charles P. Darnell
Exxon Corporation

ABSTRACT

Fuel cells offer the potential of ultra-low emissions combined with high efficiency. Recent, rapid advances in the past few years in fuel cell technology have resulted in a vast increase in fuel cell research and development directed towards wide spread application in vehicle and stationary power. The PEM (proton exchange membrane) fuel cell proposed for future vehicles requires hydrogen as a fuel. Supplying hydrogen, as either hydrogen gas or a hydrogen-rich reformate, is a critical issue. A large number of fuel sources can be used to provide the hydrogen, each has advantages and disadvantages. For example, hydrogen provides the simplest and easiest on-board system, but requires the largest infrastructure hurdles. Methanol has some advantages in producing usable reformate, but several other issues ranging from performance on cold-start and transients, to acute toxicity and maintaining purity in transport would need to be addressed. Similarly, gasoline-like liquids present opportunities to capitalize on existing infrastructure and customer familiarity, but may increase system complexity and require longer hardware development time. This paper will explore the fuel choice issue, highlighting the advantages and disadvantages of hydrogen, methanol reformers, and hydrocarbon (gasoline) reformers. Infrastructure, safety & health issues, system complexity and feasibility will also be addressed.

INTRODUCTION

The past 30 years have seen rising public concern with air quality and fuel economy. For the automobile and energy industries, this has led to a series of changes in automotive hardware and fuel composition designed to reduce pollution and improve efficiency. Hardware improvements such as catalytic converters, three-way catalysts, fuel injection and sophisticated engine management; along with fuel changes to unleaded gasoline, detergents for injectors, valves, and combustion chambers, reformulated gasoline, and new low-sulfur fuels have eliminated well over 90% of the precontrol emissions and doubled fuel economy. Since the mid-1980s, increasing consumer demand for light trucks, increased horsepower, and vehicle size have stalled progress on fuel economy. In addition, light trucks have not been subject to the same emissions standards as passenger cars.

Despite dramatic improvements in air quality, standards have not been achieved in many areas. Additionally, increasing emphasis on energy efficiency and CO_2 emissions have pressured automakers and energy suppliers to produce even cleaner and more efficient vehicles. These drivers have led to a proliferation of new engine and powertrain designs. Direct injection for spark and diesel engines, variable valve timing, and improved emissions technologies such as NOx and particulate traps, promise to reduce emissions and improve efficiency for conventional powertrains. New vehicles utilizing these advancements, and several announcements of future vehicle have been made recently. [1-4]

Powertrains based on electric power are now under development to compete with these advanced engines. Battery electric vehicles – with the advantage of zero tailpipe emissions - introduced first into the market, have not been successful due to high cost, limited range and lack of recharging infrastructure. Hybrid electric vehicles have only recently been introduced, but have proven much more successful, and are now considered a major option at many automakers [5-10]. Though they cost more to produce than a conventional vehicle, hybrids do offer improved fuel economy, low emissions, and take advantage of the existing fuel infrastructure. Fuel cell vehicles, offering potential for up to twice the efficiency, and near-zero emissions, will compete against these advanced technologies. To be successful, fuel cells will have to provide emission and efficiency advantages while competing on cost and consumer appeal. Table 1 lists the emerging vehicle technologies, energy efficiency improvement, and potential fuels.

Table 1. Competing Technologies

Technology	Potential Efficiency Improvement	Fuel
Direct Injection	15-40%	Gasoline Diesel
Hybrid Electric	40-90%	Gasoline Diesel
Fuel Cell	30-100%	Gasoline Diesel Methanol Hydrogen

FUEL OPTIONS FOR FUEL CELLS

Two options are now under consideration for supplying hydrogen to PEM fuel cell powered vehicles: Hydrogen, and fuel reforming of liquid hydrogen sources such as methanol or gasoline. Each material has advantages and disadvantages related to technical feasibility, cost, and infrastructure.

HYDROGEN GAS OR LIQUID – The simplest and most efficient vehicle use hydrogen gas, supplied from a liquid or gaseous reservoir on-board the vehicle. Most of the earliest fuel cell vehicles have used compressed hydrogen as a fuel source.

Due to its low energy density, hydrogen is very expensive to transport and store. In addition, no infrastructure is currently available for distributing hydrogen to the motoring public. Studies by the Department of Energy, California Air Resources Board, and others [11-15] highlight the difficulties of this approach and estimate that the infrastructure for hydrogen production and distribution would cost $100 billion or more to produce 1 million barrels of oil-equivalent energy per day – just 10 percent of today's U. S. road-fuel energy requirement. The cost of building a hydrogen infrastructure has been debated in numerous publications (e.g., [16-18]). Also, the use of hydrogen in passenger vehicles causes safety concerns that must be addressed. Hydrogen would require a new safety infrastructure where it is dispensed and transported. Dispensers would likely work at either high pressure or cryogenic temperatures, neither of which is within the normal range of experience of most vehicle owners. The public, long accustomed to the use of conventional fuels such as gasoline, diesel, or propane, would have to be "trained" in the use of hydrogen which behaves far differently and poses different dangers.

Problems with supplying hydrogen to the public will probably limit the use of hydrogen fuel to niche fleet applications, such as inner-city buses, or where true zero-emission vehicles are required (e.g., vehicles used in enclosed spaces, such as current applications for electric-powered forklifts). A recent assessment prepared for the California Air Resources Board (CARB) concluded that hydrogen "is not considered a technically and economically feasible fuel for private automobiles now or in the foreseeable future" [12].

Vehicle range could also suffer greatly as the energy density of hydrogen is a small fraction of the hydrogen abstractable from liquid fuels as shown in Figure 1. Large tanks, or storage as liquid hydrogen – which incurs a large energy efficiency penalty – would be required to achieve acceptable vehicle range. Advanced technologies such as storage in metal hydrides, carbon nanotubes, ultra-light tanks, or other high-density media, are yet to be proven commercially viable. Should these technologies move from the laboratory to commercial practice at competitive cost, hydrogen storage on-board vehicles may become more competitive with liquid fuels, but it would not solve the large financial hurdles of developing a hydrogen infrastructure. As a result, the balance of this paper will concentrate on liquid fuels as sources of hydrogen for fuel cells.

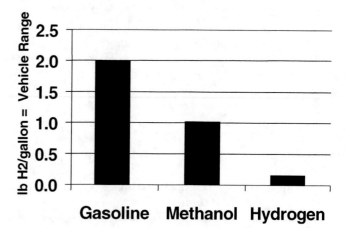

Figure 1.

HYDROGEN FROM LIQUID FUELS

On solution to storing hydrogen at high density on-board a vehicle to extract hydrogen from a liquid fuel via a fuel reformer. This allows the use of a wide variety of fuel sources for a fuel cell including gasoline, methanol, naphtha, ethanol and other bio-derived fuels.

Fuel reformers currently under consideration all share some common features. All feature a primary reformer, responsible for taking the liquid fuel and converting it to a gaseous reformate which contains a large amount of hydrogen; though it will also contain some level of unwanted impurities, most particularly CO. Following the primary reformer one or more cleanup stages are employed. These are designed to convert CO to CO_2 by reaction with water (e.g., as in water-gas shift), or reaction with oxygen (e.g., as in preferential oxidation, PROX), reducing the CO concentration to 5-100 ppm. Final mitigation of CO poisoning is often done by oxygen addition to the fuel cell where reports of up to 500 ppm tolerance to CO have been made recently [19].

It is in the primary reformer that most of the key differences between fuels are found. Three basic designs are being evaluated: steam reforming, partial oxidation, and autothermal reforming.

PARTIAL OXIDATION – Partial oxidation (POX) reformers combine fuel with oxygen to produce hydrogen and carbon monoxide. This approach generally uses air as the oxidant, and results in a reformate which is diluted with nitrogen. Typical stoichiometry for iso-octane (a typical gasoline molecule) is shown below.

$$C_8H_{18} + 4O_2 + 16N_2 \Rightarrow 8CO + 9 H_2 + 16N_2 + 659 \text{ kJ/mol}$$

Partial oxidation reformers require subsequent water-gas shift reactor(s) to capture more hydrogen from H_2O by conversion of CO to CO_2. The reaction is exothermic which has the advantage of very fast response to transients, but requires efficient heat integration between the high temperature POX reactor and subsequent processing steps.

STEAM REFORMING – Steam reforming (SR) combines fuel with water to produce hydrogen and carbon monoxide. The concentration of hydrogen in the reactor effluent is higher than in POX because no nitrogen diluent flows through the reactor and the only reactant is water. Typical stoichiometry for iso-octane and methanol are shown below.

$$C_8H_{18} + 8H_2O + 1.28x10^3 \text{kJ/mol} \Rightarrow 8CO + 17 H_2$$

$$CH_3OH + H_2O + 122 \text{ kJ/mol} \Rightarrow CO_2 + 3 H_2$$

The highly endothermic steam reforming reaction must be powered by burning some of the fuel, typically 15-25% of the total fuel heating value. While the steam reforming reaction produces a higher concentration of hydrogen, it is more difficult to deliver power through a sharp upward transient. As more fuel is fed to the reactor, the endothermicity drives the reaction temperature down, reducing the reaction rate. This in turn requires the heat source for the reactor to respond quickly in order to maintain reactor temperature. Though there is some heat which can be supplied by burning the anode tail gas (the residual hydrogen from the fuel cell stack which is not completely utilized), the heat produced by this stream is unlikely to be synchronized with the heat requirements for transient response in the steam reformer.

AUTOTHERMAL REFORMING – Autothermal reformers (ATR) combine fuel with both air and steam so that the exothermic heat from the partial oxidation reaction is balanced by the endothermic heat of the steam reforming reaction. The stoichiometry for isooctane is shown below.

$$C_8H_{18} + 2.64O_2 + 9.9 N_2 + 2.72H_2O \Rightarrow 8CO + 11.7 H_2 + 9.9 N_2 \text{ (no net heat input/production)}$$

Autothermal reforming produces a more concentrated hydrogen reformate than straight POX, though less concentrated than pure steam reforming. Heat transfer between the high temperature POX and SR reaction allows for better heat integration and lower operating temperature of the primary reformer, but does require a catalyst(s) to perform both reactions. While not as fully developed as the others, ATR offers the most flexibility in heat management and thus potentially higher efficiency than POX with better transient response than SR.

In general any of the fuels considered as hydrogen carriers (e.g., gasoline, methanol, ethanol, LPG) can be used in any of the three reformer designs. Differences in the chemical nature of the fuels, however can favor one design over another.

The qualitative advantages and disadvantages of fuel processors for methanol and gasoline, described below, are shown in Table 2.

Table 2. Relative Advantages of Fuel Reformers

	Gasoline (POX or ATR)	Methanol (SR)
Development Progress		X
Reformer Simplicity		X
Heat Integration		X
Fuel Flexibility	X	
Start-up / Transients	X	
Vehicle Range	X	
Emissions	X	X

METHANOL FUEL PROCESSOR

Steam reforming is the leading design for conversion of methanol in a fuel cell vehicle. Methanol's unique chemical structure allows conversion at temperatures of only about 250C. The reaction is generally done over ZnO supported Cu, which is also the lead catalyst for water-gas shift reactors. As a result, the reactor effluent is generally close to the equilibrium composition of CO, CO_2, H_2, H_2O. CO concentrations of 1-5% are typical, and can be further processed by preferential oxidation (PROX) or other CO cleanup stages downstream. A lower temperature steam reformer would favor even lower CO concentrations (the equilibrium concentration of CO decreases with decreasing temperature), but the reaction rate would be too slow.

ADVANTAGES – The low temperature methanol steam reformer is the simplest of the three primary reformer designs, and can have as few as two reaction steps before the fuel cell stack. By operating at temperatures close to those of the fuel cell stack, methanol reformer systems do not require much additional equipment to remove excess heat. Another plus is that the low-temperature operation produces nearly zero nitrogen oxide emissions, while the use of steam as an oxidant results in a high concentration of hydrogen in the fuel cell feed gas.

In its simplest form, the methanol steam reformer offers the potential for the lowest cost and smallest-sized engine, although some of the methanol system's disadvantages could result in the need for additional components and further product development.

DISADVANTAGES – The methanol steam reformer is endothermic, thus it requires heat input to operate. To provide that heat, an additional reactor or burner is required to burn some of the fuel and/or exhaust gases from the fuel cell stack. Also, the heat required to vaporize methanol is nearly four times that of gasoline.

The heat required to power the steam reforming reaction and the high heat of vaporization for methanol results in long start-up times before operating temperatures are reached. Current design take up to 30 minutes to start producing power, and much current research is focused on mitigating this problem. Adding a battery to run the vehicle until the reformer and fuel cell stack are producing power could solve this problem, but will add considerable cost and complexity, especially if the start-up time cannot be reduced below several minutes.

Another disadvantage of the methanol steam reformer stems from methanol's chemical uniqueness. Only methanol, and not any other commonly proposed simple oxygenate or hydrocarbon (e.g., ethanol or propane) can be processed by a methanol steam reformer. This leads to two problems. First, the system does not offer any fuel flexibility. It would preclude the use of conventional hydrocarbon sources as well as alternative fuels derived from petroleum, gas, or biomass sources not converted to pure methanol. Secondly, most impurities, organic or inorganic (e.g., salts), are poisons to the catalyst system. This imposes severe purity constraints on the fuel, and will preclude the use of many additives which might be needed for safety or operability requirements.

Concerns over transient performance and fuel specificity have been addressed by moving to an autothermal design such as in Johnson Matthey's HotSpot reactor. [20] This approach may improve response, but at the expense of some the advantages cited for low-temperature steam reforming.

GASOLINE FUEL PROCESSOR

Partial oxidation and ATR are the leading designs for conversion of gasoline in a fuel cell vehicle. They are also the leading designs for conversion of other hydrocarbon and alcohol fuel sources such as methane and ethanol. POX reactors generally run at high temperatures, typically around 800-1100C and produce hydrogen and carbon monoxide (syngas). The conditions in the POX reactor, while much hotter than the methanol SR, do not produce significant amounts of NOx

from the nitrogen in the air. Syngas generation technology is widely practiced on an industrial scale for several processes. It is a key step in synthetic fuel production via Fischer-Tropsch synthesis.

ADVANTAGES – The highly exothermic nature of the gasoline POX system allows a much faster start-up of the primary reformer. Typically only a few seconds are needed for the POX reactor to "light-off" and begin complete conversion of the reactants to CO and hydrogen. Bench scale processors have achieved warm-up in as little as 2 minutes and progress towards faster start-up is likely in the near term [11,13]. The high temperature of the primary reformer allows use of many fuels, including methane, LPG, naphtha, ethanol, methanol, and potentially even kerosene or jet fuel. The process is less sensitive to small impurities, as many species are converted to gaseous components which may be "scrubbed" for the reformate if required, before reaching the more sensitive catalytic steps used to convert CO. In addition, liquid fuels like gasoline have a much higher hydrogen density than methanol producing up to twice the miles per gallon.

DISADVANTAGES – There are two main disadvantages of the gasoline (or other hydrocarbon fuel) processor. First, a POX or ATR design is likely to need additional process steps, most likely water-gas shift reactor(s), to completely convert CO to CO_2 and H_2. Secondly, the higher temperature present in POX and ATR will demand greater heat integration than the lower temperature methanol SR reactor, which complicates the design. This is likely to make a gasoline reformer larger and more complex than a methanol system. – though not necessarily less efficient. The efficacy of the heat integration design will have a large impact on the overall processor efficiency.

Another consideration is that today's gasoline grades are specifically designed for internal combustion engines using spark ignition. Gasoline contains small amounts of sulfur, a poison to fuel cell stacks in the part-per-million range, and other heteroatoms may be present at trace levels. Fuel cell vehicles may require a new grade of gasoline with very low sulfur levels and possibly other compositional changes. This grade could replace one of the existing grades now available in the fuel distribution system. In a section below we will examine the opportunity and costs associated with producing a new fuel cell grade gasoline.

DEVELOPMENT PROGRESS – At present, methanol SR systems are further advanced in development than gasoline POX or ATR designs. General Motors, Daimler-Chrysler, Toyota, and Nissan all have operating fuel cell test vehicles running on methanol. The first of these was developed in late 1997. Both General Motors and Epyx Corporation have demonstrated integrated gasoline systems in test laboratories, but vehicle demonstration is not expected before 2000.

Although fuel cell engine designs have made significant technical progress, there are further hurdles to overcome before they are ready for consumers. Each design must meet a competitive level of cost, reliability and size while providing performance which competes favorably with traditional internal combustion engines, and recently introduced battery-IC engine hybrids. At the present time, given the uncertainties of technology development, it is too early to assess when these vehicles will become truly competitive, and which system will emerge as the leading technology.

PRODUCING A HYDROCARBON FUEL FOR FUEL CELLS

Over 500 refineries around the world currently produce nearly 30 million barrels (5 billion liters) of gasoline and diesel fuel per day [21]. Refineries may produce upwards of 10 different gasoline blends, each made from multiple individual streams. The refinery infrastructure feeds associated pipelines, marine terminals, trucks, and ultimately retail stations of which there are about 190,000 in the United States alone. A fuel cell "gasoline" would take advantage of this infrastructure, minimizing the cost of producing, and easing the distribution of a new fuel.

Researchers at Epyx [11,13] have demonstrated that a flexible POX processor can be used on current gasoline in short-term tests, though it was more difficult than some simpler fuels such as ethanol. A POX based processor has the advantage of fuel flexibility, and may allow for wider distribution of the fuel cell system. A specially formulated gasoline could improve on the performance of current gasoline, and be produced at an attractive cost by eliminating some of gasoline's ingredients.

FUEL CELL GASOLINE – Current gasoline is tailored to meet the demands of today's engines. Foremost among these demands is the need for high octane values to provide knock-free operation. Octane, however, is unlikely to improve operation of a fuel cell reformer. Likewise, expensive ingredients like oxygenates that are now blended into gasoline to lower emissions (and under debate as to the overall environmental benefit), may not be needed.

Sulfur is likely to be the most important fuel quality issue. Sulfur can act as a poison in the water-gas shift reactor, preferential oxidation stage and in the fuel cell stack. Although the acceptable level of sulfur is not currently defined, it is likely that it will need to be reduced to just a few ppm or lower for long-term performance of the fuel cell system.

In today's gasoline, sulfur levels vary between about 30 and 300 ppm in industrialized countries. Sulfur could be removed on-board the vehicle – adding additional cost and complexity to the reformer – or in the fuel itself. The sulfur currently found in gasoline is concentrated in specific compounds found in specific process streams blended to make gasoline. Several refinery processes

are sensitive to sulfur, and the feeds to these processes are currently extremely low in sulfur content.

For example, catalytic reforming – a process which converts low octane molecules to high octane aromatics – has both a feed and product which often contain <1 ppm sulfur. Alkylation and isomerization, which produce high octane branched paraffins, have feeds and products which are typically a few ppm sulfur. Similarly the product (though not the feed) to a hydrocracker is very low in sulfur. Conversely, catalytic cracking, which produces a high percentage of high-octane olefins, does contain substantial amounts of sulfur in both feed and product. Each of these streams are present in millions of barrels per day quantities – more than enough to meet the projected needs of fuel cell vehicles in the foreseeable future. Examples of current and future fuels which may be used for fuel cells are shown in Table 3.

Table 3. Potential Gasoline Blendstocks for Fuel Cells

Source	Blendstock	Sulfur, ppm
Crude oil	Unleaded Gasoline	30-300
Crude oil	Hydrotreated virgin naphtha	<1
Crude oil	Hydrocrackate	<10
Crude oil	Alkylate/isomerate	<10
Natural gas	Gas-to-liquids (Fischer-Tropsch) naphtha	0
Natural gas	Hydrotreated condensate	<1 to <10

Judicious use of current low sulfur materials already present in the refinery may satisfy the demands of fuel cell vehicles. If additional sulfur removal is needed in selected refineries, the investment required for removing sulfur from, for example, a low-octane naphtha stream is less than 1 percent of the investment required to produce and supply a new fuel such as methanol. A likely candidate for a fuel cell gasoline is naphtha – the direct, hydrotreated, light boiling range material extracted from crude oil. Naphtha is easily desulfurized to very low levels. Currently naphtha requires additional, expensive processing to increase its octane level. Such steps would be unnecessary with a fuel cell gasoline, reducing its cost, and improving the energy efficiency and CO_2 emissions of producing fuel.

Other fuels with excellent potential for fuel cells are derived from natural gas such as Fischer-Tropsch liquids. These liquids contain no sulfur at all, and are very high in paraffins and hydrogen content. Though not currently available in significant quantities, these fuels could become available just as the fuel cell vehicle industry begins to mature. Another potential fuel is condensate – the liquid co-produced with natural gas. These streams are typically low in octane requiring processing to become conventional gasoline, but are easy to desulfurize and cost less than current gasoline.

Sulfur could also be removed on-board by traps in the fuel processor system. Long-known, and inexpensive materials like ZnO have been used in other applications to achieve very low sulfur levels. Epyx has demonstrated sulfur trapping to less than 1 ppm using current pump gasoline.

USING METHANOL AS A FUEL CELL FUEL

Worldwide, about 600,000 barrels of methanol are produced each day (energy equivalent of about 300,000 barrels of oil) [22,23]. For comparison, the world transportation demand for road-transportation fuels is about 100 times as large as shown in Figure 5. Given a low initial rate of market penetration for fuel cell vehicles, current excess methanol production capacity would be sufficient to meet demand [23]. However, if methanol fuel cell vehicles begin to penetrate beyond selective, niche applications, a significant amount of new production and distribution capacity would be needed to supply the fuel.

PRODUCTION, DISTRIBUTION AND COST – The vast majority of the world's methanol is made from natural gas. Methanol can also be made from crude oil, coal, and renewables such as biomass and wood. However, synthesis from natural gas is significantly less expensive than from any alternative starting material. Without a substantial change from today's energy market, natural gas is likely to remain the source of virtually all new methanol production capacity [24].

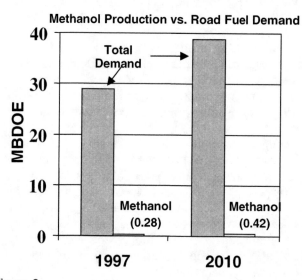

Figure 2.

Due to the significant differences between methanol and gasoline, they cannot share the same distribution system. The high miscibility of water in methanol can result in methanol which is highly corrosive. Methanol also requires different installations because it presents a greater flammability hazard than gasoline when exposed to air within a confined space – as it might be in distribution terminals, service stations, or vehicle fuel tanks. While methanol can be handled safely, the equipment involved will be different than the current

gasoline infrastructure provides. Lastly, before methanol can be widely used as a transportation fuel, it requires a dedicated network of distribution facilities. This would be necessary in order to maintain the high fuel purity requirements of the methanol steam reformer. Each retail store which sells methanol will need a separate storage tank and fuel dispenser.

Methanol Cost and Prices – New methanol pipelines, terminals, ships, barges, storage tanks, and dispensers can be built, but at a significant cost. Constructing a infrastructure to produce, distribute and market methanol is expected to cost from $15 billion to $23 billion for each million barrels per day (MBD, approximately 160 million liters/day) of capacity. Methanol production now costs about $14-15 billion per MBD, but that could drop as much as 30 percent as new technologies and additional economies of scale come about. The cost for an infrastructure to distribute and market methanol would add $4-$8 billion per MBD of methanol produced [24]. The capital cost requirements per million barrels per day of oil equivalent energy production are summarized in Table 4.

Though these costs are high, during initial introduction of fuel cell vehicles, volumes will be low enough to use excess production capacity and investment costs should be fairly low [23]. Were methanol fuel cell vehicles to penetrate broadly into the market, however, meeting demand for fuel would take a very large investment. If, for example, methanol fuel cell vehicles were to capture half of today's gasoline market, the investment would be hundreds of billions of dollars to duplicate the existing gasoline infrastructure.

Table 4.   Capital Cost Requirement

Cost/million barrels/day	Low	High
Production	11 billion	15 billion
Marine	2	3
Distribution	1	3
Retail	1	2
TOTAL	15	23
oil-equivalent MBD	30	46

The history of gasoline and methanol prices over the past 10 years, shown on an energy equivalent basis, is plotted in Figure 3. Methanol prices, excluding distribution, marketing, and taxes, have average more than $1 per gallon ($4/liter) on a energy equivalent basis, some 85 percent higher than the wholesale price of gasoline over the same period. Excluding the price spike which occurred in 1994, methanol's price has averaged 60 percent higher than regular gasoline.

Historical Price Comparison: Oil-Equivalent Energy Basis

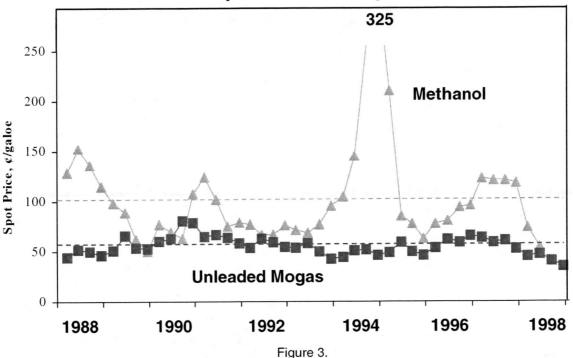

Figure 3.

A recent study by major U.S. automakers and energy companies [23] under the joint Auto/Oil Air Quality Improvement Program projected the retail price of methanol transportation fuel. The study predicted that M85 (85% methanol/15% gasoline) would be 50-100% higher than regular gasoline on an energy equivalent basis - in line with the historical differences in retail prices shown in Figure 3. The manufacturing cost of methanol may decline in the future as a result of economies of scale and new technologies. However, in a tax-neutral environment, it will be difficult for methanol prices to compete with gasoline prices on an energy equivalent basis.

EFFICIENCY AND EMISSIONS

Gasoline and methanol fuel cell vehicles hold the promise of substantially better energy efficiency than today's internal combustion engines. Methanol and gasoline fuel processor systems have not been developed to the point where overall system efficiency can be reliably estimated over standard automotive driving cycles. Vehicle efficiency will be determined by both the efficiency of the individual fuel processing steps, as well as the success of integrating those components into an overall, efficient power-train design.

Currently fuel cell stacks can achieve a maximum of 50-60 percent efficiency under normal loads. Well integrated fuel processors have been reported to have 78-85 percent efficiency under steady-state conditions [11,13]. The American Methanol Institute recently estimated a 38% vehicle efficiency (processor + fuel cell + motor/electronics) for methanol fuel cell vehicle [23], in line with this estimate. Though these numbers provide a

useful reference, efficiency losses during start-up, transient and low-load conditions, and real world driving cycles has not been well quantified. Each system needs further development before the efficiency differences between them can be accurately quantified.

Methanol and gasoline fuel cell vehicles should also provide significant emission benefits over today's vehicles. Since the fuel reformers, even the high temperature POX process for gasoline, operate at much lower temperatures than combustion in an internal combustion engine, NOx production is not a concern. In addition, emissions of CO and hydrocarbons have been reported to be well below the proposed SULEV standard for California. [11,13,23]

PRODUCTION EFFICIENCY – On average, hydrocarbon fuels retain 85 percent of energy originally contained in the crude oil. About 5% is lost during extraction of the oil and an additional 10% is consumed in the refinery, primarily to provide heat to the refinery processes. Production efficiency has some potential to improve for fuel cell fuels since some candidate fuels, such as hydrotreated virgin naphtha, would require less processing than conventional gasoline. Current refineries use several energy intensive processes to produce high octane components which would not be required. In addition, oxygenated fuels would not be used saving further on energy efficiency.

Methanol manufacturing consumes more energy than it takes to produce gasoline, diesel, naphtha and other hydrocarbon fuels. In comparison to the 90% efficiency of crude oil refineries, process limitations hold the efficiency of even the newest natural gas-based methanol plants to about 68%-70% [23]. If coal is used as the energy source, the process is less efficient, only about 60% [25]. A substantial amount of this inefficiency is unavoidable. Methanol is, in essence, partially oxidized methane, and the loss in heating value per molecule in transforming methane to methanol is not recoverable.

Delivering methanol to the market is also less efficient than delivering the energy in hydrocarbon fuels. Gasoline has twice the energy density of methanol, so twice the volume of methanol must be transported through the system in order to supply the same amount of energy. While comparing a methanol fuel cell car, with twice current efficiency, to a current gasoline car may seem equivalent; the proper comparison – a methanol fuel cell vs. a gasoline fuel cell – finds distribution of the methanol is half as efficient. Thus, by the time methanol reaches consumers, only about 63% of the energy originally contained in the natural gas is retained.

WELL-TO-WHEELS EFFICIENCY – Current, mid-sized passenger cars are about 18% efficient, i.e. 18% of the energy in the fuel is converted to work driving the wheels. When this efficiency is multiplied by the production and distribution efficiency of 85% for gasoline, about 15% of the energy of the crude oil in the ground in converted to useful work at the wheels. This is the well-to-wheels efficiency for a current internal combustion energy. All comparisons to current systems must be made on this basis, including production of the resource, conversion to a useful fuel, distribution, and vehicle efficiency. In addition, it is only fair to compare systems measuring vehicle efficiency under standard fuel economy test conditions such as the FTP cycle or equivalent European or Japanese fuel economy standards. Approximate efficiencies and emissions are shown in Figure 4 for current engines, fuel cell alternatives, and other

competitive technologies including methanol from coal [26].

Advances in internal combustion engine technology, such as direct injection spark and diesel designs, lean-burn operation, and variable valve timing, have the potential to increase the efficiency of today's engine to greater than 20%. Hybrid engines, utilizing both a battery and small internal combustion engine such as the Toyota Prius and upcoming Honda Insight have claimed to improve efficiency to up to double the current engine (36%), though more driving cycle measurements need to be completed to get a broad comparison to today's engine.

For comparison, the efficiency of future fuel cell engines was assumed to be twice current engine technology, about 36%. This is a combination of the efficiency of the fuel cell and the fuel processor over typical driving cycles. Since fuel cell efficiency is typically around 50%-60%, and steady-state processor efficiencies have been published of around 78-85% (for an idealized steady-state efficiency of around 40-50%), this assumption seems reasonable. A fuel cell vehicle will also lose efficiency due to losses in the electric motor and power electronics (around 5-10%), and due to the less efficient test cycle conditions which will include start-up and transient driving conditions. The actual efficiency of a fuel cell vehicle can, at best, be a rough estimate at this time, and production vehicles may differ from these estimates.

From Figure 4 we see that advanced designs may improve efficiency to about 20%, with hybrid gasoline-electric vehicles achieving around 31%, based on current claims from recently introduced hybrids. A hydrocarbon fuel cell vehicle should be similar to the gasoline hybrid in energy efficiency assuming the fuel processor system will approach the performance currently indicated by steady-state measurements. Methanol energy efficiency will be

Fuel Production/Distribution		Vehicle		Overall System	
VEHICLE	EFF. %	EFF. %	Emissions Potential	EFF. %	CO$_2$ g/km
Mogas IC (Current)		18	LEV	15	220
Mogas IC (Advanced)		23	ULEV/SULEV	20	170
IC Hybrid	85		ULEV/SULEV	31	110
Hydrocarbon FC		36	SULEV	31	110
MeOH (From NG) FC	63		SULEV	23	110
MeOH (From Coal) FC	54		SULEV	19	220

Figure 4.

lower, due to the poor production efficiency – especially if the source for the methanol is not natural gas. Though it is possible that methanol processors will achieve higher efficiency within the vehicle than gasoline processors, they would need to achieve nearly 50% overall vehicle efficiency to approach the well-to-wheels energy efficiency of the gasoline system. Since we assume the fuel cell stack to have at best an efficiency of $\cong 50\%$, the processor would have be at or above 100% efficiency to accomplish this, an unrealistic assumption.

WELL-TO-WHEELS CO_2 EMISSIONS – As with energy efficiency, the global issue of CO_2 emissions should be considered not only in terms of vehicle emissions, but also on a well-to-wheels basis. This would include the amount of CO_2 released during extraction, production and delivery of the fuel. As with energy efficiency, the emissions of CO_2 will only be accurately determined, and compared to current designs, when fuel cell vehicles are tested under identical circumstances – production vehicles running standard driving cycles.

A mid-sized gasoline-powered vehicle sold today with an energy efficiency of about 18% produces well-to-wheels CO_2 emissions of about 220 gm/km. Hybrid vehicles, based on gasoline powered engines and with advertised fuel economy of approximately double today's vehicles would cut that to 110 gm/km. Some non-hybrid designs based on much smaller vehicles for European markets have advertised CO_2 emissions of 90-130 gm/km on a vehicle-only basis (e.g., [1]).

Fuel cell vehicles should be competitive with the best of the alternative vehicle designs based on internal combustion engines. We estimate that both methanol and hydrocarbon fuel cells have the potential to halve well-to-wheels CO_2 emissions. If methanol is not produced from natural gas, these emissions would be substantially higher. Should the methanol fuel efficiency be improved over our estimates, CO_2 emissions could drop further from the values in Figure 4. If either system is unable to obtain the vehicle efficiencies we have assumed, it is likely that advanced hybrid designs could provide lower CO_2 emissions. Methanol would be expected to have an advantage over other fuels because it starts from a fuel, methane, with the highest H:C ratio; however, this is offset by very low production and distribution efficiency. Similarly, even hydrogen fuel produced from methane would have substantial CO_2 emissions, as production and distribution energy efficiency for hydrogen would be below 60%.

LOCAL VEHICLE EMISSIONS – In contrast to efficiency and CO_2 emissions, production of unwanted hydrocarbon, CO, NOx and particulate matter can be considered a local, vehicle only emission. For these emissions, fuel cell vehicles may provide the largest advantage over conventional, advanced, and hybrid designs. In eliminating combustion, fuel cell vehicle designs are fundamentally different from all internal combustion based designs. Lower temperature operation, coupled with catalytically based processes nearly eliminate all local emissions. Early tests suggest that local emissions will be well below California's SULEV specifications. While neither system is technically "zero" emissions, producing barely measurable amounts of engine emissions along with evaporative emissions, they would pass the most stringent proposed regulations.

SAFETY, HEATH AND ENVIRONMENT

Gasoline and methanol have both advantages and disadvantages. Both have been used safely in industrial settings, and substantial data does exist on flammability and toxicity. While the public is accustomed to the dangers of using gasoline through long experience, methanol has not seen wide distribution and use by the public. Issues regarding fire safety, toxicity, spills and groundwater contamination must be adequately addressed to ensure the safe handling of methanol in this extended environment.

FIRE SAFETY – Fire safety issues for gasoline and methanol are shown in Table 5. Methanol has a lower risk of open fire than gasoline. It would be less likely to ignite in an open atmosphere such as an automotive accident because methanol's volatility is well below its lower flammability limit. Thus we would expect a reduction in vehicle fires from this source. The EPA equates the fire risk for methanol as similar to diesel fuel, which, compared with gasoline, has fewer instances of fire [27-29]. In addition, methanol fires are substantially less severe, with a heat release rate of only 22% of the heat release from gasoline. This results from the combination of methanol's low heat of combustion and high heat of vaporization.

Table 5. Fuel Fire Safety Comparison

	Gasoline	Diesel	Methanol
Ignition: Open	Highest	Lowest	Middle
Ignition: Closed	Lowest	Middle	Highest
Relative Heat Release Rate	4.9	4.5	1.0
Flame Luminosity	High	High	Nearly Invisible

In a closed environment, such as a gasoline tank, methanol poses a greater fire risk. The vapor pressure of methanol overlaps its flammability range, forming a combustible fuel/air mixture in a closed tank. For diesel fuel the fuel/air mixture in a tank is too lean to burn, while for gasoline the mixture is too rich. Shipping and storing methanol requires precautions, such as blanketing the fuel with a layer of nitrogen or inert gas, keeping a roof on the tank to minimize air space, or other preventative measures.

Flame luminosity is very low for methanol, as pure methanol flames are difficult to detect, and with a low heat release rate, it is easy to get within close proximity of a methanol fire without detection. Generally, this is not a problem for the methanol fuel mixtures sold today such as M85. The 15% gasoline blended into the fuel provides ample flame luminosity [28]. For compounds without visible flames, such a methanol, one would typically add a luminosity additive [29]. While this is not an issue for a combustion engine, the catalytic processes used at low temperature to reform methanol are unique to methanol, and will not work effectively on luminosity agents currently being used. This leads to a dilemma where either a safety aspect of the fuel, flame luminosity, must be sacrificed, or the fuel processor may have to be changed to effectively process the additives.

TOXICITY – Both methanol and gasoline are toxic and can injure or kill through ingestion. Injury from gasoline injection is quite rare as a substantial amount needs to be ingested and gasoline has significantly objectionable odor and taste. Methanol is significantly more toxic with just 2 to 4 ounces (60-120 ml) needed for a fatal dose. Methanol is odorless, colorless, and tasteless. It is easily ingested when mixed with water at moderate concentrations [30-32]. A study by the National Poison Control Center suggests that without preventative measures, the accidental ingestion of methanol fuel creates a significant added risk over gasoline [32].

The two main sources for ingestion are siphoning and children drinking from a fuel container at home. Thus, anti-siphoning devices will be required, and additives to provide taste, odor, and/or color would normally be added (such as in windshield-washer fluid). However, as with flame luminosity, these additives would not reform under the mild, low-temperature conditions present in a methanol fuel cell processor. Again, an acceptable safety solution, balancing the chemical needs of the methanol reformer catalyst, with the health dangers of unadditized methanol would have to be reached. The EPA has suggested restricting the use of methanol for nonvehicle applications, such as lawnmowers, chain saws, weed cutters and other tools [27].

SPILLS – Large marine spills of crude oil or methanol have severe immediate environmental impact. Recovery from a methanol spill, however, would be much more rapid and complete than from a spill of crude oil. Methanol quickly dissolves in water and evaporates, it biodegrades relatively quickly and completely in both aerobic and anaerobic environments. Concentrations in an ocean would reach non-toxic levels quickly, and will typically be low enough for recovery to begin within days, with nearly complete recovery in a few months. No surface or shoreline impact would be lasting

For groundwater spills, the information is less complete. Methanol biodegrades faster than gasoline, and much faster than MTBE, a widely used gasoline oxygenate.

Thus, long-term damage from methanol spills will be minimal. In addition, low concentrations of methanol do not appear to have long-term health effects.

On the other hand, methanol's miscibility with water may cause acute effects in the short term. Methanol spills will dissolve into groundwater, rivers, or other fresh water sources. Since it is odorless, colorless, and tasteless, no simple way exists to detect groundwater contamination in the event of a large spill into an aquifer. This may cause immediate heath consequences for those using these water supplies without adequate detection measures. Thus, the EPA has stated that the use of additives to impart a color, odor or taste to the fuel is essential for methanol to be detected in groundwater [30]. Gasoline is not miscible in water, and is easily detected by taste and odor. Components like MTBE are extremely easy to detect, as the taste is highly objectionable even in minute, non-toxic concentrations. Of course, the effects of a gasoline spill can be long-lived and the long-term cleanup more difficult than with methanol spills.

CONCLUSION

The choice of a fuel to power the fuel cell vehicles of tomorrow is critical, and a large number of factors must be weighed. Both natural gas and crude oil based energy supplies are abundant [33,34] and either can produce fuels suitable for fuel cell use for the foreseeable future. Fuel cell vehicles must compete with other emerging technologies, and be competitive on cost and performance. Hydrogen, while an ideal fuel for the fuel cell stack, currently has disadvantages as a transportation fuel. Current production and distribution technology results in a very high cost, low energy density, and safety concerns. Only with substantial gains in production efficiency, storage, and infrastructure costs will hydrogen become more competitive with liquid fuels.

Methanol and gasoline have both advantages and disadvantages as sources of hydrogen for fuel cells, and both should continue to be developed before a commercialization choice is made. Production prototypes running standardized driving cycles, and true well-to-wheel analysis will determine which is the best overall choice. This choice needs to consider not simply the fuel processor technology, but safety and health considerations, overall infrastructure costs, fuel cost on a tax neutral basis, and acceptance by the public of new technologies.

REFERENCES

1. Automotive Environment Analyst, "VW Launches Lupo 1 Year Early", October 1998.
2. Business Week, "Enviro-cars: The Race is On", 2/8/99, p74.
3. National Research Council, "Review of the Research Program of the Partnership for a New Generation of Vehicles, Fourth Report," 1998, p.7

4. White, Gregory L., "GM, Isuzu Investing $320 Million to Build Advanced Diesel Engines for GM Pickups", Wall Street Journal, 9/9/1998

5. Buss, Dale L., "Holistic Hybrid", Automotive News Insight, 2/23/1998, p.44.

6. The Clean Fuels Report, "Diesel-Powered Audi Duo Hybrid Reaches Market Maturity", November 1997, p. 173

7. The Clean Fuels Report, "General Motors Pledges Production-ready Hybrids by 2001," April 1998 p.103.

8. Reuters, "New Nissan Models Could Include Hybrid," 4/15/1998.

9. Simison, Robert L., Fort to Start Building Prototypes of 'Super Car'," Wall Street Journal, 3/18/1997 p. A4.

10. Toyota Press Release, "Toyota Introduces new Passenger Vehicle Hybrid System," http://www.toyota.co.jp, 3/25/1997.

11. Bentley, J. M., "Fuel Processors for Fuel Cell Power Systems," Arthur D. Little, Inc., Presentation to CPPI/TSTF and AITF, Toronto, Canada, 1/26/1998.

12. Kalhammer, Fritz, et. al., Status and Prospects of Fuel Cells as Automobile Engines: A Report of the Fuel Cell Technical Advisory Panel," Prepared for the State of California Air Resources Board, July 1998.

13. Arthur D. Little, Inc., "Fuel Processors for Fuel Cell Power Systems,", Presented at the IQPC Fuel Cell Technology Conference, 9/14-15/1998.

14. Wang, M. et. al., "Assessment of PNGV Fuels Infrastructure: Phase 2 Report: Additional Capital Needs and Fuel-Cycle energy and Emissions Impacts," Argonne National Laboratories, August 1998.

15. Moore, R. B. et. al., "Hydrogen Infrastructure for Fuel Cell Transportation," Air Products and Chemicals Inc., Int. J. Hydrogen Energy, Vol. 23, #7, pp.617-620, 1998.

16. C. E. Thomas et. al., "Societal Impacts of Fuel Options for Fuel Cell Vehicles," SAE publication 982496, 1996.

17. J. M. Ogden et. al., "Fuels for Fuel Cell Vehicles: Vehicle Design and Infrastructure Issues," SAE publication 982500, 1998.

18. Mark, Jason, "Infrastructure Requirements for a Hydrogen-Vehicle Future," Energy Resources Group, Union of Concerned Scientists, May 1996.

19. Zawodzinski, et. al., "R&D on Optimized Cell Performance for Operation on Reformate and Air," DOE Fuel Cells for Transportation Program Annual R&D Meeting, Argonne National Laboratory, June 1999.

20. Golunski, S. "HotSpot Fuel Processor", Platinum Metals Rev. 1 (1998) p.2

21. American Petroleum Institute, "Basic Petroleum Data Book: Petroleum Industry Statistics,", Vol. XVIII, No. 2, July 1998

22. 1997 World Methanol Conference

23. American Methanol Institute, "the Promise of Methanol Fuel Cell Vehicles," AMI 1999, http://www.methanol.org

24. Hahn, R. W., "The Economics of Methanol,", Auto/Oil Air Quality Improvement Research Program, January 1992

25. Ford Motor Company, "Economic, Environmental and Energy Life Assessment of Coal Conversion to Automotive Fuels in Shanxi Province,", July 1997

26. DeLuchi, M. A., "Emissions of Greenhouse Gases From the Use of Transportation Fuels and Electricity," Argonne National Laboratory, November, 1991

27. Environmental Protection Agency, "Analysis of the Economic and Environmental Effects of Methanol as an Automotive Fuel," September 1989, p. 70

28. Machiele, P. A., "Methanol Fuel Safety: A Comparative Study of M100, M85, Gasoline, and Diesel Fuel as Motor Vehicle Fuels," EPA-AA-SDSB-90-01, November, 1990.

29. Machiele, P. A., "Summary of the Fire Safety Impact of Methanol as a Transportation Fuel," SAE 90113, U. S. EPA, May 1-4, 1990.

30. American Petroleum Institute, "Transport and Fate of Dissolved Methanol, Methyl-Tertiary-Butyl-Ether, and Monoaromatic Hydrocarbons in a Shallow Sand Aquifer," American Petroleum Institute Publications No. 4601, April 1994.

31. Heath Effects Institute, "Automotive Methanol Vapors and Human Health: An Evaluation of Existing Scientific Information and Issues for Future Research," May 1987

32. Litovitz, T., MD "Acute Exposure to Methanol in Fuels: A Prediction of Ingestion Incidence and Toxicity," National Capital Poison Center, October 31, 1998.

33. Oil and Gas Journal, "Worldwide look at reserves and production," Vol. 95, 12/29/1997, pp.38-9.

34. McCabe, Peter J., "Energy Resources – Cornucopia or Empty Barrel?" American Association of Petroleum Geologists, November 1998.

Societal Impacts of Fuel Options for Fuel Cell Vehicles

C. E. Thomas, Brian D. James, Franklin D. Lomax, Jr. and Ira F. Kuhn, Jr.
Directed Technologies, Inc.

ABSTRACT

Methanol, gasoline and hydrogen are the primary fuel options under consideration for fuel cell vehicles (FCV). The ideal fuel would eliminate local air pollution, substantially reduce greenhouse gas emissions and oil imports, cost no more than current transportation fuels per mile driven, and require little investment in new infrastructure. In addition, the fuel used for future fuel cell vehicles should be suitable for near-term hybrid electric vehicles using internal combustion engines, to avoid the need for introducing more than one new motor fuel in the 21st century.

All three primary FCV fuels have limitations: gasoline is the most difficult to reform to produce hydrogen and produces the least improvement in the environment and yields the least oil import reductions. Methanol is cleaner, easier to reform onboard the FCV, eliminates oil imports, but requires some new infrastructure. Hydrogen provides the best environmental and oil import improvements, but requires the largest infrastructure investment. This paper will provide a comprehensive comparison of FCV fuels and their societal impact on local air pollution, greenhouse gases and oil imports. In addition, we will estimate the likely infrastructure costs for each fuel, and compare their societal impact if used as a fuel for hybrid electric vehicles.

INTRODUCTION

As society places increased emphasis on the environment, climate change and energy sustainability, conventional fuels such as gasoline and diesel and conventional spark ignition and compression ignition engines may be replaced by entirely new drivetrains and alternative fuels. The choice of an alternative vehicle to meet societal objectives will depend on some combination of vehicle performance and cost, fuel availability and cost and the emphasis consumers and society place on the capability of these fuels and vehicles to reduce local air pollution, greenhouse gases, and oil imports.

In this study we have assumed a Ford AIV (aluminum intensive vehicle) Sable as the glider for all vehicles. This vehicle weighs about 270 kg less than the commercial Ford Sable or Taurus, but otherwise has the same aerodynamic and rolling resistance characteristics. By keeping a common vehicle glider, the study permits the direct comparison of the societal attributes of various alternative fuel/drivetrain combinations.

Three fuels were analyzed for each of three types of vehicle drivetrains in this study: conventional ICEVs (gasoline, natural gas and "hythane"[1] -- a mixture of 30% hydrogen and 70% natural gas), hybrid vehicles (natural gas, hydrogen and diesel fuel), and fuel cell vehicles (hydrogen, methanol and gasoline). Three types of hybrids were also analyzed for each fuel: a parallel hybrid and two types of series hybrid operation (thermostat "on/off" mode and load-following mode.)

For each fuel/vehicle combination, we have estimated the vehicle weight, fuel economy, mass production cost, local emissions of criteria pollutants, well-to-wheels greenhouse gases and oil consumption. We have also evaluated the possible costs of the two alternative fuels (hydrogen and methanol), both in terms of fuel industry investment costs required and in terms of fuel cost per mile in each vehicle.

ALTERNATIVE VEHICLES

Three types of fuel cell vehicles were analyzed:

DIRECT HYDROGEN FUEL CELL VEHICLE – The direct hydrogen FCV was designed under a DOE/Ford contract. This FCV stores hydrogen onboard the vehicle in 5,000 psi carbon fiber wrapped composite tanks (3.6 kg of stored hydrogen). The fuel cell system is designed for 38-kW maximum continuous power, with a peak power battery bank providing an additional 40-kW of peak power. The estimated FCV test weight after replacing the ICE drivetrain and four-speed automatic transmission with the fuel cell drive train including electric motor and one speed transmission is 1,291 kg, compared to 1,304 kg for the baseline AIV Sable ICEV. Alternatively, with PNGV body parameters and 1,032 kg test weight, this FCV architecture scaled down in size would have 610 km (380 mile) range and a fuel economy of 46.8 km/liter of

---

[1] "Hythane" is a registered trademark of Hydrogen Components, Inc. of Littleton, Colorado.

gasoline equivalent (110 miles per gallon of gasoline equivalent-- mpgge[2]) on the EPA 55% urban/45% highway schedule [1].

METHANOL FUEL CELL VEHICLE – The methanol FCV includes an onboard steam reformer to convert methanol into a hydrogen-rich fuel stream. After water gas shift reactor(s) and a preferential oxidation (PROX) system to reduce carbon monoxide down to less than 20 to 100 ppm (since CO is a poison for the fuel cell anode catalyst), the methanol reformate gas stream contains about 75% hydrogen with the remainder mostly carbon dioxide. Daimler-Benz has demonstrated a test FCV with an onboard methanol reformer, and has announced plans to start mass production of these vehicles by 2004 [2]. Since we do not have experimental data on methanol FCVs, we have assumed a probable case and a best case methanol FCV. The parameters of these two systems are summarized in Table 1. The fuel cell system peak power will probably decrease when operating on 75% hydrogen instead of pure hydrogen. Scientists at the Los Alamos National Laboratory have measured a fuel cell maximum power drop of 10 to 12% operating on simulated methanol reformate, using a Ballard fuel cell stack [3]. In addition, not all of the hydrogen can be utilized in the anode of the fuel cell, since large gas exhaust flow is required to remove the CO_2 from the system. As a result, some of the hydrogen leaves the anode exhaust. We assume that 75% of the lower heating value of this hydrogen is recovered in a burner used to raise steam for the methanol reformer.

Table 1.   Methanol Fuel Cell Vehicle Parameters

	Best case	Probable case
Ratio of Fuel Cell Maximum Power on Methanol to Maximum Power on Pure Hydrogen (Power Loss Due to CO_2 Dilution)	0.91	0.89
Hydrogen Utilization	90%	83.3%
Methanol Reformer Efficiency (H_2-LHV/MeOH -LHV)	84.5%	80%
Anode Exhaust Gas Heat Recovery	75%	75%
Methanol Reformer Weight (kg)*	46	60
Total Vehicle Weight Increase (Relative to H_2 FCV) (kg)	99	121
Vehicle Test Weight (kg)	1,390	1,413

*Before vehicle weight compounding

The total vehicle weight increase for the methanol FCV relative to a direct hydrogen FCV listed in Table 1 includes the effects of weight compounding -- the fuel cell system weight must increase to provide the necessary vehicle acceleration and all other powertrain components must also increase in size to maintain vehicle performance. The weight compounding includes a 15% structural weight increase, but most of the increase is due to growth of other power train components.

GASOLINE FUEL CELL VEHICLE – The gasoline FCV includes a partial oxidation (POX)/autothermal reformer (ATR) to produce approximately 40% hydrogen in a gas stream after gas cleanup to convert most CO to CO_2. A. D. Little (now Epyx) has demonstrated a gasoline reformer in the laboratory, operating with a PROX system made by the Los Alamos National Laboratory, driving a Plug Power fuel cell [4]. As with the methanol FCV, we have assumed a probable and best case set of parameters to characterize the gasoline FCV as summarized in Table 2. With only 40% hydrogen in the reformer system output, fuel cell maximum power loss has been measured up to 25% [3]. Several organizations are working on "CO-tolerant" fuel cells that will reduce these degradations, but so far these cells have produced lower power density, used more platinum catalyst, run with 1.3 (77% hydrogen utilization) or higher hydrogen stoichiometry or required up to 6% air bleed into the anode gas stream. Each of these remedies could increase the cost of the fuel cell stack. In any case, Los Alamos has now defined "CO tolerance" as less than 20% drop in maximum power, with at most a 5% drop in power at 0.7 volts [5]. While the fuel cell generally runs at much less than peak power, the fuel cell vehicle designer must still provide the required fuel cell maximum power to maintain the required vehicle hill climbing characteristics.

Table 2.   Gasoline Fuel Cell Vehicle Parameters

	Best case	Probable case
Ratio of Fuel Cell Maximum Power on Gasoline Reformate to Maximum Power on Pure Hydrogen (Loss Due to Dilution by CO_2 and N_2)	0.83	0.74
Hydrogen Utilization	90%	83.3%
Gasoline POX Efficiency (H_2-LHV/Gasoline -LHV)	75%	70%
Anode Gas Heat Recovery	70%	0
Gasoline POX Reformer Weight (kg)	55	100
Vehicle Weight Increase (kg)	96	184
Vehicle Test Weight (kg)	1,387	1,475

Three types of hybrid vehicle were also analyzed for each of three fuels (natural gas, hydrogen and diesel fuel):

---

[2]   A FCV with 46.8 km/liter gasoline equivalent fuel economy would travel 46.8 km on that amount of hydrogen with the same lower heating value as one liter of gasoline.

SERIES HYBRID IN THE THERMOSTAT MODE – An electric motor supplies all mechanical power to the wheels. The ICE plus generator are used to provide electrical energy to the motor and to recharge the batteries. The ICE is turned on at a fixed power level when the battery state of charge (SOC) reaches a pre-set minimum (40%), and turned off when the SOC exceeds a maximum (60%), independent of the electric motor power requirements.

SERIES HYBRID IN THE LOAD-FOLLOWING MODE – The ICE is turned on when the power demand exceeds a threshold, and turned off whenever the vehicle stops, the battery SOC exceeds the maximum, or the power demand falls below a lower threshold. Both upper and lower power thresholds for turning on the ICE are a function of battery SOC to maintain the battery SOC within a narrow range. ICE output power varies over a limited range to optimize system efficiency by minimizing the energy passing into and out of the battery. Since the ICE can be turned on quickly in this mode, the battery peak power capacity can be reduced substantially compared to the thermostat mode, where the battery must be able to provide the full electric motor power rating for fast accelerations.

PARALLEL HYBRID – The ICE and a supplemental traction motor are both mechanically coupled to the transmission. A computer algorithm determines when to turn the ICE on, what power level to produce, and when to turn the ICE off. The performance of both the load-following series hybrid and the parallel hybrid vehicles depends on the control strategy used to turn the ICE off and on. For a detailed description of the hybrid control algorithms, see reference [6].

These six types of alternative vehicles were compared with both conventional gasoline ICEVs and with ICEVs running on natural gas and hythane. All spark ignition engine maps are based on the Ford Taurus 3.0 liter engine that has a peak efficiency of 33% operating on gasoline. For the alternative fuels, we adjusted peak engine efficiency to 38% on natural gas and 42% on hydrogen, based on estimates from Smith and Aceves[7]. The diesel engine has a peak efficiency of 43% with the engine efficiency map modeled on the VW compression ignition, direct injection engine.

VEHICLE FUEL ECONOMY

CHOICE OF DRIVE CYCLES – Directed Technologies, Inc. (DTI) has developed a spreadsheet computer program to calculate the fuel economy of alternative fueled vehicles on various driving schedules. Data from this program have matched actual measured fuel economies for conventional ICE vehicles, and they have also matched estimates made by the Ford Motor Company driving simulation programs for fuel cell vehicles. The choice of driving schedule is crucial when comparing FCVs with conventional vehicles, due to their much different efficiency maps. An ICE has a peak efficiency at an intermediate torque level at moderately high RPM. Efficiency falls off away from this "sweet spot" if either torque or engine speed are increased or decreased. If an ICEV is driven over a slow speed urban driving cycle, below the engine efficiency sweet spot, then fuel economy can be improved by driving the vehicle more aggressively at higher speeds.

A fuel cell system, on the other hand, generally has a monotonically decreasing efficiency for power draws above a low level, as illustrated in Figure 1. The fuel economy of the FCV will almost always decrease on faster driving schedules. This fuel cell system operates with variable air pressure and flow -- the cathode air pressure is set at 3 bar at maximum power, but the compressor is turned down to 1.2 bar at low power draw to reduce parasitic losses, keeping the air stoichiometry fixed at 2.0. If the air pressure were maintained at 3 bar at all power levels, then the efficiency curve would be even steeper - variable air pressure operation effectively flattens out the efficiency vs. power curve over the mid to high power levels.

The characteristics of several common driving schedules are summarized in Table 3. The first five schedules shown are in common use: the US Federal Urban Driving Schedule (FUDS), the Federal Highway Driving Schedule (FHDS) and the combined schedule that includes 55% urban and 45% highway driving, along with the European and Japanese 10-15 mode schedules. These driving schedules are rather anemic compared to actual US driving habits. The last two driving schedules are meant to more closely reflect real-world driving conditions: the 1.25 times combined cycles where the speed at each time segment of a 55/45 combined schedule is multiplied by 1.25, and the US06 schedule proposed by EPA. The 1.25 times combined cycle provides about the same vehicle fuel economy as the proprietary Ford customer driving cycle. For each driving schedule, we list the average and peak power required from the fuel cell and battery of a direct hydrogen FCV based on the AIV Sable glider with a 1,291 kg test weight, 0.33 drag coefficient, 2.14 m^2 cross sectional area and 0.0092 rolling resistance. The inverter, controller and electric motor maps are all based on the Ford Ecostar electric vehicle drive train. The average power required from the fuel cell/battery system over the cycle is probably the best indicator of the degree of robustness of a cycle -- the 1.25 times combined and US06 cycles have greater average power draw, which better represents real driving conditions.

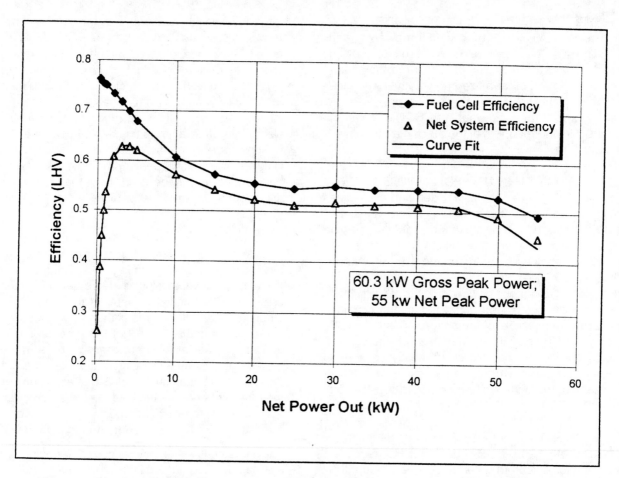

Figure 1. Fuel Cell Stack Efficiency and System Efficiency as a Function of Net Output Power

Table 3.   Driving Schedule Characteristics

	Average Speed (mph)	Peak Speed (mph)	Average Cycle Power (kW)	Peak Cycle Power (kW)
FUDS	19.5	56.7	4.93	31.5
FHDS	48.6	60	10.1	29.8
55/45 Combined	32.6	60	7.26	31.5
European	20.5	74.6	5.2	46.9
Japanese 10-15 Mode	14.4	43.5	3.59	21.4
1.25 X Combined	40.8	75	11.24	47.1
US06	48.1	80.3	16.1	84.2

The effect of drive cycle on relative fuel economy is shown in Figure 2 for methanol and direct hydrogen FCVs, all normalized to the estimated fuel economy on the given cycle for the gasoline-powered ICEV. The direct hydrogen FCV would have a 3.6 times greater fuel economy on the rather anemic Japanese 5-10 mode driving cycle, dropping to a 2.9 advantage on the European cycle and to 2.6 to one on the EPA combined 55% urban and 45% highway schedule. But on the more realistic 1.25 times faster combined schedule, the direct hydrogen FCV fuel economy advantage drops to 2.2 to one, which closely matches the advantage on the proprietary Ford customer driving schedule, and the advantage falls even further to 1.8 to one on the proposed EPA US06 schedule. The methanol FCVs have a fuel economy advantage in the range from 1.2 to 1.6 to one compared to the ICEV on these more realistic driving schedules. These data illustrate that the choice of driving cycle is essential when comparing FCVs to other vehicles. For this study, we use the 1.25 times faster combined cycle for all fuel economy comparisons.

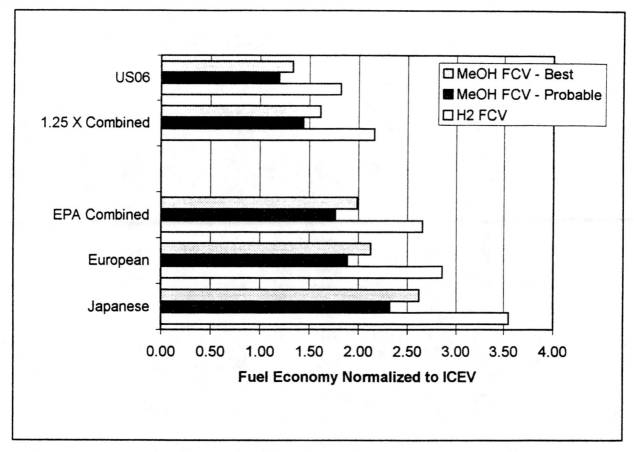

Figure 2. Fuel Economies of Fuel Cell Vehicles Normalized to ICEVs on Various Drive Cycles

VEHICLE WEIGHT – The fuel economy of each vehicle will depend in part on weight. We have estimated the weight of each alternative vehicle drivetrain. Representative drivetrain weights are presented in Table 4 for the direct hydrogen and probable cases for the gasoline and methanol FCVs. The test weights should be compared with the ICEV test weight of 1,304 kg.

Representative component weights are shown in Table 5 for the parallel hybrid vehicles.

VEHICLE FUEL ECONOMY – The estimated fuel economies for the various alternatively fueled vehicles are shown in Figure 3, including the best and probable cases for the FCVs with onboard reformers, and the three different varieties of HEVs. All fuel economies are based on the 1.25 times faster EPA 55/45 combined cycle. Surprisingly, the diesel parallel hybrid has almost the same fuel economy as the direct hydrogen FCV.

Table 4.   Fuel Cell Vehicle Weight Estimates (kg)

	Direct H$_2$ FCV	Methanol FCV (Probable)	Gasoline FCV (Probable)
Fuel	4.71	41.4	22.7
Fuel Tank	34.9	20.0	14.0
Fuel Cell System	76.2	91.9	115.5
Fuel Processor		64.8	112.1
Motor/Controller	75.3	81.0	83.9
Battery	71.9	78.4	81.6
Transmission	27.0	27.0	27.0
Glider*	865.0	878.5	882.2
Curb Weight	1,155	1,283	1,339
Test Weight	1,291	1,419	1,475

*"Glider" includes all components of the AIV Sable common to the ICEV and the alternative vehicle, including extra structure to carry additional weight of the alternative vehicles.

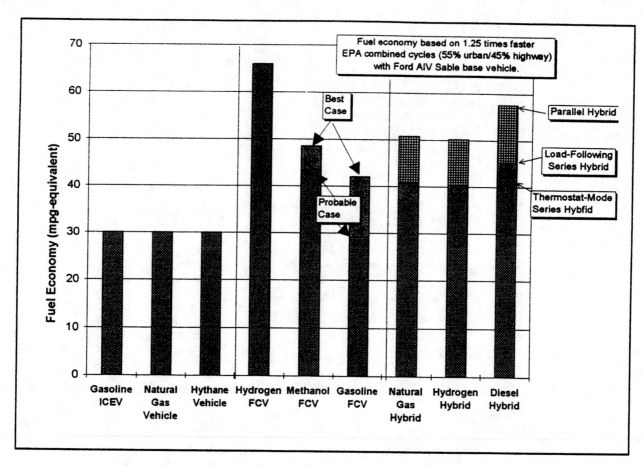

Figure 3. Estimated Fuel Economy for Conventional, Fuel Cell and Hybrid Vehicles

Table 5.　Parallel Hybrid Weight Estimates (kg)

	H2 Parallel Hybrid	NG Parallel Hybrid	Diesel Parallel Hybrid
Fuel	5.92	14.8	15.0
Fuel Tank	45.8	15.7	14.0
ICE	71.9	71.0	98.2
Motor/Controller	41.2	40.6	41.1
Battery	64.0	62.6	63.9
Transmission	44.0	44.0	44.0
Glider	725.1	722.7	693.5
Curb Weight	1,111	1,083	1,109
Test Weight	1,247	1,219	1,245

VEHICLE MASS PRODUCTION COST ESTIMATES

We have estimated the incremental mass production cost of each alternative vehicle compared to a conventional gasoline ICEV. The mass production cost of the proton exchange membrane (PEM) fuel cell system was estimated by a detailed bottom-up design and costing procedure, based on Ford costing methodology. For each fuel cell component, we analyzed the least costly material and least costly process to produce that component, with emphasis on reducing parts count and minimizing assembly costs during manufacturing. Four different bipolar plate designs were evaluated, including a three piece metallic separator, a unitized metallic plate, a solid amorphous carbon separator and a carbon-polymer composite plate. The unitized metallic plate and the carbon-polymer composite yielded the lowest cost fuel cell stack, C_s, given by:

$$C_s = \$298 + P_N \ x \left[\frac{5.34 + 27 x\, L_p}{p_d} \right] \quad \text{(Eq. 1)}$$

where P_N = the net fuel cell maximum output power in kW,

L_p = the total cell platinum loading (anode plus cathode) in mg/cm^2, and

p_d = the cell maximum power density in W/cm^2.

For a modern PEM fuel cell with 0.25 mg/cm^2 of total platinum loading and a cell power density of 0.65 W/cm^2, a 70-kW stack would cost about $1600 or $22.96/kW in mass production. A lower power fuel cell would cost more per kW, since some components of the stack do not scale down in size -- connectors, nuts, valves, wiring, plumbing, safety devices, etc., will cost about the same for a 70-kW stack as for a smaller fuel cell stack. Thus an 18-kW stack would cost about $634 or $35.26/kW. This cost excludes the ancillary system components such as the air compressor/expander, humidification system, controls, etc. Adding these components brings the total fuel cell system mass production cost up to:

$$C_{System} = 1,073 + P_N \times \left[18.70 + \frac{5.34 + 27 \times L_p}{p_d} \right] \quad \text{(Eq. 2)}$$

or, with our previous assumptions of 0.25 mg/cm^2 platinum loading and 0.65 W/cm^2 yields:

$$C_{System} = 1,073 + 21.97 \times P_N \quad \text{(Eq. 3)}$$

The 70-kW fuel cell system would then cost $2,610 or $37.30/kW, while the 18-kW system would cost $1,468 or $81.56/kW. This clearly illustrates that the use of a constant $/kW rate for all fuel cell systems is inappropriate. Even with the same component cost assumptions, the 18-kW system costs more than twice as much as the 70-kW system on a $/kW basis.

A similar detailed, bottom up cost analysis was made for the gasoline fuel processor system, resulting in a cost estimate of between $29/kW (best case) and $40/kW (probable case). However, A.D. Little, the company developing the gasoline POX processor, has estimated a possible cost of $20/kW[8], so we use this value for the best case gasoline processor in this study. We have not yet analyzed the methanol processor in detail, but it should cost less than the gasoline processor given the much lower temperatures required to reform a simple alcohol fuel (260°C) compared to the 800°C or more required to reform the many heavy hydrocarbons that make up gasoline. The U.S. Department of Energy has set a cost goal of $10/kW by 2004 for a flexible fuel processor that could process gasoline[9]. We have seen no credible design that could meet these cost goals for a gasoline processor, but we have assumed this $10/kW as the best case methanol fuel processor. Costs for other alternative fueled vehicles were taken from the literature or from the DOE cost targets (See reference [6] for cost details.)

The resulting mass production cost estimates for the conventional gasoline ICEV drivetrain is compared with the direct hydrogen fuel cell vehicle drivetrain in Table 6. With the cost assumptions used here, the FCV would cost about $2,180 more in mass production than a conventional vehicle.

The drivetrain costs for the conventional ICEV is compared with that of a natural gas parallel hybrid vehicle in Table 7. The parallel hybrid has two separate power sources, but each is lower power than the ICEV engine, resulting in a net cost increase of $1,291 for the parallel hybrid. The estimated incremental cost for all other alternative vehicles are compared in Figure 4.

Table 6.   Cost Comparison: ICEV vs. Fuel Cell Vehicle

	Conventional ICEV		Direct Hydrogen Fuel Cell Vehicle	
	Power (kW)	Cost ($)	Power (kW)	Cost ($)
Fuel Cell System			38.1	1911
ICE & Ancillaries	100	1600		
Transmission		700		200
Fuel Tank System		176	4.71 kg	760
Motor Controller			82	906
Battery System			40.3	728
Controller				150
Drivetrain Total Costs		2476		4,655
Additional Cost				2,179

Table 7.   Cost Comparison: ICEV vs. Parallel Hybrid

	Conventional ICEV		Natural Gas Parallel Hybrid	
	Power (kW)	Cost ($)	Power (kW)	Cost ($)
ICE & Ancillaries	100	1600	32.9	889
Transmission		700		700
Fuel Tank System		176	72.6 liters	334
Generator			32.9	519
Motor /Controller			34.8	533
Battery System			34.8	642
Controller				150
Drivetrain Total Costs		2476		3,767
Additional Cost				1,291

FUEL COST

The decision to buy an alternative vehicle will depend on fuel cost and availability in addition to vehicle cost and performance. No public fuel infrastructure exists for the two FCV alternative fuels: hydrogen and methanol. We need to estimate potential fuel costs to assess the relative lifecycle costs of owning and operating a FCV compared to a conventional vehicle running on gasoline or diesel fuel.

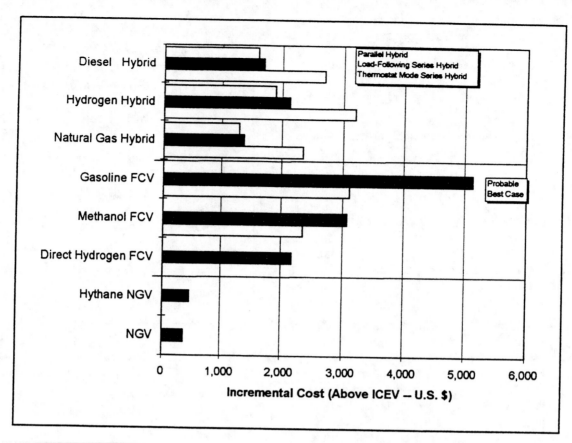

Figure 4. Additional Mass Production Cost of Alternative Vehicles

METHANOL COST AND AVAILABILITY – Methanol is available today in less than 50 public fueling stations as M-85 -- 15% gasoline with 85% methanol, and there are a few M-100 fueling facilities for buses in California. M-85 would not be suitable for an onboard methanol reformer, so new neat methanol fueling facilities would have to be added before customers would purchase a methanol FCV. The key issues are methanol cost, production capacity, and local distribution cost and capacity.

Methanol Cost – The cost of methanol today is quite variable, averaging near 50 cents/gallon over the last decade delivered to the US Gulf, with spikes to over $1/gallon in 1994, falling to current levels near 30 cents/gallon. The historical average of 50 cents/gallon of methanol would be equivalent to about $1/gallon of gasoline on an energy basis, since methanol has only half the lower heating value of gasoline. But the methanol FCV should have about 50% greater fuel economy than the ICEV on realistic driving schedules, resulting in an effective gasoline-equivalent cost per mile driven of about 65 cents/gallon, which is comparable to the wholesale cost of gasoline. Assuming that the retail markup of methanol is equal to the retail markup of gasoline, then the retail price of methanol should be similar to the price of gasoline per mile driven.

Methanol Production Capacity – Methanol production capacity should not be an issue initially, since the methanol industry has had excess capacity for many years. Global methanol production capacity is about 34 million metric tonnes per year, while consumption has been averaging only 26 million metric tonnes, or 75% plant capacity [10]. Assuming that methanol plants could operate at 90% capacity, then there is an excess useable methanol capacity of 4.6 million tonnes per year, or 1.5 billion gallons of methanol, which is enough to supply about 2.8 million FCVs. We conclude that methanol production capacity should not inhibit early introduction of FCVs.

Methanol Retail Availability – If we assume that there are about 180,000 gasoline stations in the U.S., and also assume that conversion of 10% of those stations to at least one methanol pump at a cost of $50,000 each would suffice for FCV market penetration, then a total investment of just under $1 billion would be required in retail methanol fueling facilities [11]. This investment should be compared with the estimate of $11 billion spent annually to maintain and expand the current gasoline fueling infrastructure system, including new drilling, refinery additions and maintenance, new gasoline tanker trucks, etc.[12].

HYDROGEN COST AND AVAILABILITY – Like the methanol industry, the hydrogen industry does have some excess capacity in some parts of the country, although the increased demands for hydrogen within oil refineries to produce cleaner gasoline blends is increasing the demand for hydrogen. Ogden estimates, for example, that existing industrial hydrogen excess capacity in the Los Angeles basin could support up to 100,000 FCVs [13]. Unlike methanol, however, the costs for transporting gaseous hydrogen are much greater than liquid methanol, and liquefying hydrogen consumes about 35% of the heating value of the fuel. DTI has previously analyzed various options to initiate a gaseous hydrogen fueling infrastructure, working with industrial gas partners and the Ford Motor Company under contract to the U.S. Department of Energy [14,15]. These studies suggested that hydrogen could be produced economically onsite at local gasoline stations or fleet operator's garages by either electrolysis of water or by steam reforming of natural gas. In effect, this scenario utilizes the existing natural gas infrastructure or the existing electrical power grid to provide energy to the station. Hydrogen is made locally, when and where it is needed, avoiding the cost of a national hydrogen pipeline system or a fleet of liquid hydrogen tanker trucks.

Our cost analyses show that hydrogen could be made locally at a cost that was equivalent to the cost of wholesale gasoline per mile driven, provided that small scale steam methane reformers were factory-built in quantities of 1,000 or more. This cost equivalence assumes that natural gas is purchased at $3.79/GJ ($4/MBTU) with electricity purchased at 6 cents/kWh (primarily to compress the hydrogen to 6,000 psi for overnight storage) and a plant capacity factor of just under 70%. The steam methane reformer costs are based on commercial reformers sold as part of the PC-25 phosphoric acid fuel cell system sold by the ONSI division of International Fuel Cells. This reformer system could support about 375 FCVs. We have used scale factors to estimate the likely costs for larger and smaller reformer systems plus manufacturing progress ratios to project possible cost savings at higher production volumes. Based on these scaling factors and manufacturing progress ratios, we project that a local steam methane reformer system suitable for supporting a fleet of 1,000 FCVs including compressor and hydrogen storage tanks could be manufactured for about $230,000 at the 1,000 quantity level, or an investment of about $230 per vehicle. Ogden has estimated that a similar stationary natural gas reformer system might cost $380 per vehicle served[16], based on producing a few hundred such fueling systems.

For frame of reference, we project that the cost of the methanol FCV would be between $180 (best case) to $900 (probable case) more per vehicle than a direct hydrogen FCV, suggesting that the cost of adding hydrogen infrastructure at the fueling station might be compa-rable to or even less than the cost of adding the methanol onboard reformer. Furthermore, this $230/vehicle to $380/vehicle hydrogen infrastructure estimate is significantly less than the $900 to $2,900 incremental cost per vehicle estimated for adding an onboard gasoline reformer system. Thus the total societal cost of adding a chemical processor to each vehicle could exceed the cost of installing a hydrogen fueling infrastructure.

These cost differentials between onboard and stationary reformers can be rationalized on the basis of capacity factor: the onboard reformer is used at most one or two hours per day, and even when operating, the average power is usually only 10 to 20% of peak power. The effective capacity factor for an onboard reformer is therefore less than one percent -- the cost of the mechanical equipment must be recovered operating less than one percent of rated capacity. The stationary reformer, on the other hand, can operate up to 70% of the time at full capacity, filling tanks during the night to meet peak daytime demand. In addition, the stationary reformer can operate close to steady state, without the significant weight restrictions, fast warmup time, fast transient response requirements, and the shock, vibration and temperature extremes found on a motor vehicle.

The previous comparisons between methanol and hydrogen cost have all assumed that no new methanol production capacity is required. But if the methanol FCV sales exceeded the estimated 2.8 million vehicles that could be supported by existing excess methanol production capacity, then new capacity would have to be added. Assuming that new methanol plants would cost approximately $1 billion for a 10,000 tonne/day plant that could support 2.2 million FCVs, then the effective new investments would amount to $450/vehicle, or nearly twice the potential investment required per vehicle to support a hydrogen FCV with small steam methane reformers, compressors and hydrogen storage tanks. Some of this increased capital cost of methanol per vehicle is due to the lower onboard efficiency of a methanol FCV compared to a hydrogen FCV --about 33% more joules of methanol LHV are required per mile than joules of hydrogen LHV, and the efficiency of producing methanol from natural gas ($\approx$64% on a lower heating value basis) is slightly lower than the industrial efficiency of producing hydrogen from natural gas ($\approx$72% LHV).

In summary, we conclude that either methanol or hydrogen could be made available for FCVs at a cost comparable to gasoline per mile driven, with the proper infrastructure investments prior to vehicle market introduction. Methanol has the advantage of excess production capacity now, requiring only investments in local retail infrastructure ($\approx$$1 billion), plus the convenience and familiarity of a liquid fuel. Hydrogen requires a more substantial investment initially ($\approx$$4 billion) to install both production and dispensing capacity, but offers the potential of lower investment cost per vehicle in the long run, particularly when new methanol production capacity must be built, as summarized in Table 8. The hydrogen option must also address and overcome the safety issues or perception of safety hazards associated with high pressure onboard gas storage.

Table 8.   Potential Near-Term and Long-Term Methanol and Hydrogen Infrastructure Costs

	Hydrogen	Methanol
Initial Infrastructure for 10% Coverage	≈$4 Billion	≈$1 Billion
Incremental Vehicle Cost	-	$180 - $900
Long Term Infrastructure Costs per Vehicle	$230-$380	$450
Total Long Term Infrastructure Cost per Vehicle	$230 - $380	$630 - $1,350

Given this potential for fuel cost equity between the three primary FCV fuels, we conclude that customer acceptance may depend solely on initial vehicle cost. FCVs may eventually have lower maintenance costs than ICEVs, but we have no basis for projecting such savings now. In any case, fuel and maintenance is generally less than 15 to 20% of the life cycle costs of owning and operating a car. We therefore use the initial vehicle price as the main indicator of lifecycle costs in this study. We compare these costs with the avoided costs of air pollution and greenhouse gases.

VEHICLE AIR POLLUTION

We have estimated the "real world" emissions of volatile organic compounds (VOCs - Figure 5), carbon monoxide (CO -Figure 6), oxides of nitrogen (NOx - Figure 7) and particulate matter (PM) for the same set of vehicles.[3] By "real world" we mean the actual emissions from each vehicle averaged over the lifetime of the car, not the emissions standards or even the measured laboratory emissions used to certify new vehicles. These emission estimates are based on the work of Ross et al.[17], who have shown that the average emissions over the life of a car can be five to six times the certified emission levels when the car was new. They have also projected the likely variance between actual emissions for the years 2000 and 2010 and emission standards. We have used these projected future emissions as the basis for estimating the emissions from alternatively fueled vehicles.

---

[3]   We have assumed negligible PM emissions for all vehicles accept the diesel hybrid vehicles which emit 0.03 g/mile PM at a fuel economy of 45.6 mpg-gasoline equivalent.

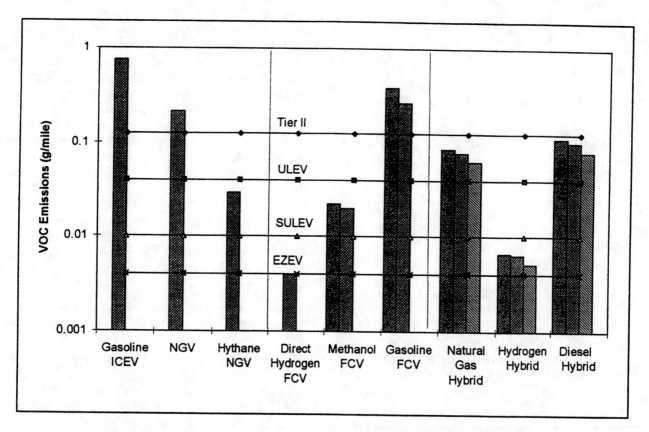

Figure 5. "Real World" Volatile Organic Compound Emissions in 2000

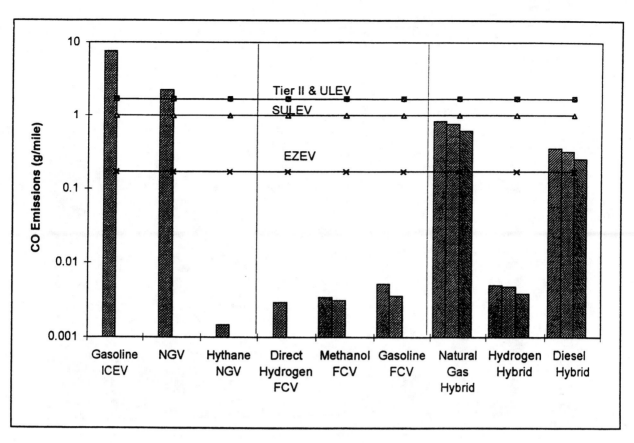

Figure 6. "Real World" Carbon Monoxide Emissions in 2000

The horizontal lines in Figures 5 through 7 indicate the emissions standards: Tier II for the Federal standards, ULEV for the California ultra-low emission vehicles, SULEV for the new standards for a super-ultra-low emission vehicle proposed by the staff of the California Air Resources Board as part of the LEV II package [18], and EZEV for the previously proposed equivalent zero emissions vehicle standard. This EZEV standard was originally proposed to approximate the emissions from in-basin power plants used to charge electric vehicle batteries, but has been effectively replaced by the less stringent SULEV proposal. In each figure, the methanol and gasoline FCVs include a probable and best case (lower emissions), and each hybrid electric vehicle is represented by a series thermostat, series load-following and parallel hybrid (lowest emissions). The parallel hybrid has lower emissions than the other hybrids here based strictly on improved fuel economy. In actual operation the start/stop cycling of the ICE may increase these emissions, but we have not modeled this transient behavior here.

VEHICLE GREENHOUSE GAS EMISSIONS

Based on the work of Delucchi [19], we have estimated the total well-to-wheels greenhouse gas emissions for each vehicle. In addition to CO_2, the main greenhouse gas, this analysis includes the other major greenhouse gases. Each of these gases is converted into a CO_2 equivalent emission by estimating the total greenhouse gas impact -- both the capacity to block infrared radiation combined with the atmospheric dwell time of the gas -- compared to CO_2. The analysis attempts to capture all sources of emissions involved with extracting, transporting, refining and dispensing the fuel, including any emissions from the vehicle itself. The resulting greenhouse gas emissions in CO_2-equivalent grams per mile are summarized in Figure 8 for each of the alternative vehicles.

The greenhouse gas emissions of Figure 8 assume that the hydrogen is generated by steam reforming of natural gas. Hydrogen could also be generated by electrolysis of water. If we used the projected utility marginal grid mix for the year 2000, then electrolytic hydrogen would generate more than twice the greenhouse gases as a conventional gasoline ICEV as shown in Figure 9, since over 50% of all U.S. electrical energy is derived from coal-burning plants. However, even replacing all coal generation with natural gas combined cycle turbines would not suffice to make electrolytic hydrogen a net greenhouse gas benefit in a FCV -- the direct hydrogen FCV would still generate 1.5 times more greenhouse gases than an ICEV with all natural gas generated electricity.

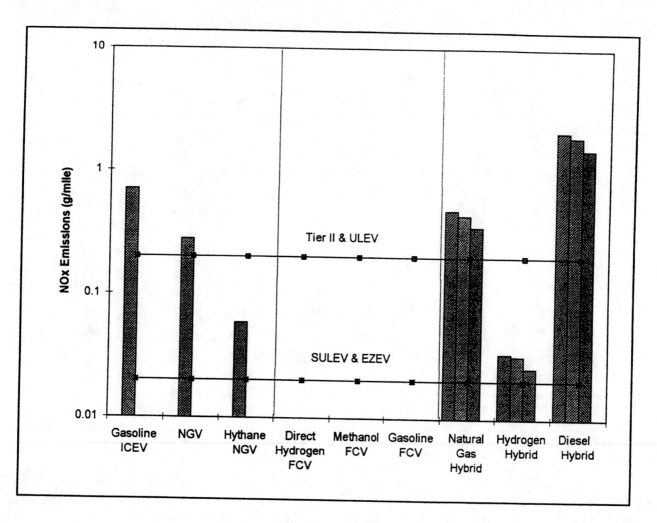

Figure 7. "Real World" Oxides of Nitrogen Emissions in 2000

Some analysts have suggested using wind or PV electricity to generate hydrogen for FCVs. However, we estimate that using renewable energy to displace existing grid electricity in the U.S. will reduce greenhouse gases by 1.8 times more than using that same renewable electricity to make hydrogen for a FCV. Eventually, when renewables and nuclear supplies a large fraction of the electricity, then electrolytic hydrogen in a FCV would reduce greenhouse gases compared to a gasoline ICEV. For example, if all coal were replaced by natural gas combined cycle turbines, and if renewable electricity provided 35% of all grid electricity, then the hydrogen FCV would produce the same greenhouse gases as the ICEV. Therefore renewable energy must exceed 35% grid penetration before the direct hydrogen FCV operating on electrolytic hydrogen provides a net reduction in greenhouse gases.

We have also assigned a monetary cost for both local and greenhouse gas emissions. Estimating the societal costs of emissions is very subjective, but several states have now assigned an avoided cost for reducing each of these emissions for utility planning purposes. The cost of alternative technologies to reduce each pollutant is taken as a surrogate for the societal or health cost of that emission. For example, if a coal plant scrubber to remove SO_2 from the flue gas costs $4,000 per kg of SO_2 removed, then this would be the "avoided cost" of that pollutant. We have selected the lowest avoided cost estimate from the literature for each pollutant gas to permit a quantitative estimate of the net cost of operating each alternative vehicle, as summarized in Table 9.

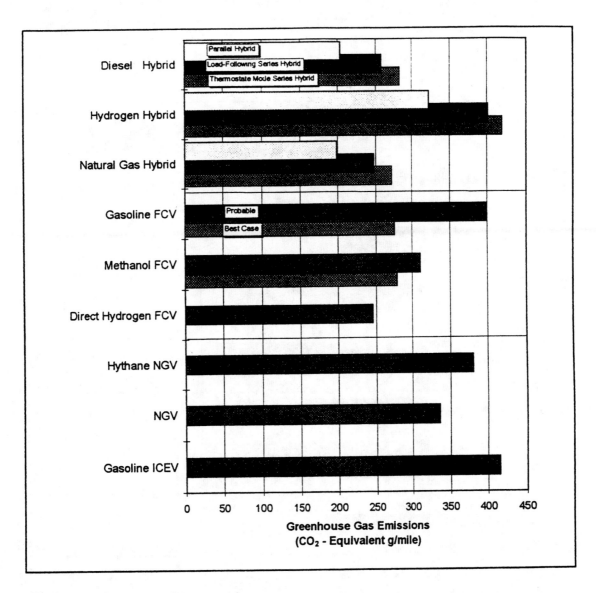

Figure 8. Vehicle Greenhouse Gas Emissions in 2000 (Hydrogen Produced by Steam Reforming of Natural Gas)

Table 9.   Avoided Costs of Emissions

Pollutant	Avoided Cost ($/metric tonne)
VOC	5,840
CO	960
NOx	7,200
PM	4,400
CO_2-equivalent	24

COST-BENEFIT PROJECTIONS

The final results of this study are shown in Figures 10 through 12, which compare the incremental mass production costs of each alternative vehicle (as a surrogate for lifecycle costs as discussed above) with the avoided costs of pollution. In each of these figures, "goodness" is in the lower left hand corner of the graph -- low pollution cost and low vehicle incremental costs.

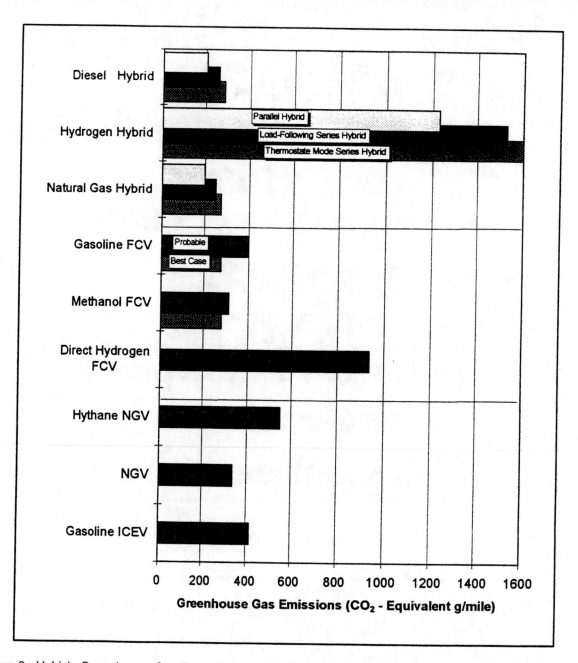

Figure 9. Vehicle Greenhouse Gas Emissions (Hydrogen Derived by Electrolysis with U.S. Grid Electricity)

Figure 10 illustrates the cost/benefit tradeoff if we take local air pollution as the only criterion. The ICEV is projected to create about $200/year of pollution. A natural gas vehicle cuts this cost to only $70/year with only a few hundred dollars of incremental vehicle cost. Adding hydrogen to the natural gas (the "hythane" vehicle) is also very cost effective -- reducing avoided pollution costs to $10/vehicle per year at only slightly higher vehicle costs. Of course the introduction of hythane fuel may be very costly, effectively requiring both a natural gas dispensing and a hydrogen production infrastructure. The direct hydrogen FCV is the only true zero emission vehicle, but the methanol FCV and the hydrogen parallel hybrid vehicle have only minor emission costs. Furthermore the hydrogen hybrid is projected to cost less than the hydro-

gen FCV, and the hybrid ICE drivetrain certainly needs less technological development.

If greenhouse gas emission is the only criterion, then the picture changes significantly as shown in Figure 11. In this case the natural gas vehicle provides only 16% reduction in greenhouse gases, and the hythane vehicle increases greenhouse gases, since the hydrogen is made from natural gas at less than 100% efficiency -- burning hythane requires more net natural gas than burning neat natural gas. The best options to reduce greenhouse gases include the natural gas and the diesel parallel hybrid vehicles. Both vehicles generate less greenhouse gas emissions than the direct hydrogen FCV, again assuming that all hydrogen is derived from natural gas.

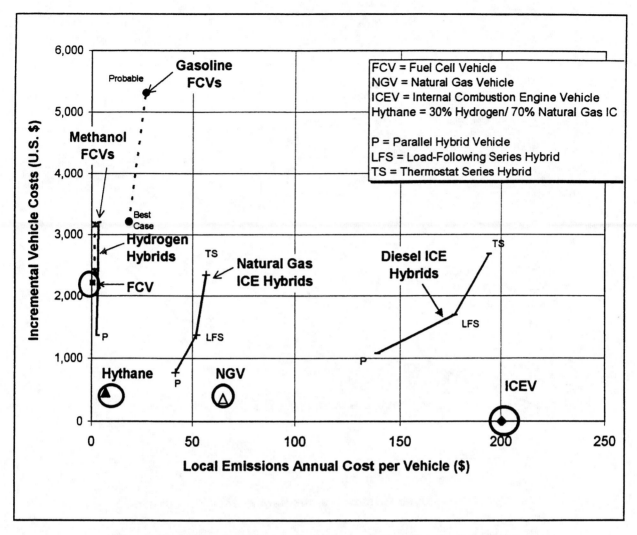

Figure 10. Incremental Vehicle Cost vs. Local Emission Avoided Costs

Figure 12 shows the effects of combining local air pollution and greenhouse gas emissions. Now the direct hydrogen FCV does emerge as the best option, but by a very narrow margin. The methanol FCV, the hydrogen parallel hybrid and the natural gas parallel hybrid vehicles all provide nearly the same net benefit as the direct hydrogen FCV, and the parallel hybrids are projected to have equal or lower incremental vehicle cost. The gasoline FCV and diesel hybrid are both far from optimum from this perspective.

OIL IMPORT EFFECTS

The other major societal goal for alternative vehicles is to reduce our dependence on imported oil. All of the vehicles considered here except the gasoline FCV and the diesel hybrid depend on natural gas as the primary fuel source. Since these two oil-dependent options are not good net environmental candidates as indicated in Figure

12, adding societal costs for imported oil would only shift these two options farther to the right, or even less beneficial to society.

CONCLUSIONS

We conclude that there are no clear winners. The optimum fuel and optimum vehicle depend on the priority given to societal objectives and (in the case of hydrogen and methanol) the willingness of industry and possibly government to invest billions of dollars in new fuel infrastructure .

If reduction of local air pollution is the sole criterion, then natural gas vehicles or hythane fuel vehicles provide the most cost effective solution. A natural gas hybrid would provide little advantage over an NGV but would cost more. Either a hydrogen hybrid, a direct hydrogen FCV or a methanol FCV would provide zero or near zero local emissions, but with added vehicle costs.

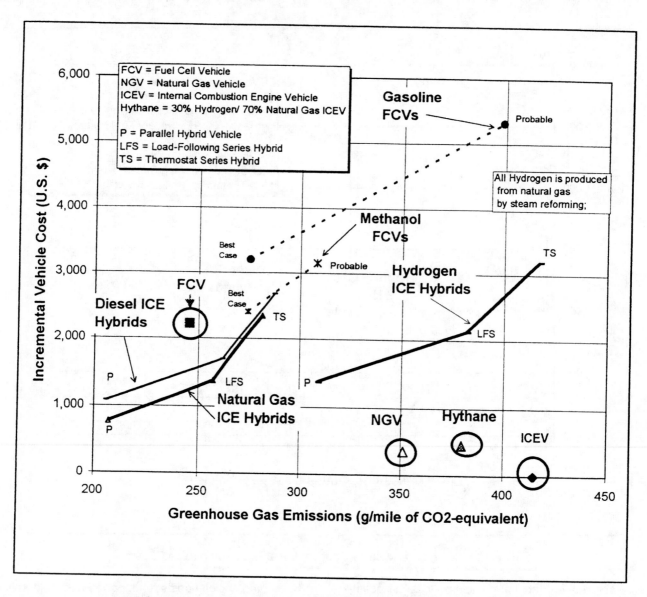

Figure 11. Incremental Vehicle Cost vs. Greenhouse Gas Emissions

If reduced greenhouse gases is the sole criterion, then the clear winners are natural gas or diesel parallel hybrids, with diesel parallel hybrids providing almost 15% lower greenhouse gas emissions than a direct hydrogen FCV, assuming that hydrogen is derived from natural gas.

If we consider both local air pollution and greenhouse gases, then any of the natural gas derived fuel/vehicle systems (direct hydrogen FCV, methanol FCV, hydrogen parallel hybrid, natural gas parallel hybrid) provide good performance. If oil import reduction is added to the goals, then this accentuates the advantage of the natural gas derived fuel systems over either gasoline FCVs or diesel hybrid vehicles.

Overall we conclude that natural gas is the preferred fuel feedstock, either used directly in a parallel hybrid vehicle or indirectly to produce methanol or hydrogen for fuel cell vehicles. Methanol requires less investment initially due to existing excess production capacity, although the added cost per vehicle to provide an onboard methanol processor could be equal to or greater than the cost per vehicle to install a distributed hydrogen infrastructure based on mass produced, small scale steam methane reformers. Hydrogen would require less infrastructure investment in the long run, once the methanol excess capacity was consumed, but hydrogen as a fuel must overcome real and perceived safety barriers before it will be accepted as a major fuel carrier for private passenger vehicles.

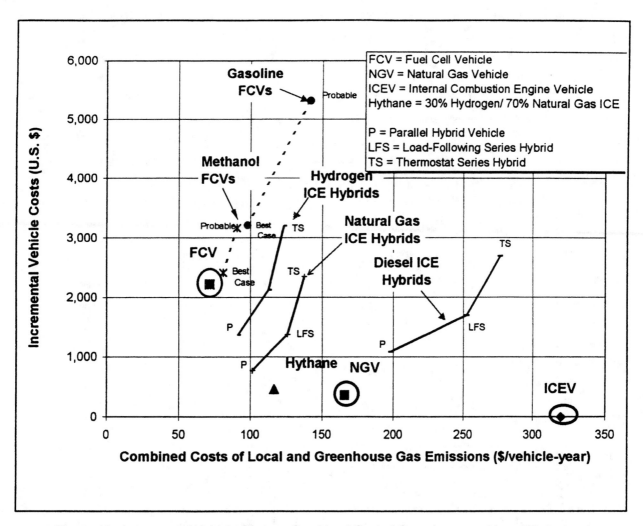

Figure 12. Incremental Vehicle Cost vs. Combined Cost of Greenhouse and Local Emissions

ACKNOWLEDGMENTS

We acknowledge the support of the U.S. Department of Energy under contract No. AXE-6-16685-01.

REFERENCES

1. Brian D. James, C. E. Thomas, George N. Baum, Franklin D. Lomax, Jr., Ira F. Kuhn, Jr., "Making the Case for Direct Hydrogen Storage in Fuel Cell Vehicles," *Proceedings of the National Hydrogen Association's 8th Annual U.S. Hydrogen Meeting,* Alexandria, Virginia, 11-13 March 1997, p. 447.
2. Stuart F. Brown, "The Automakers' Big-Time Bet on Fuel Cells," *Fortune*, March 30, 1998, p. 122.
3. Mike Inbody, Jose Tafoya, Jim Hedstrom and Nick Vanderborgh, "Fuel Cell Stack Testing at Los Alamos National Laboratory," presented at the U.S. Department of Energy Fuel Cells for Transportation Exploratory R&D Program Meeting, Washington, D.C., September 25-26, 1996.
4. Matthew L. Wald, "In Step Toward Better Electric Car, Fuel Cell Used to Get Energy from Gasoline," *The New York Times, October 21, 1997.*

5. T. Zawodzinski, B. Mueller, J. Bauman, T. Rockward, S. Savett, T. Springer, F. Uribe, J. Valero, and S. Gottesfeld, "Reformate Tolerance in PEFCs," presented at the Annual National Laboratory R&D Meeting of the DOE Fuel Cells for Transportation Program, Los Alamos, New Mexico, July 28-29, 1998.
6. C. E. Thomas, B. D. James, F. D. Lomax, Jr., and I F. Kuhn, Jr., *Integrated Analysis of Hydrogen Passenger Vehicle Transportation Pathways*, prepared for the National Renewable Energy Laboratory under Subcontract No. AXE-6-16685-01, March 1998.
7. J. Ray Smith and Salvador Aceves, "Hybrid Vehicle System Studies and Optimized Hydrogen Engine Design," DOE Hydrogen Program Review, Coral Gables, Florida, April 18-21, 1995.
8. Arthur D. Little, "Fuel Cell Powerplant System Considerations," Environmental Vehicles '96 Conference, January 22, 1996.
9. *Office of Advanced Automotive Technologies R&D Plan: Energy-Efficient Vehicles for a Cleaner Environment*, Office of Transportation Technologies, Energy Efficiency and Renewable Energy, U. S. Department of Energy Report No. DOE/ORO/2065, March 1998.

10. Jim Jordan, "Global Methanol Review," presented at the International Methanol Producers and Consumers Association, San Juan, Puerto Rico, May 4-6, 1998.

11. *Looking Beyond the Internal Combustion Engine: The Promise of Methanol Fuel Cell Vehicles,* prepared for the American Methanol Institute by Gregory P. Nowell, State University of New York at Albany, Washington, D.C.

12. Jason Mark, "Fuel Choices for Fuel Cell Vehicles: Environment vs. Infrastructure," *Proceedings of the World Car Conference,* Riverside, California, 19-22 January 1997, p. 393.

13. Joan Ogden, Adam Cox and Jason White, "Options for Refueling Hydrogen Vehicles: A Southern California Case Study," *Proceedings of the National Hydrogen Association's 7th Annual U.S. Hydrogen Meeting,* Alexandria, Virginia, April 2-4, 1996, p. 111.

14. C. E. Thomas, B. D. James, I. F. Kuhn, F. L. Lomax and G. N. Baum, *Direct-Hydrogen-Fueled Proton-Exchange-Membrane Fuel Cell System for Transportation Applications: Hydrogen Infrastructure Report,* prepared for the U. S. Department of Energy, Office of Transportation technologies by the Ford Motor Company, Report No. DOE/CE/50389-504, July 1997.

15. C. E. Thomas, I. F. Kuhn, Jr., B. D. James, F. D. Lomax, Jr., and G. N. Baum, "Affordable Hydrogen Supply Pathways for Fuel Cell Vehicles," *Int. J. Hydrogen Energy,* Vol. 23. No. 6, pp. 507-516, 1998.

16. Joan Ogden, Margaret Steinbugler, and Thomas Kreutz, "Hydrogen as a Fuel for Fuel Cell Vehicles: A Technical and Economic Comparison," *Proceedings of the National Hydrogen Association's 8th Annual U.S. Hydrogen Meeting,* Alexandria, Virginia, 11-13 March 1997, p. 469, and private communication with Joan Ogden, August 19, 1998.

17. Marc Ross, Rob Goodwin and Rick Watkins, University of Michigan, Michael Q. Wang, Argonne National Laboratory, and Tom Wenzel, Lawrence Berkeley Laboratory, "Real-World Emissions from Model Year 1993, 2000 and 2010 Passenger Cars," American Council for an Energy-Efficient Economy, Berkeley, California, November 1995.

18. *Proposed Amendments to California Exhaust, Evaporative and Refueling Emission Standards and Test Procedures for Passenger Cars, Light-Duty Trucks and Medium-Duty Vehicles, "LEV II," and Proposed Amendments to California Motor Vehicle Certification, Assembly-Line and In-Use Test Requirements, "CAP 2000,"* State of California Air Resources Board, preliminary draft staff report released June 19, 1998.

19. Mark A. Delucchi, "Summary of Results from the Revised Model of Emissions of Greenhouse Gases from the use of Transportation Fuels and Electricity," Institute of Transportation Studies, University of California, Davis, California, November 1996

ABBREVIATIONS

AIV: aluminum intensive vehicle
DOE: U.S. Department of Energy
EZEV: equivalent zero emission vehicle
FCV: fuel cell vehicle
FHDS: federal highway driving schedule
FUDS: federal urban driving schedule
HEV: hybrid electric vehicle
ICEV: internal combustion engine vehicle
LEV: low emission vehicle
LFS: load-following series hybrid vehicle
LHV: lower heating value
NG: natural gas
P: parallel hybrid vehicle
POX: partial oxidation
PROX: preferential oxidation
SOC: state of charge
SULEV: super-ultra-low emission vehicle
TS: thermostat series hybrid vehicle
ULEV: ultra-low emission vehicle
ZEV: zero emission vehicle

HYDROGEN STORAGE AND GENERATION

Performance Bench Testing of Automotive-Scale Hydrogen on Demand™ Hydrogen Generation Technology

Richard M. Mohring, Ian A. Eason and Keith A. Fennimore
Millennium Cell, Inc.

ABSTRACT

Millennium Cell has developed a novel catalytic process (called Hydrogen on Demand™) that generates high purity hydrogen gas from air-stable, non-flammable, water-based solutions of sodium borohydride, $NaBH_4$. This paper discusses initial performance bench testing of an automotive-scale hydrogen generation system based on our proprietary technology. Our system was coupled to a hydrogen flow controller system designed to simulate the hydrogen draw of a large (>50 kW net) fuel cell engine. The controller was programmed to emulate various driving cycles, and behavior of the Hydrogen on Demand™ system under realistic load transient conditions was measured. The testing indicates that the Hydrogen on Demand™ system successfully provides hydrogen under realistic load conditions, and that the data are qualitatively indistinguishable from the runs performed with compressed hydrogen gas.

INTRODUCTION

A general description of the Millennium Cell Hydrogen on Demand™ technology can be found in Reference [1]. In summary, the "hydrogen" is stored on-board at ambient conditions in a liquid fuel – a stabilized aqueous solution of sodium borohydride, $NaBH_4$. The solution is non-flammable, and can be handled similarly to common household chemicals.

The high purity, humidified hydrogen produced by this system can be used for numerous applications, addressing a wide range of power requirements. In particular, these systems can be implemented to supply hydrogen for fuel cells and internal combustion engines in automobiles.

The Hydrogen on Demand™ system releases the hydrogen stored in sodium borohydride solutions by passing those liquid fuels through a chamber containing a proprietary catalyst bed. The hydrogen is liberated in the reaction:

$$NaBH_4 + 2\,H_2O \xrightarrow{\text{catalyst}} NaBO_2 + 4\,H_2 + \text{heat}$$

A noteworthy point is that there is no heat input to the reaction. In fact, there is enough heat generated to vaporize a portion of the excess water in the fuel solution, resulting in a naturally humidified hydrogen stream. The product, sodium metaborate, can be recycled as the starting material for the generation of sodium borohydride.

In its solid form, $NaBH_4$ can store greater than 10% hydrogen by weight. In our current designs, $NaBH_4$ is dissolved in water to form a fuel solution. As an example, a 30 wt% $NaBH_4$ solution contains approximately 6.7% hydrogen by weight.

SYSTEM SCHEMATIC AND OPERATION

A general schematic for an automotive-scale hydrogen generation system is shown in Figure 1. A fuel pump directs fuel from a tank of sodium borohydride solution into a catalyst chamber. Upon contacting the catalyst bed, the fuel solution generates hydrogen gas and sodium metaborate (in solution). The hydrogen and metaborate solution separate in a second chamber, which also acts as a small storage ballast for hydrogen gas. The humidified hydrogen is processed through a heat exchanger to achieve a specified dewpoint, and is then sent through a regulator to the fuel cell or internal combustion engine, or for this work, to the fuel cell emulator (FCE) device.

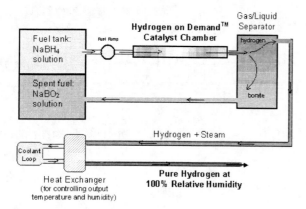

Figure 1. General schematic for a typical Hydrogen on Demand™ system

In operation, the rate at which hydrogen gas is generated is directly proportional to the rate at which the borohydride solution is pumped into the catalyst chamber. This operational simplicity translates into relatively straightforward control strategies.

For these tests, the system was run in a simple "on/off" mode using the system pressure as a process variable to control the state of the fuel pump (analogous to the way an air compressor operates). The delivery pressure setpoint for our system was 150 psig (~10 bar). This low operation pressure for Hydrogen on Demand™ is in contrast to the high pressures of 2200-5000 psig (~150-350 bar) that are present in typical compressed hydrogen tanks. In a compressed hydrogen system, the pressure is dropped through a regulator to deliver hydrogen to the fuel injectors at a much lower pressure of 75-150 psig (~5-10 bar). Operating the entire Millennium Cell system at the proper pressure for the fuel injectors eliminates the need to have highly pressurized hydrogen on board.

When the FCE demand for hydrogen gas increases, hydrogen is drawn off from the ballast tank, and the system pressure drops below the setpoint. This triggers the fuel pump to start up. As fuel reaches the catalyst chamber, hydrogen gas is generated causing the pressure to increase above the setpoint, and the fuel pump is turned off (see Reference [1]).

More sophisticated strategies will be implemented at the integration stage with the vehicle to achieve higher degrees of control. For example, in a fuel cell vehicle, a signal proportional to the electric current load demanded from the fuel cell will be used to directly control the rate at which hydrogen is produced by the generator. For an IC engine vehicle, a signal related to the already-existing instantaneous injector pulse widths can be used.

SYSTEM TESTING

The system under test was designed to power a Millennium Cell internal combustion vehicle, a Ford Crown Victoria, which originally ran on compressed natural gas (see Figure 2).

Figure 2. Crown Victoria natural gas internal conbustion engine, modified to run on hydrogen

Millennium Cell made minor modifications to the car so that it would run on hydrogen. Initially, we installed two cylinders of compressed hydrogen in the trunk area and took measurements of the gas flow rates under driving conditions. During these tests, it was noted that there was a significant occurrence of autoignition (pinging) while running the system. A simple water-injector was added to the fuel manifold, which essentially eliminated the autoignition problem.

The hydrogen generator system is shown mounted in the rear of the vehicle in Figures 3 and 4. The catalyst chamber is the near-horizontal cylinder mounted in the foreground of the photo and is easily visible in Figure 4. Sodium borohydride fuel is stored in a standard ABS plastic tank mounted underneath the vehicle in front of the trunk area (not shown) and is metered into the catalyst chamber from its right side through a flexible line by a piston pump. The hydrogen and borate solutions separate via gravity in the vessel directly behind the catalyst chamber. The borate solution is periodically pulsed (using the hydrogen pressure as a driver) from the separation vessel into an ambient pressure tank located below the trunk area.

Figure 3. Hydrogen on Demand™ system mounted in trunk of vehicle

The hydrogen is taken from the top of the separation vessel, passed through a heat exchanger, and sent into the engine (or in the bench testing, into the Fuel Cell Emulator). It should be noted that the hydrogen generation system is mounted such that the passenger compartment was not compromised, and the majority of the trunk area has been retained.

Prior to beginning the installation into the vehicle, the prototype Hydrogen on Demand™ system was run on the bench and connected to the hydrogen inlet line of the engine. The engine ran with no difficulty, in fact, an interesting consequence of the humidified hydrogen stream was that the water injection system could be removed completely (as the humidity provided plenty of moisture to defeat the autoignition effect).

The fuel cell emulator (FCE) is a device commissioned from ATC, Inc. (Indianapolis, IN) and is shown in Figure 5. The system is essentially a mass flow controller coupled via a PID control loop to a precision hydrogen mass flowmeter. A hydrogen source is connected to the inlet, and the hydrogen flow rate is controlled to meet a flow schedule that is programmed into the computer system. With proper modeling of the vehicle systems, one can convert a speed profile (for example) into realistic fuel cell or engine power demands, and further into a realistic profile for hydrogen gas demand. At that stage, various hydrogen sources can be compared identically as to their performance under these conditions.

Figure 4. Close-up of Hydrogen on Demand™ system, showing catalyst chamber and separation vessel

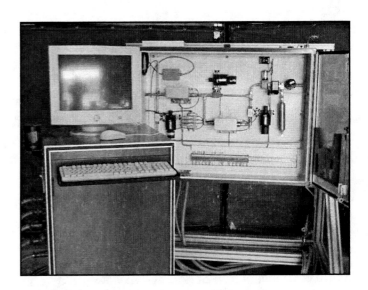

Figure 5. Fuel Cell Emulator (FCE) device with data acquisition system

This paper compares a specific FCE flow profile for both compressed hydrogen gas and the Hydrogen on Demand™ prototype system. The flow profile was manually generated to be very aggressive in order to demonstrate system response under transient and heavy load conditions, although it is not modeled for an specific vehicle. The magnitude and frequency of the transients was chosen to be representative of conditions similar to highway driving with accelerations to >60 mph (> ~95 kph) as well as a short period of aggressive stop-and-go driving. The range of flow rates was chosen to be consistent with our measured flow rates for the Crown Victoria running on compressed hydrogen.

In order to establish a baseline, we connected a manifold of six "standard" 2200 psig (~150 bar) compressed hydrogen cylinders to the FCE and ran the selected profile. Figure 6 shows our results for compressed hydrogen. The profile setpoint for hydrogen flow rate (in standard liters per minute, SLM) is shown as the solid line, while the measured flow rate is the line with hollow triangles.

Of note in Figure 6 are the over- and under-shooting of the measured flow relative to the setpoints as well as a slight delay in the response. This is simply an effect of the PID control loop of the FCE being calibrated in a slightly "underdamped" state, and bears no reflection on the ability of the hydrogen cylinders to follow the transients. As one can see, the compressed hydrogen cylinders are quite able to provide a satisfactory amount of hydrogen to meet the demand of the profile.

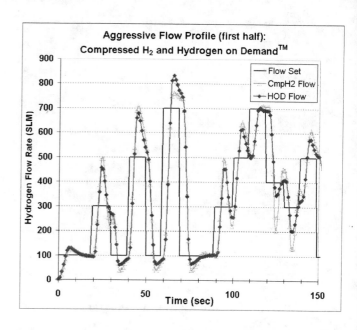

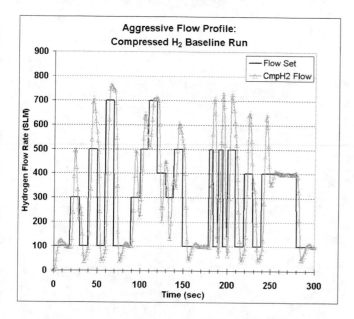

Figure 6. Baseline run of aggressive flow profile on the FCE using the manifold of compressed hydrogen cylinders at 2200 psig (~150 bar)

After establishing the baseline with compressed hydrogen gas, we removed the manifold and connected the outlet of our prototype Hydrogen on Demand™ system to the FCE. The results are shown in Figures 7 and 8, broken into two graphs (first half and second half) so as to show better detail. Our system is shown as the line with solid circles, while the profile itself and compressed hydrogen baseline are depicted as before.

Figure 7. First half of aggressive flow profile shown with data from the Hydrogen on Demand™ (HOD) system overlaid atop the flow profile and baseline run performed with compressed hydrogen gas at 2200 psig (~150 bar)

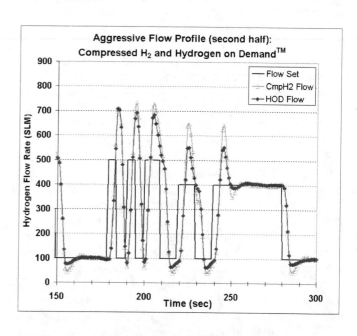

Figure 8. Second half of aggressive flow profile shown with data from the Hydrogen on Demand™ system overlaid atop the flow profile and baseline run performed with compressed hydrogen gas at 2200 psig (~150 bar)

These data are qualitatively indistinguishable from the runs made using our prototype system. As with the compressed hydrogen baseline run, the PID control on the FCE was such that there are delays and over- and under-shooting of the setpoints our the prototype runs. However, the data from the Hydrogen on Demand™ system are qualitatively indistinguishable from the baseline compressed hydrogen data.

As of this writing, we are currently completing the integration of Hydrogen on Demand™ with our Crown Victoria, and will begin in-vehicle evaluation. In addition to verifying our dynamic load performance, we expect to investigate the benefits of the humidified hydrogen stream on autoignition as well as a potential reduction of NO_x emissions.

CONCLUSIONS

In this paper, we have demonstrated on the bench that Hydrogen on Demand™ technology under a realistic power profile provides excellent delivery of hydrogen to meet demand requirements. There was no difference detected between compressed hydrogen at 2200 psi (~150 bar) and the HOD system at 150 psi (~10 bar). Hydrogen flow rates of up to ~800 SLM were demonstrated, without requiring operation at high pressure. With proper system integration with a vehicle, we expect our technology to provide sufficient response to supply the power demand of the engine or fuel cell. From a packaging perspective, Hydrogen on Demand™ systems have the potential for relatively high energy storage density (estimated to be greater than 4.5 wt% for a system as described herein) and do not suffer from the (real or perceived) safety issues associated with many technologies for storing hydrogen. The proof of concept design in our Crown Victoria potentially represents a commercially viable strategy for retrofitting existing cars to give low emissions performance. In-vehicle testing will be presented at a later time.

ACKNOWLEDGMENTS

The authors would like to acknowledge the dedicated efforts of the Millennium Cell team. Special thanks to Rex Luzader and Ying Wu for assistance and insightful commentary on the draft.

REFERENCES

[1]   Mohring, R. M., Luzader, R. E., "A Sodium Borohydride On-Board Hydrogen Generator for Powering Fuel Cell and Internal Combustion Vehicles," SAE Future Transportation Technologies Conference, Society of Automotive Engineers, August 2001, Paper Number 2001-01-2529.

CONTACT

Dr. Mohring is Program Director of the Hydrogen on Demand™ efforts at Millennium Cell, and can be reached by email at *mohring@millenniumcell.com*, by phone at 732-544-5706, or by mail at 1 Industrial Way West, Eatontown, NJ, 07724. Further information can be found at *www.millenniumcell.com*.

2002-01-0097

Investigation of Hydrogen Carriers for Fuel-Cell Based Transportation

Sebastien E. Gay-Desharnais, Jean-Yves Routex, Mark Holtzapple and Mehrdad Ehsani

Advanced Vehicle Systems Research Program Group, Texas A&M Univ.

ABSTRACT

Hydrogen is the only fuel for fuel-cells. It may be stored pure onboard the vehicle, but this approach is difficult. It is preferable to attach hydrogen to other elements to form a hydrogen carrier, which will be cracked onboard to feed the fuel-cell. This paper explains the concept of hydrogen carriers, details their preferred characteristics, and compares thoroughly the available candidates.

THE CONCEPT OF HYDROGEN CARRIERS

PURE HYDROGEN STORAGE - To this day, hydrogen is the only fuel that fuel-cells can directly process. Even direct alcohol fuel-cells oxidize only hydrogen, which they extract at the anode from the alcohol. The storage of pure hydrogen is complicated, costly, inefficient and dangerous. There are three ways to directly store hydrogen onboard a vehicle: compressed gas, cryogenic liquid, and metal hydrides.

Hydrogen has the lowest density of all elements. Therefore, compressed gas and cryogenic liquid hydrogen have very low energy densities. This is aggravated by the tank construction. The high pressure of compressed hydrogen (in excess of 300 atmospheres) requires thick, heavy tank walls for mechanical strength and to prevent the natural diffusion of the small hydrogen molecules. For cryogenic liquid hydrogen, an efficient thermal insulation is required to prevent boil-off, which takes space and adds weight, hence reducing the energy density.

The compression or liquefaction of hydrogen consumes significant amounts of energy [8], which lowers the efficiency of pure hydrogen storage and adds to its cost. Overall, both compressed gas and cryogenic liquid hydrogen are dangerous either because of the risk of tank rupture, spill of liquid hydrogen at –253 °C, and risk of leakage yielding an explosion.

Metal hydrides are a potential solution to these problems. The hydrogen is "trapped" in a metallic matrix (or sponge) made of materials such as nickel, lithium or boron. Although the whole is rather safe, it takes significant amounts of energy to desorb and extract hydrogen from the matrix. Besides, the cost of lithium (the most efficient absorbent) makes this solution unfit for most commercial or personal transportation.

HYDROGEN CARRIERS - The solution to the storage problems is to bind hydrogen to other chemical elements, thus forming hydrogen-rich, chemically stable compounds that can be stored easily onboard the vehicle as liquids or low-pressure gases. The hydrogen is then extracted at will by a reformer or cracker, purified if necessary, and then fed to the fuel-cell.

Hydrogen carriers have three origins: fossil, biochemical or synthesis. Oil and natural gas are composed mostly of hydrocarbons, which contain significant amounts of hydrogen tied to carbon atoms. Natural fermentation processes can form many organics, mostly oxygenates such as alcohols and ketones. Finally, most organics, oxygenated or not, and many hydrogen-rich molecules can be synthesized in a chemical plant.

PREFERRED CHARACTERISTICS OF HYDROGEN CARRIERS

The preferred characteristics of hydrogen carriers are highly influenced by three factors: their use in a fuel-cell powered vehicle, the need to save energy, and the presence of human beings onboard. Following is a detailed review of these characteristics.

STORAGE - The main reason why hydrogen carriers are used, instead of pure hydrogen, is their ease of storage. Hydrogen carriers must be either solid or liquid within possible operating temperatures (-40 °C to +50 °C) or gases that can easily be liquefied at low pressures. This ensures that a light, robust, and cost-effective tank can be used onboard the vehicle to contain the hydrogen carrier.

HYDROGEN EXTRACTION - Hydrogen must be easily extracted from the hydrogen carrier. Only simple, light, compact, and cheap hardware should be required. This penalizes extraction processes (reforming or cracking) that require multi-steps, purification steps, exotic catalysts, significant quantities of additives, and high temperatures. Complicated piping increases both the manufacturing and maintenance costs, and gives room for more failures. Multi-step reforming processes have a slow dynamic response, are more complex to control, and require more hardware. Obviously, if significant amounts of additives are required, the complexity, hardware cost and fuel cost increase and the vehicle weight increases. High temperatures are difficult to integrate under the hood of most vehicles and often mean long warm-up times and increased risk of fire or explosion in case of failure or accident. A warm-up time of a few seconds up to a few minutes should be considered reasonable depending on whether the drive train is a fuel-cell alone or a fuel-cell hybrid. Longer warm-up times could be allowed only in a few rare cases where the vehicle is started long before it must be operational.

HYDROGEN ENERGY DENSITY - Obviously, the added value of a hydrogen carrier is its hydrogen content or the actual amount of hydrogen that can be extracted from the carrier. Therefore, the more hydrogen that can be extracted per unit of carrier mass and volume, the better the carrier. This is achieved by minimizing the weight of the binding element(s) and their quantity. The best binding materials are found in the low atomic number region of the periodic table (second line). Elements that have higher valences - i.e. that can share, accept or lose a high number of electrons on their peripheral layer - are better candidates. Molecular structure is also important: in some cases, a longer molecule is preferable because it increases the relative quantity of hydrogen versus the binding material. This is true for common alcohols such as methanol, ethanol, propanol. Obviously, beyond a certaine molecule length, the relative quantity of hydrogen decreases again. In other cases, a shorter molecule is preferable (alkanes). Compounds that have redundant chemical bonds are also bad candidates because fewer bonds are available to tie hydrogen (cyclic hydrocarbons). Elements that do not help bind any more hydrogen lower the energy density (oxygen in alcohols).

Because the energy contained in the carrier is not entirely conserved in the hydrogen, it is useful to express the hydrogen content in terms of energy. We thus define the hydrogen energy density as the amount of energy corresponding to the hydrogen extracted from the carrier.

ONBOARD EFFICIENCY - The added value of a hydrogen carrier is its hydrogen content; therefore, it should be the only element participating to the energy content of the carrier. Throwing away the binding element(s) should not result in a loss of energy content. Another factor affecting the onboard efficiency is the energy required extracting the hydrogen, which obviously

should be as low as possible. A catalytic reforming or cracking reaction helps lowering this requirement.

WELL-TO-WHEEL EFFICIENCY - This is defined by the amount of energy that is actually transmitted to the vehicle from the initial amount of energy extracted from the well. In the case of a hydrogen carrier, this efficiency may be defined two different ways:

- The amount of energy contained in the hydrogen after extraction versus the amount of energy initially contained in the carrier or required for its synthesis.
- The amount of energy actually being transmitted to the wheels of a vehicle versus the energy initially contained in the carrier or required for its synthesis.

These two definitions are different because some fuel-cell technologies cannot accommodate certain hydrogen carriers. For the first definition, we should speak about well-to-fuel-cell efficiency. However, it is more realistic to consider the limitations induced by the fuel-cell technology on the hydrogen carrier; therefore, the second definition will be used.

The initial energy is defined as the energy contained in the primary materials or compound that are at the base of the synthesis or extraction process. Additional energy may be required such as the electrical energy needed for electrolysis. The well-to-wheel efficiency is affected by the on-board efficiency, but also by the refining or synthesis process that yields the hydrogen carrier and also by the ease of storage. It is increased if the carrier is liquid, solid, or easily liquefiable. The origin of the hydrogen also plays a role because, in most cases, the carrier is not readily available and must be synthesized either artificially or biologically from hydrogen and other materials.

RENEWABILITY - One of the primary advantages of hydrogen as a fuel is that it can be derived from renewable sources, such as wind, solar, hydraulic power, or biomass. This is opposed to fossil fuels that are not renewable at human scales. For the purpose of this paper, only renewable sources will be considered because the context chosen is that of a complete or near-complete exhaustion of fossil fuel resources. It should be noted that renewable sources have a minimal impact on the environment because the reactants and products involved are part of a closed cycle in which the only input is the sun (at the most basic level). The synthesis of some carriers from nuclear energy should be considered as this energy has a great potential to supply nearly infinite amounts of energy via super-generators and possibly fusion.

MANUFACTURING COST - This notion is seldom synonymous of well-to-wheel efficiency, but it is an economic indicator, which considers the cost of primary materials and the cost of the energy required by the manufacturing process (refining or synthesis).

SAFETY AND TOXICITY - The presence of humans onboard the vehicle requires that hydrogen carriers be very safe or that they can be easily and effectively isolated from the passengers and vehicle operators. It is often considered that the compound must be no more flammable than gasoline or diesel fuel and that its toxicity must be no worse than that of gasoline. However, it is useful to consider that some chemicals are more toxic than gasoline but that their spills or leaks can be easily managed or that the compounds can be rapidly neutralized. On the other hand, some chemicals are deadly or extremely dangerous even in trace amounts and therefore they should not be considered as hydrogen carriers.

FUEL-CELL POISONNING ISSUES - The electrodes of a fuel cell are partly made of catalytic materials that have a tendency to react more strongly with compounds other than hydrogen and oxygen. These compounds are poisons that may lead to the slow but certain destruction of fuel cell electrodes. They include, but are not limited to carbon monoxide, sulphur oxides and some hydrocarbons. Other chemicals tend to react with the electrolyte, such as carbon dioxide with potassium hydroxide in an alkaline fuel cell, or any alkali with the acidic membrane (perfluorosulphonic acid) of a PEM fuel cell. These chemicals must be removed totally from the hydrogen stream fed to the fuel cell, which often results in an increased complexity, additional weight, and increased cost.

CARBON-BASED HYDROGEN CARRIERS

One carbon atom can share four electrons, the maximum in the periodic classification. Carbon is also one of the lightest fixing elements. The greatest advantage of carbon is that it is present in all living things. Biological fermentation or decomposition (formation of petrol or natural gas) processes naturally yield a wide variety of carbon-based compounds that fix hydrogen, sometimes with other elements such as oxygen.

ALKANES – Alkanes are compounds in which the chemical formula strictly conforms to the following equation:

$$C_nH_{2n+2}$$

Because they are composed solely of a fixing element and hydrogen, alkanes have high hydrogen energy densities. They originate from two sources: fossil or biomass.

Methane, ethane, propane, and butane - These four compounds have the following chemical formulae:

Methane: CH_4

Ethane: C_2H_6

Propane: C_3H_8

Butane: C_4H_{10}

They are gaseous at room temperature and atmosphere pressure. They can be liquefied by compression at the following temperatures and sea level pressure:

Methane: -162 °C, Density: 423kg/m³

Ethane: -89 °C, Density: 545 kg/m³

Propane: -43 °C, Density: 583 kg/m³

Butane: -0.55 °C, Density: 601 kg/m³

Methane cannot be liquefied by compression at room temperature. In contrast, ethane, propane and butane can be liquefied by compression at room temperature (20 °C):

Ethane: 37.2 atm, Density: 339 kg/m³

Propane: 8.25 atm, Density: 500 kg/m³

Butane: 2.1 atm, Density: 578 kg/m³

Obviously, butane is the easiest to store, only requiring low pressures. However, propane and butane are only present in small quantities in natural gas:

Methane: 75%

Ethane: 15%

Propane and butane: 5%

From a supply viewpoint, methane is the best choice, especially because it can be produced very easily from biomass. There is no economical way to directly produce ethane, butane or propane from biomass. The problem with methane is storage because it must be stored as a gas compressed at high pressures, or as a cryogenic liquid at very low temperatures. Either way, the tank is heavy and costly. It is to be noted that in terms of safety, methane storage is comparable to gasoline: serious advances have been made by promoting natural gas for internal combustion engine powered automobiles.

Despite its storage disadvantages, methane is more often considered than any other low alkane because of its availability and ease of production. The extraction of hydrogen from methane is simple. The most commonly used reaction is steam reforming:

$$CH_4 + H_2O \rightarrow CO + 3H_2$$

Then a water-gas shift reaction is used to convert the harmful carbon monoxide into harmless carbon dioxide and extract extra hydrogen from water:

$$CO + H_2O \rightarrow CO_2 + H_2$$

The overall reaction is then:

$$CH_4 + 2H_2O \rightarrow CO_2 + 4H_2$$

The steam reforming reaction is endothermic, requiring a $\Delta H = 226$ kJ/mol. The water-gas shift reaction is exothermic and requires the presence of a catalyst such as iron oxide around 1000 °C. Other catalysts allow operating at temperatures as low as 240 °C. The water need not be stored onboard; it can be recuperated from the fuel-cell stack exhaust. Methane can also be reformed by partial oxidation:

$$2CH_4 + O_2 \rightarrow 2CO + 4H_2$$

This exothermic reaction is often followed by a water-gas shift reaction. The overall reaction is:

$$2CH_4 + O_2 + 2H_2O \rightarrow 2CO_2 + 6H_2$$

The theoretical energy efficiencies of methane reforming are shown here:

Steam reforming: 92%

Partial Oxidation: 90%

These efficiencies are theoretical and do not account for the energy necessary to bring the water and methane to operating temperatures. The water-gas shift reaction is expected to impair the efficiency slightly more because of its high operating temperature. Efficiencies between 70% and 80% are practically achieved. It should also be noted that reforming requires complex equipment that is difficult to fit under the hood of an automobile.

The hydrogen energy density of methane is shown below for cryogenic storage and partial oxidation reforming, not accounting for the tank:

Methane: 45 MJ/kg   19.0 GJ/m³

Although the energy density is high, methane is impaired by its storage conditions (high pressure, low temperature) that make the tank heavy and bulky. Most commonly, natural gas is stored as a compressed gas, which gives the following figures for hydrogen energy density:

Methane (300 atm)   45 MJ/kg   9.9 GJ/m³

Lower alkanes are not toxic, but are highly inflammable chemicals, though less than hydrogen. Methane is explosive in air for concentrations ranging from 3% to 18% [3] and its auto-ignition temperature is around 630 °C.

One problem occurring with all reforming processes is that carbon monoxide remains, even in small quantities in the hydrogen stream. This requires additional steps to completely remove this gas because it is a poison for most catalysts found in fuel-cell electrodes. Platinum is especially sensitive to CO poisoning. The carbon monoxide binds to the active sites of the platinum catalyst much more strongly than oxygen. It is difficult to recover the poisoned electrodes. Other compounds found in natural gas are also harmful for both the reformer and the fuel-cell stack. Sulphur oxides are the most frequent impurities when the gas comes from fossil sources.

Gasoline – Gasoline may be represented as octane, which chemical formulae is:

$$C_8H_{18}$$

Gasoline is found in the intermediate fractions resulting from petroleum distillation. Obviously, it is a fossil fuel, which puts it out of the scope of the present paper. However, it has been argued – with reason – that gasoline is readily available and that it might ensure a smooth transition from gasoline-ICE-based transportation to pure hydrogen-fuel-cell-based transportation.

The production, storage and transport of gasoline are very well mastered after more than 100 years of use. Gasoline is a liquid, which boils at 125 °C and auto-ignites at about 450 °C. It is explosive for small concentrations in air. It is not deadly toxic by inhalation, although it induces cancer if its vapours are inhaled too often.

It is useful to note that by its very nature, gasoline is a mixture of hydrocarbons rather than a pure compound.

Gasoline can be reformed by three methods: auto-thermal reforming (ATR), partial oxidation (POX) or steam reforming (SR). All these reformers must be followed by a water-gas shift reactor to convert the carbon monoxide into carbon dioxide and to get more hydrogen from water.

The POX reaction is written as follows [1]:

$$C_8H_{18} + 4O_2 \rightarrow 8CO + 9H_2 + 659 \text{ kJ/mol}$$

The whole reaction (including the exothermic water-gas shift reactor), is then written as:

$$C_8H_{18} + 4O_2 + 8H_2O \rightarrow 8CO_2 + 17H_2$$

Note that the water can be recycled from the exhaust stream of the fuel-cell stack. The resulting hydrogen energy density is 35.8 MJ/kg, which corresponds to about 26.8 MJ/m³. The theoretical energy efficiency of gasoline POX reforming is 81%. However, an efficiency of 70% only is practically achieved [4]. Other gasoline reforming methods achieve roughly the same practical efficiencies.

Gasoline reforming offers several advantages: gasoline is already available, the POX reformer is quick to warm-up, and the high operating temperatures allow processing a wide variety of fuels. On the other hand, the high operating temperatures require special materials and careful heat integration. In addition, all three reforming methods require a water-gas shift reaction to convert the poisonous carbon monoxide into hydrogen and carbon dioxide. This addition slows down the warm up of the reformer. Gasoline reformers require an additional step to remove the excess heat from the hydrogen stream and bring its temperature down to that of the fuel-cell stack.

Synthesized Gasoline - Gasoline can actually be synthesized by gasifying biomass and converting the resulting synthesis gas to gasoline, which remedies to the non-sustainability of gasoline. However, this process is very inefficient and makes little sense from an economical point of view. Therefore, it shall not be considered in the future.

PRIMARY ALCOHOLS – Primary alcohols have a chemical formula similar to that of alkanes, but there is an additional hydroxyl radical (OH). Hence, their formula conforms to the following equation:

$$C_nH_{2n+1}\text{--}OH$$

One of their primary interests resides in the fact that some alcohols are synthesized easily and efficiently by fermentation and that this process can very easily be reproduced and greatly enhanced in a biochemical plant.

Methanol - Methanol is the lowest primary alcohol. Its chemical formula is:

$$CH_3\text{—}OH$$

Methanol is liquid at room temperature and its boiling point is 65 °C. Its density is 800 kg/m^3. Today, methanol is mostly synthesized from natural gas (methane). However, it can be produced from methane resulting from the anaerobic decomposition of biomass, municipal waste or sewage. This ensures its sustainability as a renewable fuel.

Methanol is the only organic that can be fully reformed without the water-gas shift reaction. The steam reforming reaction is:

$$CH_3\text{—}OH + H_2O \rightarrow CO_2 + 3H_2$$

This reaction is simple, and occurs at low temperature (250 °C), which makes methanol a very good hydrogen carrier. However, the warm-up time is very long, up to 45 minutes [9] and the dynamic response is slow, although faster than gasoline. Like all hydrocarbons, methanol reforming requires an additional step to remove totally the carbon monoxide from the output hydrogen stream.

Methanol reforming is much more efficient than gasoline reforming. Theoretically, an efficiency of 85% can be achieved but practically it is estimated that an efficiency of 80% is achievable. Methanol thus yields the following hydrogen energy density:

16 MJ/kg   12.8 GJ/m^3

The greatest advantage of methanol reforming is simplicity. The reformer operates at low temperatures, which makes it easy to integrate the reformer in a fuel-cell power system and under the hood of a vehicle.

However, simple as it seems, methanol reforming suffers from many disadvantages. It requires a specific catalyst that is easily polluted by impurities. Therefore, the methanol supply must be extremely pure or a purification step must be added to the process, which costs and impairs reliability. A methanol reformer is incapable of processing any other fuel than methanol, which harms the fuel flexibility.

Methanol is toxic by ingestion: a few cm^3 cause blindness or death. There are some concerns because, unlike gasoline, methanol is miscible in water. A tank rupture could result in the pollution of water supplies. At small doses, methanol is tolerated and processed by the human body and is even present in some foods as traces. The effects of methanol do not appear until a few hours after ingestion. This leaves time to inject antidotes, which are readily and widely available. Therefore, methanol can be considered as a safe fuel.

Methanol is a flammable chemical, with an auto-ignition temperature around 430 °C. Methanol vapor concentration in air must be four times higher than gasoline in order for auto-ignition to occur [3]. It is explosive for concentrations in air ranging from 5 to 50%. Methanol vapor disperses faster than gasoline vapor, as it is a lot less dense [5]. Indeed, methanol is deemed safer than gasoline from a fire hazard point of view. There are only two difficulties with methanol: first, methanol burns with a pale blue flame, nearly colorless. This issue could be resolved by adding chemicals to make the flame visible [5]. Second, methanol is more likely than gasoline to ignite at room temperature in tanks. This is because methanol produces less vapor than gasoline under the same conditions. Because of that, the concentrations of methanol vapors in air are more likely to reach critical ranges and to ignite given a source of ignition. However, this issue is manageable through a proper tank design [5].

Ethanol - Ethanol is found in alcoholic beverages. It results from the fermentation of sugars by yeasts. Its chemical formula is:

$$C_2H_5\text{—}OH$$

Like methanol, it is a liquid at room temperature. Its density is 789 kg/m^3. Half of today's industrial ethanol is produced by fermentation of sugars by enzymes present in yeasts [6]. After fermentation, ethanol is concentrated by distillation. The alcohol not intended for drinking is produced by synthesis from acetaldehyde (itself produced from acetylene) or from ethylene (derived from petroleum). Obviously, the only way to ensure the renewability is to produce ethanol by fermentation.

Reforming ethanol is not very well documented, probably because of its disadvantages compared to methanol or alkanes. Because its molecule is longer than that of methanol, ethanol requires higher reforming temperatures. In addition, because only one of its carbons is oxygenated, carbon monoxide is produced, which requires a water-gas shift reactor. The absence of a water-gas shift reactor requirement and the low reforming temperature are the most significant advantages of methanol over alkanes. For these reasons, ethanol is not considered as a valuable hydrogen carrier.

Higher alcohols - Higher alcohols are those alcohols above ethanol. These alcohols suffer from the same disadvantages that impair ethanol as a hydrogen carrier.

OTHER HYDROCARBONS – Of all organic molecules, alkanes offer the highest hydrogen energy density. This is because other organic molecules may have more than one bond between carbons, which reduces the number of bonds available to hydrogen. Secondary alcohols, esters, etc. are partially oxidized and have less energy than alkanes. These are not valid alternatives to alkanes or methanol.

THE ENERGY PROBLEM OF CARBON-BASED HYDROGEN CARRIERS - Carbon-based hydrogen carriers are inherently inefficient because extracting hydrogen throws away part of the carbon energy.

The reforming of carbon-based hydrogen carriers requires high temperatures and results in complexity. With the exception of methanol, all reformers must be followed by a water-gas shift reactor to grasp the energy contained in the carbon. This reactor accounts for a large part for the difficulty to fit a reformer under the hood of an automobile because of its high operating temperature. In addition, the use of water steam accounts for the slow dynamics of reforming.

Because of this disadvantage in efficiency and flexibility, it has been proposed to use hydrogen carriers based on chemicals other than carbon.

NON-CARBON-BASED HYDROGEN CARRIERS

These compounds have the great advantage over carbon-based carriers of not relying on carbon as the binding material. Therefore, there is no loss of efficiency or energy density due to a waste of carbon atoms. This is true if the binding material has no energy content. It is the case with nitrogen, which two compounds present interesting characteristics: ammonia and hydrazine.

AMMONIA – Nitrogen is the closest element to carbon that is able to fix hydrogen. Nitrogen is readily available in the atmosphere and is a totally harmless element. The simplest compound of nitrogen and hydrogen is ammonia:

$$NH_3$$

Ammonia is very commonly used in the industry, mostly as a base for fertilizers; 100 million tons are produced every year [2]. Therefore, the storage, transportation and manipulation of this chemical are well mastered.

Ammonia is produced from hydrogen and nitrogen through the Haber-Bosch process. In this process, a stoichiometric mixture of hydrogen and nitrogen is compressed to high pressures at high temperatures in presence of a catalytic material. The efficiency of this process is about 88% [2]. The reaction is:

$$N_2 + 3H_2 \rightarrow 2NH_3$$

Although it is impractical to produce ammonia directly from biomass, it is possible to produce the needed hydrogen from biomass either directly or from biomass gasification. It is interesting to note that ammonia can be manufactured from urea, a very common waste product of human and animal biological activity. The reaction follows:

$$NH_2—CO—NH_2 + H_2O \rightarrow 2NH_3 + CO_2$$

The carbon dioxide emitted is not responsible for greenhouse effect because the carbon originates from biomass and will be re-absorbed as part of the normal carbon cycle.

Ammonia can be liquefied either by compression at 8.5 atmospheres or cryogenically at -33 °C. These figures are very reasonable compared to other gases, especially hydrogen. Under these conditions the following densities are obtained:

8.5 atm / 20 °C   610 kg/m^3

1.0 atm / -33 °C 680 kg/m^3

The hydrogen energy density of ammonia is then:

8.5 atm / 20 °C   21 MJ/kg   12.9 GJ/m^3

1.0 atm / -33 °C 21 MJ/kg   14.4 GJ/m^3

Although it appears clearly that cryogenic ammonia has a higher hydrogen energy density per volume, there are concerns about storing a fluid at such temperatures: boil-off, weight and volume of thermal insulation. Because of this, it is preferable to work with compressed ammonia, for which the construction of a tank is not a major challenge. The experience gained for natural gas easily translates to ammonia, which has thermodynamic properties relatively close to propane [3].

The use of ammonia raises some safety questions: ammonia is a caustic chemical that can severely burn the lungs and the eyes [2]. Ammonia can be smelt at concentrations under 1 ppm and starts to irritate around 25-100 ppm [2]. Above 1000 ppm, it strongly irritates and can kill a human being in a few minutes [2]. The large difference between the detectability threshold and the lethal concentration has proved to be an important factor of safety in the handling of ammonia [2]. Leaks are easily detected and measures can be taken to fix the situation. In addition, ammonia can be contained just as easily as propane in very solid and very reliable tanks that ensure virtually no rupture and spill in case of a wreck. The technology exists and can be adapted to the market for refuelling vehicles with ammonia. Obviously, a specific hermetically sealed hose would be necessary to prevent leaks of ammonia during the coupling and decoupling from the tank. It is believed that the dangers associated with the toxicity of ammonia are more manageable than the dangers associated with the flammability of hydrogen.

Ammonia presents safe thresholds of flammability: its auto-ignition temperature is 650 °C. For comparison, gasoline auto-ignites around 450 °C, diesel fuel around 250 °C and hydrogen around 680 °C. Ammonia has a very narrow explosive range: between 15 and 28%. By comparison, gasoline is explosive at low concentrations in air and hydrogen is explosive between 5% and 75%. Neither small leaks nor large spills of ammonia are likely to result in explosion hazard. Ammonia can then be considered as a very safe hydrogen carrier.

The extraction of hydrogen from ammonia is very easy because the synthesis reaction is reversible:

$$2NH_3 \rightarrow N_2 + 3H_2$$

This reaction is called "cracking". The required conditions are a high temperature and catalyst. The fuel-cell technology has an influence over the operating temperature of the ammonia cracker: the membrane of a PEM fuel cell is made of perfluorosulphonic acid, whereas ammonia is an alkali. Therefore, ammonia must be completely cracked into hydrogen and nitrogen, which requires high temperatures of about 900 °C. If the fuel-cell used is an alkaline fuel cell, then the operating temperature can be reduced to 500 °C, thus reducing the temperature requirement, and improving the cracker efficiency. It is estimated [3] that the cracking takes about 20% of the ammonia energy content if the fuel cell is a PEM. Other fuel-cell technologies are expected to

allow higher efficiencies: molten carbonate and solid oxide fuel cells allow using their high temperature exhaust stream to heat up the ammonia and crack it. The cracking energy is then taken from the wasted heat of the fuel cell and does not affect the hydrogen energy efficiency. However, it is not desirable to have high-temperature fuel-cells in a car. High operating temperatures mean that the warm-up thermal energy stored in the stack is lost when the fuel cell is turned off, thus reducing the energy efficiency of the vehicle. This is a particularly sensitive issue in the case of a fuel cell hybrid drive train, where the fuel cell is turned-on and off frequently. It is also a problem for conventional drive trains on short distance trips, which represent a significant share of the miles driven by automobiles in the world.

Because the direct oxidation of ammonia is possible in a low-temperature alkaline fuel-cell [10], it seems reasonable to believe that a cracker can be built that works at the stack exhaust temperature. The probable requirements would be a high-quality catalyst and a re-circulation of the anode or "tail" gas.

The cracking of ammonia is potentially a rapid reaction for two reasons: the synthesis of ammonia is reversible and there is no water-steam to produce. Warm up times are expected to be much lower than reformers. Most ammonia crackers get their heat from the combustion of a small part of the ammonia stream. Because the operating temperature of the ammonia cracker is too low, there is no production of nitrogen oxides. Obviously, there is no production of carbon monoxide or any sulphur oxide. Therefore, the only issue is the poisoning of an acidic fuel-cell (PEM or PAFC) by residual ammonia.

The onboard efficiency ranges anywhere from 80% (PEM fuel cell and cracking at 900 °C) to 100% (high-temperature fuel cell, exhaust stream used for cracking). The well-to-wheel efficiency is much more difficult to evaluate because of the wide variety of sources of hydrogen:

- From natural gas: this method suffers the same inefficiencies that carbon-based carriers do, aggravated by the inefficiency of ammonia synthesis.
- By water electrolysis: water electrolysis has an energy efficiency of 80% (from the electricity to the hydrogen). The efficiency of this route depends then on the efficiency of electricity production. Obviously, electricity produced from natural gas or coal is not desired, so the most likely sources are nuclear power, geothermal power, hydraulic power, solar, or wind power. Hydraulic and wind power are very efficient and totally renewable, but their potential is reduced. Indeed, most of the hydraulic potential has already been harnessed. Geothermal power could prove to be a valuable solution in a few regions such as Iceland. Nuclear power and geothermal power

also suffer from the inefficiency of their thermal engine (steam turbine). However, geothermal power is totally renewable so there is no concern about its extinction. The concerns are then purely economical. The same conclusions hold for solar power. However, in this case, the cost issue might prove fatal.

- From solar sources: by altering the nutrients fed to some algae, researchers have made these produce hydrogen by photosynthesis instead of oxygen. They expect this method to be economically viable [11] because the primary energy (the sun) is free.

In conclusion, once the hydrogen is paid for, the energy efficiency of ammonia follows:

- Ammonia synthesis: 80%
- Transport and storage [3]:   95%
- Cracking: 80% to 100%

Ammonia really conforms to the strict definition of a hydrogen carrier: its synthesis binds hydrogen to nitrogen atoms and the binding material has no energy content. This explains the many benefits of ammonia as a hydrogen carrier.

Ammonia is a valuable option if hydrogen is obtained from renewable sources. In such case, the energy efficiency of hydrogen production is relatively irrelevant because the primary energy (wind, sun, geothermal heat) is free and the wasted part goes back to nature. The only issue is then the cost of producing hydrogen. Ammonia synthesis, transport, storage, and cracking have to be optimized for efficiency. Although the idea of using ammonia as a fuel for fuel-cells is not new [3], it has received very little attention. A steady research effort is likely to optimize the efficiency of the ammonia chain and to enhance the safety of ammonia as a fuel.

HYDRAZINE - Hydrazine is a colourless liquid, with strong ammonia-like odor. Its chemical formula follows:

$$N_2H_4$$

The hydrogen energy content of hydrazine is lower than that of ammonia and its production is very costly. In addition, hydrazine is deadly, even at small doses [10]. Therefore, it shall not be considered for any commercial transportation.

OTHER NON-CARBON BASED COMPOUNDS - Other hydrogen-rich chemicals could be used as hydrogen carriers but none presents a good hydrogen energy density. This is because the binding materials are above nitrogen and carbon in the classification. Some of these chemicals are quickly reviewed below, for they are proposed by research teams or companies as hydrogen carriers.

Alkali hydrides and alkaline metals - These compounds dissociate water upon contact. For instance the reaction of sodium with water is:

$$2Na + 2H_2O \rightarrow H_2 + 2NaOH$$

Obviously, the energy density is low because alkaline metals are heavier than carbon or nitrogen (with the exception of lithium in which cost is a major obstacle). However, these chemicals are not hydrogen carriers. The energy does not come from the hydrogen but from the metal. The cycle can be written as follows:

Hydrogen production: $2Na + 2H_2O \rightarrow H_2 + 2NaOH$

Fuel-cell reaction: $2H_2 + O_2 \rightarrow 2H_2O$

Then the water from the fuel cell is recycled to the hydrogen production. This method of releasing hydrogen is impractical (handling of solid fuels), dangerous (alkaline metals are corrosive to the skin and so are their hydroxides), and costly.

CONCLUSION

The most significant issues influencing the choice of a hydrogen carrier are:

- Well-to-wheel and onboard energy efficiency
- Hydrogen energy density
- Reforming
- Renewability
- Safety

The well-to-wheel efficiency requirement eliminates synthetic gasoline. All fossil carriers are not to be considered because they are not renewable. Hydrazine is too dangerous. The intelligent choice is between methanol and ammonia.

Methanol certainly presents the highest well-to-wheel efficiency. The well-to-wheel efficiency of ammonia is more difficult to assess, partly because until recently ammonia has not been considered for its energy content but as a fertilizer. The production of ammonia by the Haber-Bosch process certainly is less efficient than methanol, though it can be made totally renewable.

It is much easier to extract hydrogen from ammonia than from methanol. The efficiency is expected to be higher and importantly the dynamic characteristics are much better. Methanol requires unrealistic warm-up times and its transient response is slow. In addition, the complexity of the piping in the case of methanol and the cost of the reformer are beyond compare with the simplicity of an ammonia cracker and its absence of requirement for expensive catalysts.

The hydrogen energy density of ammonia is higher than methanol in terms of weight and is similar in term of volume. Considering the tank design will not significantly impair ammonia because of the low-pressure requirement.

The safety issues of ammonia need to be addressed thoroughly in the case of an onboard storage. However, all records are showing that ammonia safety is manageable.

The present analysis did not consider all the parameters influencing the selection of a hydrogen carrier. Market issues and energy policies have at least the same weight as technical issues in this question. It is the authors' conclusion that ammonia will be a very serious technical challenger for methanol. In further publications, we shall investigate thoroughly the ammonia economy and its technical, environmental and social implications.

ACKNOWLEDGMENTS

The authors wish to acknowledge the financial support of the Snead Institute, Georgetown, TX.

REFERENCES

1. "Fuel choices for fuel-cell powered vehicles", Paul J. Berlowitz, Charles P. Darnell, SAE 2000-01-0003
2. "Ammonia fuel: the key to hydrogen-based transportation", James J. Mackenzie, William H. Avery, Proceedings of the 32nd Intersociety Energy Conversion Engineering Conference 1996, Volume 3, pages 1791-1796
3. "Ammonia as a hydrogen carrier for use in fuel-cells", Rune Halseid, Preben J. S. Vie, Rolf Jarle Aaberg, Reidar Tunold, 2000 Nordic Symposium on Hydrogen and Fuel-cell Energy
4. "Societal Impact of Fuel Options for Fuel-cell Vehicles", C.E. Thomas, Brian D. James, Franklin D. Lomax, Jr, Ira F. Kun, Jr, SAE 982496
5. http://www.epa.gov/orcdizux/08-fire.htm
6. http://www.methanol.org
7. http://www.ethanol.org
8. "Fuels for Fuel-cell Powered Vehicles", Sean Casten, Peter Teagan, Richard Stobart, SAE 2000-01-0001
9. "Development of the Jeep Commander 2 Fuel-cell Hybrid Electric Vehicle", Doanh tran, Michael Cummins, Euthemios Stamos, Jason Buelow, Christian Mohrdieck, SAE 2001-01-2508
10. "Fuel-cell Systems Explained" James Larminie, Andrew Dicks, John Wiley & Sons, Chichester, UK
11. www.iahe.org

CONTACT

Sebastien E. Gay: TheMapleLeafForever@yahoo.ca

Jean-Yves Routex: jyroutex@yahoo.com

Mark Holtzapple: m-holtzapple@tamu.edu

Mehrdad Ehsani: Ehsani@ee.tamu.edu, Department of Electrical Engineering, Texas A&M University, College Station TX 77845, (979) 845-7582

DEFINITIONS, ACRONYMS, ABBREVIATIONS

ATR: Auto Thermal Reforming

LHV: Lower Heating Value

PAFC: Phosphoric Acid Fuel-cell

PEM: Proton Exchange Membrane or Polymer Exchange Membrane

POX: Partial Oxidation reforming

SR: Steam Reforming

Compressed Hydrogen Storage for Fuel Cell Vehicles

Monterey R. Gardiner, J. Cunningham and R. M. Moore
University of California, Davis

ABSTRACT

Near term (ca. 2005) Fuel Cell Vehicles (FCVs) will primarily utilize Direct-Hydrogen Fuel Cell (DHFC) systems. The primary goal of this study was to provide an analytical basis for including a realistic Compressed Hydrogen Gas (CHG) fuel supply simulation within an existing dynamic DHFC system and vehicle model.

The purpose of this paper is to provide a tutorial describing the process of modeling a hydrogen storage system for a fuel cell vehicle. Three topics were investigated to address the delivery characteristics of H_2: temperature change (ΔT), non-ideal gas characteristics at high pressures, and the maximum amount of hydrogen available due to the CHG storage tank effective "state-of-charge" (SOC) -- i.e. how much does the pressure drop between the tank and the fuel cell stack reduce the usable H_2 in the tank.

The Joule-Thomson coefficient provides an answer to the expected ΔT during expansion of the H_2 from 5000 psi to 45 psi. The temperature change, however, was found to be negligible with regard to fuel cell thermal control issues. The departure from the ideal gas law was evaluated using the Redlich-Kwong equation of state. This provides the most accurate description of the PV=nRT relationship for simple equations of state. The pressure drop must be calculated from a number of factors such as: pipe material, bends within the pipe, length of pipe, and the number of valves (pressure regulators) the gas must pass through. The pressure drop and initial tank volume are used to calculate the remaining hydrogen – and hence the effective SOC for the CHG storage tank.

Primary results for the CHG fuel systems considered include: the temperature shows a change of ca. 13 K, the initial volume was calculated to be 264 Liters (69.7 Gallons) for 6 kg of H_2 stored at ambient temperature and 5000 psi, and the usable H_2 depends on the pressure drop within the specific fuel system design. The system was used within an existing dynamic FCV model for fuel cell vehicle analyses.

INTRODUCTION

Direct hydrogen fuel cell systems supplied by compressed hydrogen are generally more efficient and inexpensive (for similar vehicle platforms) when compared to the different fuel cell systems being investigated. The alternatives include indirect methanol, indirect hydrocarbon, and direct methanol fuel cell systems. The indirect systems reform methanol or a hydrocarbon liquid on board the vehicle to produce hydrogen to feed into a fuel cell stack. The fuel cell stack in these systems is nearly identical to the one used in a direct hydrogen system. Direct methanol systems inject liquid methanol solution directly into the fuel cell stack.

One may ask, why go to all the trouble of developing these alternatives? These systems are more expensive to produce as a result of the higher capital costs associated with a reformer at each vehicle versus a centralized fueling station. There are also reliability issues that need to be addressed due to the added complexity of onboard reformation systems. In addition, they produce emissions at the vehicle level.

The short and much simplified answer is on-board hydrogen storage. Direct hydrogen systems require on-board storage of hydrogen in one form or another. The possibilities include CHG, liquid hydrogen (LH) or hydrogen that must be combined with some exotic material. The material alternatives include metal hydrides, sodium borohydride slurry[1] or carbon nanotubes (CNT). The main obstacles related to storage include the needed volume to attain a standard range for a vehicle (380 mi for the PNGV requirement), low weight for the storage system, higher than average safety requirements to assist with the FCV consumer acceptance and most importantly, low cost.

Taking into account the limitations listed above, compressed hydrogen vehicle systems have come closest to providing an answer except for the tank volume problem. The majority of prototype FCVs use

[1] S. Amendola etal, "Fuel breakthrough offers a world of opportunity", Sustainable Development International p. 2.

either LH or CHG. Metal hydrides require high temperatures and the total system weights are not obvious. Metal slurries may require complicated recycling of the material as hydrogen is extracted and expensive catalysts may be required as well. Storage densities of 7 wt% have been claimed for both high temperature metal hydrides from Ovonic Battery Company (OBC)[2] and a sodium borohydride slurry produced by Millennium Cell[3]. The wt % from Millennium Cell does not include total system weight and it is unclear if the OBC wt% estimate includes the "Total System" weight. The quantity of hydrogen being stored in CNTs is still too small and has been limited to laboratory settings only at this point[4].

CONSTRAINTS Compressed hydrogen was chosen as the first storage method to be analyzed with respect to modeling "hydrogen storage" here at UC Davis for the Direct Hydrogen Fuel Cell Vehicle Model. The investigation was begun by trying to put constraints on the system and determining what aspects of hydrogen storage would be most likely to affect the model and therefore need to be studied more completely. The initial and most important constraints were determining the likely quantity and pressure at which the hydrogen would be stored.

HOW MUCH HYDROGEN? The quantity of hydrogen required was estimated using the UC Davis Fuel Cell Vehicle model. The auto manufacturers estimate fuel economy using both the Federal Urban driving cycle (FUDS) and the Highway driving cycle (HIWY). These values are then combined in a weighted equation.

The combined fuel economy result from the model was used to estimate the fuel requirement. A range of 380 miles (PNGV requirements) called for approximately 5 gallons of gasoline. The amount of hydrogen needed for this range was calculated with the energy content of a gallon of gasoline (121,330kJ/gallon) and a gram of hydrogen (120kJ/gram). The quantity of hydrogen is ~5 kg. This value was bumped up to 6 kg for use in modeling. This was done to account for state of charge (SOC) issues discussed later.

WHAT PRESSURE TO STORE THE HYDROGEN? The next step was to determine how large a volume this would require. The pressure at which the gas is stored affects this volume. The pressure at which the hydrogen could be stored ranges between three possible values: 3600 psi, 5000 psi and 10,000 psi. Many tanks have been constructed for compressed natural gas vehicles at

3600 psi. However, hydrogen is less dense and therefore higher-pressure tanks have been developed at 5000 psi. Research is being conducted to develop 10,000 psi tanks. IMPCO has publicly stated this.[5] 5000 psi tanks were taken as a compromise between the lower pressure tanks which would require a large volume and 10,000 psi tanks which are still in development.

TOPICS TO BE INVESTIGATED

With both pressure and quantity of hydrogen required, the next step was to determine what exactly would need to be added to the model. Many aspects of the storage system were considered for modeling. The possibilities were selected by determining the extent of their impact on the model. Those topics that were significant, therefore, required modeling. To correctly model the hydrogen storage in the vehicle model, three characteristics needed a thorough investigation: Δ T of the hydrogen gas, non-ideality of hydrogen gas and the associated pressure drop of the hydrogen gas. These factors are important to the proper operation of a fuel cell system.

To calculate the volume of the hydrogen gas and its behavior as it is expanded, a state law was required. The Ideal gas law could have been used, however the high pressures require a more complex equation of state. Next, with regards to the SOC, one must know what pressure drop exists between the tank and the fuel cell. A large pressure drop would significantly reduce the amount of "usable hydrogen" and would thus require a larger than expected tank to achieve the same mileage. The change in temperature was determined using data tables and an understanding of the Joules-Thompson Coefficient.

TEMPERATURE CHANGE

The temperature of the expanding hydrogen will actually increase slightly as shown by the Joules-Thomson Coefficient. The temperature increase from the expansion is only about 13K when expanding from 5000 psi to 45 psi. The temperature change will be neglected, however, and the basis of this assumption is verified below.

JOULE-THOMSON COEFFICIENT Hydrogen gas heats up when under-going a throttling process. The gas will only begin to get colder below negative 68C (the inversion state for hydrogen). The basis for this concept can be shown experimentally. Imagine a porous plug through which a gas may pass.

During steady state, the gas enters an apparatus at a specified temperature T_1 and pressure p_1 and expands through a plug to a lower pressure p_2, which is controlled by an outlet valve. The temperature T_2 at the exit is

[2] D. Corrigan, "Metal Hydride Technologies for Fuel Cell Vehicles", Commercializing FCVs 2000 4-12-00, Berlin Germany p. 1

[3] S. Amendola etal, "A safe, portable, hydrogen gas generator using aqueous borohydride solution and Ru catalyst", Int. J. Hydrogen Energy 25 (2000) p.1

[4] A.C Dillon, etal, "Carbon Nanotube Materials For Hydrogen Storage", Proceedings of the 2000 DOE/NREL Hydrogen Program Review NREL/CP-570-28890 5/8/00

[5] www.prnewswire.com/cgi-bin/stories.pl?ACCT=104&STORY=/www/story/11-09-2000/0001361534&EDATE=

measured. Because the gas undergoes a throttling process, the exit states are fixed by p_2 and T_2 and have the same value for specific enthalpy as at the inlet and $h_2=h_1$. By lowering the output pressure in incremental steps, one can create isenthalpic curves. The slope of the curve at any state is the Joule-Thomson coefficient at that state. States where the slope is zero is called an inversion state.[6]

$$-\frac{1}{Cp}\left[v - T\left(\frac{\partial v}{\partial T}\right)_P\right] = \left(\frac{\partial T}{\partial P}\right)_h = \mu \qquad (1)$$

TEMPERATURE CALCULATION The actual temperature was calculated using enthalpy tables for hydrogen.[7] The enthalpy value of 1890.047 BTU/lb was listed for hydrogen at 5000 psi and an initial temperature of 300K. This enthalpy was used to determine the temperature at a pressure of 4000 psi. The temperatures for the following pressures were plotted at a constant enthalpy value, and a cubic equation was curve-fitted. The pressure values were: 5000, 4000, 3500, 3000, 2400, 1000, 45 and 14.7 psi. The resulting equation is shown in the figure 1. Where Y is temperature in Kelvin and X is pressure in psi. Temperature changes with pressure drop and is shown by the graph. (Figure 1)

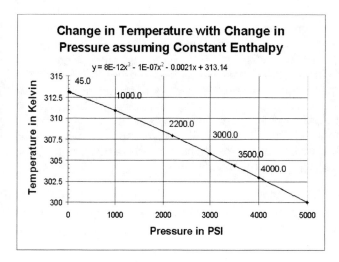

Figure 1 Temperature Increase Due to Expansion

PRESSURE CHANGE

At higher pressures the ideal gas law is not precise enough to use. Refer to Figure 2 below.

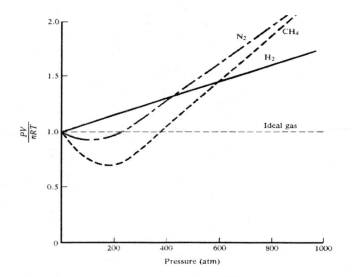

Figure 2 Non-Ideality of H_2 at High Pressure

At 340 atm (5000 psi), the compressibility factor Z is off by as much as 25%. There are two types of corrections that should be applied to the ideal gas law at higher densities corresponding to higher pressures. These corrections are associated with the forces between molecules. The first correction is concerned with forces that are repulsive over short distances. The second correction deals with the attractive forces that hold molecules together and thus reduce the pressure they exert on outside walls. This attraction depends on pairs and should be proportional to the square of the number of molecules per unit volume.[8]

Even though the Redlich-Kwong equation is somewhat more difficult to manipulate mathematically than the van der Waals equation, it is more accurate, particularly at higher pressures. In recent years several modified forms of this equation have been proposed to achieve improved accuracy. The two-constant Redlich-Kwong equation performs better than some equations of state having several adjustable constants[9] and is the most accurate equation of state that does not involve polynomials of higher degree than 2.

The Redlich-Kwong Equation of state specifies:

$$p = \frac{\overline{R}T}{\overline{v} - b} - \frac{a}{\overline{v}(\overline{v} + b)T^{1/2}} \qquad (2)$$

[6] Michael J. Moran et. al. Fundamentals of Engineering Thermodynamics, John Wiley & Sons 1992 p.498
[7] McCarty, D.R, "Hydrogen Technological Survey-Thermophysical Properties", Cryogenics Division, Institute for Basic Standards National Bureau of Standards, Boulder Colorado. Scientific and Technical Information Office, Washington, DC 1975

[8] David W. Oxtoby et. al. Principles of Modern Chemistry, University of Chicago, 1987 p. 94
[9] M.J. Moran and H. Shapiro, "Fundamentals of Engineering Thermodynamics", 1995 p. 491

$$a = .42748 \frac{\overline{R}^2 T_c^{5/2}}{p_c} \qquad (3)$$

$$b = .08664 \frac{\overline{R} T_c}{p_c} \qquad (4)$$

Table 1 Critical Values for Use in the Redlich-Kwong Equation

	Value	Units
Critical Temp	33.18	K
Critical Pressure	1315	kPa

The Solver Function in Excel was used with the Redlich-Kwong equation to calculate the tank volume at 5000 psi and 300K for 6 kg of hydrogen. An iterative process was required because the specific volume cannot be easily solved directly. The resulting volume was 264.16 liters.

PRESSURE DROP

Now that the volume has been determined, we must determine what pressure drop exists between the tank and stack to see if this volume and associated mass will be sufficient for the PNGV range of 380 miles. There are pressure losses due to the regulator and due to the head loss in the pipe between the tank and the stack inlet. The two-stage pressure regulator is used to adjust the high pressure of the hydrogen in the tank down to a lower value desired in the fuel cell stack and the pressure will be semi-continuous, The hydrogen looses energy as it travels through the various geometries within the piping. The pressure drop is associated with three coefficients, equivalent length L/D, resistance coefficient K, and the "flow coefficient Cv"[10]. The pressure drop for different system configurations can be determined from these three coefficients. However, the two-stage pressure regulator will cause the majority of the pressure drop. Quantum Technologies, a division of Impco, gave a value of 480 psi. This pressure drop can be used to calculate the residual amount of hydrogen in the tank that will not be usable by the fuel cell stack.

Below is a table of expected pressure drops and the resulting hydrogen loss as a percentage assuming a tank volume of 264.16 liters and an initial mass of 6 kg of hydrogen

---

Table 2 **Pressure Drop and Lost H$_2$**

Pressure Drop in PSI	Grams of H$_2$	% of usable H$_2$ lost
380	563.12	9.39
480	711.31	11.86
580	859.50	14.32

CONCLUSION

The mass of hydrogen and the pressure has been calculated for a typical "mid-size" FCV. These values would, of course, be lower for a smaller subcompact vehicle and higher for larger vehicle platforms such as SUVs. A value of 264.16 liters (69.78 gal) was calculated. A typical vehicle has a gas tank between 10 and 20 gallons. Therefore, a nontrivial redesign of current vehicles or a dedicated "ground-up" direct hydrogen design will be required to accommodate a fuel storage system that is 3.5 times larger. Alternatively lower weight vehicles could be developed that would lower the required amount of hydrogen required to be stored on board as a result of higher vehicle fuel economy.

The temperature change was found to be an increase of 13K. This is counterintuitive for the temperature change of an expanding gas. This temperature increase is actually helpful, however, considering the fact that a fuel cell requires humidified hydrogen to function properly. The increase in temperature allows more water to saturate the incoming flow of hydrogen. This increase is minimal compared to the advantage of recirculated hydrogen that is already humidified and much closer to the required temperature (~80 C).

To travel 380 miles using a combined FUDs and HIWY driving cycle, ~5 kg of hydrogen was calculated using the vehicle model. However, the system was sized to accommodate 6 kg. The next step for this research will be to model an alternative storage system that uses metal hydrides. The requirements for a metal hydride system will be much more complicated than those of the compressed hydrogen system due to the high temperature nature of the material and relative newness of the technology.

REFERENCES

1. S. Amendola etal, "Fuel breakthrough offers a world of opportunity", Sustainable Development International

---

[10] Engineering Dept of Crane Co. "Flow of Fluids Through Valves, Fittings, and Pipe", Technical paper No. 410, Joliet IL. 1998. p. 2-8

2. D. Corrigan, "Metal Hydride Technologies for Fuel Cell Vehicles", Commercializing FCVs 2000 4-12-00, Berlin Germany

3. S. Amendola etal, "A safe, portable, hydrogen gas generator using aqueous borohydride solution and Ru catalyst", Int. J. Hydrogen Energy 25 (2000)

4. A.C Dillon, etal, "Carbon Nanotube Materials For Hydrogen Storage", Proceedings of the 2000 DOE/NREL Hydrogen Program Review NREL/CP-570-28890 5/8/00

5. www.prnewswire.com/cgibin/stories.pl?ACCT=104& STORY=/www/story/11-09-2000/0001361534&E DATE=

6. Michael J. Moran et. al. Fundamentals of Engineering Thermodynamics, John Wiley & Sons 1992

7. McCarty, D.R, "Hydrogen Technological Survey-Thermophysical Properties", Cryogentics Division, Institute for Basic Standards National Bureau of Standards, Boulder Colorado. Scientific and Technical Information Office, Washington, DC 1975

8. David W. Oxtoby et. al. Principles of Modern Chemistry, University of Chicago, 1987

9. M.J. Moran and H. Shapiro, "Fundamentals of Engineering Thermodynamics", 1995

10. Engineering Dept of Crane Co. "Flow of Fluids Through Valves, Fittings, and Pipe", Technical paper No. 410, Joliet IL. 1998

DEFINITIONS, ACRONYMS, ABBREVIATIONS

Rbar = 8314.34kPa*m^3/(kmol*K)

T = Temperature in K

v = Specific Volume in m^3/kmol

T_c = Critical Temperature for H_2

p_c = Critical Pressure for H_2

1999-01-1320

On-board Hydrogen Generation for PEM Fuel Cells in Automotive Applications

Ian Carpenter, Neil Edwards, Sue Ellis, Jack Frost, Stan Golunski, Nick van Keulen, Mike Petch, John Pignon and Jessica Reinkingh

Johnson Matthey

ABSTRACT

In the search for clean and efficient power, PEM fuel cells have been identified as the technology that can meet our future needs for transport applications. Hydrogen-powered PEM fuel cell vehicles are perceived to give the ultimate advantage, but the complications involved with hydrogen storage and refuelling, as well as the lack of infrastructure call for a different solution. In the near term, this is almost certain to be the on-board generation of hydrogen from a readily available fuel.

At Johnson Matthey, a novel modular reformer (HotSpot^{TM}) has been developed for methanol, and has been demonstrated to have many of the qualities that are required for automotive applications. Auxiliary technologies for CO removal (Demonox) and aftertreatment have also been developed, and integrated with the reformer to form 20 kWe processor, which is currently undergoing brass-board testing. Although not designed for multi-fuel operation, HotSpot is versatile, and can be adapted for other fuels apart from methanol. A 5 kWe natural-gas processor has recently been demons-trated, in tests that have included successful coupling with a PEM fuel cell.

INTRODUCTION

PEM fuel cells are low-temperature electrochemical reactors, which generate electric power by consuming hydrogen as a fuel and oxygen or air as an oxidant [1]. For smaller and private vehicles, on-board hydrogen storage has proven bulky, and suffers from complicated "refuelling" issues. As an alternative approach, several research groups have concentrated on developing technology for on-board production of a hydrogen-rich reformate from more conventional fuels [2], i.e. alcohols, natural gas, and higher hydrocarbons. In so doing, they have identified additional requirements associated with providing suitable reformate to a fuel cell, like CO removal, anode exhaust combustion, heat integ-ration, and air compression.

The demands on each fuel cell system com-ponent for transport applications are uncom-promising. Small, mini-mal start-up time, fast dynamic response, light-weight, reliable, durable, efficient, and cheap, are the necessary properties to realise commercialisation of fuel cell powered vehicles. It is a great challenge to develop each fuel cell system component, and in particular, to ensure that the sum of the components still complies with the stringent demands made of the total system [3]. This paper summarises our progress on the development of various system components and their integration.

ON-BOARD HYDROGEN GENERATION

A PEM fuel cell system, which includes hydrogen generation from any type of carbon-based fuel, comprises several interdependent process stages (Figure 1). The first step, fuel reforming, not only requires heat input from the afterburner, but also sets the requirements for the CO removal stage. The performance of the fuel cell depends on the quality of the reformate, and dictates the anode off-gas afterburner operation.

FUEL REFORMING – There are two well-known routes to produce hydrogen from carbon-based fuels [4], i.e. steam reforming, and partial oxidation. Steam reforming is a slow endothermic process, in which hydrogen is released from both the fuel and the steam. Partial oxidation is a fast exothermic process, converting fuel to hydrogen by reaction with oxygen from a controlled amount of air. Both processes can generate mainly CO as the carbon-containing product. However, with careful design of the catalyst and reactor, the CO can be used to generate more H_2 by the water-gas shift reaction. Equations 1 and 2 show the reaction energetics of the above processes for methane, with water vapour as the reactant ($H_2O_{(l)}$ $H_2O_{(g)}$ ~ +44 kJ/mol).

$$CH_4 + H_2O_{(g)} \rightarrow CO + 3H_2 + 206 \text{kJ/mol} \qquad \text{(Eq. 1a)}$$

$$CO + H_2O_{(g)} \rightleftarrows CO_2 + H_2 - 41 \text{kJ/mol} \qquad \text{(Eq. 1b)}$$

$$CH_4 + \tfrac{1}{2}O_2 \rightarrow CO + 2H_2 - 33 \text{kJ/mol} \qquad \text{(Eq. 2a)}$$

$$CO + H_2O_{(g)} \rightleftarrows CO_2 + H_2 - 41 \text{kJ/mol} \qquad \text{(Eq. 2b)}$$

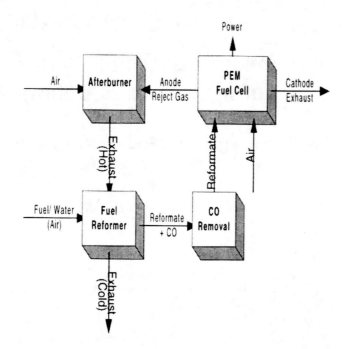

Figure 1. Interdependent process stages in PEM fuel cell system

Johnson Matthey's **HotSpot**^{TM} reformer [5] combines both processes, using the exothermic partial oxidation to drive the steam reforming reaction, which is generally called autothermal reforming. Conventional autothermal processes often have two separate reactor stages, transferring the heat generated in the combustion section to the steam reforming section. HotSpot technology distinguishes itself from other autothermal reforming processes, in that both reactions occur in a single catalyst bed, practically decreasing the heat transfer distance to zero.

Partial oxidation contributes to HotSpot's compactness, fast start-up time and dynamic response. Steam reforming raises the reactor's hydrogen concentration and efficiency, and realises excellent catalyst durability. HotSpot can be operated under various thermal conditions, i.e. the ratio of partial oxidation to steam reforming performed in the reactor can be varied making the net heat balance over the reactor exothermic or endothermic.

The standard HotSpot module for methanol-processing is presently a 0.25 liter canister. Our most advanced version of this has an internal mass of 120 g, of which 30% is catalyst. Its maximum output is 1100 standard litres of hydrogen per hour, in a reformate which when dried contains 55% H_2 and 2.5% CO. The methanol conversion is >99%, with no evidence for the formation of any organic species, apart from traces of CH_4. The start-up characteristics of the HotSpot reformer are excellent reaching 75% of the maximum output after 20 seconds upon feed-

ing the vaporised feed, while after 50 seconds the full product is obtained.

The HotSpot module has been scaled-up to produce hydrogen (>0.2 mols/s) for a 20 kWe PEM fuel cell system, which is illustrated in Figure 2. In this system methanol and water are fed to the reformer by automotive liquid injectors. The use of injectors provides an easy and accurate way to control the feed, and enhances the vaporisation as well.

Figure 2. Methanol HotSpot reformer with methanol and water injectors. (Maximum hydrogen production $\geq$ 19 m^3/h)

Although HotSpot has been developed with methanol as a fuel, it can also be applied to other fuels like higher hydrocarbons, LPG, and natural gas (NG). Recently we have developed a prototype NG processor. This first prototype produces hydrogen for up to 5 kWe from NG, and includes sulphur and CO removal steps. The dry product of this multi-step unit contains approximately 43% H_2, 40% N_2, 16% CO_2, 1% CH_4, and less than 10 ppm CO. This corresponds with a conversion of about 94%.

Similar to the methanol reformer, the NG HotSpot reformer itself produces a reformate with only 1-2% CO. The start-up of the unit has not yet been optimised, but initial results are promising, yielding suitable reformate for a PEM fuel cell within 12 minutes. Figure 3 demonstrates the performance during several hours of steady state operation, including the GC measurements and the NG conversion.

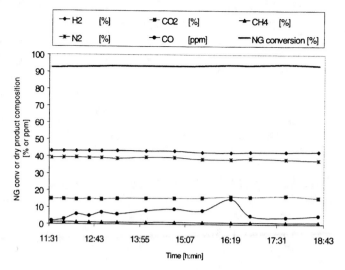

Figure 3. Conversion and GC measurements of product during steady state operation of NG fuel processor

NG is generally not regarded as a suitable fuel for private cars, but it could be considered for buses or other fleet vehicles. Compared to the conventional fuels like diesel and gasoline, NG has the advantage of being a relatively clean fuel, where at present most of the sulfur in the NG is added as an odorant. Compared to methanol, NG is fully accepted by the public and already has a widespread infrastructure. However, NG is a gas and requires compression for storage on-board the vehicle. For NG to become a fuel of choice in transport applications, the fuel processing technology needs to be developed into a compact, reliable, responsive and efficient system.

CO REMOVAL – Independent of the fuel reforming method, in the process of hydrogen generation of carbon-based fuels some carbon monoxide is produced. The dry CO concentration in the reformate typically varies between 0.5-3 during steady state operation of most reformers, and can reach values between 5-10% during start-up or transients. The PEM fuel cell, due to its low operating temperature, can not tolerate high levels of CO as it poisons the catalyst. Although recent advances in anode catalyst technology have made them more tolerant, CO levels of < 10 ppm in the reformate are still preferable [6-8].

Catalytic or physical CO removal approaches can be considered, i.e. conversion of CO into components that are not harmful to the fuel cell or separation of CO from the hydrogen fuel. Pd-membranes have the advantage of selectively separating the hydrogen from the other components in the reformate stream, producing pure hydrogen. In Pd-membranes the driving force is determined by the hydrogen partial pressure difference over the membrane. This means that, particularly for more dilute hydrogen mixtures as produced by partial oxidation or HotSpot reforming, the inlet pressure needs to be relatively high. For a 40% (wet) H_2 mixture at least a 10 bar total pressure difference is required to recover 80% of the hydro-

gen. Hence, a multi-stage compressor would be required to compress the air to the reformer, which will have an adverse impact on the system complexity and efficiency. The pressure requirements, as well as the relatively high operating temperature (300-400°C) cause the Pd-membrane to be incompatible with HotSpot technology.

Catalytic carbon monoxide removal options from a hydrogen-rich stream include the water-gas shift reaction (3), selective methanation of carbon monoxide to methane (4), and selective oxidation of carbon monoxide (5).

$$CO + H_2O_{(g)} \rightleftarrows CO_2 + H_2 \qquad (3)$$

$$CO + 3H_2 \rightarrow CH_4 + H_2O_{(g)} \qquad (Eq.\ 4a)$$

$$CO_2 + 4H_2 \rightarrow CH_4 + 2H_2O_{(g)} \qquad (Eq.\ 4b)$$

$$CO + 0.5O_2 \rightarrow CO_2 \qquad (Eq.\ 5a)$$

$$H_2 + 0.5O_2 \rightarrow H_2O_{(g)} \qquad (Eq.\ 5b)$$

The water-gas shift reaction would appear the most favorable since it produces hydrogen. However, for the hydrogen-rich streams generated in fuel reformers it is not possible to reduce the carbon monoxide concentration to levels acceptable for the fuel cell as the reaction is limited by the thermodynamic equilibrium.

Selective methanation is only an option when the CO concentration in the feed stream is sufficiently low (preferably less than 0.5%) so that hydrogen losses are minimised. For a 100% selective process, the removal of each molecule CO consumes 3 molecules of hydrogen (4a). Methanation of CO and CO_2 (4b) are both exothermic. If CO_2 is present in the reformate stream, it is important to carefully control the reactor temperature as this system could potentially run-away, generating nothing but methane and water.

Having assessed the advantages and disadvantages of the various options we ave developed a catalytic CO removal called Demonox. Demonox is based on selective oxidation, using the concept that CO can be removed most effectively in several small consecutive catalyst beds, each operating within a narrow 'window' of optimum conversion and selectivity. By contrast, in conventional clean-up reactors (containing a single bed of either CO-methanation or CO-oxidation catalyst), the heat of reaction and the removal of CO can force the reactor into a non-selective operating regime.

Since first being demonstrated in a micro reactor three years ago, Demonox has been scaled-up to match the output of each of our current HotSpot processors, which range in size from 2-20 kWe. Typically, the reformate produced by these processors contains 1-3% CO (based on dry analysis), which a Demonox unit will reduce to sub-10 ppm. During the course of development, we have substantially improved the selectivity of the units, and therefore reduced wastage of hydrogen. Initially, the ratio for H_2 lost per CO removed was >2, but this has been decreased to <1 in our most recent unit.

Each catalyst in a Demonox unit responds rapidly to changes in throughput, allowing CO-control to be maintained during transients. During start-up, the individual catalyst beds take different lengths of time to reach a steady-state, with the whole unit taking in excess of 20 minutes to stabilise. However, this does not mean that CO-removal is ineffective during this time. For a Demonox unit coupled with a HotSpot methanol-processor, the highest exit CO-concentration that we observe during cold start-up was <20 ppm (Figure 4).

Although designed to remove the CO generated by HotSpot processors, Demonox has also been adapted for use with other fuel processors. For example, a 25 kWe clean-up unit has been custom-built for coupling with the Epyx flexi-fuel processor (Figure 5). In tests at Epyx, the Demonox unit achieved the agreed CO target of sub-10 ppm under all steady-state conditions at which it was tested, irrespective of whether the processor was fuelled with ethanol or gasoline. The high selectivity of the unit to CO meant that its impact on the efficiency of the Epyx processor was very slight, reducing it by less than 2%.

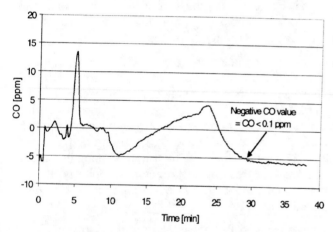

Figure 4. CO break through during start-up of 7 kWe methanol HotSpot + demonox processor

Figure 5. Demonox system tested with Epyx fuel processor (L = 53.9", W = 10", H = 10.4")

AFTERBURNER – Unlike the other catalytic components, the afterburner does not have just one function. By operating a fuel cell with a reformate feed, only a certain fraction of the hydrogen reacts on the fuel cell anode.

The rejected hydrogen can be combusted on a catalytic burner, creating a hot exhaust stream. Striving for a high system efficiency, the hot exhaust stream is used to pre-heat the feed to the reformer, allowing HotSpot to perform more steam reforming. During start-up, the catalytic burner is used to combust methanol with air. This unit combusts methanol instantaneously at ambient temperature, and so the hot exhaust product can be led through HotSpot's feed evaporation section, contributing to a fast start-up time.

FUEL-PROCESSING SUBSYSTEM – Our largest subsystem to date is a combination of a 20 kWe HotSpot methanol-reformer, Demonox plus an Afterburner, which was constructed for the EU-supported CAPRI project (in which the partners are ECN, JM, Volkswagen and Volvo). It was designed to achieve the performance targets of

- steady-state output : H_2-formed/CH_3OH is 2.3
- transient response : 10%-90% of maximum output in 10 s
- CO : sub-20 ppm of dried reformate.

The subsystem has been incorporated into the brass board fuel cell facility at ECN, where it is currently undergoing 'real life' testing. The complete fuel cell system will then be transferred to Volkswagen, where it will be integrated into a battery/fuel cell hybrid vehicle (VW Golf).

In principle, the 'catalyst loop' shown in Figure 1 is generic to many fuel cell systems. In practice, the type of fuel cell used will indeed be common to a variety of systems, but the fuel-processing subsystem will need to be carefully tailored, particularly to different fuels. Although a multi-fuel subsystem may seem highly desirable, the widely differing control and heat management issues are likely to make uniform performance for all fuels an impossibility. We anticipate, therefore, that our subsystems will continue to be created from our building-block technologies of HotSpot, Demonox and Afterburner, but that the individual components will be designed for the intended application.

FUEL CELL SYSTEM EFFICIENCIES – Fuel cell system performances and efficiencies depend on the performance of the individual components, as well as on the total system integration. During generation of electricity from a fuel with a PEM fuel cell system a series of efficiency losses should be expected.

• In the hydrogen production step losses are caused by fuel combustion instead of partial oxidation, by the production of CO, through heat losses, and due to the need to raise additional steam (more than stoichiometrically required).

• A significant loss occurs in the fuel cell itself, most of which is due to over-potential required to activate the cathode. Other losses in the fuel cell are due to cell resistance, anode activation, and at the higher current densities due to mass transport resistances.

- To operate the fuel cell system parasitic power is consumed by the control system and by auxiliary equipment, like pumps and a compressor.
- To generate useful electricity the DC power from the fuel cell is converted to AC power by an inverter.

An adequate definition of the overall fuel processor efficiency needs to account for the heat generated by the rejected hydrogen, and can be defined as the lower heating value of the hydrogen converted on the fuel cell anode, divided by the lower heating value of the fuel fed to the reformer. The hydrogen converted on the fuel cell anode is the product of the total hydrogen generated in the fuel processor and the anode hydrogen utilisation (AHU).

$$FP_{Efficiency} = \frac{LHV_{H_2} \cdot H_2 \ utilised \ on \ anode}{LHV_{CH_3OH} \cdot FP \ fuel \ feed}$$

$$= \frac{LHV_{H_2} \cdot total \ H_2 \ produced \cdot AHU}{LHV_{CH_3OH} \cdot FP \ fuel \ feed}$$

The efficiency of our methanol HotSpot reformer is 95%, while the efficiency of the CO clean-up is 94%. With an expected AHU of 85%, the overall fuel processor efficiency is 76%. The fuel processor efficiency of our NG reformer is at present still lower. We expect to raise this to 70-75% range in the near term through optimisation of the reformer air to fuel to steam feed ratio and the system heat integration.

PEM Fuel cell efficiencies can vary from 40 to 65%, depending on the operating conditions and the cell/ stack design. The auxiliary power consumption strongly depends on the system's operating pressure, since the compressor is a major parasitic power consumer. However, recently good progress has been made on the development of "ambient pressure" PEM fuel cells. The parasitic power losses are generally found in the range of 5-15%. Assuming a DC/AC inverter efficiency of about 95%, an average overall methanol to electricity efficiencies of $76\% \times 55\% \times 90\% \times 95\% = 36\%$ can be expected, but potentially efficiencies of 45% can be reached. Probably the biggest challenge in realizing efficiencies greater than 40% is to improve the fuel cell's performance at lower operation pressures. This will decrease the parasitic power consumption but requires careful system heat and water management.

CONCLUSIONS

The methanol HotSpot reformer has many of the qualities required of hydrogen generation for fuel cells in automo-tive applications. It is compact, fast starting, efficient and responsive, but requires to be coupled to a CO removal unit. We have demonstrated that our Demonox CO removal system can reliably reduce the CO from concentrations of 3% to less than 10 ppm, while during start-up maximum CO levels of less than 20 ppm were observed.

To complete the fuel processor subsystem HotSpot and Demonox are coupled to our catalytic afterburner, closing the loop between the fuel cell and the reformer. The catalytic afterburner increases the system efficiency by feeding the generated heat back into the fuel processor. Additionally, the afterburner ensures that the system does not produce harmful emissions.

An important point of discussion and disagreement has been on the fuel that should be used for automotive applications. HotSpot technology is not limited to methanol, and we have built and operated a 5 kWe NG fuel processor including sulphur removal and Demonox. We are continuously developing HotSpot technology, designing catalysts and reactors that will allow us to generate hydrogen from alternative fuels.

ABOUT THE AUTHOR

Jessica Reinkingh is a senior engineer in Fuel Cells, Catalytic Systems Division at Johnson Matthey, Wayne, PA (P/F: (610)-341-8533/(610)-341-3495, reinkj@mat-they.com), providing technical support to JM's fuel cell products in North America. Previously she worked at Johnson Matthey's Technology Center, Reading, England where she was involved in hydrogen generation and fuel cell system integration for almost three years. She received a Chemical Engineering degree (Ir) from the University of Groningen, and completed a two year graduate program "Process and Equipment Design" at Delft Technical University, both in the Netherlands.

REFERENCES

1. TR Ralph and GA Hards, Powering the cars and homes of tomorrow, *Chemistry & Industry*, 4th May 1998, pp 337-342

2. Status and Prospects of Fuel Cells as Automobile Engines, Fuel Cell Technical Advisory Panel, Prepared for CARB, Sacremento, July 1998

3. Review of the Research Program of the Partnership for a New Generation of Vehicles, Third Report, National Academy Press, Washington DC, 1997

4. MA Pena, JP Gomez and JLG Fierro, New catalytic routes for syngas and hydrogen production, *Applied Catalysis A : General*, 144, pp 7-57, 1996

5. N Edwards, S Ellis, J Frost, S Golunski, N van Keulen, N Lindewald, J Reinkingh, On-board hydrogen generation for transport application: the HotSpot[TM] methanol processor, *J of Power Sources*, 71, pp. 123-128, 1998

6. D Wilkinson and D Thompsett, and S Cooper, A Gunner, G Hoogers, and D Thompsett in New materials for fuel cell and modern battery systems II, (Eds. O Savagodo and P Roberge); ETSU F/02/00014/REP/1, ETSU F/02/00092/REP

7. T Ralph, G Hards, J Keating, S Campbell, D Wilkinson, M Davis, J StPierre, M Johnson, Low Cost Electrodes for Proton Exchange Membrane Fuel Cells, *J Electrochem Soc*, Vol 144, No 11, 1997

8. W Ernst, Reformate fuelled PEM Fuel Cells, presented at SAE Fuel Cells for Transportation TOPTEC, March 18, 1998

2001-01-1918

An Analysis of Hydrogen Production from FT Liquids for use in Fuel-Cell Systems

David Edlund, William Pledger and Brad Turnbull
IdaTech, LLC

Branch Russell
Syntroleum Corp.

ABSTRACT

Synthetic saturated hydrocarbons produced from Fischer-Tropsch (FT) processes offer potential significant advantages as fuels for fuel cell engines. Additionally, FT synthetic fuels, unlike petroleum-derived liquid fuels, are produced absent of sulfur or metals. Elimination of fuel clean-up processes to remove these threats to fuel cell systems offers the potential of significant capital and operational savings. Additionally, the flexibility of synthetic fuel production offers the potential for optimization of synthetic fuel-cell fuels for use in fuel processors.

To better assess the hydrogen carrying capacity and reformability of saturated FT fuels in a fuel cell system under development for commercial applications, IdaTech tested four synthetic fuels provided by Syntroleum, including synthetic naphtha (FC-2), synthetic diesel (S-2), synthetic light oil (S-1) and synthetic blend (S-5). Methanol was used as the comparison case because this alternative fuel is being seriously considered for use in fuel cell vehicles. All tests were conducted using a fully integrated IdaTech™ fuel processor (incorporating hydrogen purification) and rated to deliver up to 40 sLm product hydrogen. Test results show consistent high yields of hydrogen from all FT liquid fuels.

INTRODUCTION

With the recent dramatic increase in the world's cost of energy from petroleum, alternative fuels include Fischer-Tropsch (FT) synthetic fuels are receiving increased attention. Fuel cell technology could be a significant beneficiary if hydrogen-saturated FT fuels become commercially available, because hydrogen-saturated FT fuels contain none of the numerous poisons of the fuel cell catalysts that commonly occur in both conventional petroleum and natural gas. As such, Synthetic fuels produced using the Syntroleum™ Process—a cobalt-based FT process designed for natural gas feedstock—

were surmised to be advantageous for fuel cells because they are easily refined to be hydrogen saturated.

It goes without saying that for the production of FT fuels to become economic, it must ultimately be commercially competitive with both conventional petroleum liquid fuels and other available "alternative fuels" including compressed natural gas (CNG) and methanol. When oil prices are above $20/ barrel, commercial production of synthetic fuels becomes increasingly viable, particularly when the feedstock is natural gas associated with oil production that would otherwise be re-injected or flared. Early commercial projects will generally assume no significant premium for the near-zero-sulfur, hydrogen-rich characteristics of synthetic fuels. It is noteworthy that synthetic fuels in CIDI (diesel engines) appear to be at least as efficient as CNG and other "alternative fuels" on a carbon emission well-to-wheels basis[1]. With equivalent tax treatment and $20+ oil, synthetic fuels should also be competitive with EPACT "alternative fuels" on an economic basis.

REQUIREMENTS OF PEMFC AND AFC FUEL CELLS - Fuel cells offer promise for generating electricity and heat while offering several benefits relative to conventional generators including reduced noise, reduced emissions, high energy efficiency, and reduced maintenance. In particular low-temperature fuel cells, specifically proton-exchange-membrane fuel cells (PEMFCs) and alkaline fuel cells (AFCs), are being developed for applications ranging from large-scale and small-scale stationary and portable power generation to automotive and other vehicular propulsion systems. Both PEMFCs and AFCs require hydrogen as the operating fuel; preferably relatively pure hydrogen.

CHALLENGES OF FUEL CELL HYDROGEN GENERATION - Since a widespread hydrogen fuel distribution infrastructure does not exist in any substantial capacity, and it is arguably not likely to be introduced in the near term due to prohibitively high costs, a fuel processor (or reformer) is often considered to be an enabling component to a practical commercial fuel cell system. For instance, the PC-25 family of 200 kW fuel cell systems (commercially available from International Fuel Cells) utilizes a fuel processor that generates hydrogen from the feedstock, typically natural gas, propane, landfill gas, or digester gas. And Ballard's 250 kW stationary fuel cell system likewise incorporates a natural gas fuel processor to generate a hydrogen fuel stream.

Practical fuel-cell systems rated <10 kW for the stationary and portable markets, and that use conventional fuels (feedstocks), have not yet been commercialized. In many cases, the preferred fuel for these applications will be a liquid hydrocarbon fuel. Liquid fuels offer the advantage of ease of transport and storage. Candidate liquid fuels would ideally have little to no sulfur, as sulfur is a well-known poison of the catalytic processes used to generate hydrogen. Sulfur compounds, if present in the product hydrogen stream, will also poison the electrocatalysts used in the fuel cell. Fischer-Tropsch (FT) liquid fuels meet these criteria, and further offer the advantage of being produced from natural gas feedstock, which is more widely distributed than oil around the world. This paper will present the results of testing conducted using a small-scale fuel processor under development by IdaTech and FT fuels provided by Syntroleum

FUEL CELL PROCESSING

CONVENTIONAL FUEL PROCESSOR - A practical fuel cell system with integrated fuel processor is shown schematically in Figure 1.

The fuel processor comprises a device for generating a hydrogen-rich product gas from a conventional fuel such as alcohols and hydrocarbons. The hydrogen-rich reformate stream produced by the fuel processor must be purified prior to feeding the hydrogen to the fuel cell. Both the fuel processor and the fuel cell are supported by a balance of plant (BOP) that includes pumps, fans, sensors, plumbing, heat exchangers, and at least one controller.

A conventional fuel processor (see Figure 2) typically is based on either steam reforming or autothermal reforming and utilizes a series of unit operations to purify the hydrogen- rich reformate. However, the purification methods (high-temperature and low-temperature shift reactors followed by selective oxidation of carbon monoxide) only partially purify the product hydrogen[2-4]--common impurities such as sulfur compounds, unsaturated hydrocarbons, and amines, are not removed by the conventional purification methods.

IDATECH™ FUEL PROCESSOR - In contrast, IdaTech is developing a unified fuel processor that incorporates a universal hydrogen purifier as an integral part of the device (see Figure3). This universal purifier is based on a palladium-alloy membrane (hydrogen selective) that rejects all impurities in the hydrogen-rich reformate stream, essentially passing only hydrogen[5]. Hydrogen-rich reformate is produced by steam reforming, but the product hydrogen is very pure (typically >99.95% hydrogen with <1 ppm CO). More importantly, trace contaminants such as sulfur compounds, unsaturated hydrocarbons, and amines, are rejected by the hydrogen purifier to yield a product hydrogen stream of superior quality.

The high purity of the product hydrogen has been verified by analysis. Using a hydrocarbon feedstock containing about 3 ppm alkylthiols (commercial natural gas) the product hydrogen stream was found to be .>99.95% hydrogen with <1 ppm CO and <1 ppb total sulfur compounds. Hydrogen of this quality exceeds typical specifications for PEM fuel cells, typically given as <10 ppm CO and <50 ppb total sulfur. The hydrogen-selective membrane also yields the fuel-gas stream that supplies heat necessary to vaporize the feedstock and satisfy the endothermic enthalpy of the steam reforming reactions. This fuel-gas stream consists of the impurities that are rejected at the membrane along with some of the hydrogen produced by reforming.

The IdaTech™ unified fuel processor has been operated using several different FT liquid fuels supplied by Syntroleum. These liquid fuels and their properties are summarized in Table 1. Because these fuels are synthesized from desulfurized natural gas, they are exceedingly pure with respect to sulfur content and heavy metal contamination. Furthermore, the processes used to manufacture the FT fuels tested during this study ensures that the concentration of unsaturated hydrocarbons and aromatic hydrocarbons is very low. Thus, the Syntroleum™ fuels are essentially blends of paraffinic hydrocarbons.

EXPERIMENTAL RESULTS

PROCEDURE - All fuels were steam-reformed at a steam-to-carbon ratio of 3:1. This is a typical value for the steam-to-carbon ratio, and it proved to be very suitable in this work. Although further work is required to definitively identify the optimum steam-to-carbon ratio, a summary of the pros and cons of varying the steam-to-carbon ratio is presented in Table 2. Two different steam-reforming catalysts were investigated: a precious-metal steam reforming catalyst (supplied by Süd-Chemie, Louisville, Kentucky, and designated FCR-9) and a nickel-based catalyst designated G91 (also from Süd-Chemie). Both catalysts were used in a standard packed-bed arrangement. The steam-reforming temperature was in the range of 650°C to 730°C. Typically, the gas space velocity through the steam-reforming bed (corrected for temperature and pressure) was low, about 4,000 hr-1.

The Syntroleum™ fuels are compared to commercial-grade methanol which was also steam-reformed in a fuel processor of nearly identical construction to that used for the Syntroleum™ fuels. The commercial catalyst used with methanol is a low-temperature shift catalyst, a copper-zinc formulation, designated G66B and purchased from Süd-Chemie. The catalyst was used in a conventional packed-bed reactor configuration and was operated at approximately 300°C.

RESULTS - Experimental runs were conducted using various Syntroleum™ fuels, taking care to ensure that the fuel processor was operated for a sufficient time interval (usually 2-6 hours) to ensure steady state operation. An example data set obtained using Syntroleum™ FC-2 is shown in Figure 4. In this data set, the hydrogen production is at steady state for about 2 hours, although there is a small amount of fluctuation in the fuel feed rate (FC-2) during the first 60 minutes or so of operation (beginning with introduction of FC-2 at about 10:20 hours. Except for a slight upward drift in the reforming catalyst temperatures, the other operating parameters are fairly constant. The test run was stopped at about 13:00 hours.

Highly pure hydrogen was generated from all of these FT fuels. A typical analysis shows 99.95% hydrogen with about 1 ppm CO and about 3 ppm CO_2. The hydrogen yields from the Syntroleum™ fuels in comparison to the hydrogen yield from methanol is given in Table 3.

Clearly, the yields of pure product hydrogen from the saturated FT hydrocarbon fuels are significantly greater than the yields of hydrogen from methanol. Using catalyst FCR-9, the hydrogen yields per liter of FT fuel are twice the hydrogen yield per liter of commercial methanol. This result is consistent with expectations based on the balanced stoichiometric steam-reforming reactions.

Examination of the data obtained using FC-2 fuel suggests that G91 is an inferior catalyst selection compared to FCR-9. Since G91 was formulated specifically for natural gas (methane) reforming, and FCR-9 was formulated for reforming higher hydrocarbons, this result again is not surprising. The poor performance obtained with FC-2 and G91 does raise doubt concerning the hydrogen yield from the S-2 fuel, as this fuel was only tested with G91. Further testing is warranted using S-2 and FCR-9 reforming catalyst; higher hydrogen yields are expected.

We have also analyzed the emissions for the fuel processor operating on Syntroleum™ S-5 fuel, and the data are summarized in Table 4. The emissions originate with the combustion of waste gases (necessary to heat the steam reforming region of the fuel processor) that are rejected at the hydrogen purifier. Thus, impurities in the S-5 fuel and byproducts of the steam-reforming process will be combusted and can potentially contribute to undesirable emissions. For instance, sulfur contamination in the fuel would cause sulfur oxides in the combustion emissions. Not surprisingly, since fuels are essentially free of sulfur compounds, no sulfur oxides were detected in the emission. Nitrogen oxides are also

relatively low in concentration, and unburned hydrocarbons are about 10%. These results are very similar to the emission composition we observe when natural gas is the fuel.

IMPLICATIONS - The composition of the FT liquids makes these fuels especially attractive for use in a fuel processor to generate hydrogen for the following reasons:

- the lack of sulfur and heavy metals in the FT fuels means that there is no need for a sulfur removal bed or guard bed to ensure adequate performance of the steam reforming catalyst; and

- the paraffinic nature of the FT liquids, more specifically the lack of unsaturated and aromatic hydrocarbons, favors minimal coking in the steam-reforming region of the fuel processor.

Both of these attributes are expected to contribute significantly toward minimal operating (maintenance) costs and favorable overall economics.

In contrast, liquid hydrocarbon fuels derived from petroleum are typically contaminated with sulfur as well as significant concentrations of unsaturated and aromatic hydrocarbons. Extensive hydrotreating is an effective process for reducing, to acceptable levels, the concentrations of sulfur compounds, unsaturated hydrocarbons, and aromatic hydrocarbons, but the cost of deep hydrotreating is often prohibitive. However, the boiling-point ranges and other properties of these fuels are based on fuel standards established for internal combustion engines. They are not necessarily optimal for fuel cell applications.

CONCLUSION

It is important to make a distinction between fuels that are blended for combustion and fuels that are suitable for use as a feedstock in a chemical process (i.e., production of hydrogen). The relative purity of FT liquid fuels makes them more amenable for hydrogen production than are petroleum-derived hydrocarbon fuels. The lack of sulfur compounds and unsaturated hydrocarbons leads to lower maintenance costs, simpler system design, and reduced operating costs. And yet, FT liquid fuels may also be used in internal combustion engines with minimal or no modifications, thereby providing a ready bridge to the adoption of fuel cell engines in vehicles.

Hydrogen-saturated FT fuels generate higher hydrogen yields per volume than methanol. This is an important consideration for vehicular fuel-cell propulsion systems since fuel-carrying capacity on a vehicle is limited. FT liquids also deliver a significantly greater volume of hydrogen (compared to methanol) on a mass basis. Again, this is an important advantage that FT liquids offer for vehicular applications.

A comparison of hydrogen yields from FT liquid fuels and methanol could also be made on a energy basis. In this case, the FT liquid fuel is comparable to methanol; both fuel yield similar volumes of hydrogen when normalized to the energy capacity of the liquid fuel as methanol has about half the energy capacity of liquid hydrocarbon fuels. But the practical aspects of transportation applications, most especially automobile operation, argue against this comparison as a meaningless quantity. Vehicles are constrained by the volume and weight of fuel they can carry, not specifically by the energy capacity of the fuel.

Clearly, opportunity exists for optimizing FT fuels as well as fuel processor designs. Additional development and testing is needed in the overall fuel cell system design as well. These activities are ongoing and we can expect to see continued design evolution over the next several years.

REFERENCES

1. Wang, M.Q., GREET 1.5 - Transportation Fuel-Cycle Model, Vol. 1 and Vol. 2, (Aug. 1999), http://www.ipd.anl.gov/anlpubs/1999/10/34035.pdf

2. Stell, S., and J. Cuzens "Fuel-flexible, fuel processing technology" Fuel Cells Bulletin, 7(April 1999)9-12
3. McNicol, B. "The oil industry response to the challenge of fuel cells" Fuel Cells Bulletin, 9(June 1999)6-11
4. Ahmed, S., R. Kumar, and M. Krumpelt "Fuel processing for fuel cell power plants" Fuel Cells Bulletin, 12(September 1999)4-7
5. Edlund, D. "Versatile, low-cost and compact fuel processor for low-temperature fuel cells" Fuel Cells Bulletin, 14(November 1999)8-11

CONTACT

David Edlund is Senior Vice President, Chief Technology Officer & Co-Founder of IdaTech (email address dedlund@idatech.com). He has a Ph.D. in chemistry from the University of Oregon and more than 18 years experience in the field of hydrogen production and purification.

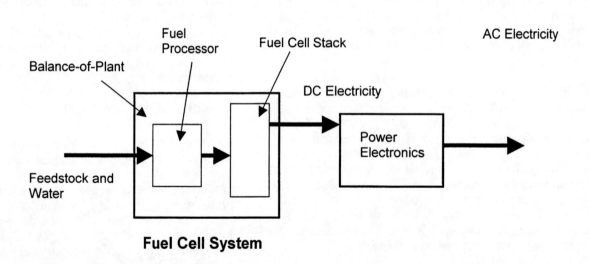

Fuel Cell System

Figure 1. Schematic representation of a fully integrated fuel cell system with a fuel processor.

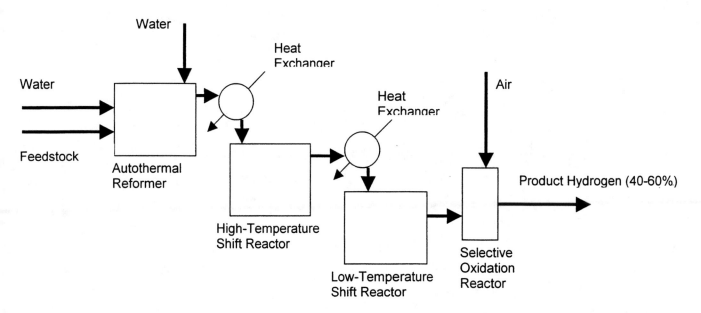

Figure 2. The conventional approach to fuel processing combines autothermal reforming with multiple reactors and heat exchangers.

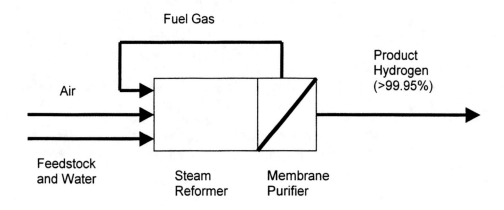

Figure 3. The IdaTech™ unified fuel processor has fewer components than conventional fuel processors and utilizes a membrane-based hydrogen purifier in a very compact package.

Physical Properties	Test Method	Units	Typical Value			
			FC-2	S-2	S-1	S-5
Flash Point	ASTM D-93	°F	>70	157	>70	>60
Specific Gravity	ASTM D-1298	60/60	0.70-0.72	0.771	0.76	0.759
Gravity, API	ASTM D-287	°API	65-70	52	55	55
Reid Vapor Pressure	ASTM D-323	psia	max 10	nil	max 10	----
aromatics	ASTM D-1319	vol %	nd	nd	nd	nd
olefins	ASTM D-1319	vol %	nd	nd	----	nd
saturates	ASTM D-1319	vol %	>99.99	>99.99	----	>99.99
sulfur	ASTM D-2622	wt%	<0.00005	<0.00005	<0.00005	<0.00005
Distillation method			ASTM D-2887 °F	ASTM D-86 °F	ASTM D-2887 °F	86 °F
	IBP, wt%		70 min	320	115	379
	10		165	390	210	387
	20		----	----	----	410
	30		234	----	336	----
	50		261	493	415	416
	70		305	----	488	----
	90		----	601	599	486
	95		350 max	----	650	----
	FBP, wt%			662	690	572
nd = not detected						

Table 1. Selected properties of four Syntroleum™ FT liquids.

Steam-to-Carbon Ratio	Anticipated Benefit	Anticipated Drawback
<3:1	Higher hydrogen partial pressure, lower energy input to make steam	Greater rate of coke formation, reduced catalyst life
>3:1	Lower hydrogen partial pressure, greater energy input to make steam	Lower rate of coke formation, longer catalyst life

Table 2. Anticipated effects of varying the steam-to-carbon ratio for steam reforming hydrocarbon fuels in the unified fuel processor.

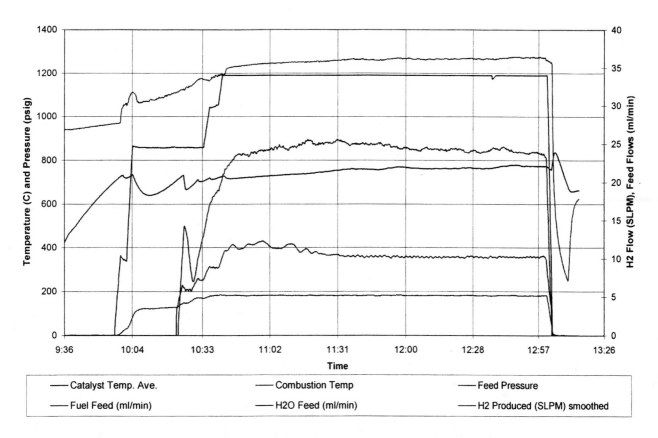

Figure 4. Example data set showing steady-state hydrogen production using Syntroleum™ FC-2 fuel.

Fuel	Catalyst	Range of Hydrogen Production (std. L/L of fuel)
Syntroleum™ FC-2	G91	1500-1900
Syntroleum™ FC-2	FCR-9	2600-3200
Syntroleum™ S-2	G91	1500-1800
Syntroleum™ S-1	FCR-9	2400-3200
Syntroleum™ S-5	FCR-9	2300-3300
Commercial Methanol	G66B	1200-1500

Table 3. Summary of hydrogen yields from different Syntroleum™ fuels using the IdaTech™ fuel processor.

SO_x	0 ppm
NO_x	6 ppm
NO	5 ppm
NO_2	1 ppm
C_xH_y	10.06%

Table 4. Average emissions from a fuel processor producing hydrogen using Syntroleum™ S-5 fuel.

A Sodium Borohydride On-board Hydrogen Generator for Powering Fuel Cell and Internal Combustion Engine Vehicles

Richard M. Mohring and Rex E. Luzader
Millennium Cell, Inc.

ABSTRACT

Hydrogen gas, H_2, is the environmentally desirable fuel of choice for powering fuel cells and internal combustion engines. However, hydrogen is extremely difficult to store, handle, and transport. One major hurdle to commercializing hydrogen-powered vehicles is providing a way to effectively and safely generate, store, and deliver the large amounts of H_2 needed to achieve acceptable vehicle range while minimizing the weight and volume of the storage system. Millennium Cell has developed a novel catalytic process that generates high purity H_2 gas from air-stable, non-flammable, hydrogen-rich water-based solutions of sodium borohydride, $NaBH_4$. This on-board system has already been used by Millennium Cell to successfully power a hydrogen powered series-hybrid sport utility vehicle [1], a full size six-passenger sedan with an internal combustion engine running on hydrogen gas, and is currently being installed into a fuel cell vehicle.

INTRODUCTION

The world has embarked on a fundamental shift in transportation technology. The decades-long acceptance of petroleum fuels has been replaced with growing public demand for cleaner air and reduced health risks. The transportation industry has become very interested in creating low- or zero- emission vehicles. Growing energy needs have prompted specialists to recognize that because traditional energy sources such as petroleum or natural gas are finite (and are becoming increasingly costly), other energy sources must be developed. In addition, issues such as environmental protection, global climate change, and energy supply security also make alternative fuel development crucial. With the depletion of fossil fuels, future world energy needs will have to be based on renewable resources. Battery powered vehicles, once seen as the solution, are not commercially viable as present batteries do not supply adequate driving power and range. Fuel cell powered and battery-fuel cell hybrid electric vehicles are therefore receiving attention from automobile manufacturers as a way to achieve zero emissions and comply with increasingly stringent governmental mandates. Hydrogen lies at the center of these emerging technologies.

HYDROGEN GENERATION

Millennium Cell has invented and developed a proprietary process called Hydrogen on Demand™ that safely generates pure hydrogen from environmentally friendly raw materials. The "hydrogen" is stored on-board at ambient conditions in a liquid "fuel" – an aqueous solution of sodium borohydride, $NaBH_4$. Sodium borohydride is made from borax, a material that is found in substantial natural reserves globally. The process supplies pure hydrogen for energy applications without the need (and associated energy penalties) for compression or liquefaction. Hydrogen produced by this system can be used for numerous applications, addressing a wide range of power requirements. In particular, these systems can be implemented to supply hydrogen for fuel cells and internal combustion engines in automobiles.

The Hydrogen on Demand™ system releases the hydrogen stored in sodium borohydride solutions by passing the liquid through a chamber containing a proprietary catalyst. The hydrogen is liberated in the reaction:

$$NaBH_4 + 2\,H_2O \xrightarrow{\text{catalyst}} NaBO_2 + 4\,H_2 + \text{heat}$$

The only other reaction product, sodium metaborate (analogous to borax), is water-soluble and environmentally benign. The reaction occurs rapidly in the presence of catalyst. Because the reaction is exothermic there is no need to supply external heat to access the hydrogen. The amount of heat generated is calculated to be approximately 75 kJ/mol of H_2 (although laboratory calorimetry suggests a smaller value, closer to 67 kJ/mol).

The heat generated is sufficient to vaporize a fraction of the water present, and as a result the hydrogen is supplied at 100% relative humidity. This co-generated moisture in the H_2 stream is an added benefit both for fuel cells (humidifying the proton exchange membrane) and for internal combustion engines (reducing autoignition, slowing the flame speed [2]).

The reaction is totally inorganic (carbon and sulfur free), producing a high quality energy source without polluting emissions. It is safe and easily controllable – hydrogen is only produced when the liquid fuel is in direct contact with the catalyst, thereby minimizing the amount of gaseous hydrogen on-board at any given time. The fuel solution itself is non-flammable, non-explosive, and safe to transport.

In its solid form, $NaBH_4$ can effectively store greater than 10% hydrogen by weight. In our current designs, $NaBH_4$ is dissolved in water to form a fuel solution. In our testing, concentrations typically range between 20 to 35 percent $NaBH_4$ by weight. The amount of hydrogen stored in solutions of varying concentrations is depicted in Figure 1. As an example, a 30 wt% $NaBH_4$ solution contains approximately 7% hydrogen by weight.

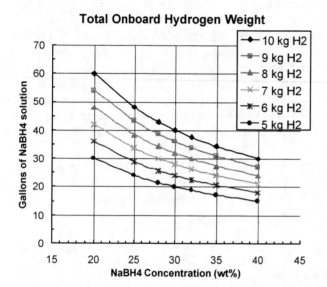

Total Onboard Hydrogen Weight

Figure 1 – Number of gallons of various concentrations of $NaBH_4$ solutions need to achieve storage of 5 to 10 kg H_2.

SYSTEM SCHEMATIC AND PERFORMANCE

A schematic for a typical system is shown in Figure 2. A fuel pump directs fuel from a tank of sodium borohydride solution into a catalyst chamber. Upon contacting the catalyst bed, the fuel solution generates hydrogen gas and sodium metaborate (in solution). The hydrogen and metaborate solution separate in a second chamber, which also acts as a small storage ballast for hydrogen gas. The hydrogen can be processed through a heat exchanger to achieve a specified dewpoint, and is then sent through a regulator to the fuel cell or internal combustion engine.

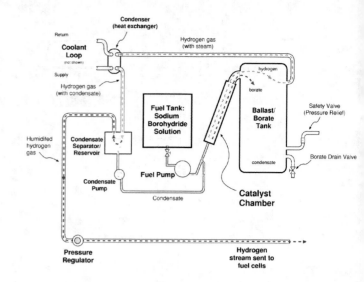

Figure 2 – Schematic of a typical Hydrogen on Demand™ hydrogen generation system.

In operation, the rate at which hydrogen gas is generated is directly proportional to the rate at which the borohydride solution is pumped into the catalyst chamber. Turning off the fuel pump stops hydrogen generation. Additionally, the delivery pressure of the generated hydrogen is variable, and is a function of the liquid pressure at which the sodium borohydride fuel is supplied to the catalyst chamber. We have successfully operated systems continuously at pressures up to 150 psig and hydrogen flow rates of 1000 SLM (standard liters per minute).

This operational simplicity translates into relatively straightforward control strategies. For example, the system can be run in a simple "on/off" mode using the system pressure as a process variable to control the state of the fuel pump (analogous to the way an air compressor operates). When the demand for hydrogen gas increases, hydrogen is drawn off from the ballast tank, and the system pressure drops below a setpoint. This triggers the fuel pump to start up. As fuel reaches the catalyst chamber, hydrogen gas is generated causing the pressure to increase above the setpoint, and the fuel pump is turned off. Experimental data showing this behavior are shown in Figure 2. Of course, more sophisticated strategies can be used to achieve higher degrees of control.

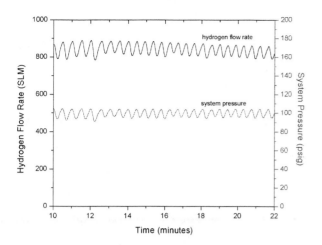

Figure 3 – Pressure and flow data from a Hydrogen on Demand™ system test stand. The pump was controlled via a simple "on/off" strategy (as described in the text), with a pressure setpoint of 100 psig. The outgoing flow was fed through a fixed regulator and hence follows the oscillations in pressure. This run produced an average output flow rate of approximately 850 SLM hydrogen gas.

PROJECT VEHICLES

In our series hybrid design, a Ford Explorer SUV (Figure 4) is powered by a DC electric motor running on conventional batteries. Our technology is used to generate hydrogen gas to run a small internal combustion engine, which drives an electrical generator that keeps the batteries charged.

Figure 4 – Hybrid SUV powered by Hydrogen on Demand™

The SUV can meet all the requisites for an emission-free electric vehicle with virtually no loss of cabin space and no material change in fully-fueled weight. To modify the SUV, we installed a fuel tank to hold the sodium borohydride fuel solution, a receiving tank for the resulting sodium metaborate solution, as well as a catalyst chamber in which the hydrogen-producing reaction takes place. The

space required for the system is approximately the same size as required for a conventional internal combustion engine.

A second project vehicle, a Ford Crown Victoria, has a conventional drive train, and was originally designed to run on compressed natural gas. Millennium Cell modified the car to run on hydrogen. We have performed load-following bench tests in which the engine was powered by hydrogen gas produced from the Hydrogen on Demand™ system, and the in-vehicle design is now being enhanced. This design potentially represents a commercially viable system for retrofitting existing cars to give low emissions performance, at minimal costs to the consumer.

The third hydrogen powered vehicle is the "Genesis", part of a project originated by the New Jersey Department of Transportation. In this case, the Hydrogen on Demand™ system is used to supply hydrogen to a proton exchange membrane (PEM) fuel cell stack that powers the vehicle.

ONGOING RESEARCH AND CHALLENGES

Ongoing research is focused on improving catalyst performance, lowering costs, and enhancing overall generator design, aimed at using higher concentrations of sodium borohydride fuel. Additionally, Millennium Cell is interfacing with major automotive manufacturers to introduce the Hydrogen on Demand™ system in fuel cell powered vehicles, hydrogen burning internal combustion engines, and hybrid applications.

Challenges ahead include validating and implementing cost effective sodium borohydride production technologies. Millennium Cell has been exploring a number of chemical and electrochemical pathways towards economic production of sodium borohydride from borax, as well as the recycling of sodium metaborate back into sodium borohydride. We are also actively seeking partnerships with organizations that will contribute to developing the infrastructure to produce, dispense, and recycle this fuel. Candidates include major energy companies, such as electric utilities, and oil and natural gas companies.

CONCLUSION

Millennium Cell has developed a novel system for storing and generating hydrogen on-board vehicles. The system has many benefits with regards to safety, hydrogen storage density, simplicity of design, and flexibility to provide hydrogen over a wide range of flow and pressure conditions. In particular, for transportation applications this system has very attractive volumetric and gravimetric hydrogen storage densities when compared to other hydrogen storage systems. Replacing the millions of point sources of hydrocarbon pollution (gasoline and diesel burning vehicles) with clean hydrogen as the energy source would be a major development in the fight against air pollution. This new hydrogen storage and generation technology can be a foundation block in the effort to create a full hydrogen economy – something that forward-thinking people have been striving to begin for decades.

ACKNOWLEDGMENTS

The authors would like to acknowledge the dedicated efforts of the Millennium Cell team: in particular, Steve Amendola, Michael Binder, Stefanie Goldman, Michael Kelly, Phil Petillo, and Ying Wu for their technical contributions, and Katherine McHale, Michael Binder, and Ying Wu for their commentary and assistance in preparing this manuscript.

REFERENCES

[1]   Amendola, S. C., et al., "SUV Powered by On-board Generated H_2", SAE Future Car 2000, Society of Automotive Engineers, April 2000.

[2]   Norbeck, J. M., et al., "Hydrogen Fuel for Surface Transportation", SAE #R-160, Society of Automotive Engineers, 1996.

CONTACT

Dr. Mohring is Program Director of the Hydrogen on Demand™ efforts at Millennium Cell, and can be reached at *mohring@millenniumcell.com*.   Mr. Luzader is Millennium Cell's Vice President, Business Development for Transportation applications, and can be reached at *luzader@millenniumcell.com*. Further information can be found at *www.millenniumcell.com*.

Hybrid Storage System: An Optimization Case

A. Di Napoli
Dept. of Mech. & Industrial Engrg., University "ROMA TRE"

G. Pede
ENEA, Centro Ricerche "Casaccia"

ABSTRACT

The paper deals with the study of an original on-board power system that combines battery, ultracapacitors (UCs) and a Fuel Cell. The project was funded by the Italian Government in the framework of the new Hydrogen Research Project and is conceived as the first step of a project involving ENEA, (Italian National Agency for New Technology, Energy and Environment, and University of Rome " Roma 3", aimed at developing a Fuel Cell-powered drive train, for a city-car. It foresees the realization of the drive train in a prototype version, powered by a 5 kW DeNora Fuel Cell, and its testing beside the ENEA laboratories in Casaccia. The report highlights the storage system design phase, emphasising the role that ultracapacitors could fulfill in a near future thanks to their peculiar characteristics, and point at an optimisation method, based on the "simplesso method", for the technical and economic sizing of such a hybrid storage system.

INTRODUCTION

Among the research and development subjects of ENEA (Italian National Agency for New Technology, Energy and Environment) an important theme is the study of innovative vehicles with high energy efficiency and low emissions [1], also with the employment of fuel cell and ultracapacitors, [2] [3].

As a matter of fact, proton exchange membrane (PEM) fuel cells are being increasingly accepted as the most appropriate power source for future generation vehicles. This acceptance is evident through the formation of a new global alliance for the commercialisation of this technology and by the growing participation of major automotive manufacturers in producing sophisticated demonstration vehicles.

With the rapid rate of technical advances in fuel-cells (FCs), and with major resources of the automotive industry being directed at commercialisation of FC propulsion systems for transportation it looks more promising than ever, that FCs would soon become a viable alternative to internal combustion engine technology. One issue that has been driving the development of FCs for automotive applications is their potential to offer clean and efficient energy without sacrificing performance or driving range.

In the case of a PEM FC-V, realising this potential means, first of all, to assure braking energy recovery [4], then to ensure that the complete FC system operates as efficiently as possible over the range of driving conditions that may be encountered.

To sum up, a storage unit can be employed combined with FCs in order:

- To recover braking energy
- To downsize the FC, the most expensive component, thus reducing the drive train total cost
- To achieve the operating voltage-current point of maximum efficiency for the FC system

According to this concept, FCHEV prototypes (Fuel Cell Hybrid Electric Vehicle) have been designed and produced by the world's major car makers, like Honda, Opel, Nissan, Renault, Fiat, etc. The majority of these vehicles has a storage system composed by a high power battery, in two cases, at least, (Honda FCX-V3 and Mazda Demio FC-EV) there are ultra capacitors (UCs). There are technical, economic and practical reasons to suggest one or another type of storage technology.

In the paper, we will try to explain the reasons behind such choices, and to propose an innovative hybrid (batteries + UCs) storage system, describing an experimental application of such a power train, for a city car application, developed by ENEA and University of Roma " Roma 3", [5][6].

WHY AN HYBRID STORAGE SYSTEM FOR HEV ?

STORAGE SYSTEMS REQUIREMENTS

In a hybrid electric vehicle (HEV) the storage system must be able:

1. to recover the otherwise dissipated braking energy,
2. to level demanded peak power (load levelling), enabling the generator to deliver an equal power to the average demanded power (such power could eventually be adequate, with very slow transitories if the primary source is a motorgenerator, to the demand fluctuation throughout time)
3. To deliver the whole time the additional power

A correctly designed system will therefore have to guarantee at the same time a maximum power sufficient to compensate the Δ between that one of the generator and the maximum forecast (required storage power) and an energetic content sufficient to avoid the complete discharge during every phase of demand for power (required storage energy). The values of these two dimensioning parameters depend on two main factors, the driving cycle and the hybrid system configuration, or, more specifically for series hybrids like FCHEV (Fuel Cell Hybrids Electric Vehicles), by the "Hybridisation level", expressed by the relationship between two installed powers, the generation power and the traction power. An analogous parameter, suitable for parallel hybrid, is the "Fraction of electric power"[7], estimated by dividing total peak motor power by combined peak ICE and motor power, Fig.1.

The concept is pointed out by the OAAT's [8], that in a specific program, the "Vehicle High-Power Energy Storage Programme", states specific objectives for energy storage requirements, different for the different types of hybrid electric vehicles, whose FCHEV are a particular case.

Speaking of HEV classification, the continuum of Series HEV, Fig. 1, starts from the "Range extender", an electric vehicle with a small generator, to full power fuel cell vehicles (FCEV) and conventional vehicles with electrical transmission (diesel–electric).

Parallel Hybrid vehicle continuum starts from ICE vehicles where braking energy recovering and a minimum torque integration is assured by a small electric machine, directly mounted on the crankshaft that also works as starter and generator (like the Siemens ISG, Integrated Starter Generator or the Toyota THS-M), up to "Power assist" and " Dual mode" hybrids.

The OATT report considers only two categories:

- Power-assist hybrid vehicles (like Honda Insight and Civic IMA Sedan, Ford Prodigy, General Motors ParadiGM etc), that are vehicles with small electrical propulsion systems relative to their engines
- Dual mode hybrid vehicles (Toyota Prius and Estima minivan, GM Precept), that have a limited driving range using only batteries

The energy storage requirements for the two cases are reported in Table 1 :

Table 1 : Technical Targets: High power Storage Requirements

PNGV Goals Characteristics	Units	Power Assist Minimum	Dual Mode Minimum
Pulse Discharge Power (18 s)	kW	25 (for 18 s, 0.125 kWh)	45, for 12 s (0,15 kWh)
Max. Regen Pulse (10 s)	kW	30 (for 10 s, 0.08 kWh)	35 (for 10 s, 0.1 kWh)
Total Available Energy	kWh	0.3	1.5 (for a ZEV range of ≈ 10 km
Max. Weight	kg	40	100
Specific Cost (at 2004)	$/kWh	940	320
Production Cost x unit (100 ku/y)	$	300	500

P/E RATIO AND STORAGE SYSTEMS

The relationship P/E (Power/Energy ratio) between two mission-dependent parameters, pulse discharge power and total available energy, is useful to correctly design the vehicle storage system. For both hybrid categories considered by OATT, this ratio is calculated and is equal to 83 W/Wh for the "power-assist" and the 27 W/Wh for the "dual mode"[8].

Moreover, the same ratio can be calculated for each kind of storage device, by the ratio between its specific power, W/kg, and its specific energy, Wh/kg. For instance, available traction batteries have a P/E ratio of 1-4, good for battery EV, but far too low for hybrid applications. High power storage system (high-power lead-acid battery or Ni-MeH batteries) have a more adequate ratio, approximately equal to 10 W/Wh (specific power and specific energy are approximately 300 W/Kg and 30Wh/kg for the former, 4500W/kg and 50-70 Wh/kg for the latter. Conceptually speaking, a storage system whose P/E ratio reflects the relationship between power and energy requested by the considered hybrid vehicle's typical mission, would have all the power in order to satisfy the cycle power peaks, without having more energy than the necessary one.

To be precise, to estimate the P/E ratio we consider the ratio between "max. regeneration pulse" and "max. regeneration available energy", since the accumulated energy must at least be double of the one used, to limit the normal storage cycling between 30% and 80% of total available energy (State of Charge fluctuation : 30% < SOC < 80%).

The P/E ratio is shown in the figure below for the whole range of EV and HEV:

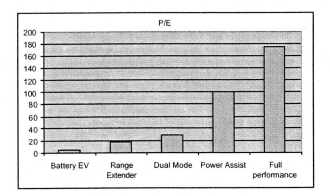

Fig. 1. P/E ratio vs. HEV class

UCs benefit from a P/E ratio much higher, more than 100 W/Wh (specific powers is very high, up to 1000 W/kg and specific energy is less than 5 Wh/kg).

Therefore UCs seem to be more suitable than batteries for "power assist" hybrids to satisfy in a balanced way the two requirements, energy and power, whereas for "dual mode" hybrids, where the energy content must be sufficient to assure a long time discharge period, a high-power battery like a NiMeH or a Li Ion seems to be a forced choice. For instance, Continental ISAD Electronic Systems GmbH & Co KG has opted for UltraCaps from EPCOS [9] and GM/Allison for Maxwell Technologies Microelectronics [10], whilst Toyota and Honda hybrids are equipped with NiMeH batteries.

BATTERIES AND UCS : OTHER PRO AND CONS

Indeed, for a proper choice of HEV storage systems, there are also other problems to take in account, related to the system reliability, life and total cost. As a matter of fact, in the battery, a high peak power rate implies an increase of losses and temperature and so a decrease of lifetime, which results in repeated replacement throughout the life of the vehicle. Replacement costs include those associated with the purchase and installation of a new battery and the removal and disposal of the old battery. Moreover, batteries have difficulty functioning in cold weather, so they create significant inconveniences, whereas UCs can

operate successfully in wide temperature ranges, including those as low as -40° C [11].

Thereby, it could be desirable to combine the characteristic of UC and batteries, load leveling the generator–battery system by means of UCs, which have high power density; it is then possible to obtain regeneration energy at high efficiency during deceleration and supplying the stored energy during accelerations in order to reduce the peak power requirements for the FC–battery unit [12]. The UC tank must supply all the power required in excess of the generator–battery system rated power, if its state of charge is greater than a specified minimum. When the power required to operate the vehicle is lower than the generator-battery rated power, the UCs can be charged with the power in excess. Whenever regenerative braking operations occur, energy is put into the UC tank, provided this device is not fully charged yet otherwise the braking energy is recovered into the battery. Such an hybrid storage system could achieve an optimum value for the P/E ratio, at a lower cost than a NiMH battery and at a lower weight than a high-power lead acid battery.

COSTS AND PERFORMANCES RELATED CONSIDERATIONS

In the table, performance and cost for prototype (Saft) or commercially available HEV storage systems are listed below [13][14]:

Table 2 : High Power Batteries Characteristics

	W/kg	Wh/kg	P/E	Specific Cost
Ni MeH (Ovonic 12HEV60)	550	68	8	1000 $/kWh
Li Ion (Saft 12 Ah)	370	125	15	'/
Pb (Genesis 13 Ah)	250	25	10	250 $/kWh

For UCs, there are not commercially available HEV storage systems, because current costs are too high for this application. For example, a prototype city car storage system (total capacity 67 F, nominal voltage 120 V), composed by EPCOS B48710 UCs, at the moment costs about 8000 $. The economic feasibility of this solution, innovative regarding that usual for the hybrids, is so largely tied to the forecast of reduction of the costs of the UCs, currently very expensive, but for which manufacturers like Epcos and Maxwell PowerCache, anticipate a drastic reduction. As a matter of fact, the basic materials used in their construction (carbon cloth or carbon particulate for the electrodes and a organic electrolyte) pose no significant barriers to affordable cost in quantities typical of the automotive market (according to

---

1 "the cost ratio is almost 2-to-1 , lithium-ion being more expensive than nickel metal hydride", M.Anderman, A.A.Batteries, EV World Update 2.11.

PowerCache and Epcos, approximately $0.01÷$0,02 per Farad, by 2004, at volumes in the millions, corresponding to 10÷20.000 $/kWh [13]). The relationship between the P/E ratio and the storage hybridisation parameter Mb/MUC, namely the ratio between Lead-acid batteries weight and UCs weight in the storage, is as following :

$$M_b/M_{UC} = \frac{(Es_{UC} \times P/E - Ps_{UC})}{(Ps_b - Es_b \times P/E)}$$

where Es and Ps are specific energy and specific power for batteries and UCs 2.

Table 3 : Muc/Mb vs. P/E

P/E	M_{UC}/M_b
104	9,85
87	7,45
52	3,62
29	1,60
20	0,91
17	0,71
13	0,39

The figure below, shows how the specific cost of a hybrid storage system increases with the P/E ratio:

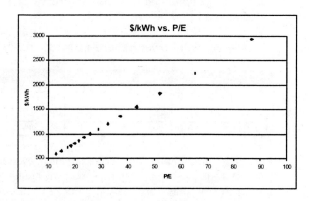

Fig. 2. Specific cost vs. P/E ratio

It is apparent that, at these costs (foreseen), a hybrid storage system would be convenient for "Range extender" and "Dual mode " hybrids, with respect to Ni MeH battery.

FUEL CELL VEHICLES AND P/E RATIO

If we move on considering the case of FCHEV, hybrid vehicles in which the primary generator is

---

2

	$/kWh	Ps (W/kg)	Es (Wh/kg)
UCs :	15000	400	1,83
Battery :	250	250	25

constituted from a fuel cell, we can also classify them according to the aforementioned HEV categories. The FIAT 600 FC has two different versions, the existing "range extender" with a 5 kW cell, and a second version, under development, with a 15 kW cell, comparable to a "power assist" hybrid, because its P/E ratio is about 100 W/Wh. As a matter of fact, simulating the current homologation European cycle, NEDC, though limited in the maximum speed by the maximum power of the traction motor (Vmax = 100 km/h), the dynamic model gives, in the two cases, the following values for peak power and energy :

Table 4 : FCHEV simulation results

Storage requested performance	Units	5 kW	15 kW
Pulse discharge Power	kW	26	20
Maximum ΔEnergy	kWh	0,7	0,1

Since the accumulated energy must at least be double of the used one, to limit the normal storage cycling between 30% and 80% of total available energy (State of Charge fluctuation : 30% < SOC < 80%), the aforesaid values for energy must at least be doubled, Tab.5.

Table 5 : FCHEV storage characteristics

Storage characteristics	Units	5 kW, R. extender	15 kW, Power assist
Storage Power	kW	26	20
Storage Energy	kWh	1,4	0,2
P/E	/	18	100

For a "Range extender" FCHEV, the P/E ratio is in the range of convenience for hybrid storage solution.

For a " Power assist" FCHEV, the ratio P/E is far closer to the typical value of Ultracapacitor than batteries, like in power assist HEV. Moreover batteries like high-power lead-acid or Ni-MeH would get an excess of accumulated energy, perfectly useless in a vehicle that is ZEV also when the generator is "on", and also harmful, because the resulting extra weight, in case of use of lead-acid batteries.

On the other hand, the P/E ratio is not in the range of convenience for hybrid storage solution, so the practicability of this application depends on a strong cost reduction (or specific energy increase) for UCs.

AN APPLICATION EXAMPLE

As an example of the application of these concepts, we decide to consider a real case, the already mentioned Fiat 600 FC, a 5 kW fuel cell powered, integrated by a 30 kW storage system, that is, at present, a lead-acid battery very heavy (261 kg), while in the future it will be a Ni-MeH.

The cost of the new storage will increase very much, so it is interesting to verify the technical-economic feasibility of the above described innovative hybrid storage system. The study is conceived as the first step of a project involving ENEA, (Italian National Agency for New Technology, Energy and Environment), and University of Rome " Roma 3 ", aimed at developing a FC-powered drive train, of hybrid type, for a city car, like the Fiat 600 Electric. The project, funded by the Italian Government in the framework of the new FC Research Project, foresee the realisation of the drive train in a prototypic version, powered by a 5 kW DeNora Fuel Cell. The proposed propulsion system comprises a 3-way power electronic converter (PEC) used to interface the FCs, the battery storage and the UC bank. Thanks to the availability of fuel cell generation system, developed by ENEA using a 5 kW De Nora PMFC stack of this power, Fig. 4, the traction system has been effectively realised and tests are on going at the ENEA laboratories in Casaccia, a full testing facility for hybrid vehicles equipped with a set of testing sections for each subsystem (i.e. electricity generation, energy storage, driving motors).

POWER INTERFACE DESIGN

The batteries load-leveling problem throughout the UCs use has been faced by the authors in the past: now our studies are concentrated on the interface power structure, in order to couple the two power sources. In all the possible configurations a bi-directional DC/DC converter is used in order to adapt the UCs voltage versus the battery voltage. It is important to observe that particularly in the derived configuration, it is possible to manage the UCs SOC through the duty cycle the dc/dc converter is being operated. This paper presents a power electronic interface (PEI) that allows managing separately three power sources being operated at different voltage levels. The proposed structure is shown in Fig. 9.

As shown in the figure, the three power sources are connected through DC/DC converters to a common node whose voltage is suitably regulated to be maintained rigorously constant during the drive train operations; the traction motor inverter is connected to the same node, eventually through an additional dc/dc converter. The converters associated to the battery and the UC tank are both bi-directional step-up converters; by regulating their duty-cycle it is possible to control the battery and UC tank input/output currents in order to achieve the desired SOC values. On the contrary, the FC-fed converter is a conventional step-up converter and thereby it does not allow current inversion during regenerative braking operations of the drain train.

STORAGE SYSTEM DESIGN

For the storage system design, first of all it needs to find the value of the maximum specific power that can be made available from the UCs during the acceleration and deceleration phases. Indeed, previously considered value (1,83 Wh/kg, 400 W/kgs), refer to the device's total stored energy. In the real case, instead, it needs to take into account two important discharge aspects:

1 Discharge profile
2. Initial conditions

Discharge profile
To get a constant value for the accelerating and braking torque, it is necessary to linearly adjust the power to the vehicle's speed.
Therefore the device will be discharged (in acceleration) and charged during braking, following a linearly increasing power characteristic.

Initial condition
Alongside feeding wich will always be possible, the device needs to be always in an intermediary equilibrium state (voltage) between the maximum the minimum admissible values.
For such purpose, the control system will have to recharge the UC after a deep discharge and to discharge it after a braking that brought it to the maximum voltage level.

Simulation
The two phases have been simulated considering the equivalent electric circuit constituted by a condenser (generating) ideal and from a resistance in series.
The device's characteristics are reported below:

Table 6 EPCOS UCs data sheet

Epcos B48710		
Capacity	1200	F
Operating range	2,3 – 1,1	V
Imax	300	A
Equivalent Resistance	0,0012	Ohm
Weight	0,4	kg

Minimum voltage was calculated considering the used electronic, which can work well with a duty cycle values up to 0,75 ÷ 0,8. Then minimum voltage (Vscmin) is calculated by:

$$Vlink/Vsc = 1/(1-D)$$

where D is the maximum admissible duty cycle (Dmax = 0,75). In this way.
Nevertheless the device can be discharged down to 0 V without damaging, whilst the allowed maximum voltage is equal to 2,6 Vs, is bigger of the used one, 2,3 V.

Equilibrium voltage has been chosen in a way such that the useful Pmax in braking is bigger than Pmax in acceleration, and equal to 1,9 V. Operational range is therefore comprised between 1,9 and 2,3 in charging and 1,9 and 1,1 Vs in discharging.

Used equations are the following:

1. $Q(t) = C\,Uc(t)$
2. $Uc(t) - I(t)\,Rint = V(t)$
3. $P(t) = V(t)\,I(t) = r\,t$

where $r = dP/dt$ is the storage power rate that, in the pre-set time, brings to the minimum voltage starting from the equilibrium voltage.

The simulation gives the followings results:

Table 7. Simulation Results

	Units	Discharge	Charge
Time	s	12	10
AV	V	1,9 - 1,1	1,9 - 2,3
dP/dt max.	W/s	15	21
Max. power	W	180	210
Max. Ps	W/kg	450	525
Imax	A	160	109
Es	Wh/kg	0,8	
Efficiency		93%	96%

In the following diagrams the calculated discharge voltage and current profile are reported :

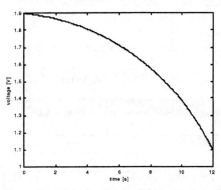

Fig.3. UCs discharge voltage profile

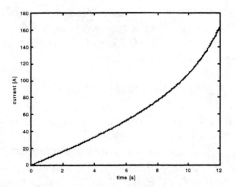

Fig.4. UCs discharge current profile

PRELIMINARY STORAGE SIZING

Once we defined the traction system and also introduced a model for the storage control system, it is now possible to design the storage.

In the calculation, the UC'S specific power value in discharge was set equal to 400 W/kgs, to take into account electronic interface efficiency.

For the considered "range extender configuration" the relationship P/E is equal to 19. A mixed configuration Batterie/UC satisfies this value when the value of the relationship between the weights of UC and batteries is 0.7, relationship 1). Since the necessary power is of 26 kW, the weight of the batteries is obtained by resolving the equation:

$$Ps_b\,x + 0.7\,Ps_{UC}\,x = 26\ kW$$

where x is the unknown weight, Ps_b and Ps_{UC} represent batteries and UCs specific power. We have therefore:

$$M_b = 26.000 / (Ps_b + 0.8 * Ps_{UC}) = 50\ kg$$
$$M_{UC} = 0,7 * 50\ kg = 35\ kg$$
$$Total\ weight = 85\ kgs$$

To get a solution that can offer an immediate application to a eal storage system, we have to take into account the commercial availability of batteries and Ucs. A possible solution for the two systems could be as following:

· n. 11 12V/13 Ah batteries, 120 Vtots
· n. 83 1200 Farad UCs, 154 Vtots

The voltage levels detected will be verified in order to be consistent with the interface design, and eventually modified.

HYBRID STORAGE OPTIMISATION

The storage system sizing could be done by taking into account previously additional constraint and more general considerations.

For instance it is possible to accept a weight slightly greater than the minimum corresponding to a precise value for the relationship P/E, in order to have a cheaper system, or to consider a acceptability range for the relationship P/E and not a precise value.

Besides, the calculation value for P/E is related to the reference mission or to other mission constraints, like the pure electric range, that could be imposed and by whom the E value depends. Speaking of reference mission, for instance, the same vehicle, driven on an urban cycle of homologation UDDS, used in the USA, has a value of the relationship P/E> 19, because this cycle is more severe than the European one.

To sum up, the problem is an optimisation problem, and can be faced in a more general way with the classical operational research methods.

As a matter of fact, the optimal technical-economic storage sizing requires the maximisation, subordinate to the respect of precise technical constraints, of a linear function Z that expresses the total cost, in presence of a series of functional constraints represented by the following positions:

- Total weight has to be smaller than a fixed value G
- Maximum power has to be equal or greater than reference power P
- Accumulated energy has to be equal or greater than the reference energy E

Functional constraints correspond to the performance invariability for the reference vehicle. In the following we will call:

	Weight	Power	Energy
Batteries	G1	P1	E1
UCs	G2	P2	E2

The three aforesaid conditions become:

- G1+ G2 <= G
- P1 + P2 >= P
- E1 + E2 >= E

The hybrid system cost is a function of weights and costs (life cycles costs) for batteries and UC.

To get a solution that can offer an immediate application to a real storage system, we have to take into account the commercial availability of batteries and UCs, whose weights and tensions obviously vary in a discontinuous way, and the need that the battery pack nominal voltage has to be maintained in a certain field, determined by the correct interface sizing criterion.

So, we have to to set another condition, related to the battery pack voltage:

$$Vbatt >= Vmin$$

and re-phrase the aforementioned equations system.

We set therefore:

1. $N1 M_1 + N2 M_2 <= G$
2. $N1 M_1 Ps_1 + N2 M_2 Ps_2 >= P$
3. $N1 M_1 Es_1 + N2 M_2 Es_2 >= E$
4. $N2 V_2 >= Vmin$

where other term meanings are expressed by the following table :

	Module number	Module weight	Module voltage	Specific power	Specific energy
Batteries	N1	M_1	V_1	Ps_1	Es_1
UCs	N2	M_2	V_2	Ps_2	Es_2

The aforementioned linear inequality system can be graphically represented on the plan N1, N2 , Fig.5, and can have no solutions, only one solution or countless solutions; in the latter case, we have to choose those ones corresponding to a minimum, for the total weight or the total cost. Because total weight and total cost are linear functions of N1 and N2, it is possible to apply the "simplex method", related to the minimum points for multi-variable linear functions. According to the theorem, the point of minimum for a linear function:

$$Z = Z(N1, N2) = N1 M_1 c_1 + N2 M_2 c_2$$

defined in the plan N1, N2 , lie on the vertexes of the figure detected by the line corresponding to the constraint conditions.

In this case the lines correspond to the equations:

$N2 *(G) = (G - N1 M_1)/M_2$
$N2 **(P) = (P - N1 M_1 Ps_1)/ M_2 Ps_2$
$N2 ***(E) = (E - N1 M_1 Es_1)/ M_2 Es_2$
$N2 ****(V) = V min /V_2$

CALCULATION HYPOTHESIS

The hybrid storage system will be constituted by lead-acid batteries from Genesis (13 Ah/12 V) and ultracapacitors by Epcos (1200 F).

So, N1 is the number of such battery modules and N2 is the number of such UCs modules, Table 10.

From the manufacturer's catalogues we drawn the coefficients M_1, Ps_1, Es_1, M_2, Ps_2, Es_2 :

Tab.10. Batteries and UCs data sheet

	M_1, M_2 module weight	Ps_1, Ps_2 specific power	Es_1, Es_2 specific energy
Batteries	4.5 kg	200 W/kg	27 Wh/kg
UCs	0.4 kg	400 W/kg	0.8 Wh/kg

Unitary costs are:
· c1 (batteries) = 10 $/kg, according with HESA (5 $/kg for 10.000 pz/year, the actual cost is twice because Ni-MH life is 50.000 km vs. 25.000 km)
· c2 (UCs) = 30 $/kg, according with the Maxwell Technologies Microelectronics forecast, 10 $/pz, for a 1000 F UC, (100.000 pz/year), corresponding to 16.000 $/kWh and to 30 $/kg.

From dynamic cycle analysis, see Table 4, and general design related considerations, we draw the following numerical values for known terms of the aforesaid equation system:

G = 100 kg
P = 26 kW
E = 1400 Wh
V min = 120 V

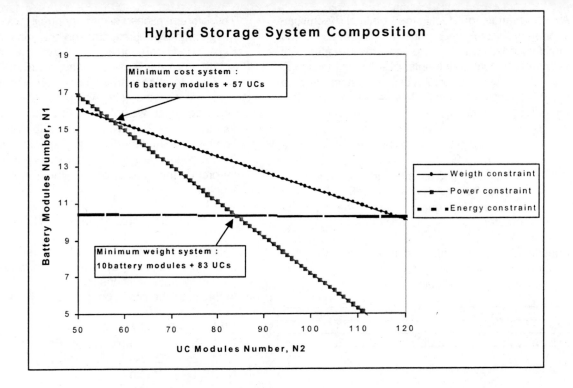

Hybrid Storage System Composition

Minimum cost system :
16 battery modules + 57 UCs

Minimum weight system :
10battery modules + 83 UCs

Weigth constraint
Power constraint
Energy constraint

Battery Modules Number, N1

UC Modules Number, N2

Fig.5

As concerns the reference system, its characteristics are pointed out in the table:

Table 11 : Ni-MH Reference System

Battery type	Modules number	Total voltage	Total weight	Total energy	Total cost
Ovonic 12HEV12	16	192 V	52 kg	2.4 kWh	2400 $[3]

The Ni-MeH battery life has been assumed to correspond to 50.000 km, while for the lead-acid, it has been assumed corresponding to 25.000 km.

RESULTS

The diagram reported in Fig. 5 detects every possible solution in the plan N1, N2. The figure corresponding to solutions is the triangle comprised between the three straight lines corresponding to the first three functions. The minimum energy straight lines condition nearly corresponds, casually, to the minimum voltage straight line, (4th equation) in correspondence to N1 = 10 modules.

A rapid verification allows us to detect the minimum weight solution in the central vertex, because, as the cost, also the weight is linear function of the two variables N1 and N2.

The point corresponds also to the exact solution, already found in the paragraph, because it is the intersection point of the two straight lines that represent the constraint conditions for power and

energy. The top-left point corresponds to the hybrid solution of lesser general cost,

The following table summarises the characteristics of both hybrid systems so detected, that of minimum weight and that one of minimum cost (per kWh):

Table 12 : Range Extender Hybrid Storage System Solutions

	Minimum weight system	Minimum cost system
Ps_{UC}, W/kg	400	400
Es_{UC}, Wh/kg [4]	1,8	1,8
N1	10	16
N2	83	57
Batt.Weight	49	78
Ucs Weight	35	24
Batt.Energy	1323	2117
UCs Energy	28	19
Total Weight	84	102
Total Power	25	26
Total Energy	1351	2136
P/E	19	12
Total cost	1536	1502
$/kWh	1137	703

---

[3] 1000 $/kWh, the Ovonic forecast is 250 $/kWh

[4] We'll consider the manufacture's value, because it is homogenous with the forecast values cited below

404

Comparing weights and costs of the hybrid solution of lesser cost with the reference one, the cost difference is around 900 $ for the two solutions, and it would be enough to cover the greatest cost of the most complex interface for UCs [15]. We underline that, for the same total energy (about 2 kWh) the hybrid storage weight is only 50 kg bigger than the NiMeH one.

HYBRID STORAGE SYSTEMS AND THE PNGV OBJECTIVES

Thanks to the simplex method, it is possible to perform a rapid analysis of the problem that underlines as the proposed OATT's objectives, Tab.1, are attainable, in the limits of weight pointed out (that are 100 kgs. for the " dual mode"and 40 kgs for the " power assist "), only if the UCs specific power arrives to 1100 W/kgs, and specific energy to 8 Wh/kgs. These new values for specific power and energy are used in Fig.10 and Fig. 11.
The table below shows both system compositions and characteristics:

Table 13 : Hybrid Advanced Storage System Specifications

	Power Assist Hybrid	Dual mode Hybrid
Ps_{UC}, W/kg	1100	1000
Es_{UC}, Wh/kg	5	8
N1	3	10
N2	50	88
Batt.Weight	15	49
Ucs Weight	20	35
Batt.Energy	397	1323
UCs Energy	100	211
Total Weight	35	84
Total Power	25	45
Total Energy	497	1534
P/E	50	29
Total cost	747	1546
$/kWh	1503	1008

Are these specifications (1000-1100 W/kgs, 5-8 Wh/kgs) attainable for UCs ?
About specific power, there are no problems, because at the moment the UCs are energy limited, not power limited. As a matter of fact, the pulse power capabily at the moment ranges between 1400-3900 W/kg.
About specific energy, according to Epcos :
" By increasing the Cell Voltage from 2.3 V auf 2.5 V, we think that in about 1 year we will get an increase of 18 %, which will then enter into 4.58 Wh/kg", and the same is true for other manufacturers, as reported in the last papers

presented at the EVS-18 in Berlin, see references. To cite Andrew Burke [16] "R&D efforts by several groups promise of improving energy density of UCs to 6-8 Wh/kg".
About the costs, the "Dual mode system" system specific cost would be twice the cost objective pointed out by the OATT. However, an equivalent system with NiMeH batteries would attain it only if the specific cost objective 250 Wh/kg was reached, (it is the objective declared by the Ovonic for 2002).
For the "Power assist" storage system, an UCs system plus a 36 V battery could cost about 750 $, twice the cost of a pure battery system. Efficiency and life-cycle related considerations could justify the higher costs.

CONCLUSIONS

We have shown the possibility and the convenience, under certain conditions, of using storage systems for hybrid vehicles, constituted by UCs and batteries. The report shown how such a hybrid system can be designed, starting from the study of the mission.
A storage system for a FC vehicle in a " range extender " configuration has been chosen to appraise its convenience in comparison with the same vehicle equipped with a high power NiMeH battery system.
Finally, a planning methodology is also shown, based on the simplex method, that allows us to quickly analyse the different possible solutions and to appraise its costs and weights. The results of the technical economic analysis are encouraging, provided that the cost reductions foreseen by the manufacturers are verified. More specifically, the economic convenience results demonstrated, for our " range extender ", for a UCs cost of 0,01 $/F, corresponding to 16.000 $/kWh. As to the possibility that such systems could satisfy the weight and cost objectives set by PNGV, this also depends on an increase of 5÷6 times of their specific energy and on a further cost reduction.

REFERENCES

[1] "Vehicle Testing in ENEA Drive-train Test Facility", G. Lo Bianco, G. Pede, E. Rossi A. Puccetti, ENEA, G.Mantovani, ALTRA, Advanced Hybrid Vehicles Powertrains, SAE SP-1607, Detroit, March 2001

[2] "Employment of Ultra-Capacitors for Power Leveling Requirements in EV: a State of the Art", F. Brucchi, G. Lo Bianco, P. Salvati, F. Giulii Capponi, L. Solero, Proc. of the 32th ISATA, 1999, pagg. 371-379.

[3] " Flow field design and testing for PEM bipolar plate", S. Galli, L.Giorgi, A. Pozio and F. Scola, VII Grove Fuel Cell Symposium - September 11-13, 2001 - London (U.K.)

[4] "Development of Fuel Cell Hybrid Vehicle", Tadaichi Matsumoto, Nobuo Watanabe, Hiroshi Sugiura, Tetsuhiro Ishikawa, Toyota Motor Corporation, EVS 18, Berlin, October 2001

[5] "Power Converter Arrangements with Ultracapacitor Tank for Battery Load Leveling in EV Motor Drives", Di Napoli, F. Giulii Capponi, L. Solero, Proc. of the 8th European Conference on Power Electronics and Applications, 1999, cd-rom.

[6] "Ultracapacitor and Battery Storage System Supporting Fuel-Cell Powered Vehicles", A. Di Napoli, F. Crescimbini , L. Solero, G. Pede, G. Lo Bianco, M. Pasquali, EVS 18, Berlin, October 2001

[7] ""Evaluating Commercial and Prototype HEVs", Feng An et alii, Argonne National Laboratory, SAE SP-1607, Detroit, March 2001

[8] "Recent Accomplishment of the Electric and Hybrid Vehicle Storage R&D Programs at the U.S. Department of Energy: a Status Report ", R. Sutula, K.L. Heitner, S.A.Rogers, Tien Q.Dong, R.S. Kirk, OATT, V.Battaglia, G.Henriksen, Argonne N.L., F.McLarnon, Lawerence Berkeley N.L., B.J.Kumar, C.Schonefeld, Energetics Inc., EVS-17, Montreal, October 2000

[9] "ISAD opts for UltraCaps For Next-Generation Vehicles" EPCOS Components, Issue 4, October 2000

[10] " Maxwell to supply ultracapacitors for GM/Allison hybrid systems ", Automotive Engineering International, March 2001

[11] "Ultracapacitors and the Hybrid Electric Vehicle", Bobby Maher, Applications Engineer, PowerCache

[12] "Combining Ultra-Capacitors with lead–acid Batteries", B.J.Arnet, L.P.Haines, Solectria Corporation, EVS-17, Montreal, October 2000

[13] "Current Status Report on U.S. Department of Energy Electric and Hybrid Vehicle Energy Storage R&D Programs", ", R. Sutula K.L. Heitner, J.A. Barnes, Tien Q.Dong, R.S. Kirk, V.Battaglia, B.J.Kumar, C.Bezanson, EVS-18, October 2001, Berlin

[14] "Ovonic NiMH Batteries: the Enablin Technology for Heavy Duty Electric & Hybrid Electric vehicles", G.Fritz, N.Karditsas, S.Venkatesan, D.Corrigan, S.K. Dahr and S.R.Ovshinsky. Ovonic Battery Comany, SAE Paper 2000-01-3108

[15] "A Progress Report of the Capacitor Hybrid System - ECS", Michio Okamura, EVS-18, October 2001, Berlin

[16] "Update of UC technologies and Hybrid Vehicles Applications: Passenger Cars and Buses" Andrew Burke, Marshall Millers, Univerisity of California - Davis, EVS-18, October 2001, Berlin

DEFINITIONS, ACRONYMS, ABBREVIATIONS

UCs	Ultracapacitors
PEM	Proton exchange membrane
FC	Fuel cell
FC-V	Fuel cell vehicle
FCHEV	Fuel cell hybrid electric vehicle
HEV	Hybrid electric vehicle
ICE	Internal combustion engine
OATT	Office of Advanced Autom. Technologies, DOE
IMA	Integrated Motor Assist
THS-M	Toyota Hybrid System-Mild
Ni-MeH	Nickel metal hydrides
SOC	State of charge
P/E	Power/Energy ratio
EV	Electric vehicle
E_s	Specific energy (per kg)
P_s	Specific power (per kg)
E_{s_b}	Battery specific energy
P_{s_b}	Battery specific power
$E_{s_{UC}}$	Ultracap. specific energy
$P_{s_{UC}}$	Ultracap. specific power
M_b	Battery mass
M_{UC}	Ultracapacitor mass
NEDC	New European driv. cycle
Q	Electric charge (Coulomb)
C	Electric capacity (Farad)
Rint	Internal Resistance
UDDS	Urban dynam. driv. schedule
Z	Total storage system cost
G	" " " " " weight
P	" " " " " power
E	" " " " " energy
$/pz	Unitary cost (for a 1000 Farad UC)

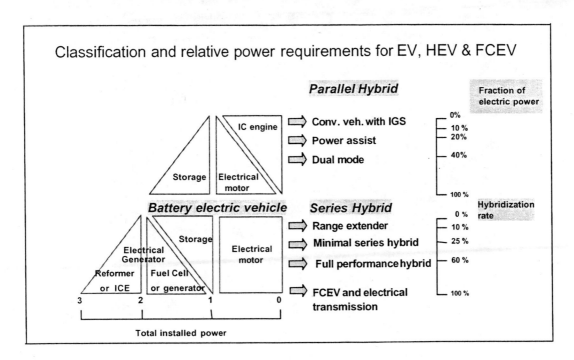

Fig. 6 Classification and relative power requirements for EV, HEV & FCEV

Fig. 7: Epcos B48710 67F/40 V   280x185x172mm

Fig. 8 Fuel cell generation system, developed by ENEA using a 5 kW De Nora PMFC

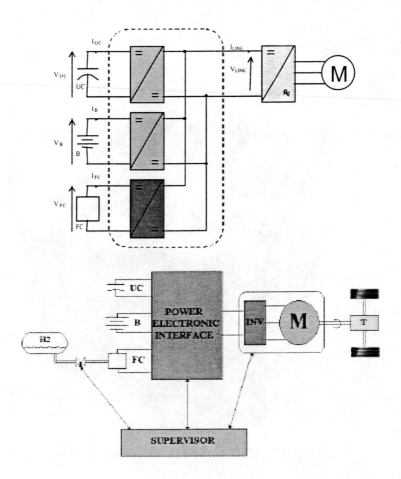

Fig. 9: Power Converter Structure

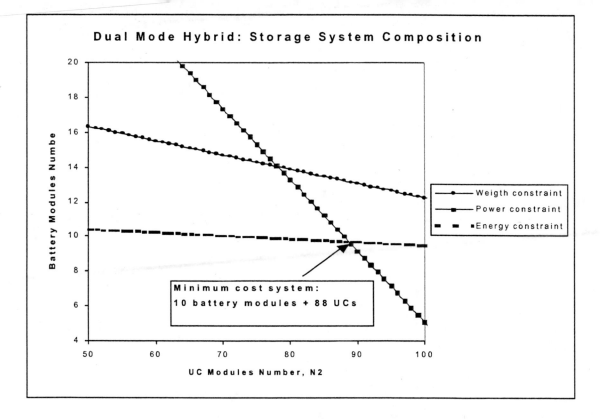

Fig. 10: Dual mode Hybrid Storage System

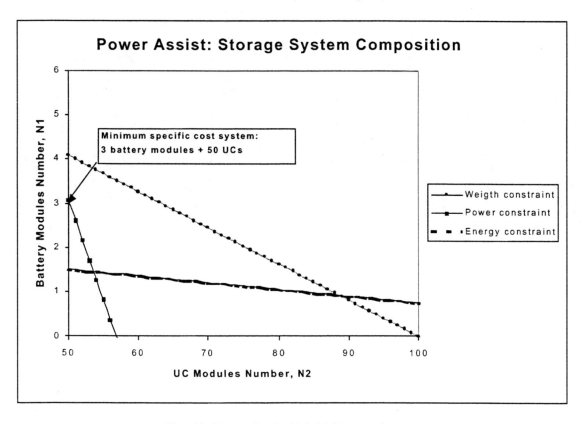

Fig. 11: Power Assist Hybrid Storage System

About the Editor

Daniel J. Holt holds a Masters of Science degree in Mechanical Engineering and a Masters of Science degree in Aerospace Engineering. He is currently the Editor-at-Large for SAE's *Automotive Engineering International* magazine. For 18 years Mr. Holt was the Editor-in-Chief of the SAE Magazines Division where he was responsible for the editorial content of *Automotive Engineering International*, *Aerospace Engineering*, *Off-Highway Engineering*, and other SAE magazines.

He has written numerous articles in the area of safety, crash testing, and new vehicle technology.

Prior to joining SAE Mr. Holt was a biomedical engineer working with the Orthopedic Surgery Group at West Virginia University. As a biomedical engineer he was responsible for developing devices to aid orthopedic surgeons and presented a number of papers on crash testing and fracture healing.

Mr. Holt is a member of Sigma Gamma Tau and a charter member of West Virginia University's Academy of Distinguished Alumni in Aerospace Engineering. He is also a member of SAE.